Cell Biology
of *Physarum* and *Didymium*

VOLUME I

Organisms, Nucleus,
and Cell Cycle

This is a volume in
CELL BIOLOGY
A series of monographs

Editors: D. E. Buetow, I. L. Cameron, G. M. Padilla, and A. M. Zimmerman

A complete list of the books in this series appears at the end of the volume.

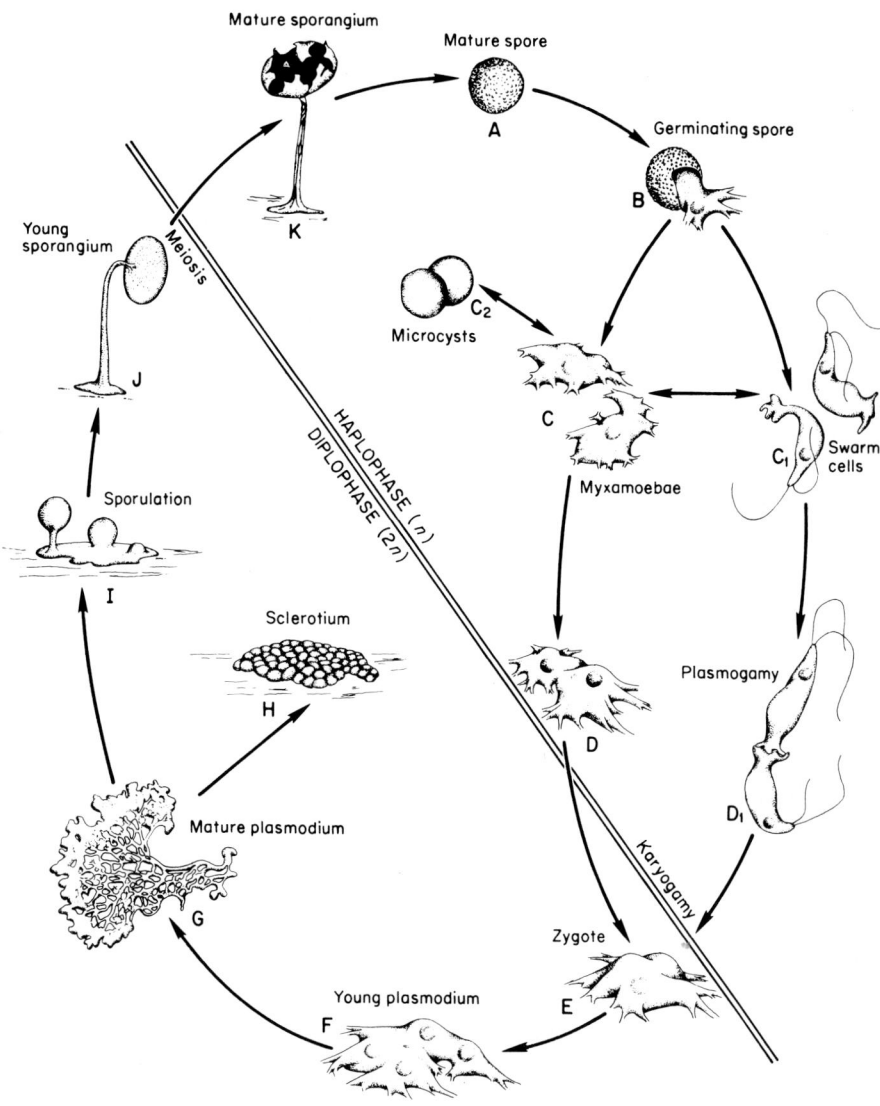

Life cycle of typical Myxomycete. From C. J. Alexopoulous and C. W. Mims (1979). "Introductory Mycology," 3rd ed., p. 69. Reprinted by permission of John Wiley & Sons, Inc.

Cell Biology of *Physarum* and *Didymium*

VOLUME I
Organisms, Nucleus, and Cell Cycle

Edited by

HENRY C. ALDRICH

Department of Microbiology and Cell Science
University of Florida
Gainesville, Florida

JOHN W. DANIEL

Department of Radiology
Division of Radiation Biology
Case Western Reserve University
Cleveland, Ohio

1982

ACADEMIC PRESS
A Subsidiary of Harcourt Brace Jovanovich, Publishers
New York   London
Paris   San Diego   San Francisco   São Paulo   Sydney   Tokyo   Toronto

COPYRIGHT © 1982, BY ACADEMIC PRESS, INC.
ALL RIGHTS RESERVED.
NO PART OF THIS PUBLICATION MAY BE REPRODUCED OR
TRANSMITTED IN ANY FORM OR BY ANY MEANS, ELECTRONIC
OR MECHANICAL, INCLUDING PHOTOCOPY, RECORDING, OR ANY
INFORMATION STORAGE AND RETRIEVAL SYSTEM, WITHOUT
PERMISSION IN WRITING FROM THE PUBLISHER.

ACADEMIC PRESS, INC.
111 Fifth Avenue, New York, New York 10003

United Kingdom Edition published by
ACADEMIC PRESS, INC. (LONDON) LTD.
24/28 Oval Road, London NW1 7DX

Library of Congress Cataloging in Publication Data
Main entry under title:

Cell biology of Physarum and Didymium.

(Cell biology)
Includes bibliographies and index.
1. Physarum. 2. Didymium (Fungi) 3. Fungi--
Cytology. I. Aldrich, Henry C. II. Daniel,
John W. III. Series. [DNLM: 1. Physarum--Cytology.
2. Myxomycetes--Cytology. QW 180.5.M9 C393]
QK635.P5C44   589.2'9   81-20483
ISBN 0-12-049601-1   AACR2

PRINTED IN THE UNITED STATES OF AMERICA

82 83 84 85   9 8 7 6 5 4 3 2 1

Contents

Contributors ix

Preface xi

Part I. Introduction to the Organisms

1. Morphology, Taxonomy, and Phylogeny
CONSTANTINE J. ALEXOPOULOS

I.	Introduction	3
II.	The Order Physarales	4
III.	General Taxonomy	17
IV.	Phylogeny	18
	References	21

2. *Didymium iridis* in Past and Future Research
O'NEIL RAY COLLINS and DONALD A. BETTERLEY

I.	Introduction	25
II.	Laboratory Cultivation and Maintenance	28
III.	Isolates Studied and Their Origin	28
IV.	Reproductive Cycles	28
V.	Chromosomal Numbers and Nuclear DNA Contents	35
VI.	Speciation and Evolutionary Diversification	38
VII.	Research on the Plasmodial Stage	41
VIII.	Sporophore Stage	50
IX.	Amoeboflagellate-Microcyst Stage	51
X.	Concluding Remarks	52
	References	53

Part II. Characteristic Biological Phenomena

3. Periodic Phenomena in *Physarum*

JOHN J. TYSON

I.	Introduction	61
II.	Nuclear Division Cycle	62
III.	Periodic Enzyme Synthesis	80
IV.	Other Periodic Events of the Mitotic Cycle	92
V.	Shuttle Streaming	95
	References	101

4. Chemotaxis in Plasmodia of *Physarum polycephalum*

TETSUDO UEDA and YONOSUKE KOBATAKE

I.	Introduction	112
II.	The Experiment	112
III.	Response to Carbohydrates	117
IV.	Hydrophobicity and Heterogeneity of Functional Membranes	119
V.	Electrostatic Interactions in Relation to Chemoreceptive Thresholds	122
VI.	Correlation between Tension Generation and Tactic Movement	127
VII.	Manipulation of Intracellular Components in the Plasmodial Strands	128
VIII.	Intracellular Control of Chemotaxis	132
IX.	Rhythmicity of the Plasmodial Contractile System	134
X.	Effects of Temperature on Membrane Activities	136
XI.	Summary	138
XII.	Glossary of Symbols	140
	References	141

Plasmodial Structure and Motility

DIETRICH KESSLER

I.	Introduction	145
II.	Plasmodial Morphology	151
III.	Mechanism of Protoplasmic Streaming in the Plasmodium	169
IV.	An Oscillating Regulatory System Controls Shuttle Streaming	185
V.	A Preliminary Model for the Molecular Mechanism of Shuttle Streaming in the Plasmodium	192
	References	196

Part III. Genetics

6. Genetics of *Physarum polycephalum*

JENNIFER DEE

I.	Genetic Control of the Life Cycle and of Plasmodium Development	212
II.	Genes Affecting Processes Other Than Development	235
III.	General Comments on Methods	244
	References	247

7. Ploidy throughout the Life Cycle in *Physarum polycephalum*

JOYCE MOHBERG

I.	Introduction	253
II.	Methods of Estimating Ploidy	254
III.	Ploidy throughout the Life Cycle	263
IV.	Factors Affecting Ploidy	266
V.	Uses of Cultures with Different Ploidy	268
	References	269

8. Genealogy and Characteristics of Some Cultivated Isolates of *Physarum polycephalum*

JOYCE MOHBERG and KARLEE L. BABCOCK

I.	Introduction	273
II.	Origin of Isolates	274
III.	Characteristics of Isolates	278
	References	281

Part IV. The Plasmodial Cell Cycle

9. Nuclear Organization during the Cell Cycle in the Myxomycete *Physarum polycephalum*

JEAN-G. LAFONTAINE and MONIQUE CADRIN

I.	Introduction	287
II.	Structure of the Chromosomes during the Mitotic Cycle	288
III.	The Nucleolar Cycle	301
IV.	The Mitotic Apparatus	309
	References	312

10. Chromosome Organization and Chromosomal Proteins in *Physarum polycephalum*

HARRY R. MATTHEWS and E. MORTON BRADBURY

I.	Introduction	318
II.	Nucleosome Structure	318
III.	Nucleolar Chromatin	325
IV.	Histones	330
V.	Postsynthetic Modifications of Histones	341
VI.	Nuclear Histone Kinase	353
VII.	Non-Histone Chromosomal Proteins	359
	References	361

11. Organization and Replication of DNA in *Physarum polycephalum*

THOMAS E. EVANS

I.	The Groundwork	371
II.	Early Advances	372
III.	Current Research Areas	374
	References	386

12. RNA Metabolism

RICHARD BRAUN and THOMAS SEEBECK

I.	Introduction	393
II.	Properties of DNA and RNA	394
III.	RNA Synthesis	400
IV.	RNA Degradation	413
V.	Transcription in the Mitotic Cycle	416
VI.	Transcription in Differentiation	425
VII.	Concluding Remarks	428
	References	429

Index   437

Contributors

Numbers in parentheses indicate the pages on which the authors' contributions begin.

Constantine J. Alexopoulos (3), Department of Botany, University of Texas, Austin, Texas 78712

Karlee L. Babcock (273), McArdle Laboratory for Cancer Research, University of Wisconsin, Madison, Wisconsin 53706

Donald A. Betterley (25), Department of Botany, University of California, Berkeley, California 94720

E. Morton Bradbury (317), Department of Biological Chemistry, School of Medicine, University of California, Davis, California 95616

Richard Braun (393), Institute for General Microbiology, University of Bern, CH 3013 Bern, Switzerland

Monique Cadrin (287), Department of Biology, Faculty of Sciences and Engineering, Laval University, Quebec G1K 7P4, Canada

O'Neil Ray Collins (25), Department of Botany, University of California, Berkeley, California 94720

Jennifer Dee (211), Department of Genetics, University of Leicester, Leicester LE1 7RH, England

Thomas E. Evans (371), Division of Radiation Biology, Case Western Reserve University, Cleveland, Ohio 44106

Dietrich Kessler (145), Department of Biology, Haverford College, Haverford, Pennsylvania 19041

Yonosuke Kobatake (111), Faculty of Pharmaceutical Sciences, Hokkaido University, Sapporo, Japan

Jean-G. Lafontaine (287), Department of Biology, Faculty of Sciences and Engineering, Laval University, Quebec G1K 7P4, Canada

Harry R. Matthews (317), Department of Biological Chemistry, School of Medicine, University of California, Davis, California 95616

Joyce Mohberg (253, 273), College of Arts and Science, Governors State University, Park Forest South, Illinois 60466

Thomas Seebeck (393), Institute for General Microbiology, University of Bern, CH 3013 Bern, Switzerland

John J. Tyson (61), Department of Biology, College of Arts and Sciences, Virginia Polytechnic Institute and State University, Blacksburg, Virginia 24061

Tetsudo Ueda (111), Faculty of Pharmaceutical Sciences, Hokkaido University, Sapporo, Japan

Preface

It is now more than twenty years since Dr. Daniel, working with Harold Rusch at the University of Wisconsin, developed the axenic culture methods and chemically defined media that enabled us to grow plasmodia of *Physarum polycephalum* in liquid shake culture. Since that time, the organism has become firmly established as an important experimental tool in cell biology. In the genetics area, a related organism, *Didymium iridis,* has also assumed importance because of its ease of manipulation in culture. This two-volume treatise summarizes important experimental research using these two organisms for developmental and cellular studies.

Because of the natural synchrony of the cell cycle events in the plasmodium, *P. polycephalum* has been used for dissecting the events of DNA, RNA, and protein synthesis. More recently, fundamental studies on the organization of chromatin have focused on the nucleus in the plasmodium. Studies on the spectacular shuttle streaming in the plasmodium have contributed greatly to our understanding of contractility and motility in nonmuscle systems. These examples suggest that these two species are experimental tools whose potential has just begun to be exploited.

This treatise was planned with several audiences in mind. It should serve as a frequent, single reference source to brief cell biologists on the primary research to date on *Physarum* and *Didymium*. To accomplish this aim, we have encouraged authors to organize their chapters as comprehensive reviews insofar as possible.

We frequently encounter cell biologists who are intrigued with the research possibilities of plasmodial slime molds but lack the familiarity with the basic biology of the organisms to handle them intelligently. To meet the needs of such scientists, we have included a general introductory chapter by the eminent taxonomist–morphologist C. J. Alexopoulos and a number of shorter chapters on experimental methods at the end of the second volume. The interest in these experimental methods chapters shown in our own laboratories indicates that they will be of utility to researchers more familiar with the organisms as well.

The volumes will be a good source for graduate students in cell biology and perhaps may even be of use in other graduate courses. The contributors have not only reviewed work to date in their areas but have also pointed out areas and topics likely to be most fruitful for future research. This approach should prove stimulating to students searching for suitable dissertation and thesis topics.

We are great believers in plasmodial slime molds as research tools. Professors W. F. Dove and H. P. Rusch have recently published a volume on "Growth and Differentiation in *Physarum polycephalum*" in which they exhibit this same type of enthusiasm. Wider use of these organisms as research tools in cell biology will benefit us all.

We wish to acknowledge with gratitude the initial encouragement of Ivan Cameron, who urged us to organize this undertaking, and the aid of the staff of Academic Press in producing this work. We are also grateful for the cooperation and understanding of our families and laboratory associates during the time we have been occupied with the preparation of this book. All scientists active in the *Physarum* research group in the United States many of whom are chapter authors, have been generous with suggestions concerning the organization of this treatise. We thank them all!

Henry C. Aldrich
John W. Daniel

PART I

Introduction to the Organisms

CHAPTER 1

Morphology, Taxonomy, and Phylogeny

CONSTANTINE J. ALEXOPOULOS*

I.	Introduction	3
II.	The Order Physarales	4
	A. Family Physaraceae	4
	B. Family Didymiaceae	12
III.	General Taxonomy	17
IV.	Phylogeny	18
	References	21

I. INTRODUCTION

The Myxomycetes are organisms with a unique life cycle. This consists of sporophores that bear the spores which, after meiosis takes place, give rise to haploid myxamoebae or anteriorly flagellate cells devoid of cell walls. Such cells behave as gametes, fusing in pairs to form zygotes. The zygote then grows into a free-living, multinucleate, diploid mass of protoplasm—the plasmodium—devoid of cell walls but generally enveloped by a slime sheath. The plasmodium feeds phagotrophically and eventually produces sporophores characteristic of the species. Many variations to this life cycle are known. Some strains, for example, are apogamic, completing the entire life cycle in a haploid condition; of the sexual strains, some are homothallic and some heterothallic, their gametes consisting of two mating types. For a comprehensive summary of the biology of Myxomycetes, see Alexopoulos (1966, 1973), Gray and Alexopoulos (1968), and Collins (1979).

Myxomycetes have been traditionally classified on the basis of their

*This chapter was partly written while the author was Visiting Professor of Botany at the University of Florida in Gainesville.

sporophore characteristics alone, but in recent years plasmodial characters and type of sporophore development have become important considerations in delimiting subclasses.

II. THE ORDER PHYSARALES

The genera *Physarum* and *Didymium* both belong to the order Physarales of the subclass Myxogastromycetidae of the class Myxomycetes. The order Physarales was established in 1922 by Thomas Macbride in the second edition of his monograph "The North American Slime Moulds." It corresponds to Lister's (1925) suborder Calcarineae, which is equivalent to Rostafinski's (1875–1876) subcohort Calcarineae.

The order, as now defined, contains myxomycete species with various types of sporophores, which are always of subhypothallic development (Alexopoulos, 1969, 1973), each covered by a typically calcareous peridium, except in *Protophysarum*, and with lime also often present in the hypothallus, stalk, columella, pseudocolumella, and capillitium when these structures are present. The spores, in mass, are dark purple-brown to black; by transmitted light, they appear purple-brown, brown, or violaceous, rarely pallid. The trophic stage is a phaneroplasmodium (Alexopoulos, 1960, 1969) of various colors, which is generally extensive but which may remain minute until it sporulates. The distinguishing feature of the order, along with the purple-brown spores, is the almost universal presence of lime in the peridium. This characteristic, however, is influenced by the environment, so that sporophores that are typically calcareous may sometimes be devoid of lime when they develop under certain conditions (Gray, 1961). The order consists of two families: the Physaraceae, of which *Physarum* is the largest genus, and the Didymiaceae, in which *Didymium* predominates.

A. Family Physaraceae

In the family Physaraceae, described by Rostafinski (1873), to which *Physarum* belongs, the peridial lime is granular except in *Protophysarum*, which is totally devoid of lime but which is placed here because of its subhypothallic development, its phaneroplasmodium, and its capillitial network, albeit devoid of lime nodes. The distinguishing feature of the family, which separates it from the Didymiaceae, is the presence of lime in the capillitium. This will be discussed in greater detail later.

Sporophores of the Physaraceae vary from minute, sessile spheres, through plasmodiocarps pendent on slender stalks, to stipitate simple or multilobed sporangia. Such sporangia may be densely massed, forming pseudoaethalia. In the genus *Fuligo*, the sporangia are aggregated into an aethalium covered by a relatively thick calcareous cortex below a common peridium.

The peridium of the Physaraceae is typically sprinkled or covered with granular lime but varies from genus to genus and indeed even within genera in its makeup from a nearly limeless, thin, iridescent membrane; to a two-layered covering, the inner layer membranous, the outer calcareous; to a three-layered structure, as in *Leocarpus*, with a calcareous middle layer sandwiched between a cartilaginous outer and a membranous inner layer. In *Cienkowskia*, the membranous or cartilaginous peridium is densely covered with lime. In *Physarella* the peridium bears spinelike, calcareous trabeculae pointing inward.

Calcareous, rarely limeless, the capillitium in most genera of the Physaraceae (*Erionema, Fuligo, Physarella, Craterium,* and *Physarum*) typically consists of a network of limeless slender tubules connecting generally numerous, but sometimes few, limy nodes. In *Badhamia*, the capillitium is made up of more or less uniform calcareous tubules. In *Leocarpus* and *Cienkowskia*, it is of a duplex nature with limy, platelike partitions present in the plasmodiocarps of the latter genus. The aforementioned calcareous, spinelike tarbeculae extending inward from the peridium of *Physarella* have also been interpreted by some (Lister, 1925; Martin and Alexopoulos, 1969; Farr, 1976) as constituting one portion of a duplex capillitium, but it is probably better not to regard them as such. Capillitial lime nodes are often aggregated near the center of the sporophore, forming a globose or rod-shaped, calcareous pseudocolumella that, however, may be missing even in species of which it is considered to be characteristic.

There is nothing to distinguish the spores of the two families. In both they are typically globose, purplish-brown, violet or pallid, with variously ornamented walls. In the Physaraceae they vary in diameter from as small as 5 μm in *Physarum penetrale* Rex to as large as 22 μm in *Fuligo megaspora* Sturgis.

GENUS *PHYSARUM*

a. General Characteristics. The genus *Physarum* was described by Persoon (1794). The chief characteristic of the genus as now delimited is the capillitium, which consists of lime granules of different types connected in a network by fine, tubular filaments. Another characteristic of the genus is the absence of pseudocapillitium from the sporophores of any species. The lime deposits on and in the sporophores are granular or amorphous. "The infrequent occurrence of subcrystalline and partly crystalline lime is believed to be due to alternate wetting and drying," according to Farr (1976, p. 112). The sporophores may be sporangiate, plasmodiocarpous or, rarely, pseudoaethalioid. No species with truly aethalial sporophores is included in the genus.

b. The Plasmodium. Up to 1960, plasmodia of all Myxomycetes were generally regarded as essentially similar in form, and that of the much studied *Physarum polycephalum* was, in the minds of biologists in general, the prototype of all myxomycete plasmodia. To be sure, de Bary (1887) had observed that the plasmodia of *Stemonitis*, of the Trichiaceae, and of *Lycogala* were much more finely granular than those of the Physaraceae; Zukal (1893) had described the species "exceptional among Mycetozoa" (Lister, 1925, p. 185), plasmodium of *Licea (Hymenobolina) parasitica* (Zukal) Martin, and Celakowski (1892), Miller (1898), and Thom and Raper (1930 had pointed out how the growing plasmodium of *Stemonitis* differs from the "typical," i.e., physaraceous plasmodium, but it was not until 1960, when Alexopoulos described three general types of plasmodia—phaneroplasmodium, aphanoplasmodium, and protoplasmodium—and signaled the existence of a fourth type—that of the Trichiales, which he did not name—that the widespread occurrence of several major plasmodial types was emphasized and later generally accepted.

The plasmodium of *Physarum* is a phaneroplasmodium (Fig. 1), as, indeed, it is in all Physarales examined up to now. It consists of a usually robust, fairly extensive network of thick viens, each separate from the others or embedded in a continuous sheet of protoplasm that terminates in one or more fleshy fans. Each vein consists of an outer tubular gel layer (ectoplasm) in which no streaming can be observed, enclosing a core of endoplasm that exhibits more or less rhythmic, reversible streaming. The protoplasm of a phaneroplasmodium is very granular, and the ectoplasm and endoplasm are very distinct. The advancing fleshy fans are very conspicuous and have definite margins. The phaneroplasmodium is enveloped in a gelatinous sheath, or "envelope," as de Bary (1887) called it, containing microfibrils, that is shed as the plasmodium creeps over the substratum, leaving traces behind that are easily visible in nature on dead leaves or in the laboratory on the agar surface.

Physarum plasmodia may or may not be pigmented. In nature, the most conspicuous plasmodia are bright yellow to orange or white. Indeed, of the 84 species of *Physarum* listed in Martin and Alexopoulos (1969), 24 are described as having yellow plasmodia and 18 white, with those of most of the remaining species described as ochraceous, greenish, olive-green, orange, red, scarlet, maroon, violet, purple, blue, gray, and black or combinations of these colors such as orange-red, yellowish-green, and grayish-black. The plasmodia of 24 *Physarum* species are listed as unknown. Within a certain range, plasmodium color is more or less stable, but sometimes it varies. The plasmodium of *P. roseum* Berk. & Br., for example, is described as maroon or bright red, but in some cultures it is pink. Also, plasmodial color sometimes changes from colorless or white to pigmented (usually yellow) on exposure to light, as in *P. gyrosum* Rost. (Fergus and Schine, 1963; Koevenig, 1964). The nature of plas-

1. Morphology, Taxonomy, and Phylogeny

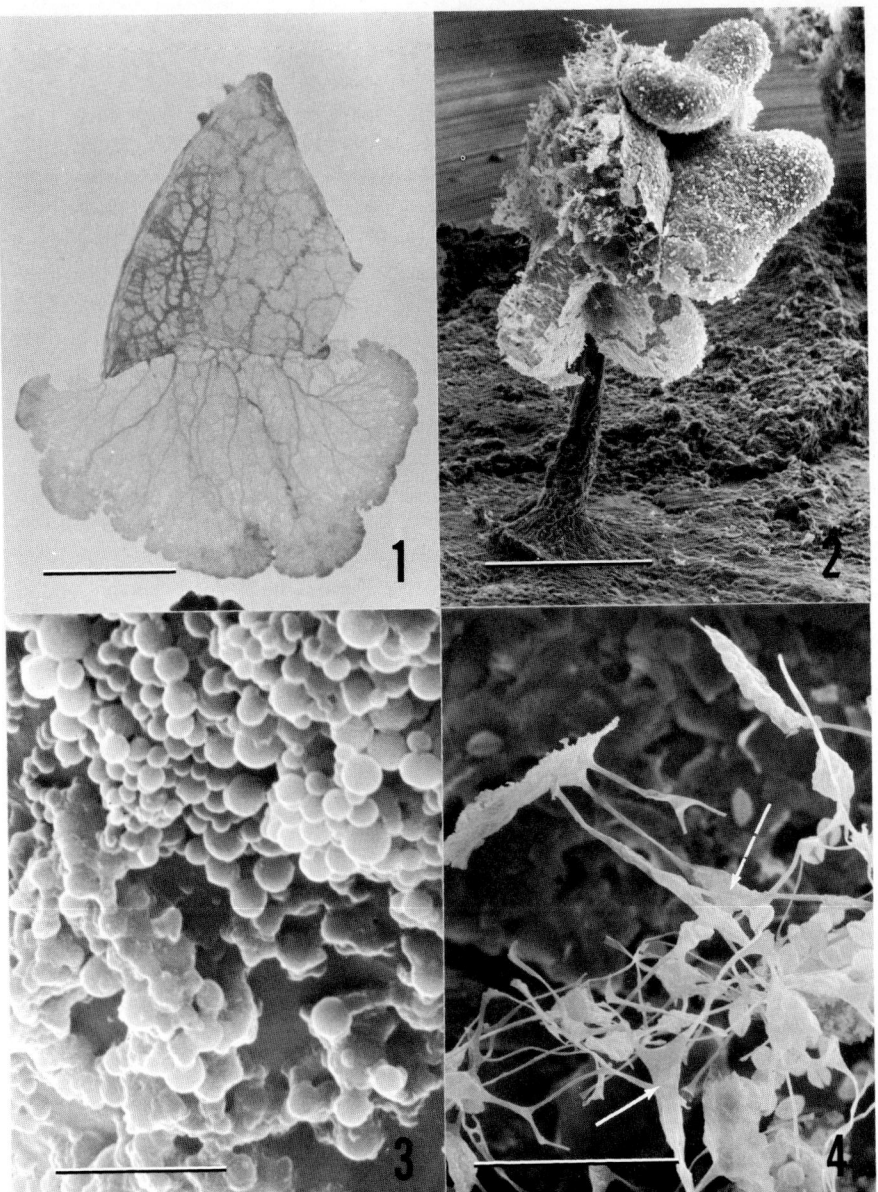

Figs. 1–4. *Physarum polycephalum*. Fig. 1. Phaneroplasmodium. Bar = 1 cm. Fig. 2. Sporangium with many heads on a single stalk. Bar = 500 μm. Fig. 3. Peridium covered with lime granules. Bar = 5 μm. Fig. 4. Capillitium showing fusoid, elongated lime nodes (arrows). Bar = 50 μm.

modial pigments remains unknown, but it appears that at least some pigments are confined in pigment granules (McManus, 1965).

In some species of *Physarum*, the plasmodium is capable of reaching a great size and may cover areas a square meter or more in extent on a well-watered lawn (*P. cinereum* Schum.) or a considerable portion of a large, decaying log [*P. polycephalum* Schw., *P. viride* (Bull.) Pers]. In other species, however, the plasmodium never attains a large size in nature, sporulating while still small (*P. roseum*). Whether the size a plasmodium is capable of reaching is genetically or environmentally controlled has not been determined, partly because only a small percentage of species have been cultured in the laboratory from spore to spore.

c. The Sporophore. *Physarum* sporophores vary from sessile to stipitate sporangia, which in *P. polycephalum* bear many heads (Fig. 2); to sessile plasmodiocarps to those with a weak, stalklike extension of the hypothallus; to pseudoaethalia consisting of crowded sporangia, as in some collections of *P. gyrosum* and *P. polycephalum*. In size, sporophores vary from the very small, sessile sporangia of *P. lateritium* (Berk. & Rav.) Morgan, 0.3–0.7 mm in diameter, to stipitate sporangia of *P. tenerum* Rex, which reach a height of 3 mm, and the pseudoaethalia of *P. gyrosum*, which attain a size greater than 3 mm across.

d. The Peridium. The peridium of *Physarum* is typically membranous but is sometimes cartilaginous (*P. nasuense* Emoto) or rugose [*P. melleum* (Berk. & Br.) Massee]. It is often double, consisting of a delicate inner wall covered by a thickly calcareous outer wall (*P. bivalve* Pers.), or sometimes triple, as in *P. bogoriense* Racib., with the two outer layers calcareous and closely attached. Typically the peridium bears lime granules (Fig. 3) and may be completely encrusted with granular lime. The amount of lime, however, varies even in different sporophores from the same collection and is probably influenced by the microenvironment at the time of sporulation. Species whose peridia are typically limy may be devoid of lime in humid atmospheres (Gray, 1961). The chemical composition of the membranous or cartilaginous walls is not known. The calcareous covering is lime, with some pigments incorporated in many species. *Physarum* sporophores "display spectacular array of colors and shades," as Collins (1979) puts it for Myxomycetes in general. Many paintings and color photographs of such sporophores have been published. Among the best known are the paintings in Crowder (1926), Lister (1911, 1925), Hattori (1935, 1964), Martin and Alexopoulos (1969), and the most recent and spectacular ones in Emoto's (1977) beautiful book. Color photographs have been published by Alexopoulos (1973). Sporophore pigments appear to be incorporated in the lime in some unknown way. The nature of the pigments is not known in any species. Of interest in this connection is Henney's (1968) study of *P. globuliferum* (Bull.)

1. Morphology, Taxonomy, and Phylogeny

Pers., the sporangia of which have been described as "white" (Lister, 1925, p. 27) or "white, pale ochraceous or pinkish" (Martin and Alexopoulos, 1969, p. 303). Henney showed that the blue sporangia of the so-called *P. bilgramii* Hagelst., formed from a blue plasmodium, are nothing more than a color variation of *P. globuliferum*.

The type of peridial dehiscence varies within the genus. Although irregular dehiscence, as in *P. polycephalum* (Fig. 2), is probably the rule, some species dehisce in a characteristic manner, which aids in identification. Thus, in the plasmodiocarpous *P. bivalve*, "dehiscence is by a more or less regular, formed longitudinal fissure" (Martin and Alexopoulos, 1969, p. 288); in *P. bogoriense*, the sides of the peridium characteristically dehisce in a stellate fashion into triangular reflexed lobes, making the identity of the plasmodiocarps almost unmistakable, even with the use of a hand lens in the field. Again, in the common *P. nutans*, Pers. dehiscence is typically petaloid or annulate.

e. The Stalk. The stalk, when present, is filled with debris from the substratum which it accumulates during its subhypothallic development. Its surface may be smooth or ridged. It may be plain, frosted with lime, or entirely calcareous. In many species the stalk extends into the sporangial sac as a generally short, conical columella. The latter, however, may be long and cylindric and may reach the apex of the sporangium, as in *P. crateriforme* Petch.

f. The Capillitium. Although the capillitium in *Physarum* varies in detail from species to species, it is built on the same pattern of a usually large number of calcareous nodes interconnected by a delicate tubular network (Fig. 4). Variations that distinguish the species mainly concern the relative abundance of capillitium in a sporophore; size, shape, and color of lime nodes; arrangement of capillitial tubules; origin of the capillitial network; and elasticity. The general structure of the capillitium may be detected under high magnification with a good dissecting microscope in a mature, dry sporangium still attached to the substratum. On removal of the peridium, the lime knots, usually abundant, are immediately evident as white, yellow, orange, red, or blue conspicuous nodes. The capillitium may appear as a dense network that completely fills the sporangium; or, in a few species, it may consist of sparingly branched threads; or, as in *P. rigidum*, it consists almost entirely of slender, rodlike tubules enclosing lime granules. In some species, notably *P. gyrosum*, the capillitium is elastic and expands instantly when the peridium is removed.

The variation in size, shape, color, and abundance of lime nodes is important in distinguishing among species. Lime nodes may be globose or flat, expanded and irregular in outline, or they may be elongated with several extensions, or fusoid or rodlike; they are usually concolorous with the lime granules on the peridium, but they may vary. In some species, many calcareous nodes are massed in the center of the sporangial sac, forming a spherical or rod-shaped pseudocolumella.

g. The Spores. The spores of *Physarum* do not differ in the main from those of other members of the order Physarales. They are almost always globose, purplish-brown, violet-gray, or rarely pallid. They average 8–10 µm but may be as small as 5 µm in *P. penetrale* Rex or as large as 15 µm in *P. albescens* Ellis. The spore walls are seldom smooth, as in *P. laevisporum* Agnih. More often they are spinulose, as in *P. polycephalum* (Fig. 5) or verrucose. In a few species, such as *P. dictyosporum* Martin, they are conspicuously reticulate (Fig. 6). In *P. echinosporum* A. Lister, the spines, often united into prominent ridges, are very conspicuous.

h. Species of Physarum. Martin and Alexopoulos (1969) recognized 84 species of *Physarum*. Since then, at least 10 additional species have been described. With interest in myxomycete floristics revived throughout the world, many more will undoubtedly be described in the near future as the African continent, the tropics in general, the deserts, the far north and far south regions, and the alpine habitats are more thoroughly explored. The moist chamber method now being widely used is revealing many previously unknown species of Myxomycetes but, so far, relatively few undescribed species of *Physarum*.

i. Supposed Relationships. In the absence of fossil evidence or comparative chemical data, we must at present use morphological resemblances alone as indicating relationships. On that basis, *Physarum* is most closely related to *Badhamia* and *Craterium*. With the former, it is linked by the badhamioid capillitium of some species now placed in *Physarum* and the physaroid capillitium of some species until recently included in *Badhamia,* such as *P. decipiens* Curt. (Farr, 1961). With *Craterium, Physarum* is linked by intermediate forms which have a persistent, cuplike base characteristic of *Craterium.* Relationships could also be claimed between *Physarum* and *Physarella,* the latter often producing physaroid plasmodiocarps instead of the characteristically introverted, thimble- or bell-shaped sporangia. In 1954, Locquin suggested that *Physarum* be merged with *Fuligo* (Martin and Alexopoulos, 1969, p. 275), but this has not been accepted by myxomycete taxonomists. As for possible relationships among species of *Physarum,* one might consult Lister (1925) and Hagelstein (1944), who grouped species which resemble one another in certain characteristics. More recent monographs (Martin and Alexopoulos, 1969; Farr, 1976) list species alphabetically and do not attempt to indicate relationships. Farr's synoptic keys, however, are useful in grouping species that resemble one another. And, indeed, some recognized species merge with one another by intermediate forms, but until experimental evidence is obtained showing that such "species" interbreed, there

1. Morphology, Taxonomy, and Phylogeny

Figs. 5–8. Fig. 5. Spore of *Physarum polycephalum* with spinulose wall. Bar = 1 μm. Fig. 6. Reticulate spore of *P. dictyosporum*. Bar = 1 μm. Fig. 7. Lime crystals on the peridium of *Didymium iridis*. Bar = 5 μm. Fig. 8. Sporangium of *D. iridis* showing scattered lime crystals on the peridium. Bar = 500 μm.

is no way to establish relationships. Alexopoulos (1969) strongly advocated the experimental approach to the taxonomy of the Myxomycetes and cited crossing experiments that indicate that *P. obrusseum* (Berk. & Curt.) Rost., at first described as a distinct species and later reduced to a variety of *P. polycephalum*, deserves neither specific nor varietal status. The work of Henney (1968) with the *P. globuliferum/bilgramii* complex has already been mentioned. Such attempts to establish genetic relationships between entities that show morphological resemblances are sorely needed before the taxonomy of the Myxomycetes can be placed on a firmer basis.

B. Family Didymiaceae

As pointed out by Martin and Alexopoulos (1969), our concept of the Didymiaceae has undergone some modifications. Originally, Rostafinski (1873), who first described the family, separated it from the Physaraceae on the basis of a limeless capillitium in the former as opposed to a limy one in the latter. This distinction was abandoned by Lister (1894, 1911, 1925), who placed emphasis on the type of lime present (granular versus crystalline) rather than on its presence or absence in the capillitium. Hagelstein (1944) followed Lister in this respect, as in most others. These two different viewpoints affect the classification of at least three genera, shifting them from one family to the other. One is, therefore, presented with the problem of deciding whether the presence versus absence of lime in the capillitium is a more or less important taxonomic criterion than granular versus crystalline lime wherever in the sporophore it may be located. We have already seen that the environment sometimes determines whether a physaraceous sporophore is limy or devoid of lime. Also, it is generally believed that the repeated wetting and drying of a calcareous sporophore tends to induce formation of crystalline lime even in species whose lime is usually granular (Farr, 1976). Of interest in this connection are Schoknecht's (1975) results with X-ray microanalysis of the calcareous deposits of 16 species in four genera of the Physaraceae and 10 species in two genera of the Didymiaceae (according to the classification of Martin and Alexopoulos, 1969). Schoknecht (1975) found phosphorus in all 11 species of Physaraceae she examined but in only one of the 10 species of Didymiaceae, the one exception being *Didymium trachysporum* G. Lister. One might expect such results if the lime in all the Physaraceae were of one type and in all the Didymiaceae were of another, but such is not the case. The genus *Diderma*, which has granular lime, is placed in the Didymiaceae by Martin and Alexopoulos (1969), Nannenga-Bremekamp (1974), and Farr (1976), yet in the two specimens analyzed by Schoknecht (1975) no phosphorus was indicated, as it was also not found in seven species of *Didymium* with crystalline lime. Schoknecht's results (1975) with *Diachea leucopodia* (Bull.) Rost. further complicate the situation. *Diachea* has no lime in

1. Morphology, Taxonomy, and Phylogeny

the peridium or capillitium but has a limy stalk and columella. The genus has been shifted back and forth from the entirely limeless Stemonitales to the Physarales. Martin and Alexopoulos (1969), following Morgan (1894) and Macbride (1922), placed *Diachea* in the Stemonitales, but Farr (1974), partially basing her decision on Blackwell's (1974) conclusion that *D. leucopodia* undergoes a subhypothallic development, returned *Diachea* to the Physarales, where Lister (1925) following Rostafinski (1873) had classified it, and placed it in the Didymiaceae because of the absence of lime from the capillitium. Schoknecht (1975) found phosphorus together with calcium in the calcareous deposits of *D. leucopodia,* in which the lime is granular, as in the Physaraceae. However, some other species of *Diachea* have either crystalline or a mixture of crystalline and granular lime, according to Farr (1976). No species of this genus other than *D. leucopodia* has been subjected to X-ray microanalysis, and it would be instructive indeed to look into this situation. For the purpose of this discussion, we have separated the Didymiaceae from the Physaraceae on the basis of a limeless capillitium in the former and will tentatively accept the inclusion of *Diachea* in the Didymiaceae. This will be discussed later in Sections III and IV.

GENUS *DIDYMIUM*

   a. General Characteristics. The genus *Didymium* was described by Schrader in 1797 with *D. farinaceum* Schrad. [now *D. melanospermum* (Pers.) Macbr.] as the type species. Since then, 40 additional species have been added, 11 of which were described since the Martin/Alexopoulos monograph was published in 1969. The genus is distinguished from others in the Didymiaceae chiefly by the crystalline lime on the peridium that is not united into scales (Fig. 7). The sporophores are sporangial (Fig. 8) or plasmodiocarpous.

   b. The Plasmodium. The plasmodium of *Didymium* is a typical phaneroplasmodium. As in *Physarum,* but possibly to a larger extent, plasmodia of different species of *Didymium* vary in the size they reach before sporulating. Thus, the plasmodium of *D. iridis,* which has been studied extensively in culture, is capable of assuming considerable proportions, whereas, at the other extreme, those of *D. atrichum* Henney and Alexopoulos (1980), and *D. eremophilum* Blackwell and Gilbertson (1980), remain minute throughout their existence. Of the 22 species in which the plasmodium has been described, 12 are stated to be colorless, gray, or white and the other 10 pigmented. Of the latter, five are described as yellow or yellowish, one as pale green, and one as orange-red (Martin and Alexopoulos, 1969).

   c. The Sporophore. All species of *Didymium* have sporangial or plasmodiocarpous sporophores. In *D. atrichum* the sporangia are often crowded or heaped, forming small pseudoaethalioid clusters. Some plasmodiocarps are terete or annulate, as in *D. annelus* Morgan; others are flattened and effuse, often widely expanded, as in some collections of *D. annelus* and *D. dubium* Rost. In *D. perforatum* Yam., the effuse plasmodiocarps are reticulate, forming a dense network. The extreme case of a spreading plasmodiocarp is reached in *D. serpula* Fries, which often resembles an extensive, very thin coating, 0.01–0.15 mm thick and up to 4 cm in extent over a leaf on which the plasmodium has sporulated. At the other extreme are the stipitate sporangia of *D. intermedium* Schroet., which often form corymbose clusters up to 1.5 mm tall. It is evident from the above that some *Didymium* sporophores are sessile and others stipitate. As in *Physarum*, the stalks in *Didymium* when present are stuffed, which is a consequence of their subhypothallic development. More often than in *Physarum*, the tip of the stalk in *Didymium* is extended into the sporangial sac, forming a columella (Fig. 9). Interestingly enough, the columella sometimes differs in color and texture from the stalk. For example, in *D. iridis* (Ditmar) Fries the stalk is yellow or yellowish-brown, whereas the columella is typically white and calcareous. In fact, the color of the columella is the main character that distinguishes this species from the closely related *D. nigripes* (Link) Fries, which has a dark brown columella that is calcareous within. The stalk of neither *D. nigripes* nor *D. iridis* contains lime.

Columellae are often also present in sessile *Didymium* sporophores In such cases, the columella represents the dome-shaped, thickened base of the sporangium. In *D. clavus* (Alb. & Schw.) Raben., a very distinctive species, the sporangia are discoid and umbilicate below and also often above. The stalk is often so short as to be contained in the lower umbilicus, giving the sporophore a sessile aspect.

d. The Peridium. Two general types of peridial wall are distinguished in the genus *Didymium:* (1) a cartilaginous wall, more or less covered with lime crystals, uniquely found in *D. leoninum* Berk. & Br., which occurs mostly in East Asia but has also been found in Jamaica and Venezuela, and (2) a membranous peridium covered with lime crystals, loosely scattered or united into a calcareous crust, but never in the form of squamules or scales. This latter type of peridium is often manifested as a double wall, the inner a thick membrane, the outer a thick limy wall, as in *D. crustaceum* Fries and *D. difforme* (Pers.) S. F. Gray, for example. The membranous peridium covered with white lime crystals is exemplified by *D. iridis* (Fig. 8) and *D. nigripes*. The crystals may be found in clusters, as in *D. ovoideum* Nann.-Brem.

e. The Capillitium. In *Didymium*, the capillitium is a system of limeless threads that branch and anastomose and may bear nodular thickenings (Figs. 10, 11). No lime is typically included in the capillitium except in *D. trachysporum*,

1. Morphology, Taxonomy, and Phylogeny

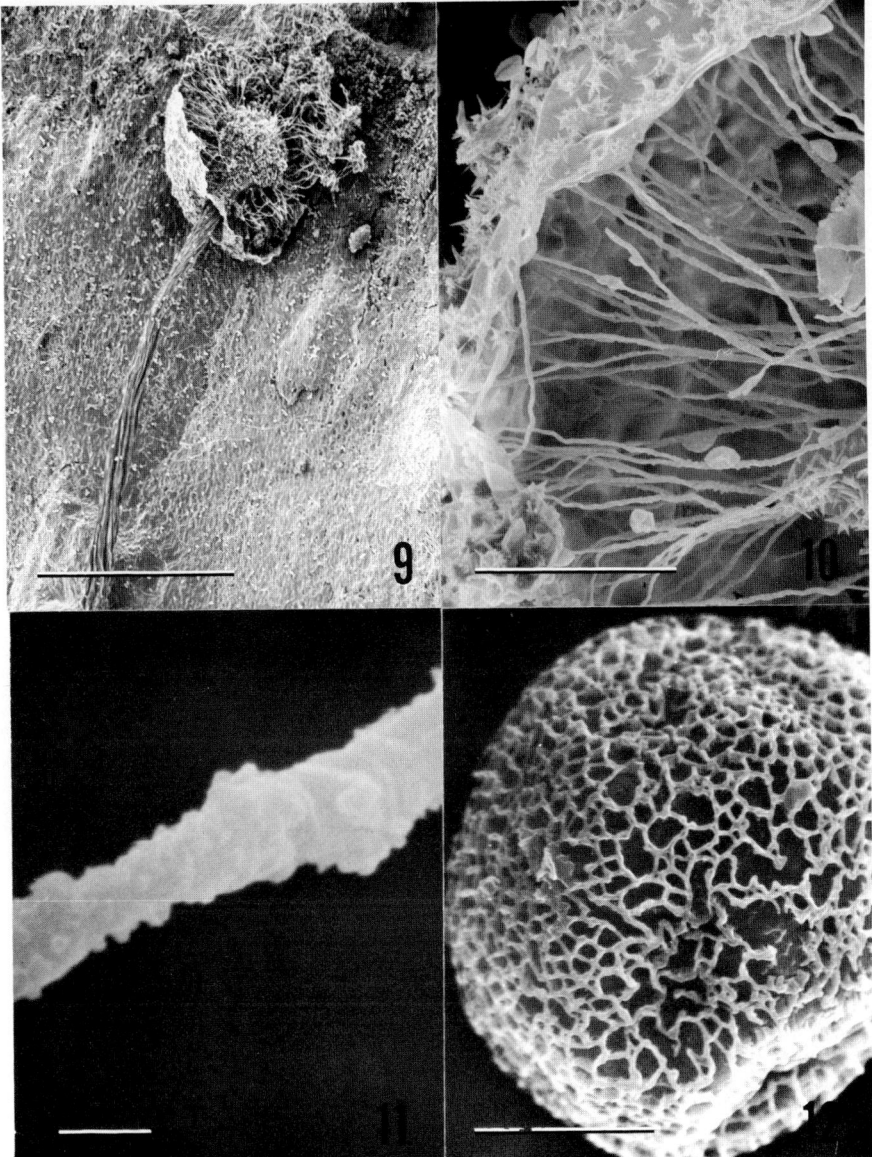

Figs. 9–12. Fig. 9. Sporangium of *Didymium iridis*, broken so as to reveal the spherical columella at the tip of the stalk inside the sporangial sac. Bar = 500 μm. Fig. 10. *D. iridis*. A portion of the sporangium showing branching capillitial threads attached to the inner side of the peridium. Note the stellate lime crystals on the outer surface of the peridium. Bar = 50 μm. Fig. 11. *D. iridis*. A portion of a capillitial thread. Bar = 1 μm. Fig. 12. Reticulate spore of *D. atrichum*. Bar = 5 μm.

the capillitial threads of which bear vesicular expansions enclosing lime crystals. It is of interest, as we have seen, that this is the only *Didymium* species out of eight studied by Schoknecht (1975) in which X-ray microanalysis revealed the presence of phosphorus in the calcareous deposits. Henney *et al.* (1980) have, however, also found phosphorus in the calcareous deposits in *D. atrichum*, which they postulated to be related to *D. trachysporum* because of a number of other similarities. The peculiar structure of the rare plasmodiocarpous *D. sturgisii* Hagelst. should be mentioned here. A columella is lacking in this species, but the thickened base of the sporophore gives rise to numerous columns enclosing white lime crystals and reaching the upper surface where they are attached. Lister (1925) considers these columns of *D. sturgisii* (*D. anomalum* Sturgis) to represent the capillitium, but Martin and Alexopoulos (1969, p. 399) state that "there is not the slightest suggestion that the trabeculae (in *D. sturgisii*) are part of the capillitial system." Such differences of opinion reflect our lack of knowledge concerning capillitium development in *Didymium* and, in fact, in the Physarales in general. Developmental and ultrastructural studies are sorely needed in this area. More species of *Physarum* and *Didymium* have been cultivated in the laboratory from spore to spore than those of any other myxomycete genera, and there should be no lack of material for such studies by interested individuals. One last statement may be warranted here. In *D. atrichum* and *D. eremophilum* there is no trace of capillitium in the sporophores.

f. The Spores. What has been said of the spores of *Physarum* holds equally well for those of *Didymium*. There are no characteristics that distinguish *Didymium* spores from those of other Physarales. The diameter of the spores varies from as little as 6 μm in *D. ochroideum* G. Lister to as much as 22 μm in *D. rugulosporum* Kowalski. In most species it falls somewhere in the middle. In color too the spores of *Didymium* are typically violaceous-brown, yellow-brown, brown, or pallid. Spore walls are variously ornamented and minutely spinulose; the spines are sometimes arranged in a reticulate pattern, as in *D. trachysporum;* warted, with the warts sometimes clustered, as in *D. nigripes;* or reticulate, as in *D. atrichum* (Fig. 12). In *D. rugulosporum,* the spores are distinctly warted with large warts and, in addition, bear a reticulum of raised ridges. The spores of *D. saturnus* Keller are unique in bearing a Saturn-like ring and those of *D. synsporon* Brooks and Keller in being borne in clusters.

g. Supposed Relationships. On the basis of the crystalline lime on its peridium, *Didymium* is grouped with *Mucilago* and *Lepidoderma*. If we accept *Squamuloderma* as a valid genus, it too could be supposed to have affinities with this group.

Concerning supposed relationships among species of *Didymium*, here again species may be grouped according to characteristics they show in common to indicate possible relationships. As with *Physarum*, this has been done again by

Lister (1894, 1911, 1925) and by Hagelstein (1944). The synoptic keys of Farr (1976) for the neotropic species, many of which are widely distributed over the earth, may be used for this purpose. There is no need to repeat these groupings here. The references cited above are readily available in any good library.

III. GENERAL TAXONOMY

A discussion of the two genera under consideration in this book cannot be meaningful without a short discussion of the taxonomy of the Myxomycetes in general. The classification of the whole group of endosporous Myxomycetes traditionally begins with the color of the spores. The dark-spored Myxomycetes, *Physarum* and *Didymium* among them, are placed in two orders: Physarales and Stemonitales. These are separated on the basis of presence or absence of lime from the peridium, lime being typically present in the Physarales and typically absent in the Stemonitales. In addition, lime is said to be absent from the capillitium of the Stemonitales. Such distinctions, although very old in the classification system and quite workable in general, are probably artificial. The separation of two major orders of Myxomycetes based solely on the presence or absence of lime in the peridium and capillitium does not seem that important.

With the establishment of the new subclass Stemonitomycetidae by Ross (1973), we may have better-circumscribed subclasses. To both Ross (1973) and Alexopoulos (1973), methods of sporophore development coupled with type of plasmodium and method of spore germination (by cracking of the spore wall or by a pore dissolved in it) seem more fundamental than presence or absence of lime in the sporophore. Olive (1975) has criticized both Ross and Alexopoulos for placing too much emphasis on the two types of stalk development, first signaled by de Bary (1887). Yet, in the same treatise (pp. 30–31), Olive himself based transfer of *Ceratiomyxa* from the Myxomycetes to the Protostelida chiefly on the manner of stalk development!

If we accept Ross' (1973) classification, as Alexopoulos (1973) and Farr (1976) do, the subclass Myxogastromycetidae includes all endosporous Myxomycetes with subhypothallic development, spore germination by cracking of the spore wall, and a trophic stage other than an aphanoplasmodium. The subclass is divided into four orders: Liceales, Echinosteliales, Trichiales, and Physarales.

The subclass Stemonitomycetidae includes endosporous Myxomycetes with epihypothallic development, spore germination by a pore dissolved in the spore wall, and an aphanoplasmodium as the assimilative stage. These are all assembled in the single order Stemonitales.

Whether the exosporous species (subclass Ceratiomyxomycetidae) belong to the Myxomycetes, where most monographers still place them, or to the Protostelida, where Olive (1975) and Ross (1979) believe them to belong, cannot be ascertained until *Ceratiomyxa* is cultured in the laboratory spore to spore and a vigorously growing plasmodium can be studied.

Accepting Ross' (1973) concepts unfortunately does not solve all our taxonomic problems. The *Comatricha/Diachea* relationship is a case in point. *Diachea,* back again in the Physarales according to Farr (1974, 1976) and Keller (1980), differs from *Comatricha,* a large genus of the Stemonitales, by the presence of lime in the stalk and columella in the former and its complete absence from the sporophores of the latter, according to all major monographs. *Comatricha caespitosa* Sturgis, however, sometimes develops lime in its columella, as Farr (1979) points out, and its transfer to the genus *Diachea* by the Listers in 1907, more or less intuitively, seems to Farr (1979) to be justified on the basis of presence of lime in the columella since (1) the sporophores of this species are usually sessile, and no developmental studies are extant, and (2) its plasmodium remains unknown. The few species of *Comatricha* that have been investigated in these respects, however, undergo an epihypothallic development (Ross, 1960; Goodwin, 1961), and the few that have been cultured have an aphanoplasmodium (Alexopoulos, 1969),* whereas *D. leucopodia,* the type species of the genus *Diachea,* undergoes a subhypothallic development (Blackwell, 1974) [a fact that Farr (1974) took into consideration, as we have seen, when she transferred *Diachea* from the Stemonitaceae to the Didymiaceae], and the plasmodium of *D. splendens* Peck is intermediate between an aphanoplasmodium and a phaneroplasmodium (Indira, 1965). On this basis, we cannot but agree with Farr (1979, p. 116) that "future evidence may well lead us to reevaluate the significance of lime and other diagnostic criteria currently considered fundamental in myxomycete taxonomy."

In conclusion, we must admit that in the Myxomycetes we are still at a stage when convenient characteristics, artificial though they may be, must still be used for identification and classification and that much more research is needed before we can develop any semblance of a natural system. Our present concepts of the origin and evolution of the Myxomycetes will be discussed next in the concluding section of this chapter.

IV. PHYLOGENY

Linnaeus (1753) included the slime molds in his "Species Plantarum," and his treatment has been designated by the International Botanical Congress (whose rules of nomenclature mycologists follow for the fungi) as the starting point for

---

*Venkataramani *et al.* (1977) state that *Comatricha suksdorfii* produces a phaneroplasmodium in laboratory culture. However, the sporophores formed in their cultures were abnormal, and their experiments should be repeated to ascertain that they were indeed culturing one of the Stemonitales.

1. Morphology, Taxonomy, and Phylogeny

myxomycete classification. Fries (1829) grouped the true slime molds with the Gasteromycetes (puffballs and similar fungi) and Link (1833) applied the name "Myxomycetes" to them, reflecting his belief that they are fungi. de Bary (1864), who initiated the serious study of these organisms, believing they have affinities with the animals rather than the plants, in which fungi were then included, named them "Mycetozoa" and classified them with the protozoa. In this he was followed by his student, Rostafinski (1873, 1875), Lister (1894, 1911, 1925) in England, and Hagelstein (1944) and Bessey (1950) in America.

Even in more modern times, the origin and relationships of the Myxomycetes or Mycetozoa continue to remain obscure. In the absence of a fossil record and of sufficient biochemical data, phylogenetic considerations must be based on morphology and physiology alone. Those who believed the Mycetozoa are more closely related to the animals (de Bary, 1864, 1887; Bessey, 1950) emphasized the absence of walls from the mobile assimilative stages and their phagotrophic nutrition. Martin (1960), who placed the Myxomycetes in the fungi, emphasized reproduction by spores borne in fungus-like sporophores and reminded us that there are some undoubted fungi with hyphae devoid of cell walls and with multinucleate coenocytic protoplasts. He also cited the fact that the plasmodium of *P. polycephalum* has been grown in axenic culture in liquid (nonparticulate) media and that therefore plasmodia are not obligately phagotrophic (Daniel and Rusch, 1956). Since then, several other species have also been grown in axenic liquid culture in both the plasmodial and myxamoebal stages (Henney and Lynch, 1969; Henney and Asgari, 1975). This view was also reflected in the classification adopted by Martin and Alexopoulos (1969), Ainsworth *et al.* (1973), Nannenga-Bremekamp (1974), and Farr (1976), all of whom designated these organisms as Myxomycetes rather than Mycetozoa. With the newer ideas of the multikingdom classification of organisms (Copeland, 1956; Whittaker, 1969; Whittaker and Margulis, 1978), the controversy about the place of the true slime molds among living organisms continues, Whittaker (1969) placing them among the fungi and Olive (1970, 1975) in the kingdom Protista. It is of interest, in this connection, that the only modern study of the chemical composition of the spore wall of a myxomycete (that of *P. polycephalum*) indicated that the chemistry of the spore wall is different from that of any fungus or protozoan (McCormick *et al.*, 1970). This brings us back to square one!

The most probable extant relatives of the Myxomycetes are the Protostelida, for as Alexopoulos (1973, p. 40) remarked, "With the discovery of the Protostelids it is no longer possible to circumscribe the Myxomycetes with a sharp line that will exclude all the Protostelida." One might speculate that some ancestral amoeboflagellate gave rise to present-day protostelids and present-day Myxomycetes, according to Fig. 13. As Olive (1975) has pointed out, the sporophores of the protostelids closely resemble those of *Echinostelium*, which might be thought of as being the most primitive of modern Myxomycetes because of its simple

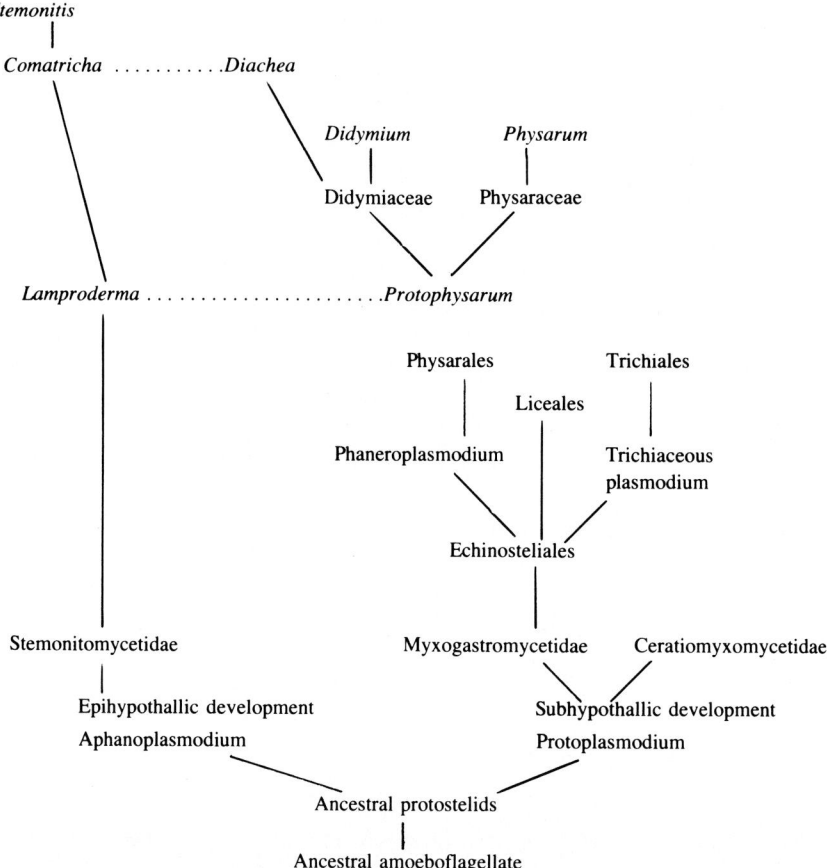

Fig. 13. Hypothetical origin and relationships of the Myxomycetes.

structure, its minute size, its apogamic life cycle (Haskins and Therrien, 1978),* and its protoplasmodium, which Alexopoulos (1960) considered to be the primitive plasmodial type. On the other hand, the plasmodia of protostelids are much more comparable to the aphanoplasmodia of the Stemonitomycetidae than to the protoplasmodia of *Echinostelium*. Furthermore, Collins (1979) regards the protoplasmodium as an advanced type, possibly a reduced phaneroplasmodium. If we accept this view and emphasize the possible evolution of the plasmodium over that of the sporophore, the Echinosteliales could be considered as an ad-

---

*These results, based on a number of strains of *E. minutum*, must be considered tentative until several other forms of the same species are studied. No other species of *Echinostelium* has been investigated for change of ploidy in the life style.

vanced (reduced) group and the aphanoplasmodial Stemonitales as a primitive group. This further emphasizes the futility of discussing phylogeny in the Myxomycetes.

If we consider an ancestral protostelid to have given rise to the Myxomycetes, we must probably postulate a triple origin of the three subclasses separately: (1) the Ceratiomyxomycetidae to have originated from ancestral forms resembling *Ceratiomyxella* or *Protosporangium;* (2) the myxogastromycetidae from an ancestral nonplasmodial, *Echinostelium*-like prototype; and (3) the Stemonitomycetidae from an ancestral plasmodial protostelid. The first line culminates with *Ceratiomyxa;* the second proceeds from *Echinostelium* to the Liceales and Trichiales in one branch and to the Physarales in another; and the third give rise to the Stemonitales through *Lamproderma, Comatricha,* and *Stemonitis*. Whether the Didymiaceae or the Physaraceae originated first we cannot know. What do resemblances between *Protophysarum* and *Lamproderma*, and between *Diachea* and *Comatricha*, mean? How significant is crystalline versus granular lime; presence or absence of phosphorus from the calcareous deposits; presence or absence of lime from peridia and capillitia? There are no answers to these questions, and it is probable that none will ever be developed.

ACKNOWLEDGMENTS

I am grateful to Professor Henry C. Aldrich for providing the phtographic illustrations for this chapter and to him as well as to Drs. Marie L. Farr, Meredith Blackwell, and Raymond W. Scheetz for help with some of the literature cited.

REFERENCES

Ainsworth, G. C., Sparrow, F. K., and Sussman, A. S. eds. (1973). "The Fungi: An Advanced Treatise," Vol. 4A, p. 5; Vol. 4B, p. 5. Academic Press, New York.

Alexopoulos, C. J. (1960). Gross morphology of the plasmodium and its possible significance in the relationships among Myxomycetes. *Mycologia* **52,** 1-20.

Alexopoulos, C. J. (1966). Morphogenesis in the Myxomycetes. *In* "The Fungi: An Advanced Treatise" (G. C. Ainsworth and A. S. Sussman, eds.), Vol. II, pp. 211-234. Academic Press, New York.

Alexopoulos, C. J. (1969). The experimental approach to the taxonomy of the Myxomycetes. *Mycologia* **61,** 219-239.

Alexopoulos, C. J. (1973). The Myxomycetes. *In* "The Fungi: An Advanced Treatise" (G. C. Ainsworth, F. K. Sparrow, and A. S. Sussman, eds.), Vol. 4B, pp. 39-60. Academic Press, New York.

Bessey, E. A. (1950). "Morphology and Taxonomy of Fungi." McGraw-Hill (Blakiston), New York.

Blackwell, M. (1974). A study of sporophore development in the myxomycete *Protophysarum phloiogenum*. *Arch. Mikrobiol.* **99,** 331-344.
Blackwell, M., and Gilbertson, R. L. (1980). A new myxomycete from the Sonoran desert of Arizona. *Mycologia* **72,** 791-797.
Celakowski, L. (1892). Ueber die Aufnahme lebender und todter verdaulicher Körper in die Plasmodien der Myxomyceten. *Flora* **76,** 182-244.
Collins, O. R. (1979). Myxomycete biosystematics: Some recent developments and future research opportunities. *Bot. Rev.* **45,** 145-201.
Copeland, H. F. (1956). "The Classification of Lower Organisms," 158 pp. Freeman, San Francisco, California.
Crowder, W. (1926). Marvels of Mycetozoa. *Nat. Geogr.* **49,** 421-443.
Daniel, J. W., and Rusch, H. P. (1956). Growth of a plasmodial slime mold in pure culture on a soluble medium. *Fed. Proc. Fed. Am. Soc. Exp. Biol.* **15,** 513.
de Bary, A. (1864). Die Mycetozoen (Schleimpilze). Ein Beitrag zur kentniss der niederen Organismen. 132 pp. Englemann, Leipzig (cf. Martin and Alexopoulos, 1969).
de Bary, A. (1887). "Comparative Morphology and Biology of the Fungi, Mycetozoa, and Bacteria" (Engl. transl.), 525 pp. Oxford Univ. Press (Clarendon), London and New York.
Emoto, Y. (1977). "The Myxomycetes of Japan," 263 pp. Sangyo Tosho Publ. Co. Ltd., Tokyo.
Farr, M. L. (1961). *Badhamia decipiens* reinstated in *Physarum*. *Britonnia* **13,** 339-345.
Farr, M. L. (1974). Some new myxomycete records from the neotropics and some taxonomic problems in the Myxomycetes. *Proc. Iowa Acad. Sci.* **81:** 37-40.
Farr, M. L. (1976). Myxomycetes. Flora Neotropica Monogr. No. 16. 304 pp. New York Bot. Gard., New York.
Farr, M. L. (1979). Notes on Myxomycetes II. *Nova Hedwigia* **31,** 103-118.
Fergus, C. L., and Schine, R. D. (1963). Light effects on fruiting of *Physarum gyrosum*. *Mycologia* **55,** 540-548.
Fries, E. M. (1829). Systema mycologicum, sistens fungorum ordines, genera et species. *Myxogastres* **3,** 67-199 (cf. Martin and Alexopoulos, 1969).
Goodwin, D. C. (1961). Morphogenesis of the sporangium of *Comatricha*. *Am. J. Bot.* **48,** 148-154.
Gray, W. D. (1961). The laboratory cultivation of *Physarum flavicomum*. *Am. J. Bot.* **48,** 242-243.
Gray, W. D., and Alexopoulos, C. J. (1968). "The Biology of the Myxomycetes," 288 pp. Ronald Press, New York.
Hagelstein, R. (1944). "The Mycetozoa of North America," 288 pp. Published by the author, Mineola, New York.
Haskins, E. F., and Therrien, C. D. (1978). The nuclear cycle of the myxomycete *Echinostelium minutum*. I. Cytophotometric analysis of the nuclear DNA content of the amoebal and plasmodial phases. *Exp. Mycol.* **2,** 32-40.
Hattori, H. (1935, 1964). "Myxomycetes of Nasu District," 1st and 2nd eds., Tokyo (text in Japanese). Edited by Biological Laboratory Imperial Household.
Henney, H. R., Jr., and M. Asgari. 1975. Growth of the haploid phase of the myxomycete *Physarum flavicomum* in defined minimal medium. *Arch. Microbiol.* **102,** 175-178.
Henney, H. R., Jr., and T. Lynch. (1969). Growth of *Physarum flavicomum* and *Physarum rigidum* in chemically defined minimal media. *J. Bacteriol.* **99,** 531-534.
Henney, M. R. (1968). Mating type systems in the Myxomycetes *Physarum globuliferum* and *Physarum bilgramii*. *Am. J. Bot.* **55,** 720 (Abstr.).
Henney, M. R., Alexopoulos, C. J., and Scheetz, R. W. (1980). *Didymium atrichum*, a new myxomycete from south-central Texas. *Mycotaxon* **11,** 150-164.
Indira, P. U., (1965). *In vitro* cultivation of *Diachea splendens* Peck. *Curr. Sci.* **34**(21), 601-602.
Keller, H. W. (1980). Corticolous Myxomycetes VIII: *Trabrooksia*, a new genus. *Mycologia* **72,** 359-403.

Koevenig, J. L. (1964). Studies of life cycle of *Physarum gyrosum* and other Myxomycetes. *Mycologia* **56,** 170-184.
Link, J. F. H. (1833). Handbuch zur erkennung der nurtzbarsten und am haufigsten vorkemmenden Gewachse. 3. Ordo Fungi, Subordo Myxomycetes. 405-422; 432-433. Berlin (cf. Martin and Alexopoulos, 1969).
Linnaeus, C. (1753). *Species Plantarum* **2,** 561-1200. Holmiae (cf. Martin and Alexopoulos, 1969).
Lister, A. (1894, 1911, 1925). "Mycetozoa," eds. 1-3. British Museum (Nat. Hist.), London.
Lister, A., and Lister, G. (1907). Synopsis of the orders, genera, and species of Mycetozoa. *J. Bot.* **45,** 176-197.
Macbride, T. H. (1922). "North American Slime Moulds," 2nd ed., 347 p. Macmillan, New York.
McCormick, J. J., Blomquist, J., and Rusch, H. P. (1970). Isolation and characterization of a galactoseamine wall from spores and sphaerules of *Physarum polycephalum*. *J. Bacteriol.* **104,** 1119-1125.
McManus, M. A. (1965). Ultrastructure of myxomycete plasmodia of various types. *Am. J. Bot.* **52,** 15-25.
Martin, G. W. (1949). "The Myxomycetes," N. Am. Flora 1, pt. 1, pp. 1-190. New York Bot. Gard., New York.
Martin, G. W. (1960). The systematic position of the Myxomycetes. *Mycologia* **52,** 119-129.
Martin, G. W., and Alexopoulos, C. J. (1969). "The Myxomycetes," 561 pp. (incl. 41 col. pls.). Univ. of Iowa Press, Iowa.
Miller, C. A. (1898). The aseptic cultivation of Mycetozoa. *Q. Jr. Microsc. Sci.* **41,** 43-71.
Morgan, A. P. (1894). The Myxomycetes of the Miami Valley, Ohio. *J. Cinc. Soc. Nat. Hist.* **3,** 127-156.
Nannenga-Bremekamp, N. E. (1974). De Nederlandse Myxomyceten. Nederl. "Naturrhist," 440 p. Ver Zutphen (Netherlands).
Olive, L. S. (1970). The Mycetozoa: A revised classification. *Bot. Rev.* **36,** 59-89.
Olive, L. S. (1975). "The Mycetozoans." Academic Press, New York.
Persoon, C. H. (1794). Neuer Versuch einer systematischen Eintheilung der Schwämme. *Neues Mag. Bot.* **1,** 63-128.
Ross, I. K. (1960). Sporangial development in *Lamproderma arcyrionema*. *Mycologia* **52,** 621-627.
Ross, I. K. (1973). The Stemonitomycetidae, a new subclass of Myxomycetes. *Mycologia* **65,** 477-485.
Ross, I. K. (1979). "Biology of the Fungi," 499 p. McGraw-Hill, New York.
Rostafinski, J. T. (1873). Versuch eines Systems der Mycetozoen. Inaugural-Dissertation, IV, 21 pp. Univ. of Strasbourg, Strasbourg, France.
Rostafinski, J. T. (1875-1876). Sluzowce (Mycetozoa). Monografia, Paryz. 432 p.
Schrader, H. A. (1797). Nova genera plantarum. I-VIII and 32 p. "Pars prima," but no later part published (cf. Martin and Alexopoulos, 1969).
Schoknecht, J. D. (1975). SEM and X-ray microanalysis of calcareous deposits in myxomycete fructifications. *Trans. Am. Microsc. Soc.* **92,** 216-233.
Thom, C., and Raper, K. B. (1930). Myxamoebae in soil and decomposing crop residues. *J. Wash. Acad. Sci.* **20,** 362-370.
Venkataramani, R., Kalyanasundaram, I., and Kalyanasundaram, R. (1977). A simple technique for inducing sporulation in cultures of Myxomycetes. *Trans. Br. Mycol. Soc.* **69,** 320-322.
Whittaker, R. H. (1969). New concepts of kingdoms of organisms. *Science* **163,** 150-160.
Whittaker, R. L., and Margulis, L. (1978). Protist classification and the kingdoms of organisms. *BioSystems* **10,** 3-18.
Zukal, H. (1893). Ueber zwei neue Myxomyceten. *Oesterr. Bot. Z.* **43,** 73-77; 133-137 (cf. Martin and Alexopoulos, 1969).

CHAPTER 2

Didymium iridis in Past and Future Research

O'NEIL RAY COLLINS and DONALD A. BETTERLEY

I.	Introduction	25
II.	Laboratory Cultivation and Maintenance	28
III.	Isolates Studied and Their Origin	28
IV.	Reproductive Cycles	28
	A. Heterothallic Cycles	29
	B. Haploid Facultative Apomicts (or "Selfers")	31
	C. Diploid Apomixis and Homothallism	32
	D. Apogamic Amoebal Mating-Type Heterozygotes	34
V.	Chromosomal Numbers and Nuclear DNA Contents	35
VI.	Speciation and Evolutionary Diversification	38
VII.	Research on the Plasmodial Stage	41
	A. Development and the Mating-Type Locus	41
	B. Morphology and Pigmentation	44
	C. Incompatibility Systems	45
	D. Senescence	49
VIII.	Sporophore Stage	50
IX.	Amoeboflagellate-Microcyst Stage	51
X.	Concluding Remarks	52
	References	52

I. INTRODUCTION

Didymium iridis (Ditmar) Fries is a well-known true slime mold whose proper taxonomic placement is in the class Myxomycetes, order Physarales, family Didymiaceae (Martin and Alexopoulos, 1969). Originally, it was given the binomial *Cionium iridis* (Ditmar, 1813), but that was changed to its present name by Fries (1829). It is a cosmopolitan species whose sporophores most commonly develop on dead fallen leaves in forests, but other substrates include twigs, dead logs, straw, and mosses. Sporophores are typically in the form of

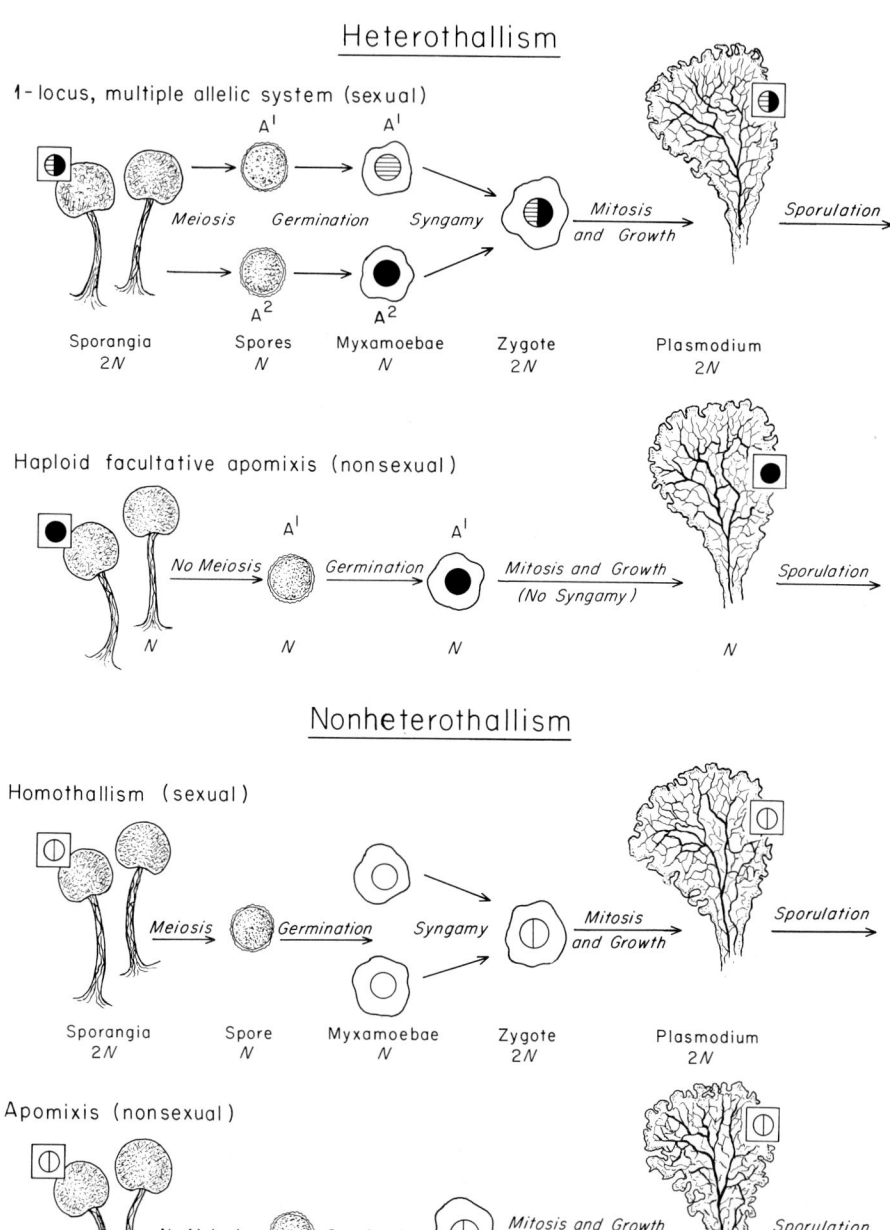

Fig. 1. Life cycles of *Didymium iridis*. Heterothallism—1-locus, multiple allelic system (sexual) and haploid facultative apomixis (nonsexual) or "selfing." Nonheterothallism—homothallism (sexual) and apomixis (nonsexual).

2. *Didymium iridis* in Past and Future Research

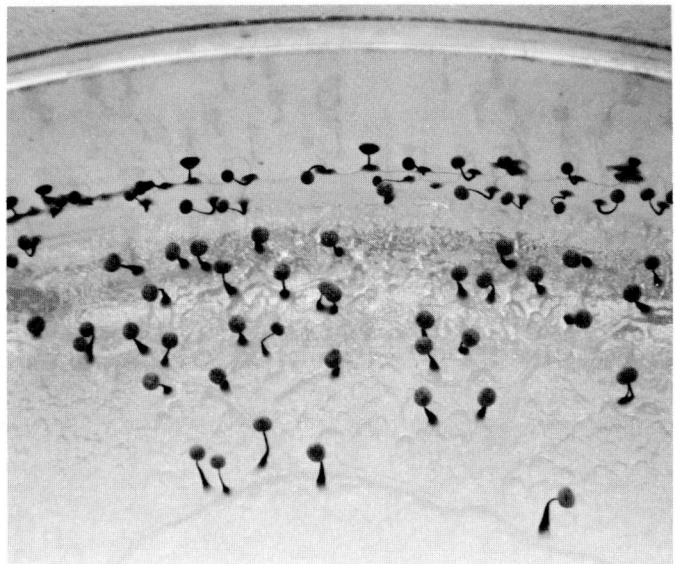

Fig. 2. Sporophores of *Didymium iridis* on agar. ×3.

sporangia. These have yellowish translucent stalks which rarely exceed 1.5 mm in height, whereas the whitish, lime-covered sporangium proper has a maximum diameter of no more than 0.7 mm. Each sporangium contains about 500,000 dark brown spores, a single relatively large, centrally located white columella, and numerous colorless, branching, threadlike capillitia. The plasmodial stage is typically brown and not easily noticed in the field against its usually dark substrates. When compared to species producing very large phaneroplasmodia, such as *Physarum polycephalum* Schw. and *Fuligo septica* (L.) Wiggers, the plasmodium of *D. iridis* is small, reaching a maximum diameter of no more than a few inches. In addition to macroscopic sporophores and plasmodia, the microscopic and colorless amoeboflagellates are important life cycle entities (Fig. 1). These emerge from the spores upon germination and give rise to plasmodia, which in turn produce more sporophores (Fig. 2).

The current prominence of *D. iridis* as a research organism derives from the relative ease with which it can be grown in monoxenic agar cultures, and especially from the discovery of multiple allelic heterothallism in the species (Collins, 1961, 1963). Knowledge of the existence of a series of mating types made controlled crosses and genetical analyses a practical and worthwhile undertaking. Since 1961, we have accumulated a considerable amount of data on the organism. This, and related information on other myxomycete species, will be presented and interpreted in the context of myxomycete biosystematics and evolutionary implications.

II. LABORATORY CULTIVATION AND MAINTENANCE

When specimens are collected from the field or from moist chambers, cultures are started from the spore contents of whole sporangia. These are called "mass-spore cultures" and theoretically reflect the genetic heterogeneity of specimens from nature. Each isolate (collection) tends to have its own peculiar traits, and these must be studied and taken into account in laboratory manipulations.

Techniques for cultivation of *D. iridis* amoebae and plasmodia in crude and monoxenic cultures are essentially the same as for *P. polycephalum*. Those procedures are covered adequately in Chapter 5 of Volume 2. *Didymium iridis* is most often maintained with the bacterium *Klebsiella pneumoniae* (Schroeter) Trevisan. Axenic growth of amoebae on agar has been achieved (Kerr, 1963a; Balamuth and Gong, 1975), but most work to date has been with monoxenic cultures. Plasmodia of *D. iridis* have not been grown successfully in axenic culture, and there are no reports of axenic growth of amoebae or plasmodia in liquid shake culture on defined or semidefined media.

Mass-spore cultures and amoebal clones are stored on half-strength corn meal agar slants or preferably lyophilized, as described earlier (Davis, 1965; Kerr, 1965b) and summarized by Collins (1979).

III. ISOLATES STUDIED AND THEIR ORIGIN

The Hon 1 isolate of *D. iridis* in which heterothallism was first demonstrated (Collins, 1961) was collected by Lindsay Olive in Lancetilla, Honduras. Subsequently, 42 additional isolates from many locations have been examined for the existence of mating types, and well over half of these have been shown to be nonheterothallic (Table I). No other myxomycete species has been so extensively sampled. Costa Rican and Panamanian specimens were obtained from banana peels in moist chamber cultures, so the places of origin of the bananas from which they were derived are listed as their geographic origins, but we cannot be certain of their actual derivation. Future geographic and ecological studies should begin with specimens collected directly from the field.

In addition to *D. iridis,* several other myxomycete species have been examined for mating types. Those for which at least three isolates have been studied are listed in Table II.

IV. REPRODUCTIVE CYCLES

It is useful to subdivide reproductive cycles in *D. iridis* and other Myxomycetes into heterothallic and nonheterothallic categories (Fig. 1). The first group is

TABLE I

Heterothallic and Nonheterothallic Isolates of *Didymium iridis*[a]

Heterothallic			Nonheterothallic			
Isolate	Geographic origin	Mating types	Isolate	Geographic origin	Isolate	Geographic origin
Hon 1	Honduras	A^1A^2	CR 3	Costa Rica[b]	Mo 1	Missouri
Ia 1	Iowa	A^1A^2	CR 4	Costa Rica[b]	Ga 1	Georgia
Pan 1	Panama	$A^3A^4(A^{4m-1})$	CR 7	Costa Rica[b]	Tx 1	Texas
CR 1	Costa Rica[b]	A^5A^6	Dom 1	Dominica	Ks 1	Kansas
CR 2	Costa Rica	A^5A^6	Pan 4	Panama	Wa 2	Washington
CR 6	Costa Rica	A^7	Pan 5	Panama	Mn 1	Minnesota
Pan 2	Panama	A^7A^8	Pan 6	Panama[b]	Mn 2	Minnesota
Pan 3	Panama	A^9A^{10}	Pan 7	Panama[b]	Mn 3	Minnesota
CR 5	Costa Rica	A^2A^{11}	Ph 1	Philippines	Mn 4	Minnesota
			Ha 1	Hawaii	Mn 5	Minnesota
Wa 1	Washington[b]	+, −	Ca 1	California	Mn 6	Minnesota
			Ca 2	California	NC 1	North Carolina
Ga 2	Georgia	A^aA^b	Ca 3	California	Tr 1	Trinidad
			Ca 4	California	So. Afr. 1	South Africa
Ky 1	Kentucky	A^xA^y	Ca 5	California	Kerr	France
			Ca 6	California		

[a] List updated from Collins (1976) by Betterley (1981).
[b] Spores are no longer viable.

characterized by the known existence of mating types, whereas the second is not known to possess them and may be either apomictic or homothallic.

A. Heterothallic Cycles

In a recent compilation of available data (Collins, 1979, p. 150, Table 1), it was pointed out that in 14 of 38 species examined, at least one isolate was known to possess mating types. This number has been increased to 15, and as shown in Table II, in 9 of these 15 species nonheterothallic isolates are also reported. Probably the earliest valid report of heterothallism in a myxomycete was that of von Stosch (1935) for *D. nigripes*. He concluded that there were two mating types, designated (+) and (−). For the next 25 years the subject lay dormant, but independent reports of + and − mating types, one in *P. polycephalum* (Dee, 1960) and the other in *D. iridis* (Collins, 1961, also in a paper presented at the 1960 Annual Mycological Society meeting in Stillwater, Oklahoma), signaled the beginning of sustained research on myxomycete heterothallism, especially in *D. iridis* and *P. polycephalum,* the two species for which the greatest amount of information is now available. Whereas studies on heterothallism in *P.*

TABLE II

Summary of Distribution of Heterothallism versus Nonheterothallism in the Myxomycetes for Species in Which Three or More Isolates Have Been Examined[a]

Species	Heterothallic isolates	Nonheterothallic isolates
Physarales		
Didymiaceae		
Didymium difforme	1	8
D. iridis	9 + 1 + 1 + 1[b]	31
D. nigripes	1, 1[c]	4
D. squamulosum	0	4
Physaraceae		
Badhamia utricularis	1, 1	1[d]
Fuligo septica	4	3
Physarum cinereum	1	2
P. compressum	0	3
P. flavicomum	3 + 2?[e]	0
P. globuliferum	3?	0
P. polycephalum	6, 1	1[f]
P. pusillum	3?	5[g]
P. rigidum	4?	0
Echinosteliales		
Echinostelium minutum	0	7
Stemonitales		
Stemonitis flavogenita	1[h]	5

[a] Modified from Collins (1979) and Clark and Collins (1976).
[b] A plus mark between numbers indicates that these are separate allelic series.
[c] A comma between numbers indicates that the isolates have not been tested for allelism.
[d] Betterley, 1981.
[e] A question mark after a number indicates that spores from interisolate crosses are nonviable.
[f] This isolate of *P. polycephalum* has been considered homothallic but is an extreme example of a facultative apomict (Collins, 1979, p. 156).
[g] One additional isolate reported by Collins, this chapter.
[h] This isolate was previously designated a nonheterothallic (Collins, 1979), but mating types are now known.

polycephalum have emphasized genetical fine structure of the mating locus and identification of additional genes and factors affecting plasmodial development (Dee, Chapter 6, this volume), studies of heterothallism in *D. iridis* have focused on its usefulness in understanding speciation and evolutionary relationships in the Myxomycetes.

As indicated in Table I, 12 of 43 isolates of *D. iridis* have been reported as heterothallic and 31 as nonheterothallic. Characteristically, a heterothallic isolate is endowed with two mating types which segregate during meiosis as an allelic

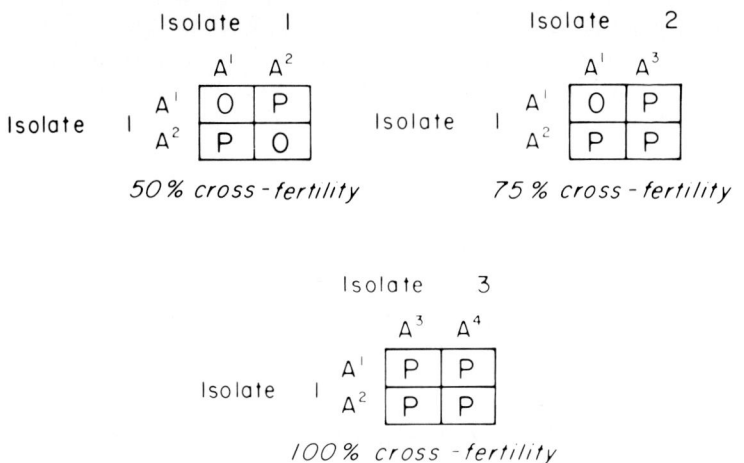

Fig. 3. Patterns of fertility in intraisolate and interisolate crosses.

pair residing at one locus. This is apparent from the pattern of intraisolate plasmodial development when clones of amoeboflagellates are crossed in all pairwise combinations. Approximately 50% of these crosses yield plasmodia, and the rest do not. On the basis of such results, each clone is assigned a mating type, e.g., A^1 or A^2. However, although 50% of random intraisolate crosses are fertile (i.e., yield plasmodia), because of the one-locus, multiallelic mating system, interisolate cross-fertility may be either 50, 75, or 100% depending on whether the two isolates used have two, one, or no mating type alleles in common (Fig. 3). Among the 12 heterothallic isolates studied (Table I), 12 different mating types belonging to a single multiple allelic series are known. A pair of mating types is known for each of three additional isolates (Table I), but since only intraisolate crosses are sexually compatible, mating types of these isolates cannot be assigned to the multiple allelic series. The three isolates (Ky 1, Ga 2, Wa 1), therefore, are reproductively isolated from each other as well as from all others studied.

B. Haploid Facultative Apomicts (or "Selfers")

In *D. iridis,* as in all heterothallic Myxomycetes which have been extensively cultured, some transfers of amoeboflagellates produce plasmodia clonally (Fig. 1). These have a dual capacity, however, and they retain their mating-type specificity, as exemplified by their yielding plasmodia in appropriate crosses. In fact, plasmodia are consistently produced in such crosses, whereas they develop more sporadically and are generally less vigorous in clonal cultures. Our understanding of what causes (or permits) clonal plasmodial development is limited.

However, Collins and Ling (1968) showed that in *D. iridis* only one mating type is recovered in the F_1 generation, indicating that clonal plasmodial development did not result from mutations to new mating types followed by syngamy between mutant and wild-type amoebae. This evidence, coupled with the demonstration that clonally produced plasmodia have haploid nuclei (Yemma and Therrien, 1972; Therrien and Yemma, 1975), clearly suggests that plasmodia develop apogamically and that true meiosis does not occur in the F_1 spores. Clones displaying a capacity to undergo this kind of life cycle may be designated "haploid facultative apomicts" rather than "selfers," because selfing refers to a kind of sexual reproduction. In *P. polycephalum*, mutations which increase the frequency of clonal plasmodial production have been described (e.g., Shinnick and Holt, 1977), and these are designated *gad* for "greater asexual differentiation." From this, it seems reasonable to suppose that in all heterothallic forms, mutations in a variety of different genes might result in clonal plasmodial formation. Further, the large number of individual haploid cells in a clonal population of amoebae is ideal for expressing mutations which in multicellular and/or diploid organisms would likely go unnoticed. Just how mutations or other factors override the mating locus' normal suppression of plasmodial development in clones is not understood, and whether haploid facultative apomixis plays a role in myxomycete evolution is yet to be determined. At present, we know the condition only as a laboratory phenomenon. In this connection, Ling and Moore (1979) found that one of their *D. iridis* clones produced giant multinucleate microcysts, and this was correlated with an unusually high selfing rate, although isolated individual giant cysts did not yield plasmodia. Further studies may provide an explanation for their interesting observation.

C. Diploid Apomixis and Homothallism

As shown in Table I, 31 of 43 *D. iridis* isolates have been classified as nonheterothallic. This label is primarily one of convenience, however, because in most instances we do not know whether the organisms are sexual (homothallic) or nonsexual (apomictic). As part of an effort to distinguish between those two alternatives, nuclear DNA contents have been measured in 16 isolates (see Table V). Whereas some of the data (Therrien and Yemma, 1974; Therrien *et al.*, 1977) appeared to distinguish clearly between homothallic and apomictic cycles, subsequent studies (Yemma *et al.*, 1978; Mulleavy, 1979; Collins, 1980) indicated that nonheterothallic isolates may be apomicts which can spontaneously shift from a nonsexual to a sexual life-cycle mode. If this is correct, all available data on nonheterothallism can be harmonized.

At present, the evidence supports the conclusion that the shift is apomictic–heterothallic rather than apomictic–homothallic. It should be understood that single-spore analyses are usually made from freshly collected specimens, and that the nonheterothallic designation has been typically based on results from such studies. Because the nonheterothallic specimens did not display mating types, they were only rarely used in further genetic experimentation, and ordinarily there was no reason to make further single-spore analyses in subsequent years. For these reasons, shifts from apomixis to heterothallism went undetected until it was noted that nuclear DNA measurements from one specimen gave $n-2n$ results, whereas a second specimen from the same isolate yielded $2n-2n$ data. At this point, the two specimens were genetically analyzed. The one yielding $2n-2n$ DNA data displayed no mating types (apomictic), and the other possessed a pair of mating types (heterothallic). Further checks verified that the two specimens did indeed have a common apomictic origin.

The convertibility hypothesis might also explain why Aldrich and Carroll (1971) found synaptonemal complexes in the Ph 1 isolate (obtained from Collins), whereas Therrien and Yemma (1974) reported a $2n-2n$ cycle, or *diploid apomixis*, and in an abstract, Yemma et al. (1978) reported recovery of one mating type in the same Ph 1 isolate. It is reasonable to assume that chromosomal synapses do occur in reversible apomicts and that reduction divisions are blocked, resulting in the production of $2n$ spores and amoebae. Mutations which arrest meiosis after DNA replication, but before reduction division occurs, have been reported (Klapholz et al., 1978; M. S. Esposito, personal communication) in the yeast *Saccharomyces cerevisiae,* an organism whose haplophase is similar to that of Myxomycetes in that it is composed of large populations of uninucleate cells.

As indicated above, experiments and observations on the Pan 4 isolate of *D. iridis* revealed that a heterothallic line was spontaneously derived from an apomictic line (Collins, 1980). The heterothallic line displays two mating types and grows less well than the apomictic line. At present, the simplest explanation for the conversion is that at some point in its history in nature a heterothallic entity gave rise to some offspring with a capacity to bypass both syngamy and meiosis while retaining the potential for reversion to a heterothallic sexual state. Selection pressures which would favor one cycle over the other have not been studied in the Myxomycetes, but facultative $2n$ apomicts are known in angiosperms (Nygren, 1967; Baker, 1978). It is easy to imagine that in certain habitats apomicts might have at least a short-term selective advantage, inasmuch as all of their offspring would be genetically identical, whereas their sexual counterparts would produce offspring of varied genetic backgrounds, some of which might not be well adapted to the same habitat as that of their parents. It now seems entirely conceivable that in nature there are heterothallic and apomictic forms only, perhaps with true homothallism nonexistent. In this connection, it is of particular interest that Kerr (1961) showed a positive correlation between isolated fusing pairs of amoeboflagellates and plasmodial development. Subsequent work (N. S. Kerr, 1967) with his nonheterothallic isolate* of *D. iridis* showed that one strain can form plasmodia without prior syngamy, and S. J. Kerr (1968) reported that ploidal levels in amoebae and plasmodia were the same in two of Kerr's strains. Further, S. J. Kerr pointed out that "although nuclear and cytoplasmic fusion may precede plasmodial initiation, such fusion is not necessary for plasmodial initiation (p. 14)". This work lends further support to the hypothesis that homothallism may not exist in the Myxomycetes.

D. Apogamic Amoebal Mating-Type Heterozygotes

It has been shown (Mulleavy and Collins (1979) that amoebal mating type heterozygotes produce plasmodia apogamically. The heterozygotes were created by first crossing $2n$ amoebal clones homozygous for mating types $A^2A^2 \times A^5A^5$. Upon sporulation, the resulting tetraploid plasmodia yielded $2n$ F_1 meiospores, about two-thirds of which were mating type heterozygotic (A^2A^5), and the remainder were homozygotic (A^2A^2 and A^5A^5). Whereas single-spore-derived clones heterozygous for mating types yielded plasmodia as well as amoebal populations, homozygous ones did not progress past the amoebal stage. Further, in the apogamic clones almost every amoeba eventually yielded plasmodia, and the number of plasmodia produced per plate was far greater than that typically produced by compatible crosses and greater than that found in naturally occurring nonheterothallic forms. However, the heterozygous state was not maintained in the F_2 generation. Instead, F_2 amoebal clones were haploid and nonapogamic, indicating the occurrence of normal meiosis. Thus, the artificially created apogamic lines are a one-generation phenomenon, with development occurring without syngamy but not without meiosis. However, it is easy to imagine obtaining mutations in the apogamic lines which would block meiosis. These mutations would effectively create $2n$ apomicts resembling those collected from nature.

The obvious question of whether $2n$ mating type heterozygotes are obligately or facultatively apogamic has been asked, and appropriate experiments have been performed. The results of Mulleavy and Collins (1981) show that mating type heterozygotes (A^7A^8) are capable of undergoing syngamy when presented with a partner homozygous (A^5A^5) for a third mating type. Thus far, it is not known whether mating can also occur between heterozygotes identical for mating type alleles (e.g., $A^7A^8 \times A^7A^8$), with one in common ($A^7A^8 \times A^5A^7$), or with none in common ($A^7A^8 \times A^2A^5$). If the yeast work of Mortimer and Hawthorne (1969) is applicable to the Myxomycetes, heterozygotes carrying like alleles will not mate.

*Kerr's *D. nigripes* = *D. iridis*. Work on the species has also been published by various authors under the names *Didymium xanthopus* (Ditmar) Fries or *Didymium nigripes* var. *xanthopus* (Ditmar) A. Lister.

2. Didymium iridis in Past and Future Research

It is clear from the work of Mulleavy and Collins (1981) that mating type heterozygosity does not prevent mitotic cell division in *D. iridis*. This suggests that zygotes produced in sexually compatible crosses may also have the capacity to multiply, contrary to present assumptions. There is a need to understand more fully the parameters determining whether and when a cell can function only vegetatively, both vegetatively and sexually, or only sexually. Some quantitative work (Ross, 1967; Ross and Cummings, 1970; Ross and Shipley, 1973; Ross *et al.*, 1973; Shipley and Ross, 1978) on *D. iridis* and similar work by Pallotta *et al.* (1979) on *P. polycephalum* is of interest in this regard, because it gives promise of future understanding of the physiology and kinetics of plasmodial development under a variety of experimental cultural conditions. This is needed to complement and extend genetic investigations.

V. CHROMOSOMAL NUMBERS AND NUCLEAR DNA CONTENTS

Historically, myxomycete researchers have experienced difficulty making good chromosome spreads (Collins, 1979, pp. 159–163). This, coupled with the small size and large number of chromosomes, has often impeded progress, especially in the realm of myxomycete biosystematics. In this regard, *D. iridis* is a typical myxomycete. As indicated in Table III, as many as 81 chromosomes have been reported for the haplophase, although the most recent work suggests that 37 may be a good approximation of the haploid number. In any case, the problem of chromosome counting is compounded by the common occurrence of polyploidy (S. J. Kerr, 1968, 1970; Collins and Therrien, 1976; Collins *et al.*, 1978; Mulleavy, 1979). For some kinds of biosystematic investigations, it would be useful to have information on chromosomal numbers in representatives of the 38 species of the genus *Didymium* in addition to *D. iridis*, but as shown in Table

TABLE III

Chromosomal Numbers in the Genus *Didymium*

Species	Haploid chromosomal number	References
D. iridis	ca. 81	von Stosch, 1935
	ca. 6–16	Ling and Collins, 1970b
	ca. 35	S. J. Kerr, 1968
	ca. 37	Mulleavy, 1979
D. nigripes	8	Schünemann, 1930
	4	Cadman, 1931
	ca. 24	von Stosch, 1935
D. squamulosum	8	Jahn, 1911

TABLE IV

Feulgen Nuclear DNA Measurements in Heterothallic Isolates of *Didymium iridis*

Isolate	Myxamoebae	Plasmodia	References
Hon 1	n	$2n$	Therrien, 1966; Therrien and Yemma, 1974, 1975; Collins and Therrien, 1976; Collins *et al.*, 1978
1–7	$n, 2n$		Mulleavy and Collins, 1979
1–2 "selfer"	n	n	Yemma and Therrien, 1972
Pan 1	n	$2n$	Collins and Therrien, 1976; Mulleavy and Collins, 1979
CR 5	n	$2n$	Collins and Therrien, 1976; Collins *et al.*, 1978
CR 2–25	$n, 2n$		Collins and Therrien, 1976; Collins *et al.*, 1978;
2–25*	$4n$		Mulleavy and Collins, 1979, 1980
2–63	$2n, 4n, 8n$		Mulleavy, 1979
2–55	$2n$		Mulleavy, 1979
Pan 2 (B1P)	n	$2n$	Collins and Therrien, 1976; Mulleavy, 1979
2–4 "selfer"	n	n	Therrien and Yemma, 1975
2–4	n	$2n$	Therrien and Yemma, 1974
2–6	$n, 2n, 4n$		Mulleavy and Collins, 1980
2–16	$n, 2n$		Mulleavy, 1979; Mulleavy and Collins, 1980
2–33 "selfer"	n	n	Yemma and Therrien, 1972
Pan 3	$n+, 2n+$	$4n$	Collins and Therrien, 1976
Ky 1	n	$2n$	Collins and Therrien, 1976

III, there is very little available. If a simple and effective technique for spreading chromosomes were developed that the average researcher could use, this might have a profound impact on the future of myxomycete biosystematics.

In the meantime, microspectrophotometric measurement of nuclear DNA contents is helpful for the resolution of nuclear cycles. The main advantage of using nuclear DNA measurements to determine ploidal level is that with proper facilities available, many determinations can be made quickly and expressed quantitatively as picograms of DNA per nucleus. Of course, DNA measurements do not tell us anything directly about chromosome numbers, and it is necessary to know whether nuclear DNA is in the replicated or unreplicated state at the time measurements are made. Further, whether increases in DNA are caused by polyploidy, cryptopolyploidy, or polyteny cannot be directly determined. For these reasons, it is always better to have both the nuclear DNA amount and the chromosomal number available, if possible.

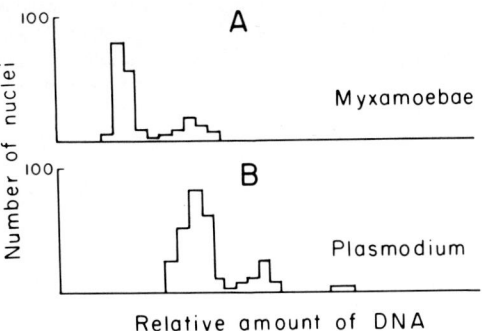

Fig. 4. Histograms of Feulgen-DNA content of myxamoebal and plasmodial nuclei of *Didymium iridis* showing the occurrence of major and minor peaks in a heterothallic isolate. (A) Myxamoebal nuclei—haploid and diploid peaks. (B) Plasmodial nuclei—diploid, tetraploid, and octaploid peaks. Redrawn from Mulleavy and Collins (1979).

Nuclear DNA measurements in several heterothallic isolates of *D. iridis* have been made (Table IV), and from these, relative ploidal levels of amoebae and plasmodia have been assigned. For purposes of simplification, only those ploidal levels corresponding to major DNA peaks are given, but it should be pointed out that small peaks reflecting other values in some nuclei were present in some cases (Mulleavy, 1979; Mulleavy and Collins, 1979). An example of the occurrence of major and minor peaks is given (Fig. 4A,B). However, although DNA values characteristically vary for nuclei from clonal amoebae as well as for nuclei from individual plasmodia, in most instances these fall within a narrow range and display a unimodal distribution, as reflected by single peaks in histograms (Fig. 5A,B).

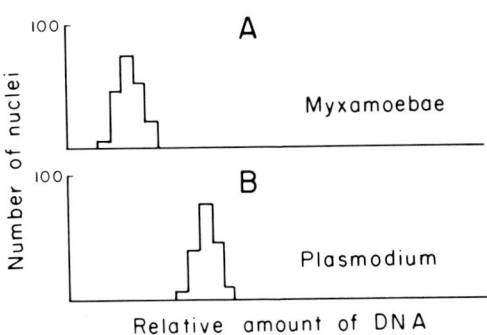

Fig. 5. Histograms of Feulgen-DNA content of myxamoebal and plasmodial nuclei of *Didymium iridis* showing typical unimodal distribution in a heterothallic isolate. (A) Myxamoebal nuclei—haploid peak. (B) Plasmodial nuclei—diploid peak. Redrawn from Collins and Therrien (1976).

TABLE V

Feulgen Nuclear DNA Measurements in Nonheterothallic Isolates of *Didymium iridis*

Isolate	Myxamoebae	Plasmodia	Reference
Pan 4	n, $2n$	$2n$	Mulleavy, 1979; Collins, 1980
	n	$2n$	Therrien et al., 1977
Pan 5	n	$2n$	Therrien et al., 1977
Kerr	n	$2n$	Therrien, 1966
	$2n$	$2n$	S. J. Kerr, 1968 (based on chromosomal numbers)
Tx 1	$2n$, $4n$	$2n$, $4n$	Mulleavy, 1979
Ga 1	$2n$	$2n$	Therrien et al., 1977; Mulleavy, 1979
Ha 1	$2n$	$2n$	Therrien et al., 1977
Ph 1	n	$2n$	Yemma et al., 1978
	$2n$	$2n$	Therrien and Yemma, 1974; Mulleavy, unpublished
Ca 1	n	$2n$	Therrien et al., 1977
	$2n$, $4n$		Mulleavy, 1979
Ca 2	$2n$	$2n$, $4n$, $8n$	Mulleavy, 1979
Mo 1	n	$2n$	Therrien et al., 1977
	$2n$, $4n$	$2n$	Mulleavy, 1979
Ks 1	$2n$, $4n$	$2n$	Mulleavy, 1979
Mn 2	$2n$, $4n$	$2n$, $4n$	Mulleavy, 1979
Mn 3	$2n$, $4n$	$2n$, $4n$	Mulleavy, 1979
Mn 5	$2n$	$2n$	Betterley, this chapter.
Wa 2	$2n$	$2n$	Betterley, this chapter.
So. Afr. 1	$2n$	$2n$	Betterley, this chapter.

Table V shows ploidal designations based on nuclear DNA measurements for several nonheterothallic isolates.

VI. SPECIATION AND EVOLUTIONARY DIVERSIFICATION

As described in Section IV, some isolates of *D. iridis* are heterothallic; others are nonheterothallic. As far as is known, organisms belonging to one group do not interbreed with those belonging to the other. Further, although nine heterothallic isolates are known to interbreed, three others do not do so with one another or with any of the other nine. The cause of the reproductive barriers is not understood, and whether gene flow occurs even indirectly across the barriers seems doubtful. Among noninterbreeding isolates, the barriers appear to represent initial steps toward their divergence into genetically distinct and reproductively isolated populations, possibly corresponding to different clonal populations or

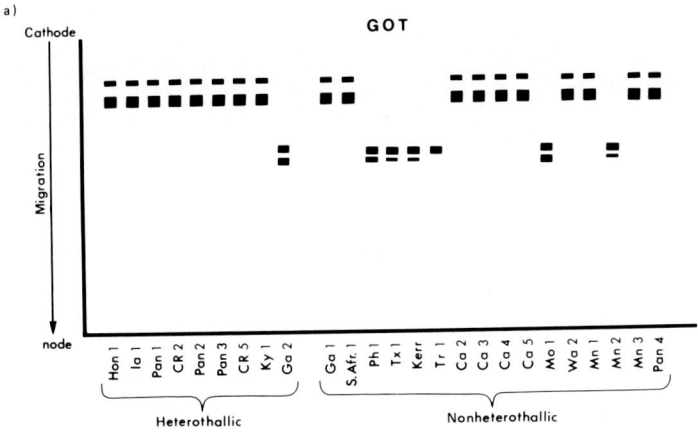
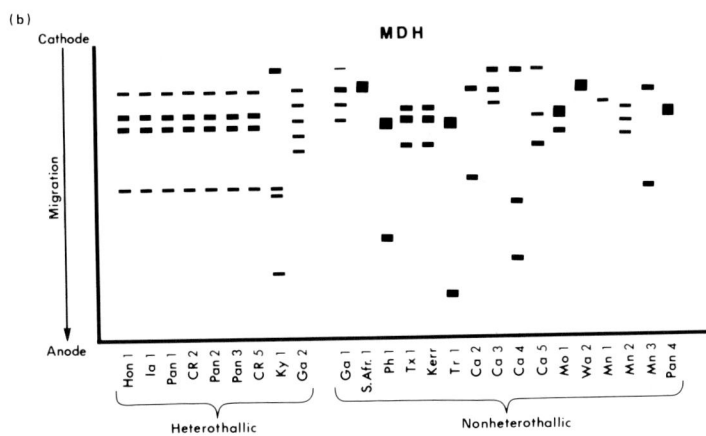
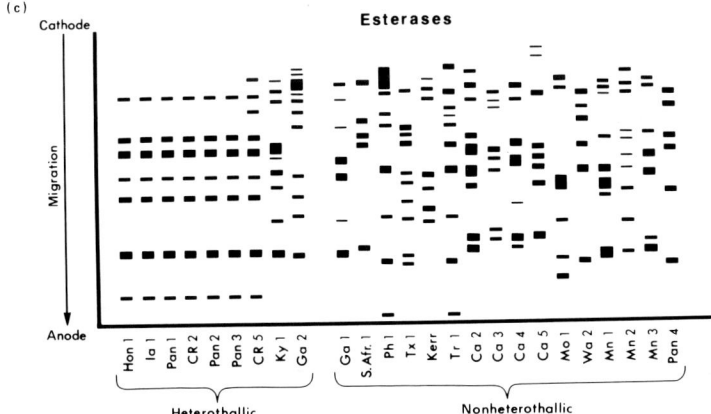

Fig. 6. Isozymes of heterothallic and selected nonheterothallic isolates of *Didymium iridis*. Composite diagrams are of polyacrylamide gels stained for (a) GOT (glutamate oxaloacetate transaminase), (b) MDH (malate dehydrogenase), and (c) α-esterases. From Betterley (1981).

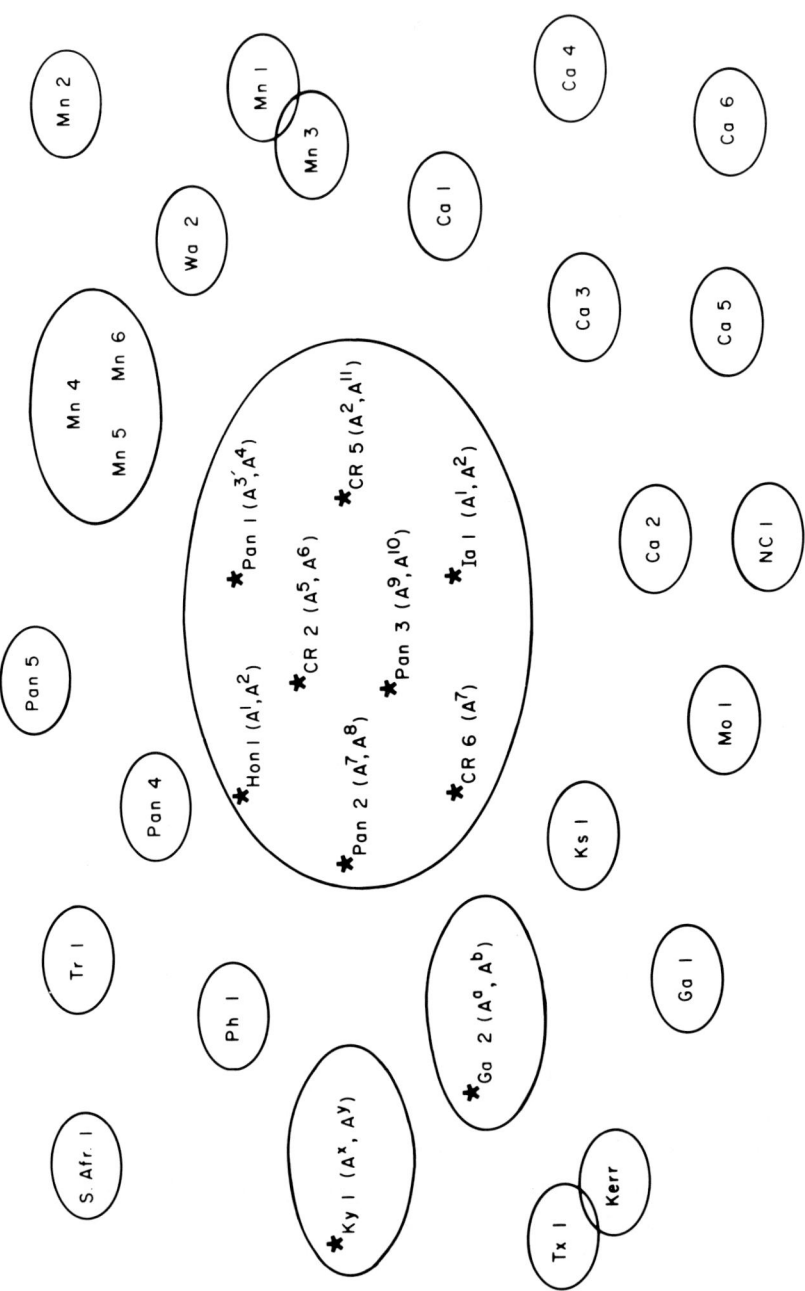

Fig. 7. Summary of relationships among isolates of *Didymium iridis* based on investigations of reproductive systems, somatic incompatibility, and isozymes of nine enzymes. Overlap of the ellipses indicates that isolates are identical for many of the enzymes studied. The order and distance between ellipses are arbitrary and do not reflect a quantitative estimation of genetic distance. Each ellipse apparently represents a reproductively isolated group in nature. From Betterley (1981). *Heterothallic; unmarked, nonheterothallic.

"biological species." This tentative conclusion is supported by studies (Betterley and Collins, 1979; Betterley, 1981) of isozyme variation.

Sample composite diagrams of polyacrylamide step-gradient gels stained for GOT (glutamate oxaloacetate transaminase), MDH (malate dehydrogenase), and α-Est (α-esterases) are shown in Fig. 6. It should be noted that there is little variation among the group of interfertile heterothallic isolates (Hon 1, Ia 1, Pan 1, CR2, Pan 2, Pan 3, and CR 5), but two of the isolates (Ky 1, Ga 2) which are not capable of interbreeding with the others display different patterns. Further, comparison of isozyme profiles from nonheterothallic isolates shows that there is much variation among them, and collectively they are also different from those shown for heterothallic isolates. The trends shown in Fig. 6 are consistent for nine enzymes studied.

In addition to 34 isolates of *D. iridis* studied, isolates from seven different species of the genus *Didymium* were examined: *D. difforme*, *D. laxifila*, *D. megalosporum*, *D. nigripes*, *D. saturnus*, *D. squamulosum*, and *Didymium* sp. Isozyme patterns for isolates of these species generally fall outside the range of those exhibited by *D. iridis* isolates.

When all *D. iridis* isolates assayed are compared with one another, their relationships can be summarized as depicted in Fig. 7. The three different heterothallic breeding groups are indicated by asterisks. Among nonheterothallics, each isolate appears to correspond to a clonal population in nature, apparently incapable of interbreeding with either heterothallic or other nonheterothallic isolates. This suggests that the nonheterothallics are indeed apomicts, an idea which was advanced in Section IV,C. Further, genetical and isozymal data are consistent with the view that in *D. iridis* there exist in nature only two basic reproductive cycles, heterothallic and apomictic. Organisms displaying other cycles are assumed to be laboratory variants or rare natural forms.

VII. RESEARCH ON THE PLASMODIAL STAGE

A. Development and the Mating-Type Locus

Factors controlling plasmodial development and function are of basic biological importance, but, our knowledge of these is still rudimentary. In *D. iridis*, development has been observed in many isolates and in thousands of cultures. From this we know that plasmodial development occurs (1) always in mass-spore cultures carrying two mating types, (2) characteristically in crosses of sexually compatible amoebal clones, (3) occasionally in uncrossed amoebal clones, (4) invariably in amoebal clones from nonheterothallic isolates, and (5) always in clones of amoebal mating type heterozygotes. Aside from the influence of extrinsic environmental factors, plasmodial development appears to be most directly under the control of the mating locus. Although at the level of genetical transmission the mating locus is fairly well understood, the manner in which it exerts its controlling influence is still a mystery. Analysis of experiments and observations on plasmodial development, however, offers the opportunity for gaining some insight into the functions of the mating locus as well as the functions of other intrinsic factors. If, for example, the assumptions are correct that specimens of *D. iridis* are either heterothallic or apomictic and that apomicts typically carry disguised mating types, this would suggest the universal involvement of the mating types in all instances of plasmodial development. This interpretation eliminates the need to explain homothallic (absence of known mating types) plasmodial development and permits a narrower focus on developmental controls.

From an evolutionary viewpoint, it seems simplest to consider that when an ancestral amoeboflagellate line acquired the capacity to mate (perhaps through mating type-producing mutations), this resulted in plasmodial formation. The development of this new plasmodium-producing capacity was a key event in the ancestral amoeboflagellates' divergence into a line which later evolved into Myxomycetes. Inasmuch as available data suggest the existence of a single mating system in all extant heterothallic entities, it is not difficult to imagine that they had a common amoeboflagellate ancestor possessing a mating type. This hypothesis has the advantage of being simple and, therefore, relatively amenable to testing. Of course, it is conceivable that today's mating types did not have a monophyletic origin and that they arose independently and repeatedly in the various myxomycete taxa, but theoretically at least this seems unlikely.

It would be useful to have a myxomycete mating type model which incorporates available information on the subject. Collins and Ling (1968, p. 866) suggested that "insofar as the sequence of events leading to plasmodial development is concerned, individual clones, or common mating type combinations of these, would remain in the repressed condition, whereas heteroallelic combinations would bring about derepression, leading to plasmodial production." Further, it was supposed that in clonal cultures, "mutations at modifying, or suppressor, loci could conceivably lead to plasmodial production, perhaps by a derepression mechanism." In *D. iridis,* evidence in support of these views is mainly inferential, and the species has not been subjected to intensive genetic analysis of the mating locus itself. However, some emphasis has been placed on understanding the cause(s) of clonal plasmodium formation (Collins, 1965; Collins and Ling, 1968; Yemma *et al.*, 1974), and there is evidence of a single mutation at the mating locus and evidence suggesting the existence of cytoplasmic modifying factors which act differentially on the various mating types.

The one-locus, multiple-allelic mating system combines some of the advantages of homothallism (outbreeding and inbreeding potential 100%) with one-locus, two-allele heterothallism (outbreeding and inbreeding potential 50%). That is, in a one-locus, multiple-allelic system, outbreeding can theoretically approach

100% (depending on the number of alleles), but inbreeding is restricted to 50%. This implies that the system should have a selective advantage over alternative sexual modes. Also, the existence of many mating type specificities suggests that the locus may be complex and that mutations in it can give rise to new specificities. Apparently, one such mutation was discovered by Collins (1965) in *D. iridis*. Perhaps a concerted effort to create and isolate new alleles through mutagenesis would prove successful. Information from such studies should be helpful for further interpretation of the structure and function of the mating type locus.

In *P. polycephalum*, a large amount of genetical work has been directed toward understanding the role of the mating locus in plasmodial development, and the results and conclusions have been summarized (Honey *et al.*, 1979). In this organism, 13 (14?) allelic specificities are known (Collins, 1975; Collins and Tang, 1977).

In *D. iridis*, as in all Myxomycetes, differentiation of a uninucleate, colorless amoeba or zygote into a multinucleate, coenocytic, pigmented plasmodium represents a fundamental shift in genomic expression. Inasmuch as this differentiation can occur apogamically, syngamy obviously cannot be the only controlling factor in plasmodial development. As suggested above, underlying controls probably reside in the mating locus and in portions of the chromosome on which the mating locus is located. These may be influenced by the occurrence of syngamy, the existence of modifying genes, and the size and age of amoebal populations, as well as by many external environmental factors. Some of the latter include pH, concentration of amoebae, presence of divalent cations, bacterial food source, and chemical composition of the medium (Kerr and Sussman, 1958; Kerr, 1961).

The fundamental nature of the shift from amoeba to zygote is reflected in the nuclear membrane behavior during mitosis. Whereas the division spindle is open in amoebae, with the nuclear membrane breaking down in late prophase, it is intranuclear in the zygote and plasmodial nuclei, with the nuclear membrane persistent throughout the division process (see the review by Collins, 1979; Hinchee and Haskins, 1980a,b). Ross (1968) emphasized that even in binucleate and multinucleate amoebae the nuclear membrane breaks down, whereas uninucleate zygotes retain their nuclear membrane. It is clear, therefore, that the shift in nuclear membrane behavior is not a simple function of change in ploidy or the number of nuclei per cell. Ross (1968) appropriately pointed out that a persistent nuclear membrane during mitosis is characteristic of some acellular or predominantly coenocytic organisms.

The studies by Shipley and Ross (1978) may provide clues to the control mechanisms. They found that plasmodium formation is affected by exposure to three proteases, but only during the first 2 hours after compatible amoebae are crossed.

B. Morphology and Pigmentation

In most characteristics, the phaneroplasmodium of *D. iridis* is typical of that produced by members of the Physarales but quite different from the protoplasmodium and aphanoplasmodium produced by some other Myxomycetes (Alexopoulos, 1960; Collins, 1979). The microscopic morphology and streaming mechanisms are dealt with by Kessler in Chapter 5, this volume.

Close observations of *D. iridis* plasmodia reveal a certain amount of morphological variability which is correlated with individual isolates and with particular crosses between different clones. Some of these morphological variants differ from one another in subtle, yet unmistakable, ways. An especially interesting example is the "patchy" plasmodium often associated with certain crosses of the Pan 1 isolate. In this example, the plasmodium is made up of alternating sheetlike and veiny regions. In some cases, plasmodia are distinguishable on the basis of the relative coarseness of their veins, their different shades of wild-type brown, the speed at which they migrate over solid surfaces, the readiness with which they sclerotize, and their longevity in continuous laboratory culture. Of course, all of these characteristics reflect expected genetic variability within the species. They are worthy of note, however, because they can be useful in research and should not be overlooked.

In addition to the wild-type brown plasmodium, cream-colored mutants have been identified in *D. iridis* (Collins and Clark, 1966; Collins, 1969). Collins and Erlebacher (1969) showed that the two known color mutations, b^1 and b^2, partially block production of a red pigment, and that these apparently affect a single biosynthetic pathway because plasmodia carrying both recessive mutations produced less pigment than those with one. Both b^1 and b^2 are "leaky," and small amounts of red pigment are produced by mutants possessing either. So, the wild-type plasmodium is brown as a result of the presence of several pigments. When the red pigment is sufficiently diminished, yellowish pigments are unmasked and the plasmodium is cream-colored. The red pigment is localized in granules, apparently has a molecular weight of 470, and when chromatographically purified, it has absorption peaks at 245, 350, and 550 nm (Collins and Erlebacher, 1969). Mutations blocking the production of pigments other than the red one have not been found, but genetical studies on all of the pigments could reveal information of value to biosystematists as well as to physiologists and biochemists. Whether any of the pigments which are obvious in extraction and chromatographic studies (Collins and Erlebacher, 1969) are photoreceptors necessary for initiating sporulation is unknown, but it would be fascinating to learn what photoreceptor was involved in Ling's (1968) sporulation experiments on *D. iridis*. He reported up to 100% sporulation in all plasmodia exposed to whole light for 1 minute at 30 ft-c, and an exposure as short as 5 seconds at 30 ft-c resulted in 23% sporulation. On the other hand, preliminary experiments indicate that there is no absolute requirement of light for sporulation in *D. iridis*. We observed that single-spore cultures of some nonheterothallic isolates, or crosses of heterothallic isolates maintained in total darkness can sporulate.

Factors inducing sporulation in *P. polycephalum* have been studied, and Gorman and Wilkins (1980) have summarized much of the relevant literature. A color mutation rendering the plasmodia white instead of the wild-type yellow-orange has also been reported (Anderson, 1977) in *P. polycephalum*.

In one naturally occurring cream-colored and nonheterothallic isolate of *D. iridis*, Kerr (1966) and Kerr and Waxlax (1968a) described a yellow, cycloheximide-resistant plasmodial variant which developed from yellow amoebae, but the genetics of the yellow trait are not understood. However, yellow color and cycloheximide resistance consistently occurred together, and the pigment tended to disappear in yellow + wild-type heterokaryons.

Laboratory observations on 39 different plasmodia, including that of *D. iridis*, were undertaken by Kambly (1939), who concluded that they exhibited considerable variation in color when maintained in moist chamber cultures. However, the variability he observed was caused by external factors such as the presence or absence of pigmented bacteria, so this is consistent with our present knowledge that plasmodial color is genetically determined. Like any inherited trait, it can be affected by environmental causes. Under standard conditions, plasmodial color is in fact remarkably consistent.

C. Incompatibility Systems

The plasmodium is one of two assimilative stages in the diplohaplontic life cycle of *D. iridis*. In the absence of a good food supply, the plasmodium migrates rapidly over solid surfaces, a characteristic which enables it to encounter possible new or better grazing areas. This capacity to migrate is useful to investigators in a variety of research areas, and it has been extensively taken advantage of in studies on the genetics of plasmodial incompatibility in *D. iridis*. For example, transfers of plasmodia can be placed side by side on an agar surface and allowed to migrate toward each other. When they collide, they may fuse and become a single entity, or they may remain discrete units regardless of how intimately they make physical contact (Fig. 8). Further, Kerr and Waxlax (1968b) showed that fusion is not affected by the presence of a variety of bacteria and yeasts or by a variety of media.

Historically, fusion tests were seen as a means to distinguish between different myxomycete species, a concept later shown to have no validity (see Clark and Collins, 1973a, for a review). Of course, it is true that plasmodia from different species do not fuse, but it is also true that the probability of fusion between random combinations of plasmodia from the same species is remote. At the time

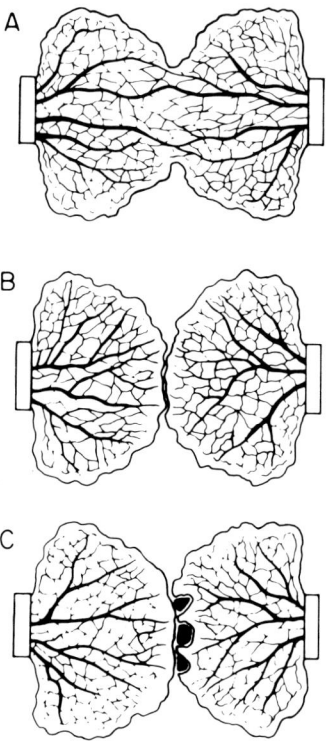

Fig. 8. Plasmodial fusion tests showing (A) fusion, with establishment of common veins, (B) nonfusion, and (C) a cytotoxic reaction with killed portions of cytoplasm. Redrawn from Collins and Clark (1968) and Clark and Collins (1973b).

plasmodial incompatibility studies were first undertaken in *D. iridis*, the species concept had been modified and nonfusion was seen as evidence for assigning plasmodia to different races (Alexopoulos and Zabka, 1962). Experiments of Collins (1963) and Mukherjee (1965) involving plasmodia produced from crosses of many clones indicated that there was a definable genetic basis for incompatibility. Later, drawing upon more extensive studies, Collins (1966) proposed that plasmodial fusion was controlled by two fusion loci, each with co-dominant alleles, plus the mating locus. This conclusion was later modified by Collins and Clark (1966, 1968), who indicated that the system was polygenic. They demonstrated the existence of five fusion loci in the Hon 1 isolate, each with alternative dominant or recessive alleles, and indicated linkage between the mating locus and one of the incompatibility loci. Fusion occurred when plasmodia were phenotypically identical for all five loci (e.g., *CC DD EE FF GG* fuses with *Cc Dd Ee Ff Gg*), but it did not occur if the two plasmodia differed for one or more of the five. With the 1968 results, all of the basic features of the incompatibility system had emerged, and this was verified by Ling and Collins (1970a,b) who identified six loci and three linkage groups in a second *D. iridis* isolate, Pan 1. Subsequently, Collins and Ling (1972) analyzed progeny from a Hon 1 × Pan 1 cross and identified a total of eleven fusion loci.

Studies intended to define the functions of individual loci led Clark and Collins (1973b) to conclude that phenotypically dissimilar plasmodia sometimes undergo temporary fusions at the points of contact; these fusions are usually quickly terminated by cytotoxic reactions. The region of protoplasmic mixing produces a clear zone of coagulated protoplasm, and the size of this region is a function of both the strength of individual loci and the number of loci for which the contacting plasmodia differ. Each locus allows a characteristic amount of mixing. When two or more loci act together, the cytotoxic reaction is more rapid and the amount of mixing is reduced. A strong locus is defined as one that is fast-acting and permits very little mixing, whereas a weak one is slow-acting, permitting a large clear zone to form. Incompatibility loci can be rank-ordered on this basis, and it has been shown (Clark and Collins, 1973b) that the lethal reactions producing clear zones are unidirectional for each locus, with killing occurring in the plasmodium with the recessive phenotype. Conclusions arrived at by Ling and Ling (1974) and Ling and Upadhyaya (1974) were similar to those of Clark and Collins, except that they considered that fusion and clear zone loci are responsible for separate kinds of functions. Ling and Upadhyaya (1974) also reported recovery of some live protoplasm from clear zones, whereas Clark and Collins (1973b) assumed (on the basis of appearance) that clear zones contained dead material only. An EM study of clear zones was carried out by Upadhyaya and Ling (1972), but they could not determine how damage was actually initiated.

In addition to their work on clear zones, Clark and Collins (1973b) reported that in some instances when fusion occurred and did not result in any noticeable killing reaction, the heterokaryons produced reverted to a homokaryotic state within 24 hours. They interpreted this to mean that there are some unidentified weak incompatibility loci which are so slow-acting that much protoplasmic mixing is permitted before the death of one phenotype takes place. Further, Clark (1980a,b) has presented evidence in support of the conclusion that knowledge of the strength of incompatibility loci can be used to predict which plasmodial phenotype will survive in mass-spore cultures or in multiple-clone crosses. In those situations, only one phenotype characteristically survives, even though theoretically there could be several kinds. It survives because it carries the largest number of dominant somatic incompatibility alleles, and these cause destruction of the other phenotypes through killing reactions. Earlier, Collins (1966) and Collins and Clark (1968) assumed that survival of the one phenotype in mass-spore and multiple-clone crosses depended on its representation by larger numbers of developing plasmodia than those of other phenotypes. This was thought to result in their gaining ascendancy by fusing with one another, thus producing a larger plasmodium than would be produced by each of the other phenotypic classes following fusion. Clark's (1980a,b) explanation is supported by a considerable amount of experimental work and now stands as the best explanation for survival of only one phenotype in situations where several develop in mass-spore or multiple-clone crosses.

Although the genetics of plasmodial incompatibility have been intensively analyzed only in heterothallic isolates, its existence has been demonstrated (i.e., intraisolate combinations fuse, but interisolate ones do not) in numerous nonheterothallic forms as well (Collins, 1966; Collins and Clark, 1968; Betterley, 1981). In fact, it can be logically assumed that the polygenic barrier to compatibility is a universal feature of the *D. iridis* plasmodium. Its function is probably similar to that of other somatic incompatibility systems known in a wide range of eukaryotes. That is, it apparently functions primarily as a self-recognition system, enabling the organism to maintain its biological integrity by rejecting foreign matter, including genetically dissimilar sibling plasmodia. However, Collins' initial interest in the system resulted from his observation that sporangial specimens carrying more than a pair of mating type factors in the multiple allelic series were never encountered. The rationale was that some of them should have more than two mating type factors if viable plasmodial heterokaryons developed between individuals heterozygous for different mating types. On the basis of what is now known, it is clear that the polygenic somatic incompatibility system is an effective barrier against the production of viable plasmodial heterokaryons. The integrity of the mating system as well as that of the entire organism is thus maintained through the plasmodial self-recognition system.

Still unexplained is the curious disappearance of a genetic marker following plasmodial fusion, reported by Kerr (1965a). Because the fused plasmodia were apparently identical for incompatibility factors, these cannot be involved as an explanation.

Carlile and Gooday (1978) reviewed the literature on somatic incompatibility in Myxomycetes, including work on *P. polycephalum, D. iridis, Badhamia utricularis,* and *P. cinereum.* A paper by Lane and Carlile (1979) on nuclear elimination in *P. polycephalum* is especially interesting in connection with the mechanism involved in lethal reactions.

Finally, from the work of Clark and Hakim (1980a), it appears that the toxic principle(s) produced in incompatible reactions are not present in the cytoplasm. In their experiments, two incompatible plasmodia were separated by filters which permitted cytoplasmic but not nuclear contact. The cytoplasm remained undamaged in repeated tests, so the nuclei appear to be the site of the as yet undefined toxic factor(s).

D. Senescence

For several years, *D. iridis* workers have observed that following prolonged laboratory cultivation, some individual plasmodia lose vigor, and even if regularly transferred to fresh media, they eventually die. When such a plasmodium is re-created by again crossing its parental amoeboflagellate clones, the new entity is again vigorous and easy to maintain in culture for another period of time, but it too will gradually lose vigor and die. Clark and his associates at the University of Kentucky have studied this phenomenon, and Lott and Clark (1980) showed that different plasmodia obtained from crossing the same two parental clones display an average life span that is characteristic for that particular genome. However, they also reported that multiple sublines, or transfers, from one plasmodium showed a remarkable likeness in the length of their life spans, much more so than those from replicate crosses. It thus appears that genomic as well as nongenomic factors are involved in aging.

When genetically identical homokaryotic plasmodia of different ages were fused to produce "age" heterokaryons by Clark and Hakim (1980b) 54% of the heterokaryons died at the time of death of the oldest homokaryon, although the remaining 46% gave variable results. These data suggest that the genome of the older homokaryon produces an aging factor which has an important effect on the heterokaryon life span. At this point, no genes have been identified, but the heterokaryon results imply that genetical transmission of aging factors may be complex. Further, Clark and Hakim (1980a) showed that, when polyploid nuclei were removed from aging diploid plasmodia, they lived longer than those from which polyploid nuclei had not been removed. Removal of polyploid nuclei was accomplished by permitting plasmodia to migrate through nucleopore filters whose pore size allowed passage of diploid but not polyploid nuclei. During investigations in inheritance of polyploidy (Collins *et al.*, 1978), we observed that compared to haploid amoebae, polyploid amoebae were as easy to maintain in culture, but they migrated faster and divided more slowly. On the other hand, polyploid plasmodia, compared to diploid ones, grew and migrated more slowly and died earlier. It seems that polyploidy may be a secondary effect of aging rather than a primary cause of it. For example, young, newly formed polyploid plasmodia resemble "old" plasmodia inasmuch as they are generally less vigorous than their corresponding $2n$ counterparts, and they also display other aging characteristics. There may be aging factors which act upon $2n$ nuclei, causing them to become polyploid, in which case the polyploid condition would be a direct result of aging. Kerr and Waxlax (1968a) studied a cycloheximide-resistant, yellow variant of *D. iridis* and noted that aging was exemplified in their mutant by a loss of the ability to fruit, to fuse with other plasmodia, and to grow. Work on plasmodial aging is in the initial stages, and further studies are indicated.

VIII. SPOROPHORE STAGE

A remarkable and fundamental event in the myxomycete life cycle is "fruiting," the conversion of the protozoan-like plasmodium into fungus-like sporophores. This process has been repeatedly observed by many workers, and it has been described for *D. iridis* in some detail (Lucas *et al.*, 1968) as well as recorded on film (N. S. Kerr, 1970b). As in other Physarales, development is subhypothallic as opposed to the epihypothallic type of the Stemonitales. Six stages in the subhypothallic pattern were recognized: (1) advancing front stage, (2) accumulation stage, (3) mounding stage, (4) rising stage, (5) white fruit stage, and (6) pigmented stage. The physical mechanisms responsible for first segmenting the plasmodium into many units and then gradually shaping them into stalked sporangia are not understood. A thorough study of this fascinating developmental process is badly needed, and the information gained would probably have relevance to myxomycete systematics as well as to the broader realm of developmental biology. It does seem, of course, that microfilaments must be the primary physical entities responsible for the movement involved in the process, but even this has not been demonstrated.

In a study primarily devoted to understanding development of the columella and capillitia, Welden (1955) observed and illustrated gross aspects of sporangial development (Fig. 9) in *D. iridis*, beginning with the "sporangial initials." (These initials correspond to the early part of the rising stage mentioned above.) In describing columella formation, Welden points out that during the downward expansion of the sporangium, a portion of the stipe is enclosed and later develops into the columella. A region just above the enclosed stipe becomes separated from the rest of the sporangium by a thickened membrane. This space is at first hollow, but by the time the columella is mature (which is after the capillitia develop), it is filled with calcium carbonate. Welden indicated that capillitia develop from tubular invaginations originating from the peridium as well as from the "columella space." These invaginations anastomose with one another, causing a branching effect in the mature capillitia. On the other hand, Schuster's (1964) EM work would seem to indicate that capillitia develop in cleavage furrows created by fusion of vesicles inside the cytoplasm. These furrows divide the protoplasm into relatively large units. In any case, at maturity the sporangium contains abundant capillitia, which in Schuster's EM cross sections are seen to have a dense outer region and a concentrically layered inner core. Gaither (1968) reported that a capillitium is formed by the deposition of a system of layered tubules which encloses portions of the cytoplasm.

Cleavage of the protoplasm into uninucleate prespore units begins soon after synchronous mitoses in the sporangial protoplasm. Meiosis occurs in the spores, as evidenced by the presence of synaptonemal complexes in prophasic nuclei (Aldrich and Carroll, 1971). Walls develop around the prespores, the outer electron-dense layer forming first, followed by the less dense but wider inner

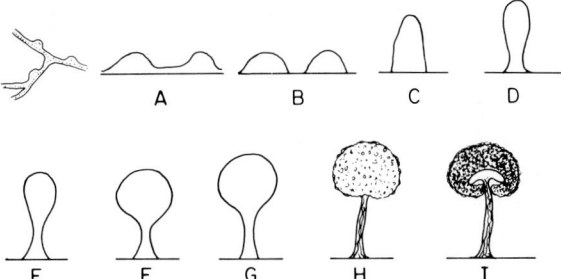

Fig. 9. Diagrams of developing sporangia of *Didymium iridis*. (A) Mounds of protoplasm still connected by plasmodial veins. (B) Discrete mounds of protoplasm. (C) Rising protoplasm to form columnar initials. (D) Expansion of the apical region of the columnar initial; stipe region beginning to constrict. (E) Further construction of the stipe region and further expansion of the sporangial region. (F-G) Outward and downward expansion of the developing sporangium; columella begins to form within. (H) First appearance of calcium carbonate crystals on the peridium—stage at which capillitium is mature and the sporangium begins to darken. (I) Longitudinal section of a mature sporangium showing the columella, spores, capillitium, and umbilicate attachment. Modified from Welden (1955).

layer (Schuster, 1969). Upon germination, a wedgelike crack develops in the wall and typically a single amoeboflagellate emerges.

Mature sporophores of *D. iridis* display a range of variability, depending on individual isolates, crosses, and culture conditions. Stalks may vary in thickness, length, or color; sporangial heads vary in size, shape, and single (typical) versus double structure. The amount of lime present on the peridium and the columella also differs from one specimen to the next. Although columella size and shape are variable, its color is consistently white in well-developed specimens, a feature separating *D. iridis* from two close relatives with a darker columella, *D. nigripes* and *D. megalosporum*. Because in the laboratory stalk color is so often a function of plasmodial color (e.g., deep brown plasmodia tend to yield dark-stalked sporangia, and cream-colored plasmodia pale-stalked ones), its usefulness as a taxonomic criterion is limited. On the other hand, spore color and size are fairly constant, and an SEM study by Gaither and Collins (1979) has shown that spore ornamentation is also relatively constant in *D. iridis*. Close examination of spores of different ploidal levels would probably reveal significant size differences. In any taxonomic work on Myxomycetes, it is essential that well-developed, mature sporophores be used, and whenever practical, observations on all life cycle stages should be recorded.

IX. AMOEBOFLAGELLATE–MICROCYST STAGE

Amoeboflagellates are characteristically uninucleate cells with a capacity to exist in two active states, either as nonpolarized amoebae or as polarized, biflagellate (heterokont) swarm cells. Bacteria are apparently their primary natural food source, and these are phagocytized by cells in either the amoebal or the flagellate state. However, mitosis is known to occur only in amoebae, probably because in the swarmer centrioles function as flagellar basal bodies and are thus unavailable for mitosis. Schuster (1965) and S. J. Kerr (1967) pointed out that centrioles of the amoebal stage are not retained throughout the life cycle and, extrapolating from the work by Aldrich (1967) on *P. flavicomum,* it is assumed that centrioles in *D. iridis* arise in the spores immediately following meiosis. In any case, amoebae display nonsynchronous, open-spindled, centriolar mitosis, whereas in plasmodia and developing sporangia, nuclear divisions are synchronous and intranuclear with no centriolar involvement. This situation is apparently characteristic of all Myxomycetes.

Detailed descriptions of amoebae and swarmers of *D. iridis* at the light microscopic and ultrastructural levels, as well as of their transition from one state to another, are available (N. S. Kerr, 1960; Schuster, 1965; S. J. Kerr, 1972a,b). In addition, N. S. Kerr's (1970a) film on *D. iridis* records various aspects of amoebal behavior, including their emergence from spores, their feeding, mitosis, encystment, and excystment, and their morphogenesis into swarmers (no doubt involving rearrangement of microtubules and microfilaments). Each of the three states (amoebal, swarmer, and microcyst) is distinctly different morphologically and is believed to correspond to a separate role in the survival of the species (Collins, 1979, p. 166). However, the amoebal state is probably the ancestral and basic one of the three, inasmuch as neither the swarmer nor the cyst is necessary for completion of the life cycle, whereas the amoeba is.

When an amoeba is inoculated onto a lawn of bacteria in an agar plate, plaques develop on the lawn as the amoeba and its mitotic descendants devour bacteria from the point of inoculation outward. Characteristically, within a few days a plaque is composed mainly of a central area of microcysts surrounded by an outer ring of densely packed feeding amoebae. However, mutant plaques with distinctive features have been described (Kerr, 1963b), and although the inheritance of "plaque-type" markers has been studied (Kerr, 1965a), no genes were identified. Many such variants exist in *D. iridis,* but generally these have only been casually observed, so little is known about them so far. One especially interesting observation, however, is that polyploid amoebae display a faster speed of migration and a slower rate of nuclear division than haploid ones. Systematic study of these and other observations on amoebal and plaque variability could provide useful information on evolutionarily significant amoebal traits.

X. CONCLUDING REMARKS

Discovery of mating types in *D. iridis* over 20 years ago signaled the beginning of its use in sustained laboratory research on the genetics of sexual and somatic incompatibility systems.

2. *Didymium iridis* in Past and Future Research

In addition, a series of comparative studies of reproductive systems in over 40 isolates of the species have made it the leading model for the study of myxomycete speciation and evolution. These comparative studies have emphasized genetical, cytological, and biochemical approaches.

ACKNOWLEDGMENTS

Supported by NSF grants DEB 75-21171 and DEB 80-05999.

REFERENCES

Aldrich, H. C. (1967). The ultrastructure of meiosis in three species of *Physarum*. *Mycologia* **59**, 127-148.

Aldrich, H. C., and Carroll, G. C. (1971). Synaptonemal complexes and meiosis in *Didymium iridis:* A reinvestigation. *Mycologia* **63**, 308-316.

Alexopoulos, C. J. (1960). Gross morphology of the plasmodium and its possible significance in the relationships among Myxomycetes. *Mycologia* **52**, 1-20.

Alexopoulos, C. J., and Zabka, G. G. (1962). Production of hybrids between physiological races of the true slime mold *Didymium iridis*. *Nature (London)* **193**, 598-599.

Anderson, R. W. (1977). A plasmodial colour mutation in the myxomycete *Physarum polycephalum*. *Genet. Res.* **30**, 301-306.

Baker, H. G. (1978). Invasion and replacement in California and neotropical grasslands. *In* Plant Relations in Pastures (J. R. Wilson, ed.), pp. 368-384. C.S.I.R.O., East Melbourne.

Balamuth, W., and Gong, T. (1975). Growth of the slime mold *Didymium iridis* in axenic culture media. *J. Protozool.* **22**, 12A-13A (Abstr.).

Betterley, D. A. (1980). Reproductive systems, enzyme polymorphism, and speciation in the myxomycete *Didymium iridis*. Ph.D. Dissertation, Univ. of California, Berkeley, California.

Betterley, D. A., and Collins, O. R. (1979). Enzyme polymorphism and reproductive systems in the slime mold *Didymium* (Myxomycetes). *Mycol. Soc. Am. Newsl* **30**, 18 (Abstr.).

Cadman, E. J. (1931). The life history and cytology of *Didymium nigripes* Fr. *Trans. Roy. Soc. Edinburgh* **57**, 93-142.

Carlile, M. J., and Gooday, G. W. (1978). Cell fusion in myxomycetes and fungi. *In* "Membrane Fusion, Cell Surface Reviews (G. Poste and G. L. Nicholson, eds.), Vol. 5, pp. 219-265. Elsevier, Amsterdam.

Clark, J. (1980a). Competition between plasmodial incompatibility phenotypes of the myxomycete *Didymium iridis*. I. Paired plasmodia. *Mycologia* **72**, 312-321.

Clark, J. (1980b). Competition between plasmodial incompatibility phenotypes of the myxomycete *Didymium iridis*. II. Multiple clone crosses. *Mycologia* **72**, 512-522.

Clark, J., and Collins, O. R. (1973a). Further studies on the genetics of plasmodial incompatibility in a Honduran isolate of *Didymium iridis*. *Mycologia* **65**, 507-518.

Clark, J., and Collins, O. R. (1973b). Directional cytotoxic reactions between incompatible plasmodia of *Didymium iridis*. *Genetics* **73**, 247-257.

Clark, J., and Collins, O. R. (1976). Studies on the mating systems of eleven species of Myxomycetes. *Am. J. Bot.* **63**, 783-789.

Clark, J., and Hakim, R. (1980a). Nuclear sieving of *Didymium iridis* plasmodia. *Exp. Mycol.* **4**, 17-22.

Clark, J., and Hakim, R. (1980b). Aging of plasmodial heterokaryons in *Didymium iridis*. *Molec. Gen. Genet.* **178**, 419-422.

Collins, O. R. (1961). Heterothallism and homothallism in two Myxomycetes. *Am. J. Bot.* **48**, 674-683.
Collins, O. R. (1963). Multiple alleles at the incompatibility locus in the myxomycete *Didymium iridis*. *Am. J. Bot.* **50**, 477-480.
Collins, O. R. (1965). Evidence for a mutation at the incompatibility locus in the slime mold, *Didymium iridis*. *Mycologia* **57**, 314-315.
Collins, O. R. (1966). Plasmodial compatibility in heterothallic and homothallic isolates of *Didymium iridis*. *Mycologia* **58**, 362-372.
Collins, O. R. (1969). Complementation between two color mutants in a true slime mold, *Didymium iridis*. *Genetics* **63**, 93-102.
Collins, O. R. (1975). Mating types in five isolates of *Physarum polycephalum*. *Mycologia* **67**, 98-107.
Collins, O. R. (1976). Heterothallism and homothallism: A study of 27 isolates of *Didymium iridis*, a true slime mold. *Am. J. Bot.* **63**, 138-143.
Collins, O. R. (1979). Myxomycete biosystematics: Some recent developments and future research opportunities. *Bot. Rev.* **45**, 145-201.
Collins, O. R. (1980). Apomictic-heterothallic conversion in a myxomycete, *Didymium iridis*. *Mycologia* **72**, 1109-1116.
Collins, O. R., and Clark, J. (1966). Inheritance of the brown plasmodial pigment in *Didymium iridis*. *Mycologia* **58**, 743-751.
Collins, O. R., and Clark, J. (1968). Genetics of plasmodial compatibility and heterokaryosis in *Didymium iridis*. *Mycologia* **60**, 90-103.
Collins, O. R., and Erlebacher. B. A. (1969). Effects of two mutations on production of a red plasmodial pigment in a true slime mold, *Didymium iridis*. *Can. J. Microbiol.* **15**, 1245-1247.
Collins, O. R., and Ling, H. (1968). Clonally-produced plasmodia in heterothallic isolates of *Didymium iridis*. *Mycologia* **60**, 858-868.
Collins, O. R., and Ling, H. (1972). Genetics of somatic cell fusion in two isolates of *Didymium iridis*. *Am. J. Bot.* **59**, 337-340.
Collins, O. R., and Tang, H.-C. (1977). New mating types in *Physarum polycephalum*. *Mycologia* **69**, 421-423.
Collins, O. R., and Therrien, C. D. (1976). Cytophotometric measurement of nuclear DNA in seven heterothallic isolates of *Didymium iridis*, a myxomycete. *Am. J. Bot.* **63**, 457-462.
Collins, O. R., Therrien, D. E., and Betterley, D. A. (1978). Genetical and cytological evidence for chromosomal elimination in a true slime mold, *Didymium iridis*. *Am. J. Bot.* **65**, 660-670.
Davis, E. E. (1965). Preservation of Myxomycetes. *Mycologia* **57**, 986-988.
Dee, J. (1960). A mating-type system in an acellular slime-mould. *Nature (London)* **185**, 780-781.
Ditmar, L. P. F. (1813). Die Pilze Deutschlands. Sturm, Deutschlands. *Flora* **3**, 1-34.
Fries, E. M. (1829). "Systema mycologicum," Vol. III, 67-199.
Gaither, T. W. (1968). Capillitial development in *Didymium iridis*. *Am. J. Bot.* **55**, 719 (Abstr.).
Gaither, T. W., and Collins, O. R. (1979). A comparative SEM study of sporangial characteristics in three species of *Didymium* (Myxomycetes). *Bot. Soc. Am. Misc. Ser. Publ.* **157**, (Abstr.).
Gorman, J. A., and Wilkins, A. S. (1980). Developmental phases in the life cycle of *Physarum* and related Myxomycetes. *In* "Growth and Differentiation in *Physarum Polycephalum*" (W. F. Dove and H. P. Rusch, eds.), pp. 157-202. Princeton Univ. Press, Princeton, New Jersey.
Hinchee, A. A., and Haskins, E. F. (1980a). Open spindle nuclear division in the amoebal phase of the acellular slime mold *Echinostelium minutum* with chromosomal movement related to the pronounced rearrangement of spindle microtubules. *Protoplasma* **102**, 117-130.
Hinchee, A. A., and Haskins, E. F. (1980b). Closed spindle nuclear division in the plasmodial phase of the acellular slime mold *Echinostelium minutum*. *Protoplasma* **102**, 235-252.

2. *Didymium iridis* in Past and Future Research

Honey, N. K., Poulter, R. T. M., and Teale, D. M. (1979). Genetic regulation of differentiation in *Physarum polycephalum. Genet. Res.* **34,** 131-142.
Jahn, E. (1911). Myxomycetenstudien. 8. Der. Sexualakt. *Ber. Dtsch. Bot. Ges.* **29,** 231-247.
Kambly, P. (1939). The color of myxomycete plasmodia. *Am. J. Bot.* **26,** 386-390.
Kerr, N. S. (1960). Flagella formation by myxamoebae of the true slime mold, *Didymium nigripes. J. Protozool.* **7,** 103-108.
Kerr, N. S. (1961). A study of plasmodium formation by the true slime mold, *Didymium nigripes. Exp. Cell Res.* **23,** 603-611.
Kerr, N. S. (1963a). The growth of myxamoebae of the true slime mould, *Didymium nigripes,* in axenic culture. *J. Gen. Microbiol.* **32,** 409-416.
Kerr, N. S. (1963b). The fusion of marked plasmodia of the true slime mold *Didymium nigripes. J. Cell Biol.* **19,** 39A (Abstr.).
Kerr, N. S. (1965a). Disappearance of a genetic marker from a cytoplasmic hybrid plasmodium of a true slime mold. *Science* **147,** 1586-1588.
Kerr, N. S. (1965b). A simple method of lyophilization for the long-term storage of slime molds and small soil amoebae. *BioScience* **15,** 469.
Kerr, N. S. (1966). A yellow mutant of the colorless true slime mold, *Didymium nigripes. J. Protozool.* **13,** 7 (Abstr.).
Kerr, N. S. (1967). Plasmodium formation by a minute mutant of the true slime mold, *Didymium nigripes. Exp. Cell Res.* **45,** 646-655.
Kerr, N. S. (1970a.). Entwicklung von *Didymium* (Myxomycetes)—amöben-phase. Film C 1044 des Inst. Wiss. Film, Göttingen.
Kerr, N. S., and Inst. Wiss. Film. (1970b). Entwicklung von *Didymium* (Myxomycetes)—plasmodium-phase. Film C 1045 des Inst. Wiss. Film, Göttingen. Publikation von N. S. Kerr, 1973, 1-18.
Kerr, N. S., and Sussman, M. (1958). Clonal development of the true slime mold, *Didymium nigripes. J. Gen. Microbiol.* **19,** 173-177.
Kerr, N. S., and Waxlax, J. N. (1968a). A yellow varient of the Eumycetozoan *Didymium nigripes* which exhibits aging. *J. Exp. Zool.* **168,** 351-362.
Kerr, N. S., and Waxlax, J. N. (1968b). The fusion of plasmodia of a true slime mold grown on various nutrients and the mixing of fused plasmodia. *Trans. Am. Microsc. Soc.* **87,** 197-200.
Kerr, S. J. (1967). A comparative study of mitosis in amoebae and plasmodia of the true slime mold, *Didymium nigripes. J. Protozool.* **14,** 439-445.
Kerr, S. J. (1968). Ploidy level in the true slime mold *Didymium nigripes. J. Gen. Microbiol.* **43,** 9-15.
Kerr, S. J. (1970). Nuclear size in plasmodia of the true slime mold *Didymium nigripes. J. Gen. Microbiol.* **63,** 347-356.
Kerr, S. J. (1972a). Inhibition of flagellum morphogenesis in the true slime mould *Didymium nigripes. J. Gen. Microbiol.* **72,** 419-427.
Kerr, S. J. (1972b). Flagellum growth and regeneration in the true slime mould *Didymium nigripes. J. Gen. Microbiol.* **72,** 429-437.
Klapholz, S., Esposito, M. S., and Esposito, R. E. (1978). The integration of events of sporulation. *9th Int. Conf. Yeast Gen. Molec. Biol. Abstr,* **81.**
Lane, E. B., and Carlile, M. J. (1979). Post-fusion somatic incompatibility in plasmodia of *Physarum polycephalum. J. Cell Sci.* **35,** 339-354.
Ling, H. (1968). Light and fruiting in *Didymium iridis. Mycologia* **60,** 966-970.
Ling, H., and Collins, O. R. (1970a). Control of plasmodial fusion in a Panamanian isolate of *Didymium iridis. Am. J. Bot.* **57,** 292-298.
Ling, H., and Collins, O. R. 1970b). Linkage studies in the true slime mold *Didymium iridis. Am. J. Bot.* **57,** 299-303.

Ling, H., and Ling, M. (1974). Genetic control of somatic fusion in a myxomycete. *Heredity* **32**, 95-104.

Ling, H., and Moore, D. (1979). Multinucleate cysts of the myxomycete *Didymium iridis*. *Mycologia* **71**, 713-721.

Ling, H., and Upadhyaya, K. C. (1974). Cytoplasmic incompatibility studies in the myxomycete *Didymium iridis:* Recovery and nuclear survival in heterokaryons. *Am. J. Bot.* **61**, 598-603.

Lott, T., and Clark, J. (1980). Plasmodial senesence in the acellular slime mold *Didymium iridis* (submitted).

Lucas, S., Razin, M., and Kerr, N. (1968). Observations on the differentiation of plasmodia into fruiting bodies by the true slime mould, *Didymium nigripes*. *J. Gen. Microbiol.* **53**, 17-21.

Martin, G. W., and Alexopoulos, C. J. (1969). "The Myxomycetes." Univ. of Iowa Press, Iowa City.

Mortimer, R. K., and Hawthorne, D. C. (1969). Yeast genetics. *In* "The Yeasts" (A. H. Rose and J. S. Harrison, eds.), Vol. 1, pp. 379-460. Academic Press, New York.

Mukherjee, K. L. (1965). Plasmodial fusion reactions between three races of *Didymium iridis* and their hybrids. *J. Indian Bot. Soc.* **44**, 224-230.

Mulleavy, P. (1979). Genetic and cytological studies in heterothallic and non-heterothallic isolates of the myxomycete *Didymium iridis*. Ph.D. Dissertation, University of California, Berkeley, California.

Mulleavy, P., and Collins, O. R. (1979). Development of apogamic amoebae from heterothallic lines of a myxomycete, *Didymium iridis*. *Am. J. Bot.* **66**, 1067-1073.

Mulleavy, P., and Collins, O. R. (1981). Isolation of CIPC-induced and spontaneously-produced diploid myxamoebae in a myxomycete, *Didymium iridis:* A study of mating type heterozygotes. *Mycologia* **73**, 62-77.

Nygren, A. (1967). Apomixis in the angiosperms. *Hand. Pflanzenphysiologie* **18**, 551-596.

Pallotta, D. J., Youngman, P. J., Shinnick, T. M., and Holt, C. E. (1979). Kinetics of mating in *Physarum polycephalum*. *Mycologia* **71**, 68-84.

Ross, I. K. (1967). Syngamy and plasmodium formation in the myxomycete *Didymium iridis*. *Protoplasma* **64**, 104-119.

Ross, I. K. (1968). Nuclear membrane behavior during mitosis in normal and heteroploid myxomycetes. *Protoplasma* **66**, 173-184.

Ross, I. K., and Cummings, R. J. (1970). An unusual pattern of multiple cell and nuclear fusions in the heterothallic slime mold *Didymium iridis*. *Protoplasma* **70**, 281-294.

Ross, I. K., and Shipley, G. L. (1973). Sexual and somatic fusion in the heterothallic slime mould *Didymium iridis*. 2. Effects of actinomycin D, cycloheximide and lysosome stabilizers. *Microbios* **7**, 165-171.

Ross, I. K., Shipley, G. L., and Cummings, R. J. (1973). Sexual and somatic cell fusions in the heterothallic slime mould *Didymium iridis*. 1. Fusion assay, fusion kinetics and cultural parameters. *Microbios* **7**, 149-164.

Schünemann, E. (1930). Untersuchungen über die Sexualität der Myxomyceten. *Planta* **9**, 645-672.

Schuster, F. L. (1964). Electron microscope observations on spore formation in the true slime mold *Didymium nigripes*. *J. Protozool.* **11**, 207-216.

Schuster, F. L. (1965). Ultrastructure and morphogenesis of solitary stages of true slime molds. *Protistologica* **1**, 49-62.

Schuster, F. L. (1969). Nuclear degeneration during spore formation in the true slime mold, *Didymium nigripes*. *J. Ultrastruct. Res.* **29**, 171-181.

Shinnick, T. M. and Holt, C. E. (1977). A mutation (gad) linked to mt and affecting asexual plasmodium formation in *Physarum polycephalum*. *J. Bacteriol.* **131**, 247-250.

Shipley, G. L., and Ross, I. K. (1978). The effect of proteases on plasmodium formation in *Didymium iridis*. *Cell Diff.* **7**, 21-32.

2. Didymium iridis in Past and Future Research

Therrien, C. D. (1966). Microspectrophotometric measurement of nuclear deoxyribonucleic acid content in two Myxomycetes. *Can. J. Bot.* **44,** 1667–1675.

Therrien, C. D., and Yemma, J. J. (1974). Comparative measurements of nuclear DNA in a heterothallic and a self-fertile isolate of the myxomycete, *Didymium iridis. Am. J. Bot.* **61,** 400–404.

Therrien, C. D., and Yemma, J. J. (1975). Nuclear DNA content and ploidy values in clonally-developed plasmodia of the myxomycete *Didymium iridis. Caryologia* **28,** 313–320.

Therrien, C. D., Bell, W. R., and Collins, O. R. (1977). Nuclear DNA content of myxamoebae and plasmodia in six non-heterothallic isolates of a myxomycete, *Didymium iridis. Am. J. Bot.* **64,** 286–291.

Upadhyaya, K. C., and Ling, H. (1972). Fine structure of clear zones produced by partial intermixing of protoplasm of two phenotypically different plasmodia in a myxomycete. *Am. J. Bot.* **59** (6), 668 (Abstr.).

von Stosch, H. A. (1935). Untersuchungen über die Entwicklungsgeschichte der Myxomyceten. Sexualität und Apogamie bei Didymiaceen. *Planta* **23,** 623–656.

Welden, A. (1955). Capillitial development in the Myxomycetes *Badhamia gracilis* and *Didymium iridis. Mycologia* **47,** 714–728.

Yemma, J. J., and Therrien, C. D. (1972). Quantiative microspectrophotometry of nuclear DNA in selfing strains of the myxomycete *Didymium iridis. Am. J. Bot.* **59,** 828–835.

Yemma, J. J., Jakupcin, G. F., and Therrien, C. D. (1978). The isolation of haploid myxamoebal clones from a diploid apomictic isolate of the myxomycete *Didymium iridis. Bot. Soc. Am.* **76** (Abstr. 1978).

Yemma, J. J., Therrien, C. D., and Ventura, S. (1974). Cytoplasmic inheritance of the selfing factor in the myxomycete *Didymium iridis. Heredity* **32,** 231–239.

PART II

Characteristic Biological Phenomena

CHAPTER 3

Periodic Phenomena in *Physarum**

JOHN J. TYSON

I.	Introduction	61
II.	Nuclear Division Cycle	62
	A. Introduction	62
	B. Perturbations of the Nuclear Division Cycle	66
	C. Comparison with Other Systems	76
III.	Periodic Enzyme Synthesis	80
	A. Introduction	80
	B. Periodic Enzymes in *Physarum*	83
	C. Origin of Enzyme Periodicities	90
IV.	Other Periodic Events of the Mitotic Cycle	93
	A. Macromolecular Synthesis	93
	B. Metabolic Pools	94
	C. Miscellany	95
V.	Shuttle Streaming	96
	A. Introduction	96
	B. Biochemical Nature of the Contraction Oscillator	97
	References	102

I. INTRODUCTION

Just as a crystal is the periodic repetition of a motif, the unit cell, in space, so the history of a population of exponentially growing cells is the periodic repetition of another motif, the cell cycle, in time. In the ideal state of balanced growth, which is no more or less fictional than a perfect crystal, all the major components of a cell—e.g., protein, RNA, DNA, plasma membrane, and cell wall—increase by the same factor during one cell cycle. In the simplest and commonest case, this factor is two.

*Dedicated to the memory of Erik Zeuthen (1914–1980).

The passage of a cell through the cell cycle is marked by a number of events which bear a definite temporal relation to each other from one cycle to the next. For example, the period of DNA synthesis (S phase) is commonly followed after some delay (G_2 phase) by nuclear division (ND) and cell division. There are other, more subtle, events in the cell cycle. Certain enzymes may show peaks of activity at given phases in the cycle, and others may show abrupt doublings in their rate of synthesis at characteristic phases. In some organisms, fluctuations in pH, oxygen consumption, energy charge, and the like are correlated with the cell cycle.

In the plasmodial stage of the life history of *Physarum polycephalum*, a syncytial myxomycete, cell division does not occur. Nuclei accumulate in a common cytoplasmic environment and divide in nearly perfect synchrony once per cycle. Since nuclear division can be determined quickly and accurately by phase contrast microscopy of plasmodial smears, it serves as the most convenient reference point for cell cycle events in this organism. Natural synchrony extends to other events of the cell cycle. For instance, DNA synthesis commences in all nuclei immediately after division (i.e., in telophase), and a number of enzyme activities have been observed to fluctuate regularly from one cycle to the next.

This chapter is concerned primarily with periodic events of the mitotic cycle in *P. polycephalum*. Other useful reviews of this subject can be found in Schiebel (1973), Braun *et al.* (1977), and Holt (1980). The last section deals with a higher-frequency rhythm in *Physarum*, the periodic reversal of the direction of flow of protoplasm in the network of veins which extends throughout the plasmodium. Shuttle streaming has been reviewed in detail by Wohlfarth-Bottermann (1979) and by Kessler (this volume, Chapter 5).

II. NUCLEAR DIVISION CYCLE

A. Introduction

Well-fed *Physarum* plasmodia provide a clear example of balanced growth: The nuclei divide, and a new round of DNA synthesis is initiated each time the cell mass doubles. For instance, the protein:DNA ratio in batch cultures of microplasmodia is constant even if the ploidy of the nucleus is increased fourfold (Mohberg *et al.*, 1973) or if the growth rate is decreased fourfold (Plaut and Turnock, 1975). Another example is ribosomal DNA (rDNA), which in *Physarum* is a linear palindromic molecule located in the nucleolus. Though rDNA is independent of chromosomal DNA and replicates autonomously, the total amount of rDNA is stringently controlled during the cell cycle and during the life cycle as well (Braun and Seebeck, 1979). These examples illustrate the close coordination between nuclear division, macromolecular synthesis, and overall growth. How can we characterize this control system in *Physarum*?

3. Periodic Phenomena in *Physarum*

1. RELATION OF DNA SYNTHESIS TO NUCLEAR DIVISION

As mentioned earlier, DNA synthesis commences immediately after nuclear division in *Physarum,* which suggests that the onset of DNA synthesis is triggered by some event in the mitotic sequence. To test this idea, Guttes and Guttes (1968) transferred nuclei from S phase plasmodia to G_2 plasmodia and vice versa. The transfer was done by allowing two cells to fuse for about 15 minutes, then separating them. During the brief fusion period, nuclei and cytoplasmic materials are exchanged between the plasmodia. If nuclei from an S-phase plasmodium are transferred to a G_2-phase plasmodium, they complete DNA synthesis normally and do not trigger DNA synthesis in the G_2-phase host nuclei. On the other hand, G_2 nuclei which find themselves in an S-phase plasmodium do not reinitiate DNA synthesis; but several hours later, after undergoing a synchronous division with the host nuclei, they synthesize DNA normally. Thus, it appears that nuclear division is necessary before a new round of DNA synthesis can occur.

An apparent exception to this rule is the observation that nuclei become polyploid after a short exposure to elevated temperature shortly before metaphase (Brewer and Rusch, 1968). However, in this case only the terminal stages of nuclear division (anaphase and telophase) are aborted by the heat treatment. DNA synthesis is delayed and less intense in these undivided nuclei, but eventually they complete DNA synthesis semiconservatively, leading to a doubling of the ploidy level (Brewer and Rusch, 1968; Wolf *et al.,* 1979; Mohberg *et al.,* 1980).

Obviously DNA synthesis is tightly coupled to nuclear division in *Physarum.* It seems likely that nuclear division and DNA synthesis are triggered by the same event in late G_2. A simple picture would be:

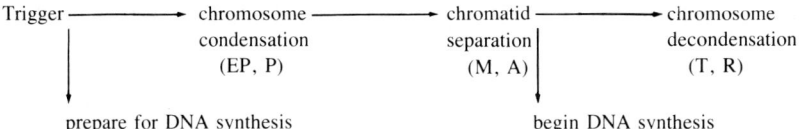

In this picture, DNA synthesis commences as soon as the chromosomes decondense. This would normally be in telophase (T), but after heat shocks in prophase (P), chromosome decondensation and DNA synthesis would occur in undivided nuclei.

We conclude that regulation of the DNA: mass ratio in *Physarum* is accomplished by the same system that controls the timing of synchronous nuclear division.

2. PLASMODIAL FUSION

If two different plasmodia are brought into contact, they will fuse spontaneously and the resulting plasmodium will share properties of the two original plasmodia. For instance, if a plasmodium in which mitosis is due at noon is fused with an equally large plasmodium in which mitosis is due at 4 P.M. the mixed plasmodia will undergo mitosis at 2 P.M. If the noon plasmodium is larger than the 4 P.M. plasmodium, then nuclear division will occur proportionately closer to noon in the mixed plasmodium (Rusch et al., 1966).

3. MODELS FOR CONTROL OF THE NUCLEAR DIVISION CYCLE

The precise synchrony of nuclear division in macroplasmodia suggests that there is some extranuclear *Zeitgeber* that triggers division in all nuclei simultaneously. The intriguing results of plasmodial fusion suggest that this substance is present in different amounts in plasmodia at different phases in the cell cycle, and these amounts are averaged according to the sizes of the plasmodia being fused.* The control substance, or substances, could monitor the nucleocytoplasmic ratio of a plasmodium and trigger mitosis and a new round of DNA synthesis each time the cell mass doubles. In theory there are a number of ways that such a control system could work. Four possibilities are discussed below.

a. The Simple Concentration Model (Fig. 1). The simplest assumption is that the cytoplasmic concentration of some substance changes monotonically during the cell cycle, and when it reaches a critical value, the substance triggers mitosis. The substance might be an activator whose concentration is increasing or an inhibitor whose concentration is decreasing. At mitosis or at the onset of DNA synthesis, a certain fraction of the activator pool, proportional to the total DNA content, is destroyed or a new aliquot of inhibitor, proportional to the total DNA content, is synthesized.

b. The Division-Protein Model (Fig. 2). Zeuthen and Williams (1969) have proposed that division is triggered upon completion of a multisubunit protein structure. The subunits are synthesized throughout the cell cycle, and the structure assembles spontaneously. The structure is unstable, so that it is continually exchanging subunits with the cytoplasmic pool. When the structure is complete, it performs some essential function in division. At division, the structure is completely dismantled and the subunits are degraded.

c. The Nuclear-Sites Titration Model (Fig. 3). Sachsenmaier et al. (1972) suggested that the concentration of a mitotic activator is regulated at a constant value. As the plasmodium grows, the total number of activator molecules increases. These molecules enter the nuclei and bind to nuclear sites, of which there are a fixed number proportional to the total DNA content of the plasmodium. After all the nuclear sites are occupied by activator molecules, nuclear

---

*Presumably a uniform concentration of mitogen is established by means of the vigorous shuttle-streaming mechanism. Even in plasmodia as large as 14 cm in diameter, mitosis is nearly synchronous (Mohberg and Rusch, 1969).

3. Periodic Phenomena in *Physarum*   65

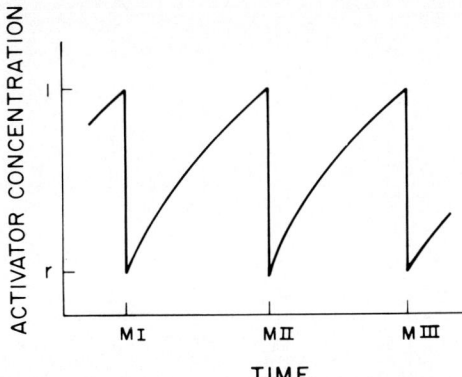

Fig. 1. Time course of activator concentration for a simple concentration model. When activator concentration reaches 1, mitosis is triggered and a fraction $(1 - r)$ of activator molecules are destroyed.

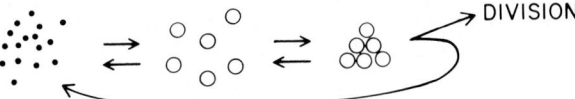

Fig. 2. A division-protein model. When the division-protein structure is complete, mitosis is triggered and the structure is disassembled.

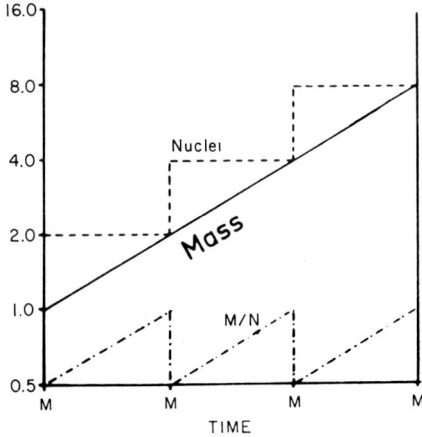

Fig. 3. The nuclear-sites titration model. The number of nuclear receptor sites is proportional to the number of nuclei in the plasmodium. The total amount of initiator is proportional to the plasmodial mass. Notice that the initiator:nuclear-sites ratio (-----) is a periodic function of time. When the ratio reaches 1, mitosis is triggered and the ratio drops abruptly to 1:2. From Sachsenmaier *et al.* (1972); used by permission of the authors and Academic Press.

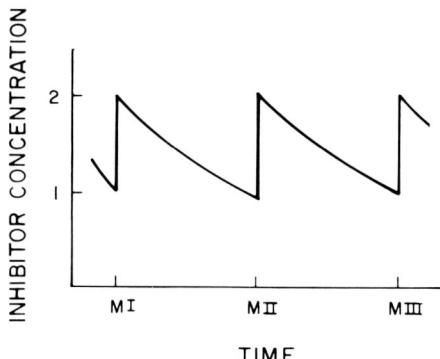

Fig. 4. Time course of inhibitor concentration for the unstable inhibitor model. When the inhibitor concentration drops to 1, mitosis is triggered and the inhibitor concentration doubles as the nuclei divide.

division commences. During division or early in the DNA synthetic period, the total number of nuclear sites doubles, producing unoccupied sites which must be filled during the next cycle.

d. The Unstable-Inhibitor Model (Fig. 4). Sudbery and Grant (1975) proposed that a mitotic inhibitor is synthesized at a rate proportional to the total amount of DNA in the plasmodium and that it breaks down spontaneously, with a half-life much shorter than the interdivision time. Thus, the total amount of inhibitor in the plasmodium is always proportional to the total amount of DNA. As the plasmodium grows, inhibitor concentration decreases because of dilution. When the concentration is reduced by a factor of two from its postmitotic value, nuclear division is triggered, and the total amount of inhibitor doubles during the next S phase.

B. Perturbations of the Nuclear Division Cycle

Of these four models, and others that might come to mind, can we decide which is most likely to be correct? Or can we at least rule out some of them because they fail to conform to certain properties of the nuclear division cycle in *Physarum?* Fantes *et al.* (1975) have pointed out that these models make divergent predictions about the outcome of certain experimental perturbations of the cell cycle. To establish this, they constructed mathematical models of the four models just described. In order to determine the timing of divisions subsequent to a perturbation of the nuclear division cycle, the mathematical equations were solved using initial conditions that represented various experimentally possible perturbations. The results are summarized in Table I. It can be seen that, whereas

TABLE I

Response Predicted by Four Models of the Mitotic Control Mechanism to Two Perturbations of the Nucleocytoplasmic Ratio[a,b]

Model	Delay mitosis[c]	Remove nuclei[d]
I. Simple concentration and division protein	$t_1 = T(1 + x)$ $t_2 = T(1 - x)$ $t_3 = t_4 = \ldots = T$	$t_1 = T \log_2 (2 - y)$ $t_2, t_3, \ldots$ lengthen asymptotically to T
II. Nuclear sites and unstable inhibitor	$t_1 = T(1 + x)$ $t_2 = T(1 - x)$ $t_3 = t_4 = \ldots = T$	$t_1 = T[1 + \log_2 (1 - y)]$ $t_2 = t_3 \ldots = T$

[a] From results of Fantes et al. (1975) and Sudbery and Grant (1975).

[b] T, normal intermitotic period; t_1, length of the intermitotic period immediately after perturbation; $t_2, t_3, \ldots$, length of the second, third, etc. intermitotic periods after perturbation.

[c] For example, heat shock. x, fraction of the cycle by which the first intermitotic period is delayed.

[d] For example, UV irradiation. y, fraction of nuclei removed. For class I models, t_{i+1} is related to t_i by $2^{t_{i+1}/T} = 1 + 2(1 - 2^{-t_i/T})$.

all four models predict the same response to a delay of mitosis, they disagree about the response to a removal of nuclei.

1. UV IRRADIATION

Sudbery and Grant (1975) used this discrepancy to test alternative models of the nuclear division cycle in *Physarum*. They chose *Physarum* not only because of its natural synchrony but also because DNA can be destroyed within a plasmodium by irradiation with UV light. If plasmodia are irradiated in early G_2, the next mitosis is delayed but otherwise normal. However, after division only a fraction of the nuclei are able to complete DNA synthesis; the rest are broken down (Sachsenmaier et al., 1970). Thus the nucleocytoplasmic ratio is abnormally small at the start of the first complete cycle after UV irradiation. As expected, this cycle is short. The test concerns the length of the second complete cycle after irradiation. Is it short as well, or is it of normal duration? Sudbery and Grant found that the second and third postirradiation cycles are of normal duration (Table II). Thus, simple concentration models and structural models, as axiomatized by Fantes et al. (1975), are not consistent with the results of these UV experiments.

A word of caution is in order here. It would be inappropriate to conclude that a division protein, of the type envisioned by Zeuthen for the *Tetrahymena* cell cycle, is definitely out of the question for *Physarum*. Fantes et al. (1975) assumed that the subunits of the division protein are destroyed at mitosis so that they cannot be reutilized in the next cycle. However, if we were to assume that

TABLE II

The Effect of UV Irradiation on Intermitotic Times of Plasmodia Growing on Vitamin-Enriched Medium Lacking Yeast Extract[a,b]

Conditions[c]	MII-MIII	MIII-MIV	MIV-MV	MV-MVI
1. UV (9)	19.5 ± 0.2	12.0 ± 0.2	15.9 ± 0.3	15.0 ± 0.4
2. Control (10)	15.4 ± 0.1	19.3 ± 0.3	16.5 ± 0.4	15.0 ± 0.4
3. UV:control	1.25 ± 0.02	0.62 ± 0.03	0.95 ± 0.04	1.00 ± 0.05
4. Class I predictions		Given	0.77 ± 0.02	0.87 ± 0.02
5. Class II predictions		Given	1.00	1.00

[a] After Sudbery and Grant, 1975; used by permission of Academic Press

[b] Plasmodia were irradiated (2000 erg/mm^2) 4 hours after MII. MIII was delayed one-quarter of a cycle due to a radiation-induced inhibition of growth. At M$_{III}$ about 20% of the nuclei were lost, causing decrease of the next intermitotic period by 38%. The two following intermitotic periods, however, were of normal duration.

[c] Lines 1 and 2 record intermitotic times of UV-irradiated and control plasmodia (the number of plasmodia in each group is given in parentheses). Line 3 is the ratio of intermitotic periods for treated and control plasmodia. Lines 4 and 5 give the value of this ratio predicted by the models of classes I and II (see Table I), given that the MIII-MIV ratio is between 0.59 and 0.65.

the subunits are conserved and divided among daughter cells (in *Tetrahymena*, daughter nuclei in *Physarum*), then the division protein model becomes equivalent to the nuclear-sites titration model, except for the question of whether the assembly of subunits occurs in the cytoplasm (e.g., the oral structure in *Tetrahymena*) or in the nucleus. This ''geographical'' difference in models cannot be resolved by perturbation experiments of the type considered by Fantes, Sudbery, Grant, and co-workers.

In general terms, the advantage of the approach taken by Fantes *et al.* (1975) is that the assumptions of a model are pinned down in mathematical equations. This allows precise predictions to be made which can then be tested. If a particular model, i.e., a particular set of mathematical assumptions, does not conform to the experiment, then we can determine which assumption is responsible for the model's failure. In this way, we learn something about the sorts of models which can or cannot account for phenomenological features of the mitotic control system.

2. CYCLOHEXIMIDE PULSES

Scheffey and Wille (1978) and Tyson *et al.* (1979) treated plasmodia at different phases in the cell cycle with pulses of cycloheximide, an inhibitor of protein synthesis, and measured the delay of subsequent mitoses (Fig. 5). For treatments early in the cycle, there is a basal delay attributable to inhibition of overall protein synthesis. For treatments later in the cycle, there is a delay in excess of the basal delay. The amount of excess delay increases as the pulse treatments are given ever later in the cycle until, about 45 minutes before mitosis, the delaying effect of cycloheximide treatment ceases abruptly.

Tyson *et al.* (1979) interpreted these results in terms of an unstable mitotic activator whose synthesis is blocked by cycloheximide. During cycloheximide treatment the activator is degraded, and after the pulse, it must be resynthesized. The later the pulse, the more catching up the plasmodium must do, so the greater the excess delay. Quantitative analysis of the experimental results indicates that the activator must have a half-life of 1–2 hours in order to account for the observed excess delay.

The cycloheximide experiments are a bit awkward for the unstable-inhibitor model to explain. If the inhibitor were a protein, then it would be degraded during the cycloheximide pulse and mitosis should be triggered prematurely. To

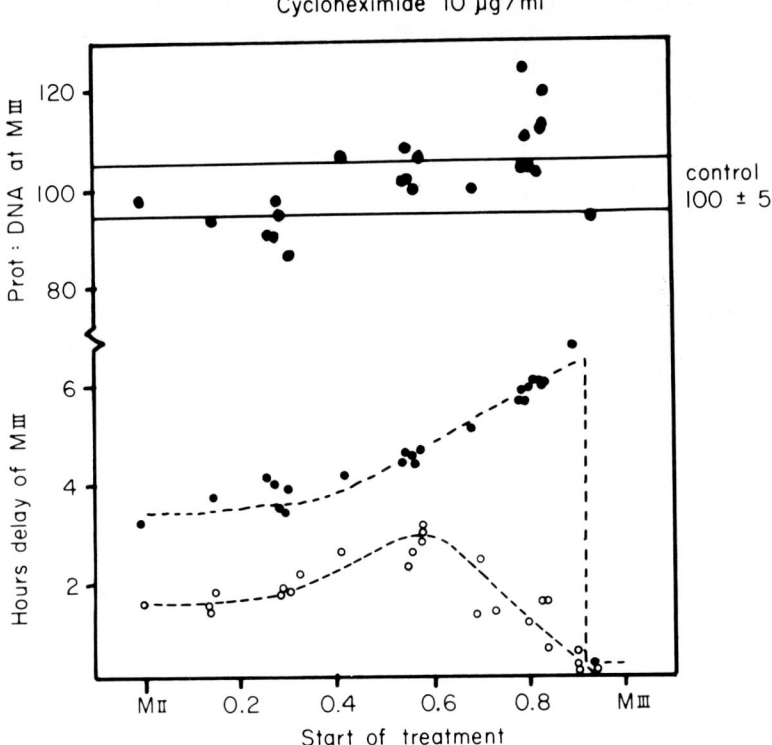

Fig. 5. Delay of mitosis (lower part) and protein : DNA ratio at the delayed mitosis (upper part) after 1.5-hour pulse treatment with cycloheximide (●, 10 μg/ml medium; ○, 0.5 μg/ml medium). Notice that the protein : DNA ratio at the delayed mitosis increases as the delay increases, indicating that pulse-treated plasmodia grow larger than untreated plasmodia during the excess delay period.

account for the opposite result, it could be assumed that the inhibitor is a metabolite whose synthesis and degradation are catalyzed by two enzymes:

$$\xrightarrow{E_1} I \xrightarrow{E_2}$$

Assuming that E_2 is unstable, one can reproduce the experimental results; however, the model is now quite similar to the nuclear-sites titration model, with E_2 analogous to the unstable activator and I analogous to the nuclear sites.

3. FUdR/FUSION

The results of the UV and cycloheximide experiments just described favor the nuclear-sites titration model. Recall that in this model the unstable activator is synthesized continuously in such a fashion that its total concentration remains constant throughout the cell cycle. What does this model predict will happen if a plasmodium is grown from the time of a synchronous mitosis on medium containing 5-fluorodeoxyuridine (FUdR), an inhibitor of DNA synthesis?

Since the plasmodium continues to synthesize protein and to grow in the presence of FUdR, we might expect that the putative activator continues to accumulate in the treated plasmodium. If the number of sites doubles in the presence of FUdR, then the signal for nuclear division should occur on schedule with an untreated control plasmodium. If the number of nuclear sites does not double in the presence of FUdR, then there will be an excess of unbound activator in the treated plasmodium and the trigger for nuclear division should be permanently pulled. In either case, however, we would not expect mitosis in the FUdR-treated plasmodium since the nuclei are unreplicated and, presumably, incapable of responding to the mitotic trigger. On the other hand, it may be that activator is synthesized only during the G_2 phase and not during the S phase. In that case, there would be very little activator in the FUdR-treated plasmodium (since activator is unstable), and the trigger for nuclear division would not be pulled as long as DNA synthesis remained blocked.

Sachsenmaier *et al.* (1972) performed an elegant experiment to test whether activator accumulates in FUdR-treated plasmodia, i.e., to test whether or not activator is synthesized during the S phase. Their idea was to fuse division-competent (replicated) nuclei into an FUdR-treated plasmodium, and to see at what time the division-competent nuclei actually divide. Their experimental protocol and results are illustrated in Fig. 6. If the FUdR-treated plasmodium synthesizes activator at the same rate as the untreated plasmodium, then mitosis in the treated/G_2 fused plasmodium should occur on schedule with the untreated/G_2 control (at least in the replicated nuclei of the treated/G_2 plasmodium). If the treated plasmodium does not synthesize activator until some time after fusion (when FUdR is no longer present and DNA synthesis can be

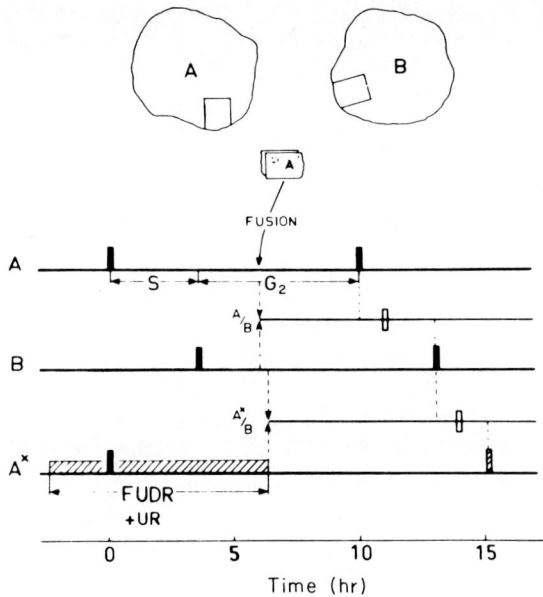

Fig. 6. Synchronous division in a mixed plasmodium prepared by fusion of an FUdR-treated plasmodium with an untreated plasmodium. The fused pair A^x/B, in which A^x has been treated with FUdR for about 7 hours between mitosis and fusion, goes through a synchronous mitosis at 14 hours, halfway between the B and A^x control cultures. Notice that the fused pair A/B, in which neither component has been treated with FUdR, goes through a synchronous mitosis at 11 hours, halfway between the two parental cultures, A and B. From Sachsenmaier *et al.* (1972); used by permission of the authors and Academic Press.

completed), then the treated/G_2 fused plasmodium should not undergo nuclear division until many hours after the untreated/G_2 control. Since Sachsenmaier *et al.* (1972) observed the latter result, they concluded that the presumptive mitotic activator is not synthesized during the S phase.

This raises a question about how the synthesis of activator is turned off and on at the beginning and end of the S phase, respectively. Tyson and Sachsenmaier (1979) have shown that if the activator gene is inducible (like the *lac* operon in *Escherichia coli*), then transcription of the gene can be turned off at the time of replication of the regulatory gene and on again at the time of replication of the operon region. Synthesis of activator only during G_2 can also be explained if there is a connection between transcription and replication in *Physarum*, as suggested by Pierron and Sauer (1980a). If the structural gene for activator is replicated at the end of the S phase and if a burst of stable mRNA is transcribed at this time, then activator would be synthesized from this message during G_2. To shut off activator synthesis at the end of G_2, we would have to assume that mRNA for activator is selectively destroyed at mitosis or in the early S phase.

4. FUSION EXPERIMENTS AND LIMIT CYCLE MODELS

Kauffman and Wille (1975) have modeled the nuclear division control system from a different point of view. They assume that two or more substances in the cell react with each other to produce sustained oscillations in the concentrations of the interacting biochemicals. These oscillations can serve as an intracellular clock, timing various events of the cell cycle; the clock, however, runs independently of the events timed. That is, in the limit cycle model, the occurrence of nuclear division or DNA synthesis does not play a role in the generation of periodicity, whereas in the models considered earlier, the occurrence of these events is essential to resetting the clock.*

Models which involve resetting can be referred to as "relaxation oscillators" because the resetting occurs over a much shorter period of time than the cell cycle period. Thus, these models appear to relax abruptly from a charged state to a discharged state. For example, in the nuclear-sites titration model, the number of nuclear sites doubles immediately after mitosis and the activator:sites ratio falls abruptly by a factor of two. This is the relaxation, or discharge, phase. During the rest of the cycle, the activator pool is recharged.

Winfree (1974) has pointed out that a limit cycle can be distinguished from a relaxation oscillator by standard plasmodial fusion experiments. The essence of Winfree's argument is that if nuclear division is timed by a continuous limit cycle oscillator, then it should be possible to find a pair of phases (ϕ_1 and ϕ_2) such that if a plasmodium ϕ_1 hours from mitosis is fused with a plasmodium ϕ_2 hours from mitosis, then the fused pair will *not* be a well-defined number of hours from mitosis. For instance, the Kauffman–Wille model predicts that mitosis in the fused plasmodium is unpredictable and very much delayed.

Tyson and Sachenmaier (1978) have carried out Winfree's analysis using data from 54 fusion experiments, and they found relaxation oscillator models to be clearly favored. Though the arguments in this paper are mathematically technical and may be difficult to follow, the fundamental objection to limit cycle models is quite easy to explain (Fig. 7). If a plasmodium shortly before mitosis is fused with one shortly after mitosis, when will nuclear division in the mixed plasmodium occur? According to relaxation oscillator models, nuclei of the mixed plasmodium should divide after about a half-cycle delay. According to limit cycle models, nuclei of the mixed plasmodium should divide either soon after fusion or after about a full-cycle delay. Fusions of this sort have been done by many different people for various reasons, and it is always found that mitosis in

---

*To illustrate this difference, think of an hourglass and an electric clock. The hourglass must be turned over each hour in order to time a repetitive sequence of events with period = 1 hour, and the turning over of the hourglass would certainly be an important event in the cycle. An electric clock, on the other hand, continues to run independently of the events it is timing.

3. Periodic Phenomena in *Physarum*

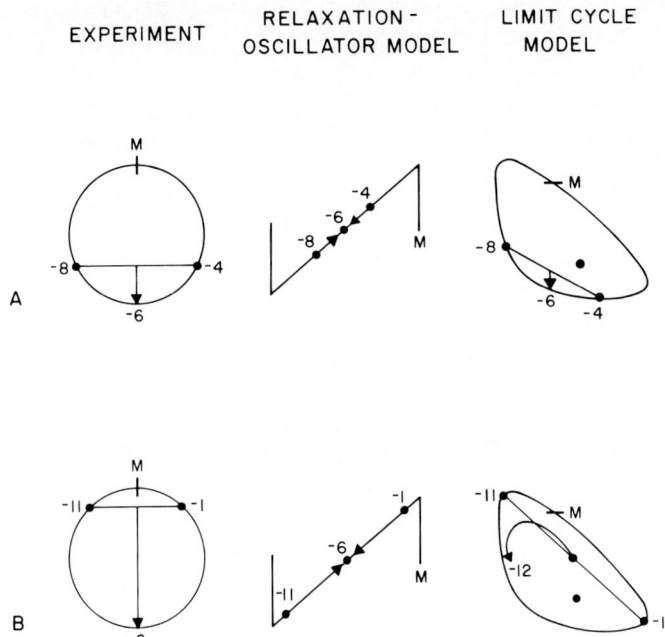

Fig. 7. Divergent predictions of a continuous limit cycle model and a relaxation oscillator model. (A) If plasmodia 4 hours and 8 hours from metaphase are fused, the mixed plasmodium undergoes synchronous nuclear division about 6 hours after fusion. This result is predicted by both models of the mitotic control system. (B) If a plasmodium about 1 hour before metaphase is fused with a plasmodium that has just gone through metaphase, the mixed plasmodium undergoes nuclear division about one-half to three-quarters of a cycle after fusion. This is consistent with the relaxation oscillator model, but the limit cycle model predicts a full-cycle delay.

the mixed plasmodium occurs after about a half-cycle delay (Sachsenmaier *et al.*, 1972; Chin *et al.*, 1972; Sachsenmaier, 1976; Loidl, 1979).*

5. HEAT SHOCKS AND TEMPERATURE SHIFTS

Brewer and Rusch (1968) transferred cultures from 26° (near-optimal temperature) to 37°C (lethal temperature) for various lengths of time at various phases in the cell cycle and determined the time of the next nuclear division in treated plasmodia in comparison to untreated controls. They found little delay induced by heat shocks early in the cycle, increasing delay in the middle of the cycle reaching a maximum about 2 hours before metaphase, and decreasing delay in

---

*Although there is no strong evidence in *Physarum* for an autogenous cytoplasmic clock keeping time for the cell cycle, there is some interesting evidence for just such a clock with the same period as the division cycle in some animal eggs (Bell, 1962; Yoneda *et al.*, 1978; Parisi *et al.*, 1978; Hara *et al.*, 1980).

the last 2 hours of the cycle. This pattern of increasing and then decreasing delay is also observed after pulse treatment with low concentrations of cycloheximide (see Fig. 5).

Wille et al. (1977) have investigated the effects of long exposures to lethal temperatures on the timing of mitoses subsequent to a return to 26°C. They interpret their results in terms of a limit cycle model. However, since no quantitative information on the effects of these long heat shocks on growth and nuclear viability is presented, it is impossible to draw any convincing conclusions from this work.

Tyson and co-workers (Tyson and Sachsenmaier, 1978; Tyson et al., 1979) have followed the protein:DNA ratio at three mitoses subsequent to heat shocks at 37°C. The ratio is 20–40% higher than normal at the first mitosis after heat shock. Subsequent interdivision periods are shorter than control periods (but never shorter than a certain minimum intermitotic period), until the protein:DNA ratio is returned to its normal value at mitosis. Thereafter, interdivision periods are normal. These results are consistent with the effects of UV irradiation on intermitotic intervals.

Wright and Tollon (1978) have followed the time of mitosis in plasmodia which have been shifted from one sublethal temperature to another at various times during the division cycle. They interpret their results in terms of two different activator substances: A, which is stable from 14° to 30°C, and B, which is synthesized at high temperature (29°–32°C) but is stable even at 22°C.

Temperature-shift experiments seem to be more informative than heat shocks at lethal temperatures. However, we will need more information on rates of protein and DNA synthesis before and after the shift in order to interpret the results of these kinds of experiments.

6. MITOTIC ADVANCES

Oppenheim and Katzir (1971) reported that when extracts prepared from late G_2 plasmodia are applied to the surface of early G_2 plasmodia, then mitosis in an early G_2 plasmodium is advanced by an hour or more relative to a control plasmodium. Blessing and Lempp (1978) have suggested an improvement in technique by using rabbit antibodies prepared against S-phase protein to precipitate these proteins from late G_2 plasmodial extracts.

Loidl and Gröbner (1981) have developed a convenient and sensitive test to determine the capacity of plasmodial extracts to accelerate nuclear division in recipient plasmodia. A high-speed supernatant prepared from homogenized plasmodia is mixed with microplasmodia, and the mixture is incubated for 15 minutes at 24°C. Then 0.2 ml of the suspension is used to start a surface culture in the usual way. The time of the first mitosis in the surface culture is determined and compared with MI in control plasmodia prepared according to the same procedure with nutrient buffer instead of plasmodial extract. The accelerating capacity of extracts undergoes a cycle dependent fluctuation with a sharp maximum (up to 30% of the control period) in late G_2. The accelerating capacity is most likely due to a protein since it is preserved during dialysis and destroyed by pronase. The substance is heat stable.

Bradbury *et al.* (1974a) have hypothesized that H1-histone kinase (HKase) is the active component of these phase-advancing mitotic extracts. This hypothesis was tested by adding kinase-HKG, isolated from Ehrlich ascites cells, to the exterior surface of plasmodia and observing the time of the next mitosis relative to controls treated similarly (Bradbury *et al.*, 1974b; Inglis *et al.*, 1976). These investigators observed maximum advances of the next mitosis (about 45-minute advances) for treatments 3-4 hours before the control mitosis.

Though the mitotic advance induced by added HKase is powerful evidence that this enzyme is part of the sequence of events which culminate in mitosis,* the role played by HKase (if any) in controlling the timing of nuclear division is still unclear. According to Bradbury *et al.* (1974a), HKase activity peaks about 2 hours before mitosis and declines rapidly 1 hour before mitosis, yet irreversible commitment to mitosis is not achieved until about 45 minutes before metaphase, as indicated by plasmodial fusion experiments.† Thus it appears that the HKase peak occurs too early to be the mitotic trigger, though we might still think of it as the trigger for chromosome condensation. In this view, chromosome condensation is a necessary, but not sufficient, condition for mitosis. A later trigger must still be pulled to initiate the mitotic sequence (nucleolar breakdown, spindle formation, chromosome movement, etc.).

Oleinick *et al.* (1981b) have repeated the assay of HKase activity during the cell cycle of *Physarum* and found that the activity peaks quite abruptly just at metaphase. Furthermore, if mitosis is delayed by γ irradiation (5.8 krad) in mid G_2, the HKase peak coincides with the delayed mitosis; and if this radiation-induced mitotic delay is reduced by caffeine treatment, the HKase peak is still coincident with metaphase. Thus, in the hands of Oleinick, Brewer, and Blank, there is a tight coupling of metaphase and the peak of HKase activity; but even if this pattern turns out to be correct, it is still unlikely that the peak of HKase activity is the mitotic trigger because, in this case, it occurs too late.

---

*Some evidence in favor of the hypothesis that HKase is the mitotic trigger is provided by the temporal correspondence between histone phosphorylating activity and the delay of mitosis induced by heat shocks at 37°C (Brewer and Rusch, 1968). If heat inactivates HKase, then the heat shock delays are accounted for nicely. But since short exposures to lethal temperatures will affect plasmodia in many ways (none of which are well characterized), this correspondence between the HKase activity peak and the heat-shock delay curve may be fortuitous.

†Plasmodia about 2 hours before metaphase can be delayed from dividing by fusion with plasmodia earlier in the cycle, but plasmodia 1 hour before metaphase can no longer be stopped by fusion with earlier plasmodia (Sachsenmaier *et al.*, 1972; Chin *et al.*, 1972; Loidl *et al.*, 1979). Since fusion takes 30-45 minutes to be effective, this puts the point of no return about 30 minutes before metaphase—long after HKase has acted. A late point of no return is also indicated by cycloheximide treatment (Tyson *et al.*, 1979).

7. SUMMARY

What have we learned about the onset of nuclear division and DNA synthesis in *P. polycephalum?* We can be sure that the control is mediated by cytoplasmic factors, and that certain appealing views of the control mechanism are not adequate. For instance, a simple model in which activator builds up in the cytoplasm until its concentration reaches a critical level (or an inhibitor is degraded by cytoplasmic factors until its concentration drops to a critical level) is not adequate to explain the precise UV-irradiation experiments of Sudbery and Grant (1975; Section II,B,1). Limit cycle oscillator models (autonomous cellular clocks that run independently of the DNA-division cycle) are also unlikely since they are not consistent with well-documented fusion experiments (Section II,B,4). On the other hand, there are at least two conceptions of the control system that are consistent with most of the evidence to date: (i) an unstable activator that titrates against nuclear sites and (ii) an unstable inhibitor that is diluted out by growth. So far, no experiment has been designed to distinguish between these two possibilities, but the cycloheximide-pulse experiments discussed in Section II,B,2 seem to favor the paradigm of an unstable proteinaceous activator.

This picture of *Physarum* is supported by compelling evidence from other cell types (to be reviewed in the next section), but we are still at a very primitive stage in our understanding of the coordination of growth, DNA synthesis, and nuclear division. Our best models are only hypotheses supported by indirect evidence. We have no knowledge of the molecular mechanism, and we hardly know where to look. We are only beginning to ferret out the relationships between the various components of the DNA-division cycle. In *Physarum,* for instance, we know that nuclear division and DNA synthesis are tightly coupled, but it is not at all clear how the mitotic cycle, the DNA synthetic cycle, the chromosome condensation cycle, the nucleolar breakdown–reconstruction cycle, the spindle cycle, etc., are interrelated. To what extent are these parallel pathways? To what extent are they dependent? Are there separate triggering events for some of these cycles? Where are the checkpoints that maintain normal phase relations among the cycles?

C. Comparison with Other Systems

1. BACTERIA AND LOWER EUKARYOTES

In prokaryotes it is widely accepted that DNA synthesis and cell division are coupled with overall growth by a cell-sizing mechanism. The most persuasive evidence for this thesis was the demonstration by Donachie (1968) that the exponential increase in average cell mass with growth rate, taken together with the Cooper–Helmstetter model for bacterial DNA replication, implies that chromosome replication is initiated at integral multiples of a critical cell mass.

Size control of the initiation of DNA synthesis and nuclear division in lower eukaryotes has been suggested quite frequently. Prescott (1956) was able to prevent cell division of *Amoeba proteus* by cytoplasmic amputation, which suggests that attainment of a critical cell mass is necessary for cell division. Frazier (1973) accomplished the opposite perturbation: By microsurgical removal of macronuclear nodes from the ciliate *Stentor coeruleus*, he induced premature DNA replication.

Amitotic division of the macronucleus of *Paramecium tetraurelia* produces daughter cells which frequently differ in macronuclear DNA content. To make up for this unequal distribution of genetic material, each cell synthesizes the same amount of macronuclear DNA regardless of its prereplication DNA content, and the amount of DNA synthesized depends on the total size of the cell (Berger, 1979). This provides an effective regulatory mechanism for maintaining an optimum relationship between gene dosage and cell size.

Compelling evidence for cell-size-related control of cell cycle events in the fission yeast *Schizosaccharomyces pombe* has been compiled in the last few years, since Nurse (1975) isolated some mutants, called *wee* mutants, that grow at a normal rate but divide at about half-normal size. From clever experiments using *wee* mutants and wild-type cells, Nurse and his colleagues have been able to determine that wild-type *S. pombe* cells have two size-related control points in the cell cycle, one at nuclear division and one at the beginning of the S phase. For instance, from nutritional shift experiments, Fantes and Nurse (1977) showed that division in wild-type cells is controlled by the requirement that cells reach a minimum size, which can be modulated by prevailing nutrient conditions, before nuclear division and cell division proceed. The execution point for this size control mechanism is shortly before nuclear division. In *wee* mutants, on the other hand, the size control mechanism seems to operate at the beginning of the S phase (Fantes and Nurse, 1978). Growth of wild-type cells from spores or under nitrogen limitation has revealed that they have the same cell-size requirement for the onset of DNA synthesis as have *wee* mutants (Nurse and Thuriaux, 1977; Nasmyth, 1979). However, under fast-growth conditions, this size requirement is cryptic since wild-type cells at birth are already larger than the minimum size for DNA synthesis. Thus, in good nutrient conditions, division is triggered when cells reach a critical size, preventing them from outgrowing the capabilities of genomic support, whereas in poor nutrient conditions, DNA synthesis is delayed until cells reach a minimum size, preventing commitment to another cycle without adequate cytoplasmic resources.

Genetic analysis of *wee* mutants, which are deficient in cell-size control over nuclear division, reveals two loci (Thuriaux *et al.*, 1978). Mutants at the *wee* 1 locus are much more common than mutants at the *wee* 2 locus, which suggests that the *wee* 1 gene product restrains nuclear division whereas the *wee* 2 gene product stimulates division. The *wee* 1 gene product could correspond to nuclear sites, but there are difficulties in assigning *wee* 2 to the activator in Sachsenmaier's model for *Physarum* (P. Nurse, private communication).

In exponentially growing wild-type *S. pombe* cells, Fantes (1977) observed no correlation between cell length at one division and cell length at the next (except for very long cells, which show positive correlation). This suggests the operation of some control mechanism which compensates in the next cycle for any deviations in cell size at division. Fantes showed that compensation is accomplished by alteration in cycle time rather than in growth rate. These results are analogous to the response of *Physarum* to UV irradiation (Sudbery and Grant, 1975) and to heat shock (Tyson *et al.*, 1979).

Evidence for size control over an early event in the cell cycle of budding yeast *Saccharomyces cerevisiae* has been reported by Johnston *et al.* (1977) and Hartwell and Unger (1977). They found that if small cells are produced by nutritional deprivation or inhibition of protein synthesis, then the delay until the onset of DNA synthesis, as evidenced by the first appearance of a bud, is inversely proportional to the initial size. The period between first and second bud formation, however, is independent of initial size. Using *cdc* (cell division cycle) mutants and alpha mating factor, they were able to determine that the size-dependent event is early in G_1 (the "start" event). Once initiated, the DNA-division cycle can be completed in the absence of growth, and cells arrest in early G_1 at start.

Sudbery *et al.* (1980) reported the isolation of two size-control mutants in *S. cerevisiae*. During exponential growth, the *whi* 1 mutants have about one-half the dry mass of wild-type cells at the stage of bud initiation. In the stationary phase, both *whi* 1 and *whi* 2 mutants are smaller (in terms of cell volume) than wild-type cells. Furthermore, in the stationary phase, *whi* 2 mutants seem to stop at a different stage in the cell cycle, since their budding index is much higher than that of wild-type cells (24 *versus* 1.6%). Sudbery *et al.* (1980) suggest that *whi* 1 is analogous to Nurse's *wee* 1 gene in *S. pombe:* The *whi* 1^+ gene product could be an inhibitor of bud initiation. The *whi* 2 gene, they conclude, ensures that the cell cycle arrests in G_1 when nutritional conditions no longer support growth.

2. MAMMALIAN CELLS

The evidence for cell-size control of the DNA-division cycle in mammalian cells is less convincing than the work just reviewed on prokaryotes and lower eukaryotes.

Studies of mammalian cell fusion, induced by inactivated Sendai virus (Rao and Johnson, 1970), have confirmed the results obtained with *Physarum* by Rusch *et al.* (1966) and by Guttes and Guttes (1968). In S/G_2 heterokaryons, an S-phase nucleus completes DNA synthesis even if the proportion of G_2 to S nuclei is large, and a G_2-phase nucleus does not reinitiate DNA synthesis even if the proportion of S to G_2 nuclei is large. Mitosis in S/G_2 heterokaryons is synchronous, and the greater the G_2 : S ratio, the earlier the initiation of mitosis in all nuclei of that cell.

Cycloheximide-pulse treatment of cultured animal cells also gives results strikingly similar to the experimental results of Tyson *et al.* (1979) on *Physarum*.

Schneiderman et al. (1971) treated Chinese hamster ovary cells with pulses of cycloheximide (1–8 hours) and observed "excess delay" in the initiation of DNA synthesis. They hypothesized that certain factors necessary for DNA synthesis decay during the inhibitor treatment and have to be resynthesized after the inhibitor is removed. From the dependence of excess delay on the duration of the pulse, they estimated a half-life of about 2 hours for the initiator factors. Rossow et al. (1979) studied the rate of growth of mouse 3T3 cells in the continuous presence of low concentrations of cycloheximide. From the lengthening of G_1 which they observed, using a model consistent with the concept of Schneiderman et al. (1971), they calculated a half-life of 2.2 hours for the hypothetical unstable initiator protein.

More direct evidence for cell-size control of DNA synthesis in mammalian cells comes from a study by Killander and Zetterberg (1965) on mouse fibroblasts. They showed that there is significantly less variation in cell mass than in cell age at the onset of DNA synthesis, which suggests that cell size is the controlling factor for the initiation of DNA synthesis. This conclusion is further substantiated by Yen et al. (1975), who showed that small cells (human lymphoid) have long generation times, principally due to extension of the G_1 period. On the other hand, Fox and Pardee (1970) observed no correlation between cell size at birth and length of the G_1 period. However, as pointed out by Nurse and Thuriaux (1977), their fractionation procedure for obtaining small cells seemed to delay all cells in entering the S phase, which may have masked a negative correlation between cell size at birth and duration of G_1.

These objections to the Fox–Pardee experiments have been circumvented by Gershey et al. (1979), who pulse-labeled a synchronized population of monkey kidney cells in the early S phase, fixed in cold glutaraldehyde, and separated by velocity sedimentation at unit gravity in 2–3% bovine serum albumin. Each fraction of the gradient was labeled to about the same extent (20%), indicating that there is no critical size requirement for entry into the S phase in these cells.

The cells of multicellular organisms are normally under strict growth control, whereas unicellular organisms grow as rapidly as nutritional conditions will allow. Thus we might expect some differences in the coordination of growth and division between protozoans and metazoans. Microorganisms that do not regulate their size at division would be at a selective disadvantage because their progeny would become too large or too small to compete with size-regulated cells. On the other hand, differentiated cells of multicellular organisms, since they divide so infrequently, may have lost this tight control over size at division. Furthermore, since single-celled organisms are subject to a more variable environment than cells in multicellular organisms, we might expect differences in nutritional control over cell division: Commitment to DNA synthesis and division in protozoans must be conditioned on adequate nutritional reserves (as determined, for example, by cell size at the onset of the S phase), but metazoan cells may have lost this checkpoint in the cell cycle (nutritional control being achieved by hormonal and neural mechanisms).

It should be kept in mind, however, that the variability in cell size and generation time is not significantly greater in metazoans than in bacteria and yeast. This observation has been used to argue that all cells are rather feckless about cell-size control and proceed through the cycle in a random fashion (Smith and Martin, 1973; Minor and Smith, 1974; Shilo *et al.*, 1976; Shields, 1978). But it could equally well be argued that mammalian cells, which maintain a constant mean mass and generation time, are likely to be size regulated just like unicellular organisms; however, this size regulation is not as apparent in cultured mammalian cells for technical reasons, especially the unsatisfactory state of mammalian cell genetics in comparison with bacterial and yeast genetics.

III. PERIODIC ENZYME SYNTHESIS

A. INTRODUCTION

Mitchison (1969) has suggested a simple classification scheme for enzyme activity patterns during the cell cycle. He distinguishes four basic patterns: continuous exponential, continuous linear, step, and peak (Fig. 8). If enzyme activity increases exponentially in an exponentially growing population of synchronized cells, then that enzyme activity is being maintained at a constant level per unit mass. Since the activity per unit mass of such an enzyme does not vary with time, it cannot be used as a marker of progress through the cell cycle. The other three patterns, on the other hand, do mark specific events in the cell cycle: the time of doubling in the rate of synthesis of a linear enzyme, the time of doubling in the amount of a step enzyme, and the time of maximum activity of a peak enzyme.

In the late 1960s and early 1970s, periodic enzyme synthesis was thought to be a common feature of the cell cycle. Many linear, step, and peak enzymes had been reported in bacteria, yeast, and mammalian cell cycles (see the reviews in Mitchison, 1971, 1977; Halvorson, 1977). One exception was *P. polycephalum*. Hütterman *et al.* (1970) had measured the activity patterns of seven enzymes involved in different areas of metabolism and found six of them to be continuous exponential enzymes.* Only glutamate dehydrogenase appeared to be a periodic

---

*The six continuous enzymes studied by Hütterman *et al.* (1970) were isocitrate dehydrogenase, glucose-6-phosphate dehydrogenase, acid phosphatase, phosphodiesterase, β-glucosidase, and histidase. Some other continuous enzymes in *Physarum* are thymidine monophosphate kinase, thymidine diphosphate kinase, uridine kinase, deoxyadenosine kinase, lactate dehydrogenase, glyceraldehyde phosphate dehydrogenase (Wolf *et al.*, 1973), thymidine monophosphatase (Hildebrandt and Sauer, 1973), succinate dehydrogenase, fumarase, and malate dehydrogenase (Forde and Sachsenmaier, 1979).

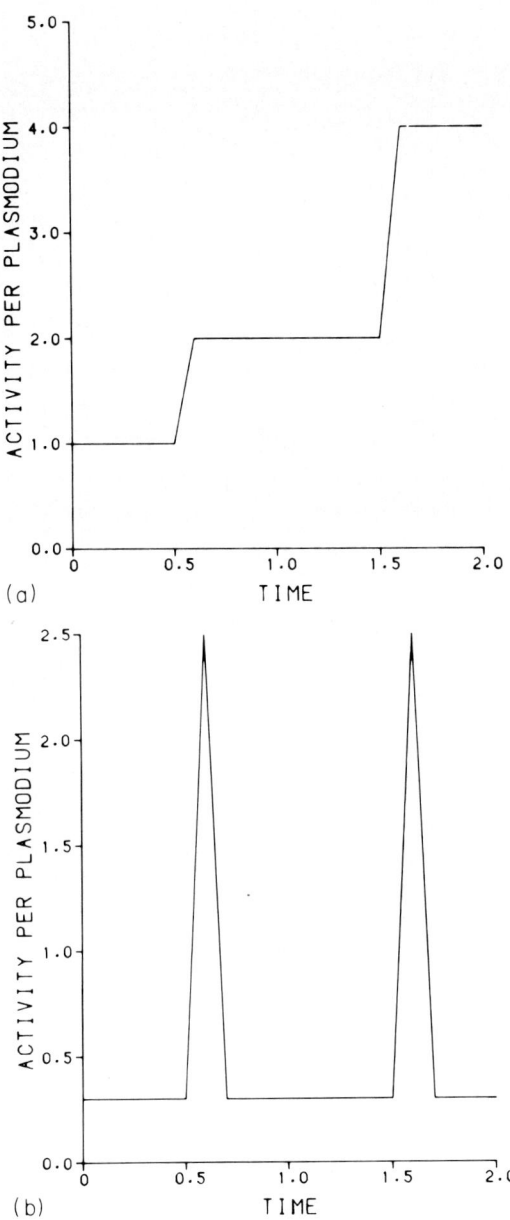

Fig. 8. Patterns of enzyme accumulation, as measured by enzyme activity, during the cell cycle. (a) Step enzyme: the amount of enzyme per plasmodium doubles at a characteristic time in the mitotic cycle. (b) Peak enzyme: high activity of the enzyme is observed only during short intervals in the mitotic cycle. (c) Linear enzyme: the rate of enzyme accumulation doubles at a characteristic time in the mitotic cycle. (d) Exponential enzyme: enzyme activity per plasmodium increases exponentially, so enzyme activity per unit mass is constant during growth and division. From Tyson (1981); used by permission of Pitman Books.

(*continued*)

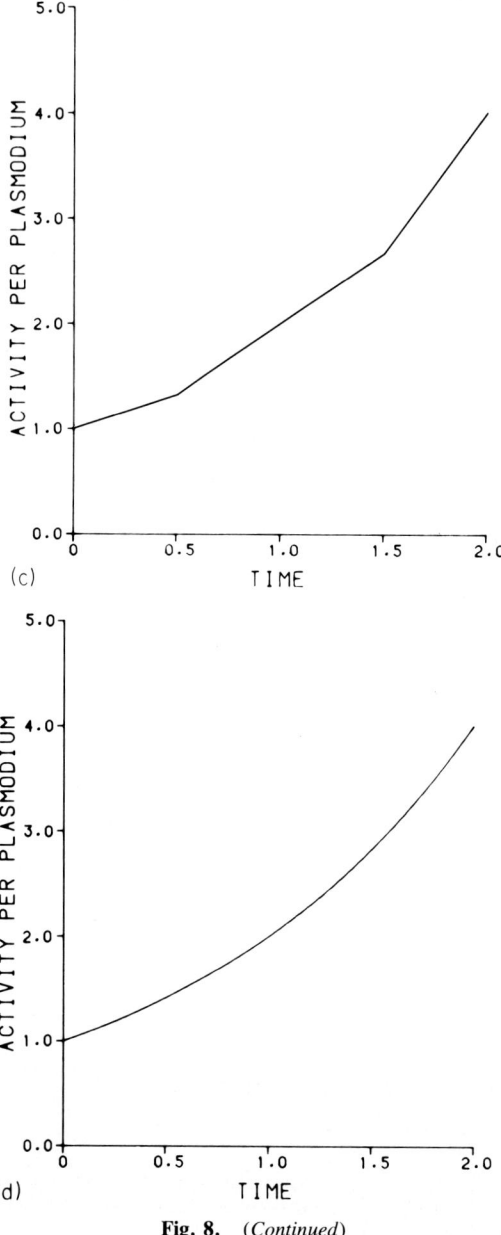

Fig. 8. *(Continued)*

3. Periodic Phenomena in *Physarum*

enzyme, increasing stepwise in late G_2* *Physarum* was a particularly troublesome exception because its cell cycle is naturally synchronous, whereas most other studies of periodic enzymes had been carried out in artificially synchronized cultures. Perhaps the periodic patterns of enzyme synthesis observed so frequently in synchronized cultures are artifacts of the synchronization procedure.

In fission yeast, *S. pombe*, synchronization is achieved by selecting small cells from an asynchronous culture that has been sorted according to size on a sucrose gradient. To control for pertubations induced by the synchronization technique, the gradient can be shaken up before inoculating cells into fresh medium. In these control cultures there is no synchrony of division, and continuous linear enzymes show continuous exponential patterns; yet alcohol dehydrogenase and homoserine dehydrogenase show stepwise increases in activity even in the shaken-up control (Mitchison, 1977). Thus, the step pattern of these two enzymes is not a characteristic of the cell cycle per se.

Further evidence against the notion that periodic enzyme synthesis is a common feature of the cell cycle comes from two studies employing two-dimensional gel electrophoresis to separate hundreds of individual cellular proteins from samples taken at different phases in the cell cycle. Using centrifugal elutriation, Elliot and McLaughlin (1978) separated cells of *S. cerevisiae* in balanced exponential growth according to their position in the cell cycle. Samples were then subjected to two-dimensional gel electrophoresis. Of 550 protein spots which could be located unambiguously on such gels, none were absent from any portion of the cell cycle. Furthermore, 111 of these spots were analyzed for rate of synthesis by pulse labeling (10-minute pulse of [^{35}S]methionine against a uniform ^{3}H background label). Each of the 111 proteins had pulse : long-term-label ratios that did not vary during the cell cycle. Lutkenhaus *et al.* (1979) have published similar results for *E. coli*. Of 750 proteins observed on their gels, none showed evidence of periodic synthesis.

In light of the present uncertainty about the role of periodic enzyme synthesis in the cell cycle, it seems particularly appropriate to review the evidence for periodic enzyme synthesis in naturally synchronous plasmodia of *P. polycephalum*.

B. Periodic Enzymes in Physarum

1. THYMIDINE KINASE

The first report of periodic enzyme synthesis in *Physarum* was by Sachsenmaier and Ives (1965), who observed a peak pattern for thymidine kinase

---

*In my laboratory, glutamate dehydrogenase in strain M_3b appears to be a continuous exponential enzyme (J. J. Tyson, unpublished observations).

(TKase), an enzyme involved in salvaging thymidine for reincorporation into DNA. From a low level in the mid-G_2 phase, TKase activity increases fourfold to a maximum value at telophase (onset of DNA synthesis). Sachsenmaier et al. (1967) investigated the effects of cycloheximide (actidion) and actinomycin C on the increase in TKase activity in late G_2. Addition of cycloheximide immediately interrupts this increase, suggesting that the increase is due to synthesis of a (relatively stable) enzyme protein. Addition of actinomycin also prevents the increase of kinase activity, but this inhibitor must be added about 1 hour earlier than cycloheximide to obtain the same effect, suggesting that the increase is also dependent on synthesis of (relatively unstable) mRNA for TKase.

Gröbner and Sachsenmaier (1976) have found by isoelectric focusing at least three variants of TKase (Fig. 9). Variant A (pI = 7.8) increases stepwise in the early S phase. Variant C (pI = 6.3) increases sharply during mitosis, reaching a maximum at telophase and dropping rapidly in the S phase. Variant B (pI = 6.7)

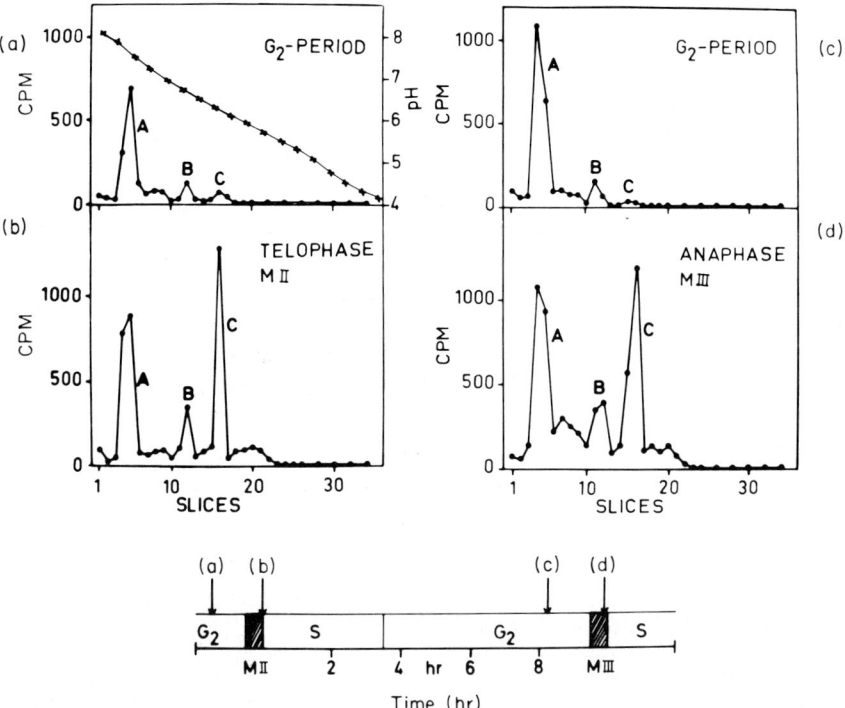

Fig. 9. Thymidine kinase enzyme variants. The three enzyme variants [peak A, pI = 7.8; peak B, pI = 6.7; and peak C, pI = 6.3] were separated by isoelectric focusing in 5% polyacrylamide gels. The times of preparation of the plasmodial extracts are given in the lowest frame. Notice that variant C, which is a major component in telophase, virtually disappears in G_2. Reprinted by permission of the authors and publisher from Gröbner and Sachsenmaier (1976); copyright 1976 by Elsevier North-Holland.

is present in small amounts throughout the mitotic cycle. The temporal relationship between variants A and C suggests that C is produced *de novo* during mitosis and is converted into A during the S phase. Variant B is probably an intermediate form. Gröbner (1979) has presented evidence that variant C is converted *in vitro* into variant A by an endogenous transforming factor which has phosphatase activity.

To determine whether the peak in TKase activity is due to de novo synthesis, Oleinick (1972) labeled newly synthesized proteins with deuterated amino acids for 3 hours, during which time TKase activity increased by a factor of two. When heavy TKase was separated from light TKase on a CsCl gradient, the two peaks were of equal size, indicating that the twofold increase in activity is accompanied by synthesis of new enzyme molecules equal in amount to the enzyme molecules present before the increase.

Wright and Tollon (1979a) have studied the activity patterns of TKase at two extreme physiological temperatures. At 32°C the enzyme activity has a half-life of about 1 hour (determined from loss of activity in the presence of cycloheximide), whereas at 22°C it is considerably more stable (half-life = 2-4 hours). As a result, total TKase activity shows a step pattern at 22°C and a peak pattern at 32°C. In a further study, Wright and Tollon (1979b) observed that on shifts from 22° to 32°C, the peak of TKase activity always occurs concomitantly with the delayed 32°C mitosis (cf. Sachsenmaier *et al.*, 1970). They suggest that the increase in synthesis of TKase and the triggering of mitosis are both controlled by the same (heat-sensitive) regulatory system. Hildebrandt and Sauer (1973), on the other hand, have argued that the TKase peak is not tightly coupled with the period of DNA synthesis.

2. ORNITHINE DECARBOXYLASE

Ornithine decarboxylase (ODCase), the first enzyme in the polyamine biosynthetic pathway, has been studied in detail by Mitchell and co-workers. During the mitotic cycle ODCase (as well as S-adenosyl-L-methionine decarboxylase) activity increases stepwise in the S phase (Mitchell and Rusch, 1973). If protein synthesis is inhibited with cycloheximide, ODCase activity decays with a half-life of 14 minutes. Theoretically, a step pattern is expected for an unstable enzyme if its rate of synthesis doubles at a certain phase in the cycle, e.g., after replication of its structural gene, and Mitchell and Rusch suggested that this might be the explanation of their observations. However, Mitchell and Sedory (1974) showed that the rapid loss in activity during exposure to cycloheximide is due to a change in the ability of this enzyme to be activated by its co-enzyme, pyridoxal 5'-phosphate (PLP). When assayed at high levels of PLP (200 μM), no loss in ODCase activity is observed after treatment with cycloheximide. Mitchell *et al.* (1976) showed that this change in activity corresponds to posttranslational modification of the enzyme protein from an active (low K_m for PLP) form to an inactive (high K_m for PLP) form. Total ODCase activity, assayed at 200 μM PLP, still shows a stepwise increase during the S phase (Sedory and Mitchell, 1977; Fig. 10), but this cannot be explained in terms of doubling the rate of synthesis of an unstable enzyme. It appears that *de novo* synthesis of ODCase is limited to a short period in the early to mid-S phase. The activity of the enzyme under physiological conditions can exhibit more complex patterns, since it is regulated by posttranslational modification. This allows extremely rapid changes in enzyme activity without the wastefulness of rapid protein turnover.

Further information on ODCase variants can be found in Mitchell and Carter (1977), Mitchell *et al.* (1978), and Mitchell and Kottas (1979).

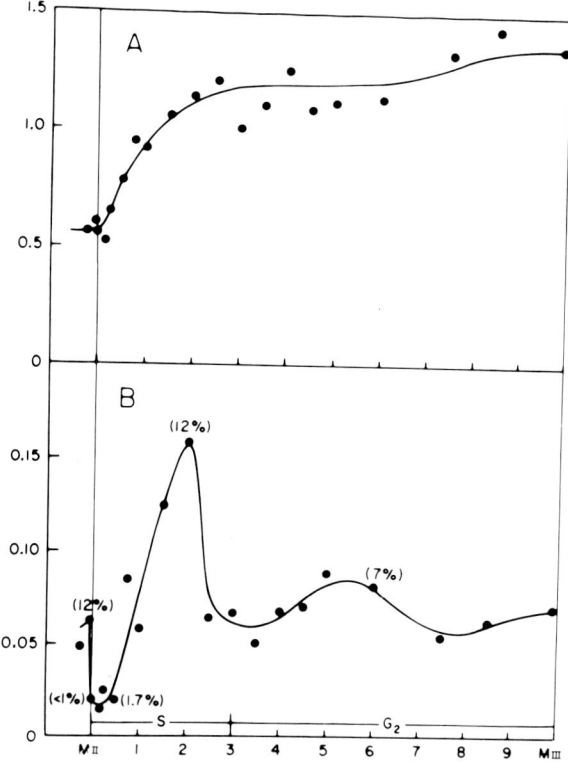

Fig. 10. Variations in ornithine decarboxylase activity during the mitotic cycle. Abscissa: time after mitosis (hours); *ordinate;* enzyme activity (μM CO_2 per hour per plasmodium). Samples were assayed at either (A) 200 μM or (B) 1 μM pyridoxal phosphate. The wild variations in activity observed at 1 μM PLP are due to the extremely rapid modification of the active, low K_m form of the enzyme to an inactive, high K_m form. From Sedory and Mitchell (1977); used by permission of the author and Academic Press.

3. Periodic Phenomena in *Physarum*

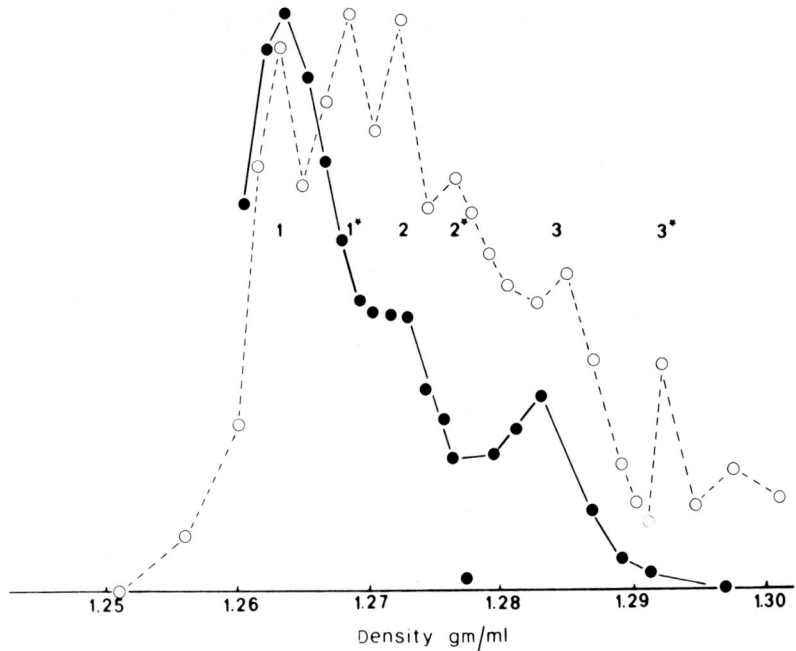

Fig. 11. Density gradient centrifugation of histone kinase. Histone kinase activity (●--●) was determined in nuclei isolated from plasmodia 6.5 hours after MII; (●) control plasmodium, transferred from normal medium to normal medium at MII; (O---O) heavy labeled plasmodium, transferred from normal medium to deuterated medium at MII. In the control plasmodium, there are 3 peaks of histone kinase activity, and in the heavy-labeled plasmodium, there are six peaks: three of normal density (unstarred) and three of higher density (starred). *De novo* synthesis of histone kinase in the 6.5 hours since MII (the deuterated molecules) cannot account for the 15-fold increase in enzyme activity observed during this interval. Reprinted by permission of the authors and publisher from Mitchelson *et al.* (1978); copyright 1978 by Elsevier North-Holland.

3. HISTONE KINASE

As mentioned in Section II,B,6, there is a nuclear histone Hl phosphorylating activity which shows a peak pattern during the *Physarum* cell cycle, but there is some disagreement about the timing of the peak with respect to metaphase. Bradbury *et al.* (1974a) reported a 15-fold exponential increase in HKase activity from early S phase to mid-G_2 (about 2 hours before metaphase) and then a rapid decrease in HKase activity in late G_2. Their peak in phosphorylating activity precedes by about 2 hours the peak in histone Hl phosphate content. Oleinick (1980b), on the other hand, found only very low levels of HKase activity until about 30 minutes before metaphase, with a very sharp peak in activity coincident with metaphase.

Is the 15-fold increase in HKase activity observed by Bradbury et al. (1974a) between telophase and mid-G_2 due to synthesis of new enzyme or activation of preexisting enzyme? Mitchelson et al. (1978) labeled *Physarum* proteins with deuterated amino acids for 6.5 hours, from mitosis (MII) to mid-G_2. A preparation of nuclear proteins was then separated by isopycnic centrifugation on metrazimide gradients, and fractions of increasing density were analyzed for HKase activity. The newly synthesized, heavy-labeled HKase appeared at a density approximately 4.5 mg/ml greater than the old enzyme (Fig. 11). Since the new and old bands were approximately equal in size, it appears that the total amount of enzyme increases by a factor of only two during the 6.5-hour labeling period. Thus the 15-fold increase in phosphorylating activity observed during this period must be due to activation of preexisting, inactive enzyme molecules.

This "activation" may only reflect changes in the extractability of HKase from nuclei. If HKase were tightly bound to chromatin, for example, then temporal changes in extractability could explain the peak of enzyme activity in nuclear lysates, and differences in extraction procedures could account for the different patterns observed in different laboratories.

Hardie et al. (1976) have separated the HKase activity of isolated *Physarum* nuclei (by ion-exchange chromatography) into two peaks. The two forms of HKase have different substrate preferences, and they peak at slightly different phases in the mitotic cycle (about 2 hours and 1 hour before mitosis).

Matthews (1978) has reported that histone phosphatase activity is low in the S phase, shows a peak in mid G_2 similar to the HKase pattern, and peaks sharply a second time at mitosis. Apparently the first peak of phosphatase activity serves to modulate the phase of histone H1 phosphorylation, and the second peak during mitosis serves to dephosphorylate H1 histone in preparation for the next mitotic cycle. Oleinick et al. (1981b), however, do not see any changes in sensitivity to the histone phosphatase inhibitor $NaHSO_3$ within 22 minutes of metaphase.

4. OTHER PERIODIC ENZYMES

There are a number of other periodic enzymes in *Physarum*, which have been studied in less detail than the three just reviewed. Two enzymes that show step patterns are glutamate dehydrogenase (Hüttermann et al. 1970) and ribonuclease (Braun and Behrens, 1969).

There are several other peak enzymes reported in *Physarum*.

1. DNA polymerase—assayed in isolated nuclei supplied with exogenous DNA and spermine, shows a peak of activity in the S phase (Brewer and Rusch, 1966). DNA polymerase has been purified to homogeneity (Choudhry and Cox, 1979; Schiebel and Baer, 1979) and ^{125}I-labeled antibodies have been prepared (Choudhry and Cox, 1979). R. A. Cox (private communication) has used a radioimmunoassay method to demonstrate that DNA polymerase is synthesized not in the S phase but in early G_2 phase.

3. Periodic Phenomena in *Physarum*

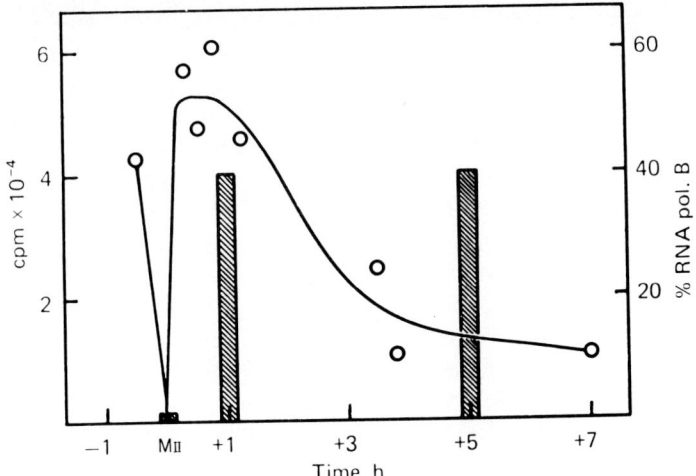

Fig. 12. RNA polymerase B activity of isolated nuclei during the mitotic cycle. The peak pattern, when assayed in 0.1 *M* KCl (○), disappears when assayed in 0.4 *M* KCl (bars). From Pierron and Sauer (1980a); used by permission of the authors and the Company of Biologists, Ltd.

2. RNA polymerase B (α-amanitin-sensitive UMP incorporation)—assayed *in vitro* from isolated nuclei at low ionic strength (0.1 *M* KCl), also shows an activity peak in the S phase (Grant, 1972; Davies and Walker, 1978; see Fig. 12). This activity peak appears to depend on activation or inhibition of the enzyme since a constant activity level is observed throughout the mitotic cycle if nuclei are assayed at high ionic strength (0.4 *M* KCl) (Pierron and Sauer, 1980a; Fig. 12). Furthermore, by titrating RNA polymerase B with tritiated α-amanitin, Pierron and Sauer (1980b) have shown that the amount of RNA polymerase B per nucleus remains constant throughout the cell cycle. Nonetheless, during mitosis (i.e., prophase → telophase) very little RNA polymerase B can be detected by activity measurements (Fig. 12) or by labeled amanitin titration. This seems to reflect release of the enzyme from condensed chromosomes rather than masking of the enzyme within condensed chromosomes (Pierron and Sauer, 1980b). RNA polymerase B has been purified to homogeneity (Smith and Braun, 1978).

3. ADP-ribosyl transferase—shows a "negative peak" in the S phase (Brightwell *et al.*, 1975).

4. NAD pyrophosphorylase—shows four times as much activity in the S phase as in G_2 (Solao and Shall, 1971).

5. Adenylate cyclase (ACase)—activity is found mainly in the low-speed pellet (10,000 *g*) of fractionated plasmodia, and the activity in this fraction peaks in late G_2 (Lovely and Threlfall, 1978). Smith and Mansour (1978) have reported activation of adenylate cyclase by its product, cyclic AMP (cAMP). However, since enzyme activation by cAMP was measured by increased cAMP production,

it could be that the added cyclic nucleotide merely provided a pool for cAMP phosphodiesterase. Smith and Mansour did not entirely rule out this possibility.

Product activation can destabilize a steady state of cAMP synthesis and degradation (Tyson, 1975) and can lead to sustained oscillation in cAMP level and ACase activity (Goldbeter and Segel, 1977); however, I would expect such an oscillation to have a period of several minutes rather than several hours.

6. Guanylate cyclase—activity is found mainly in the particulate fraction (high-speed pellet) of homogenized plasmodia and shows two peaks of activity, in mid S and late G_2 (Lovely and Threlfall, 1979). The peaks in ACase and guanylate cyclase activities reported by Lovely and Threlfall correspond to peaks in intracellular cAMP and cyclic GMP (cGMP) levels reported by the same authors; however, these reports have since been called into question by several groups of investigators (see Section IV,B).

7. cAMP and cGMP phosphodiesterase—activities, measured in isolated nuclei, are found to fluctuate during the mitotic cycle (Jeter *et al.*, 1978). Kupetz (thesis, in preparation) has also observed statistically significant fluctuations in both PDE activities in two cytoplasmic fractions (20,000 g pellet and supernatant). Lovely and Threlfall (1978, 1979), however, reported no significant fluctuations in these two PDEs in any of their cell fractions. Kinetic properties of cAMP-PDE have been investigated by Kincaid and Mansour (1979).

8. Deoxyadenosine kinase exists in two forms (Woertz and Sachsenmaier, 1979). Form I has a molecular weight of 95,000, is non-Michaelian, and has a low K_m. Form II has a molecular weight of 40,000, is Michaelian, and has a high K_m. Form I shows a peak of activity in late S and a trough in mid G_2, whereas form II shows a trough in S and a peak in G_2. In an analogy with the situation for TKase and ODCase, we might expect that form I is modified posttranslationally to the less active form II in the G_2 phase.

9. Thymidylate synthetase activity is 50% higher in the S phase than in the G_2 phase in strain M_3b and in strain TU63, which lacks thymidine kinase (Gröbner and Loidl, 1981).

For further information on periodic enzymes associated with the nucleus, see Matthews and Bradbury (this volume, Chapter 10).

C. Origin of Enzyme Periodicities

From 1965 to 1975 many examples of periodic enzymes were uncovered, but few in-depth studies were undertaken. In fact, the simplest controls were often neglected. Unsubstantiated assumptions were commonly made, and theories were elaborated thereon. As a result, the field is in some disarray at present, and it seems worthwhile to outline some of the questions that should be addressed if we are to understand the nature and significance of fluctuations in a certain enzyme activity.

1. First, is it a *bona fide* cell cycle event? Enzyme synthesis, degradation, and modification can be sensitive to nutritional changes, so that perturbations induced by selection synchrony can cause transient fluctuations in enzyme activity irrespective of the cell cycle (Mitchison, 1977). Presumably, we do not have to worry about this with *Physarum*, but it is worthwhile to investigate the pattern of enzyme activity at several temperatures (Wright and Tollon, 1979a) and under several assay conditions (Mitchell and Sedory, 1974; Pierron and Sauer, 1980a) to determine the range of variation in activity patterns for the enzyme under study.

2. Is an increase in activity due to *de novo* synthesis or to enzyme activation? This fundamental question is too rarely answered, especially in light of the fact that a number of cell cycle variations in enzyme activity are known to involve periodic activation and inhibition rather than periodic synthesis and degradation (e.g., Kuehn, 1972; Mitchell and Sedory, 1974; Mitchelson et al., 1978; Pierron and Sauer, 1980b). Some information can be gained quite easily with *Physarum* by fusing a plasmodium at a phase in the cycle of high activity with a plasmodium at a phase of low activity. If activators or inhibitors are responsible for changes in activity, then we can expect that activity in the fused plasmodium will not be an additive function of the two controls, whereas if *de novo* synthesis is responsible, activity will be additive. More direct evidence comes from density labeling, as done by Oleinick (1972) and Mitchelson et al. (1978). If the enzyme under study can be purified sufficiently, then specific antibodies can be used to monitor the amount of enzyme protein throughout the cycle (e.g., Klevecz, 1969).

3. If activity changes are due to synthesis and degradation rather than to enzyme modification, then we must ask whether enzyme synthesis is continuous throughout the cycle or restricted to certain portions of the cycle. Information from density labeling or specific-antibody precipitation can give us hints about enzyme protein synthesis, though we must remember that the total amount of enzyme present will be determined by degradation as well as by synthesis. If specific antibodies are available, then specific mRNA can be isolated from polysomes making the enzyme. Complementary DNA can be synthesized from this mRNA and used as a probe for specific mRNA during the cycle. This new technology will allow us to answer many questions about the control of enzyme synthesis.

Too often, it has been assumed that step and peak enzymes are synthesized only during some small fraction of the cycle. It was to explain this presumed restriction in enzyme synthesis that both the linear reading model and the oscillatory repression model of periodic enzyme synthesis were developed (Mitchison, 1971). However, both of these models have serious problems (Mitchison, 1971; Schmidt, 1974; Halvorson, 1977; Tyson, 1979). There is no reason to expect that one of these models must be correct, since patterns of enzyme accumulation depend as much on the degradation of enzyme protein as on its synthesis (Table III). This brings us to the final question.

4. Is the enzyme stable or unstable, or does the stability of the enzyme vary

TABLE III

Dependence of Enzyme Activity Patterns on Enzyme Synthesis and Degradation [a]

	Stability of enzyme		
Synthesis of Enzyme	1. Stable	2. Unstable	3. Unstable only during part of the cell cycle
a. Continuous, mRNA excess	Exponential	Exponential	Peak
b. Continuous, mRNA limiting	Linear	Step	Peak
c. Synthesis only during part of the cell cycle	Step	Peak	Combination

[a] The total amount of enzyme in a cell culture as a function of time, $E(t)$, depends on the rate of enzyme synthesis, $S(t)$, and the rate of enzyme degradation, $D(t)$, through the differential equation $dE/dt = S(t) - D(t)$. To derive the results of this table, I have assumed that cells are growing exponentially with doubling time T, that is, $V(t) = V(0) \exp(0.693\ t/T)$, and I have taken the following forms for synthesis and degradation:

Row (a) $S(t) = kV(t)$
Row (b) $S(t) = k$, for $0 < t < g$
  $= 2k$, for $g < t < T$

that is, the synthetic rate doubles at the time of gene replication, $t = g$.

Row (c) $S(t) = k$ for $0 < t < g$
  $= 0$ for $g < t < T$

that is, enzyme is synthesized only during the first part of the cycle.

Col (1) $D(t) = 0$
Col (2) $D(t) = qE(t)$
Col (3) $D(t) = 0$ for $0 < t < h$
  $= qE(t)$ for $h < t < T$

that is, enzyme is degraded only during last part of cycle. These functions are inserted into the differential equation $dE/dt = S(t) - D(t)$, and solutions are sought subject to the constraint that $E(T) = 2E(0)$, that is, balanced growth. Assuming for mathematical simplicity that $qT \gg 1$, that is, that enzyme degradation is rapid, we find:

(a1) $E(t) = (kT/0.693)\ V(t)$, that is, exponential increase
(a2) $E(t) = (k/q)\ V(t)$, that is, exponential increase
(a3) $E(t) = (kT/0.693)\ \{V(t) - V(0)\}$ for $0 < t < h$
  $\cong 0$ for $h < t < T$

that is, slow exponential increase during the first part of the cycle followed by an abrupt drop at $t = h$.

(b1) $E(t) = k(2T - g) + kt$ for $0 < t < g$
  $= 2k(T - g) + 2kt$ for $g < t < T$

3. Periodic Phenomena in *Physarum*

during the cycle? In the past, people relied on cycloheximide treatment to determine whether an enzyme is stable or unstable. Of course, this tells us only whether enzyme activity under certain assay conditions is stable. As Mitchell's experience with ODCase indicates, a rapid loss in activity during cycloheximide treatment may be due to rapid enzyme modification, which is reversible, rather than to irreversible enzyme degradation. Furthermore, cycloheximide is reported to have specific effects on the activity and molecular weight of glutamate dehydrogenase from *Physarum* (Wendelberger-Schieweg *et al.*, 1980), which gives us more reason for caution in interpreting inhibitor studies of enzyme "stability."

The progress which has been made in unraveling the details of TKase and ODCase synthesis in *Physarum* is encouraging, and it is time some other periodic enzymes in *Physarum* were given similar close scrutiny.

IV. OTHER PERIODIC EVENTS OF THE MITOTIC CYCLE

A. Macromolecular Synthesis

RNA synthesis during the mitotic cycle has been intensively studied (see the review by Braun *et al.*, 1977; Braun and Seebeck, Chapter 12, this volume). The rate of RNA synthesis, as measured by incorporation of [^{3}H]uridine during short pulses, appears to be low not only during mitosis but also in the early G_2 phase (Mittermayer *et al.*, 1964, 1966b; Grant, 1972). Subsequent work has revealed that whereas rRNA is synthesized continuously throughout the cycle (Hall and Turnock, 1976), mRNA synthesis occurs primarily during the S phase (Fouquet and Braun, 1974; Fouquet and Sauer, 1975; Wick, 1977; Pierron and Sauer,

---

that is, linear increase with rate doubling at $t = g$.

(b2) $E(t) = k/q$ for $0 < t < g$
           $= 2k/q$ for $g < t < T$ (step up at $t = g$)

(b3) $E(t) = kt$ for $0 < t < g$
           $= 2kt - kg$ for $g < t < h$
           $\cong 0$ for $h < t < T$

that is, linear increase during first part of cycle with an abrupt drop at $t = h$.

(c1) $E(t) = kg + kt$ for $0 < t < g$
           $= 2\,kg$ for $g < t < T$

that is, step up during the interval $0 < t < g$.

(c2) $E(t) \cong k/q$ for $0 < t < g$
           $\cong 0$ for $g < t < T$

that is, activity peak during phase of synthesis.

(c3) $E(t)$ will be a complicated function depending on the relative timing of the synthetic and degradative intervals.

1980a). No significant changes in tRNA levels are observed during the mitotic cycle of *Physarum* (Melera and Rusch, 1973; Fink and Turnock, 1977).

The rate of protein synthesis, measured by pulse labeling with [^{3}H]lysine, also shows a biphasic pattern with minima at mitosis and early G_2 (Mittermayer *et al.*, 1966a). However, the accumulation of protein, measured by an indirect isotope-dilution method, does not show a plateau at mid-cycle (Birch and Turnock, 1977).

Kuroiwa *et al.* (1978) have shown that most mitochondria divide during the nuclear S phase and that most mitochondrial DNA synthesis occurs during the nuclear G_2 phase (cf. Braun and Evans, 1969). Mitochondria have a gap (mG_1) of about 3 hours between division and mitochondrial DNA synthesis. All mitochondria divide once per cycle; there is no subpopulation of cycling mitochondria with a division time shorter than the intermitotic period.

B. Metabolic Pools

Lovely and Threlfall (1976) have reported large fluctuations in cAMP and cGMP levels, as determined by a binding-protein assay. Kupetz (thesis, in preparation) observed small (twofold) variations in cyclic nucleotide levels. However, radioimmunoassays in two other laboratories have not uncovered any significant changes in cyclic nucleotide levels (cAMP: Garrison and Barnes, 1980; cAMP and cGMP: Oleinick *et al.*, 1981a).

Threlfall and Thomas (1979) have measured chromatographically the free amino acid pools in *Physarum*. These pools are all relatively constant except for proline (20–25% of the total amino acid pool), which fluctuates by about 30%, with a maximum in early G_2 and a minimum at mitosis.

Bersier and Braun (1974) and Fink (1975) have found fluctuations in the pools of deoxyribonucleoside triphosphates; as might be expected, the pools peak at the onset of the S phase. The ribonucleotide triphosphate pools, on the other hand, do not vary during the mitotic cycle (Sachsenmaier *et al.*, 1969; Bersier and Braun, 1974).

Jeter *et al.* (1981) have measured the concentrations of Na, Mg, P, Cl, S, K, and Ca using electron-probe X-ray microanalysis of quick-frozen samples. There are significant fluctuations in all elements except Cl during the mitotic cycle. In most cases, the concentration is elevated during mitosis and reduced during the S phase. Some elements show another concentration peak in G_2. The changes in elemental concentration are not due to fluctuations in plasmodial water content, which remains constant throughout the cycle. Mg, P, and S show higher concentrations in the nucleus (chromatin) than in the cytoplasm at metaphase; otherwise, there is no evidence of unequal partitioning of these elements between nucleus and cytoplasm.

Gerson and Burton (1977) have measured the intracellular pH of cycling plasmodia. They found one alkaline shift per cycle, associated with nuclear division and the onset of DNA synthesis.

C. Miscellany

Forde and Sachsenmaier (1979) followed the rate of oxygen uptake during the mitotic cycle of *Physarum*. The rate increases in two steps, with plateaus in early G_2 and late G_2 (Fig. 13). Small fluctuations (about 15%) in cytochrome oxidase activity paralleled those in respiration rate, but the activities of other mitochondrial enzymes (succinate dehydrogenase, fumarase, and malate dehydrogenase) remained relatively constant.

Magun (1979) found a general decrease of about 30% in the phosphate content of cytoplasmic DNA-binding phosphoproteins during the G_2 phase. Several of these phosphoproteins were attenuated or disappeared during G_2.

Using a nitrocellulose membrane-filter binding assay for DNA–protein complexes, Wille (1977) has uncovered temporally specific interactions of cytosol proteins with replicating *Physarum* DNA. DNA–protein binding at about twice the nonspecific level (defined as the level of binding of G_2 cytosol proteins to either early- or late-S DNA) was found for the following pairs: early-S DNA + early-S protein, late-S DNA + late-S protein, and early-S DNA + prophase protein. Only the nonspecific level of binding was found for early-S DNA +

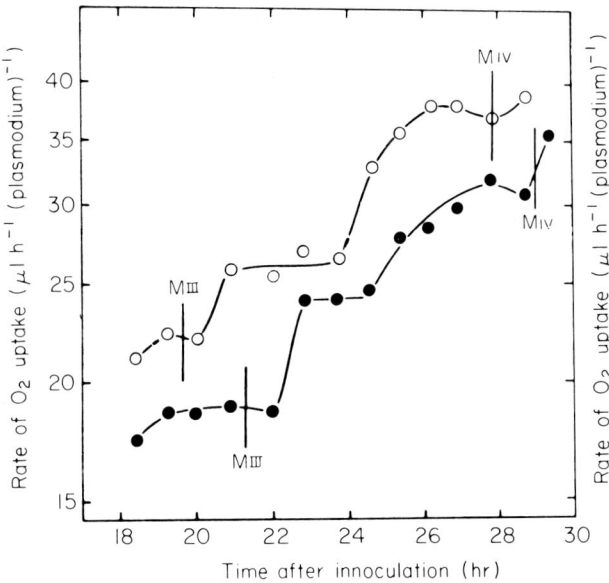

Fig. 13. Oxygen uptake during the mitotic circle. Oxygen uptake was measured manometrically at intervals on a plasmodium growing inside a Warburg flask. The two curves refer to two separate experiments. Timing of the third and fourth mitoses after preparation of the macroplasmodium was determined using sister macroplasmodia growing in an incubator at the same temperature (26°C). From Forde and Sachsenmaier (1979); used by permission of the authors and Cambridge University Press.

late-S protein, late-S DNA + early-S protein, and late-S DNA + prophase protein.

Schel *et al.* (1978) determined the nuclear pore frequency from electron micrographs of platinum-carbon replicas of isolated nuclei. They found that the number of pores per nucleus doubled between MII and MIII, with most of the increase coming in the first half of the cycle. Aldrich and Pendland (1981), counting the pores on freeze-fractured nuclei, found most of the increase in late G_2.

V. SHUTTLE STREAMING

A. Introduction

Plasmodia of the acellular slime molds are crisscrossed by a network of protoplasmic veins. In motile plasmodia, isolated veins consist of an ectoplasmic tube surrounding an endoplasmic core of transported protoplasm (Fig. 14). The ectoplasmic tube consists of deep, highly ramified invaginations of the plasmalemma (Rhea, 1966; Daniel and Järlfors, 1972) associated with actomyosin fibrils (Wohlfarth-Bottermann, 1962) running circularly around the endoplasmic core and longitudinally down the ectoplasmic tube (Wohlfarth-Bottermann, 1974). Endoplasm flows through the veins first in one direction, then in the reverse

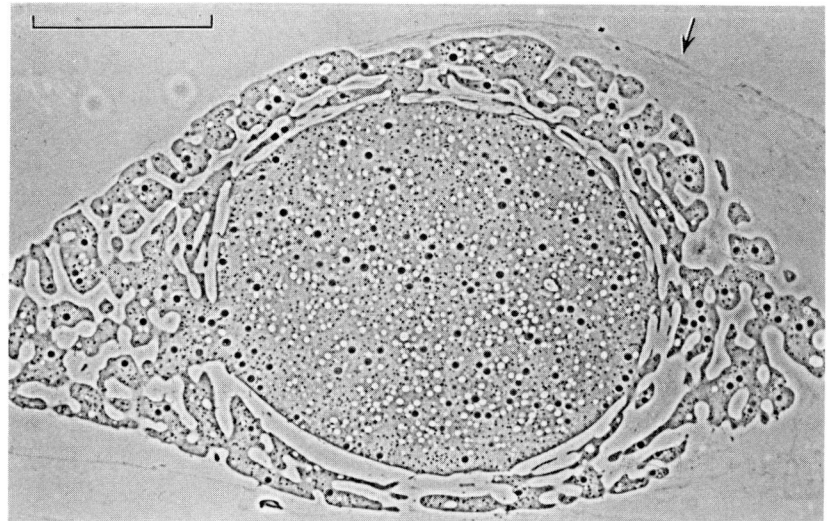

Fig. 14. Cross section through a protoplasmic strand, illustrating the endoplasmic channel and the ectoplasmic region fissured by numerous plasmalemma invaginations. Fixation: 2% OsO_4, 1% $K_2Cr_2O_7$. Phase contrast, ×315. Arrow indicates slime layer. From Wohlfarth-Bottermann (1974); used by permission of the author and the Company of Biologists, Ltd.

direction. These periodic reversals occur about once per minute, and the endoplasm flows at a velocity of about 1 mm/sec (Kamiya, 1959). This shuttle-streaming rhythm circulates oxygen, nutrients, macromolecules, and organelles throughout the plasmodium, and it is also involved in the locomotion and chemotaxis of starving plasmodia.

The study of protoplasmic streaming in *Physarum* has a long and distinguished history (Komnick et al., 1973). In the last 10-15 years, several technological achievements have led to advances in our understanding of the streaming rhythm: the measurement of longitudinal and radial contraction forces under isometric and isotonic conditions (Kamiya, 1970; Wohlfarth-Bottermann, 1975; Hülsmann and Wohfarth-Bottermann, 1978); macrocinematography of migrating plasmodia (Grebecki and Cieslawska, 1978); the replacement of endosplasm with artificial media (Ueda and Götz von Olenhusen, 1978); the production of contractile plasmodial drops without plasmalemma invaginations (Hatano, 1970; Götz von Olenhusen et al., 1979); and the measurement of oscillations in membrane potential (Rhea, 1966; Meyer and Stockem, 1979). Advances in our understanding of shuttle streaming have been reviewed by Wohlfarth-Botterman (1979), by Kessler (this volume, Chapter 5) and, in less detail, by Berridge and Rapp (1979). So here, I will only summarize briefly the conclusions that they have drawn about the nature and location of the oscillatory control system underlying the shuttle-streaming rhythm.

B. Biochemical Nature of the Contraction Oscillator

According to Wohlfarth-Bottermann (1979), there are currently four reasonable hypotheses concerning the identity of the oscillator controlling rhythmic contractions:* (a) the force-generating system (the ectoplasmic actomyosin fibrils); (b) the energy-supply system (glycolysis and oxidative phosphorylation); (c) cytoplasmic Ca^{2+} storage and release; and (d) ion transport across the plasmalemma. What can be said about the likelihood of any one, or some combination of these hypotheses being correct?

If the actomyosin fibers are intimately involved in the generation of oscillation, for example, by delayed feedback between fiber tension and contraction, then stretching veins should shift the phase and perhaps the frequency of the rhythm (Tyson et al., 1976). However, Yoshimoto and Kamiya (1978) and Nagai et al. (1978) observed no change in frequency or phase after a rapid stepwise increase in tension (accomplished by stretching the vein 10-20% of its original length). On the other hand, Krüger and Wohlfarth-Bottermann (1978) and Achenbach and Wohlfarth-Bottermann (1980) found that a strong (50%), transient stretch induces a phase shift and a small decrease in the contraction frequency (if the stretched vein remains attached to an unstretched vein, then the

---

*Each of these possibilities is known to play a role in other biological rhythms (Berridge and Rapp, 1979).

two oscillators come into synchrony). Obviously, the discrepancy between these two reports will have to be resolved before we can draw any conclusions about the role of fiber tension in the oscillator mechanism.

If fluctuations in energy supply set the pace for contractions, then the rhythm should be sensitive to perturbations of plasmodial energy-generating pathways. Again, there is conflicting evidence on this subject. In an early study by Allen and Price (1950), oxygen uptake could be reduced by up to 75% by cyanide, iodoacetate, and low O_2 tension without interfering with protoplasmic flow. Under anaerobic conditions, they observed protoplasmic flow to continue for several hours before stopping. Loewy (1950) reported that active flow could continue indefinitely under anaerobic conditions if 5% CO_2 were added to the atmosphere. Daniel (1970) reported that streaming ceases reversibly under anaerobic conditions (a high-purity nitrogen environment). Sachsenmaier *et al.* (1973) reported that shuttle streaming continues under anaerobic conditions, though the frequency is reduced by about 50%, whereas the plasmodial ATP content drops by about 40%. Wohlfarth-Bottermann (1979) found only minor changes in force output under anaerobic conditions and under inhibition of glycolysis by iodoacetate. (These discrepancies may be due to difficulties in obtaining complete anaerobiosis.)

Energy metabolism will also be perturbed by nutritional changes. Durham and Ridgway (1976) observed an increase in frequency of shuttle streaming in the presence of chemoattractants (see Fig. 16), but Ueda *et al.* (1976) reported that chemoattractants cause a decrease in amplitude but no change in frequency of the shuttle-streaming rhythm.

A possible role for calcium ions in the oscillator is quite likely, but convincing evidence is still lacking. Plasmodial actomyosin bears many structural and biochemical similarities to muscle actomyosin (Berridge and Rapp, 1979), which suggests that internal Ca^{2+} concentrations may be important in the contraction–relaxation cycle. Using ammonium oxalate precipitation, Braatz (1975) found a significant shift in the location of Ca^{2+} ions between cytoplasmic and vacuolar compartments during the contraction–relaxation cycle. However, this fluctuation may not be physiologically significant, since oxalate precipitates calcium only at high calcium concentrations. Ridgway and Durham (1976) have used the calcium-specific photoprotein aequorin to reveal changes in internal Ca^{2+} concentrations during shuttle streaming. However, the oscillations in light output that they observed might not reflect changes in internal calcium concentrations but only rhythmic changes in plasmodial thickness known to occur during shuttle streaming (Sachsenmaier *et al.*, 1973; Wohlfarth-Bottermann, 1979).

Using caffeine-derived microplasmodia (CDM), Matthews (1977) has obtained the best evidence so far that shuttle streaming is controlled by uptake and release of Ca^{2+} from an intracellular storage system. When CDM are treated with caffeine (in a Ca^{2+}-free medium containing 10 mM EGTA), Matthews observed compaction and streaming in the cytoplasm. Repeated caffeine treatments gave

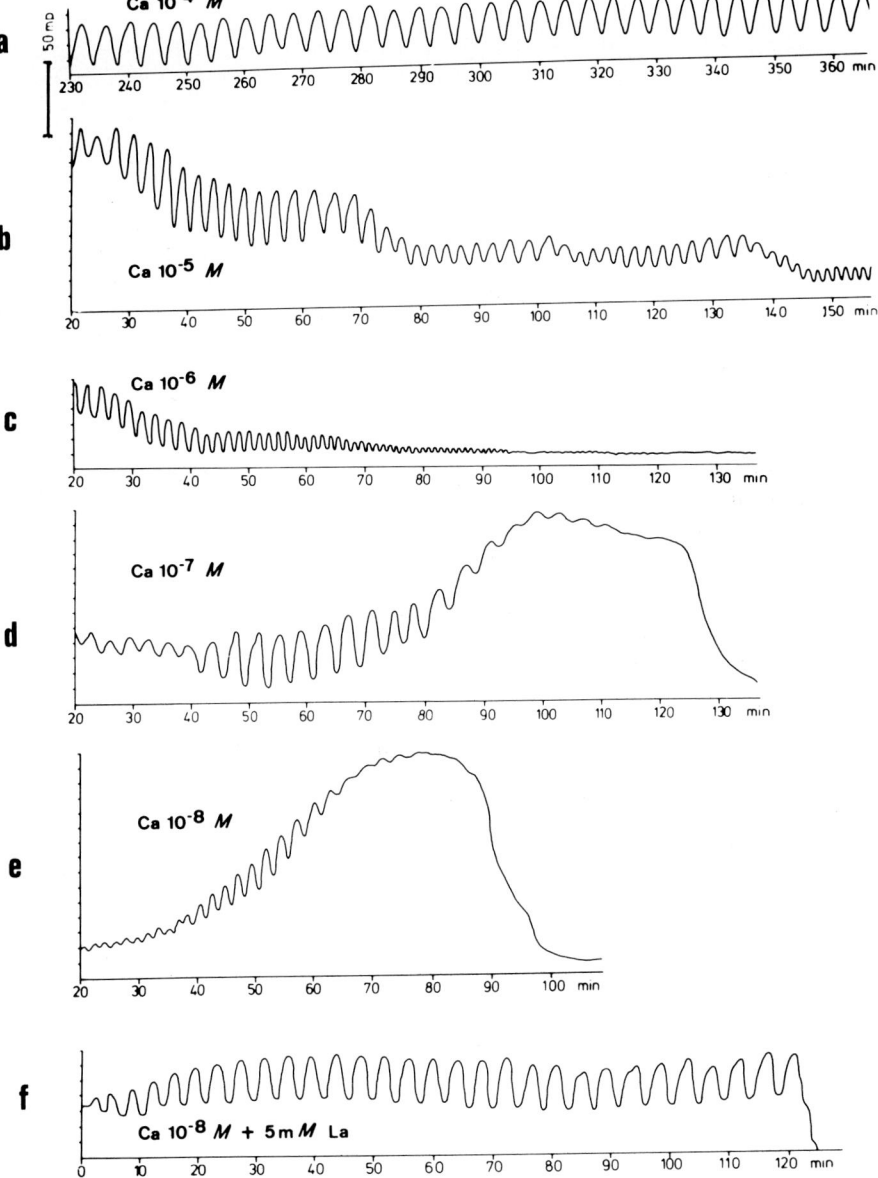

Fig. 15. Effect of external Ca^{2+} concentration on the longitudinal contraction rhythm, monitored tensiometrically. In addition to Ca–EGTA buffers, the bathing medium contained 6 mM NaCl, 3 mM KCl, and a Tris–maleate buffer (pH 7.0–7.2). Longitudinal contractions persist for over an hour even in 10^{-8} M Ca^{2+}. The eventual loss of contraction force at low $[Ca^{2+}]$ can be prevented by the addition of 5 mM La^{3+} to the external solution. External lanthanum may block the loss of internal calcium by binding to Ca-transport sites in the plasmalemma, or it may displace Ca^{2+} from the external Ca–EGTA buffer. From Wohlfarth-Bottermann and Götz von Olenhusen (1977); used by permission of the authors and Academic Press.

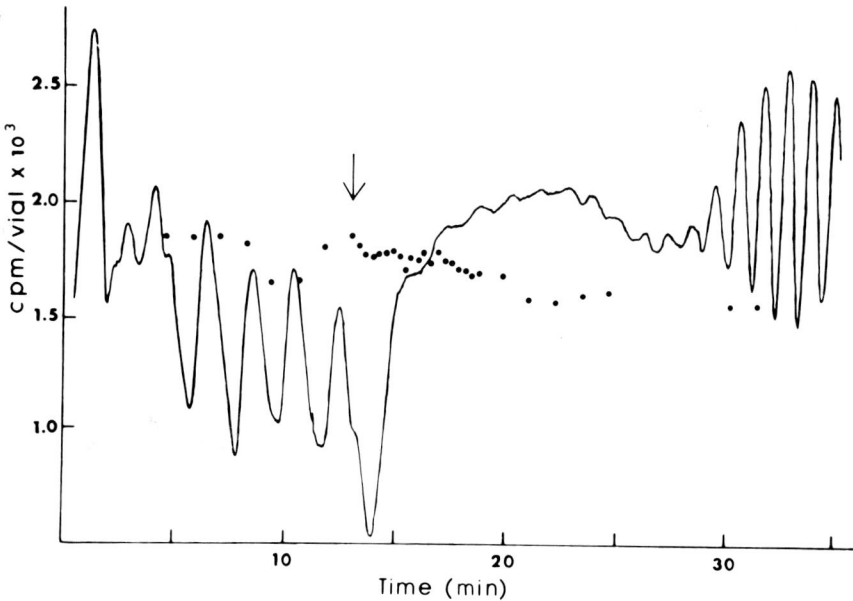

Fig. 16. Simultaneous recording of shuttle streaming (continuous line) and calcium efflux (dots). At the arrow, a plasmodium is exposed to 1 mM glucose. After a period of "shock," shuttle streaming resumes at a higher frequency, but through all this, the calcium efflux rate is unchanged. From Ludlow and Durham (1977); used by permission of the authors and Springer-Verlag.

ever weaker responses until finally no response was observed (as if some intracellular store of Ca^{2+} were depleted). Suspending these "exhausted" CDM in Ca^{2+}-containing medium restored their ability to respond to caffeine (as if the intracellular stores were replenished). Suspension in Ca^{2+}-free medium containing the ionophore A23187 also restored the ability of exhausted CDM to respond to caffeine (as if Ca^{2+} were moved from a caffeine-insensitive intracellular storage system to a caffeine-sensitive store). Matthews concluded that the intracellular Ca^{2+} storage system involved in protoplasmic streaming is membranous, nonmitochondrial, and analogous to the sarcoplasmic reticulum.

Calcium transport across the plasmalemma does not seem to be involved in the contraction rhythm since the oscillation is not affected by changes in external [Ca^{2+}] between 10^{-8} and 10^{-3} M (Wohlfarth-Botterman and Götz von Olenhusen, 1977; Fig. 15). This is a striking result, but one would like to be sure that plasmodia do indeed transport significant amounts of calcium across the plasmalemma under conditions of high pH (7.0–7.2) necessitated by EGTA buffers. Calcium transport across the plasmalemma was measured directly by Ludlow and Durham (1977), *who found no changes in efflux of* $^{45}Ca^{2+}$ from prelabeled cells during a period of dramatic changes in streaming frequency and amplitude (Fig. 16). Oscillations are affected, however, when endoplasm is replaced by artificial media containing different Ca^{2+} concentrations (Ueda and Götz von

3. Periodic Phenomena in *Physarum*

Olenhusen, 1978; Fig. 17): Oscillations appear only when $[Ca^{2+}]_{internal} > 2 \times 10^{-7}$ M (cf. Hatano, 1970).

Finally, what can be said about the involvement of the plasmalemma? Oscillating membrane potentials (Rhea, 1966) are observed even in microplasmodia devoid of shuttle-streaming activity (Meyer and Stockem, 1979), but it is not clear whether the potential oscillation is a cause or an effect of the contraction-relaxation cycle of the cytoplasmic actomyosin complex. Caffeine and D_2O, which are known to interfere with membrane-associated events, are potent inhibitors of the shuttle-streaming rhythm when applied externally but not when applied internally (Götz von Olenhusen and Wohlfarth-Botterman, 1979). This suggests that these substances affect the rhythm primarily at the plasmalemma, but it could be that they enter the ectoplasm more easily from the outside than from the endoplasm side and affect some process within the ectoplasm (Wohlfarth-Bottermann, 1979). Since neither valinomycin nor ouabain is effective in stopping the contraction rhythm, it is unlikely that a proton pump or a Na^+, K^+-ATPase pump is involved in the generation of oscillations (Wohlfarth-Bottermann, 1979; Achenbach and Achenbach, 1979). With the Ca pump ruled out as well, this does not leave much of a role for the plasmalemma. Nonetheless, membrane phenomena (ion and metabolite transport) must be able to modulate the oscillator, e.g., change its frequency, in order to account for chemotactic responses.

At present, it appears that internal Ca^{2+} storage and release is the most likely candidate for the primary oscillophore. However, this candidate seems to be the winner by default, and it is far from clear how internal, bound, and free Ca^{2+} levels interact with the actomyosin system, on the one hand, and the plas-

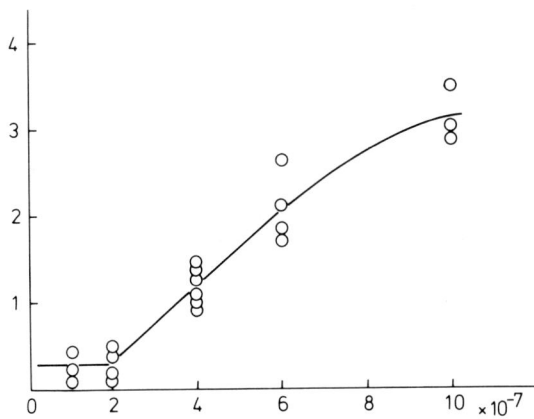

Fig. 17. Dependence of contraction activity on internal calcium concentration. Abscissa: concentration of free Ca^{2+} (moles/liter); ordinate: amplitude of radial contractions relative to control oscillation. Plasmodial strands are injected with artificial media containing Ca–EGTA buffers, and the amplitude of radial contractions after injection is compared to that of control oscillations. From Ueda and Götz von Olenhusen (1978); used by permission of the authors and Academic Press.

malemma, on the other. There is a sense of excitement and expectation in this area at present, and we can look forward to answers in the near future to many of the questions that we are left with today.

ACKNOWLEDGMENTS

Jack Daniel and Sydney Shall read this chapter carefully and made many helpful suggestions. The following individuals kindly gave me access to material prior to publication: Henry Aldrich, Larry Barnes, Jack Daniel, Ned Holt, Jim Jeter, Sam Kupetz, Peter Loidl, and Nancy Oleinick. The preparation of this chapter was supported by USPHS Research Grant GM 27629.

REFERENCES

Achenbach, F., and Achenbach, U. (1979). Oscillating contractions in protoplasmic strands of *Physarum:* Effects of externally applied ouabain, Na, K and Ca ions. *Cell Biol. Int. Rep.* **3,** 141-149.

Achenbach, U., and Wohlfarth-Bottermann, K. E. (1980). Oscillating contractions in protoplasmic strands of *Physarum:* Mechanical and thermal methods of phase shifting. *J. Exp. Biol.* **85,** 21-31.

Aldrich, H. C., and Pendland, J. C. (1981). Nuclear pores during the cell cycle in *Physarum. Tissue & Cell* **13,** 431-439.

Allen, P. J., and Price, W. H. (1950). The relation between respiration and protoplasmic flow in the slime mold, *Physarum polycephalum. Am. J. Bot.* **37,** 393-402.

Bell, L.G.E. (1962). Some mechanisms involved in cell division. *Nature (London)* **193,** 190-191.

Berger, J. D. (1979). Regulation of macronuclear DNA content in *Paramecium tetraurelia. J. Protozool.* **26,** 18-28.

Berridge, M. J., and Rapp, P. E. (1979). A comparative survey of the function, mechanism and control of cellular oscillators. *J. Exp. Biol.* **81,** 217-279.

Bersier, D., and Braun, R. (1974). Pools of deoxyribonucleoside triphosphates in the mitotic cycle of *Physarum. Biochim. Biophys. Acta* **340,** 463-471.

Birch, B., and Turnock, G. (1977). Synthesis of protein during the nuclear division cycle in *Physarum polycephalum. FEBS Lett.* **84,** 317-319.

Blessing, J., and Lempp, H. (1978). An immunological approach to the isolation of factors with mitotic activity from the plasmodial stage of the myxomycete *Physarum polycephalum. Exp. Cell Res.* **113,** 435-438.

Braatz, R. (1975). Differential histochemical localization of calcium and its relation to shuttle streaming in *Physarum. Cytobiologie* **12,** 74-78.

Bradbury, E. M., Inglis, R. J., and Matthews, H. R. (1974a). Control of cell division by very lysine rich histone (Fl) phosphorylation. *Nature (London)* **247,** 257-261.

Bradbury, E. M., Inglis, R. J., Matthews, H. R., and Langan, T. A. (1974b). Molecular basis of control of mitotic cell division in eukaryotes. *Nature (London)* **249,** 553-556.

Braun, R., and Behrens, K. (1969). A ribonuclease from *Physarum:* Biochemical properties and synthesis in the mitotic cycle. *Biochim. Biophys. Acta* **195,** 87-98.

Braun, R., and Evans, T. E. (1969). Replication of nuclear satellite and mitochondrial DNA in the mitotic cycle of *Physarum. Biochim. Biophys. Acta* **182,** 511-522.

Braun, R., and Seebeck, T. (1979). Ribosomal DNA: Extrachromosomal genes of *Physarum. In* "Specific Eukaryotic Genes" (J. Engberg, H. Klenow, V. Leick, and J. H. Thaysen, eds.), pp. 306-317. Munksgaard, Copenhagen.

Braun, R., Hall, L., Schwärzler, M., and Smith, S. S. (1977). The mitotic cycle of *Physarum polycephalum. In* "Cell Differentiation in Microorganisms, Plants and Animals" (L. Nover and K. Mothes, eds.), pp. 402-423. VEB Gustav Fischer Verlag, Jena.

Brewer, E. N., and Rusch, H. P. (1966). Control of DNA replication: Effect of spermine on DNA polymerase activity in nuclei isolated form *Physarum polycephalum. Biochem. Biophys. Res. Commun.* **25**, 579-584.

Brewer, E. N., and Rusch, H. P. (1968). Effect of elevated temperature shocks on mitosis and on the initiation of DNA replication in *Physarum polycephalum. Exp. Cell Res.* **49**, 79-86.

Brightwell, M. D., Leech, C. E., O'Farrell, M. K., Whish, W. J. D., and Shall, S. (1975). Poly (adenosine diphosphate ribose) polymerase in *Physarum polycephalum. Biochem. J.* **147**, 119-129.

Chin, B., Friedrich, P. D., and Bernstein, I. A. (1972). Stimulation of mitosis following fusion of plasmodia in the myxomycete *Physarum polycephalum. J. Gen. Microbiol.* **71**, 93-101.

Choudhry, M. K., and Cox, R. A. (1979). Purification of a DNA polymerase from *Physarum polycephalum* and the production of specific antibodies. *In* "Current Research on Physarum" (W. Sachsenmaier, ed.), pp. 71-76. Univ. Innsbruck, Innsbruck, Austria.

Daniel, J. W. (1970). Fluorescence oscillations accompanying protoplasmic flow. *J. Cell Biol.* **47**, 45a.

Daniel, J. W., and Järlfors, U. (1972). Plasmodial ultrastructure of the myxomycete *Physarum polycephalum. Tissue & Cell* **4**, 15-36.

Davies, K. E., and Walker, I. O. (1978). Control of RNA transcription in nuclei and nucleoli of *Physarum polycephalum. FEBS Lett.* **86**, 303-306.

Donachie, W. D. (1968). Relationship between cell size and time of initiation of DNA replication. *Nature (London)* **219**, 1077-1079.

Durham, A. C. H., and Ridgway, E. B. (1976). Control of chemotaxis in *Physarum polycephalum. J. Cell Biol.* **69**, 218-223.

Elliott, S. G., and McLaughlin, C. S. (1978). Rate of macromolecular synthesis through the cell cycle of the yeast *Saccharomyces cerevisiae. Proc. Natl. Acad. Sci. U.S.A.* **75**, 4384-4388.

Fantes, P. A. (1977). Control of cell size and cycle time in *Schizosaccharomyces pombe. J. Cell Sci.* **24**, 51-67.

Fantes, P. A., and Nurse, P. (1977). Control of cell size at division in fission yeast by a growth-modulated size control over nuclear division. *Exp. Cell Res.* **107**, 377-386.

Fantes, P. A., and Nurse, P. (1978). Control of the timing of cell division in fission yeast. *Exp. Cell Res.* **115**, 317-329.

Fantes, P. A., Grant, W. D., Pritchard, R. H., Sudbery, P. E., and Wheals, A. E. (1975). The regulation of cell size and the control of mitosis. *J. Theor. Biol.* **50**, 213-244.

Fink, K. (1975). Fluctuations in deoxyribo- and ribonucleoside triphosphate pools during the mitotic cycle of *Physarum polycephalum. Biochim. Biophys. Acta* **414**, 85-89.

Fink, K., and Turnock, G. (1977). Synthesis of transfer RNA during the synchronous nuclear division cycle in *Physarum polycephalum. Eur. J. Biochem.* **80**, 93-96.

Forde, B. G., and Sachsenmaier, W. (1979). Oxygen uptake and mitochondrial enzyme activities in the mitotic cycle of *Physarum polycephalum. J. Gen. Microbiol.* **115**, 135-143.

Fouquet, H., and Braun, R. (1974). Differential RNA synthesis in the mitotic cycle of *Physarum polycephalum. FEBS Lett.* **38**, 184-186.

Fouquet, H., and Sauer, H. W. (1975). Variable redundancy in RNA transcripts isolated in S and G2 phase of the cell cycle of *Physarum. Nature (London)* **255**, 253-255.

Fox, T. O., and Pardee, A. B. (1970). Animal cells: Noncorrelation of length of G1 phase with size after mitosis. *Science* **167**, 80-82.

Frazier, E. A. J. (1973). DNA synthesis following gross alterations of the nucleocytoplasmic ratio in the ciliate *Stentor coeruleus. Dev. Biol.* **34**, 77-92.

Garrison, P. N., and Barnes, L. D. (1980). Cyclic AMP in the cell cycle of *Physarum polycephalum. Biochim. Biophys. Acta* **633**, 114-121.

Gershey, E. L., D'Alisa, R. M., and Zucker, R. M. (1979). Characterization of a CV-1 cell cycle. I. No critical size for S phase entry. *Exp. Cell Res.* **122,** 9-14.

Gerson, D., and Burton, A. (1977). The relation of cycling of intracellular pH to mitosis in the acellular slime mold *Physarum polycephalum. J. Cell Physiol.* **91,** 297-304.

Goldbeter, A., and Segel, L. A. (1977). Unified mechanism for relay and oscillation of cyclic AMP in *Dictyostelium discoideum. Proc. Natl. Acad. Sci. U.S.A.* **74,** 1543-1547.

Götz von Olenhusen, K., and Wohlfarth-Bottermann, K. E. (1979). Effects of caffeine and D_2O on persistence and *de novo* generation of intrinsic oscillatory contraction automaticity in *Physarum. Cell Tissue Res.* **197,** 479-499.

Götz von Olenhusen, K., Jücker, H., and Wohlfarth-Bottermann, K. E. (1979). Induction of a plasmodial stage of *Physarum* without plasmalemma invaginations. *Cell Tissue Res.* **197,** 463-477.

Grant, W. D. (1972). The effect of alpha-amanitin and ammonium sulfate on RNA synthesis in nuclei at different times during the cell cycle. *Eur. J. Biochem.* **29,** 94-98.

Grebecki, A., and Cieslawska, M. (1978). Plasmodium of *Physarum polycephalum* as a synchronous contractile system. *Cytobiologie* **17,** 335-342.

Gröbner, P. (1979). Thymidine kinase enzyme variants in *Physarum polycephalum. J. Biochem.* **86,** 1595-1605.

Gröbner, P., and Sachsenmaier, W. (1976). Thymidine kinase enzyme variants in *Physarum polycephalum;* Change of pattern during the synchronous mitotic cycle. *FEBS Lett.* **71,** 181-184.

Gröbner, P., and Loidl, P. (1981). Thymidylate synthesis in *Physarum polycephalum. In* "Biology of Physarum: Current Problems" (L. Rakoczy, ed.), pp. 181-190. Jagiellonian Univ. Press, Cracow, Poland.

Guttes, S., and Guttes, E. (1968). Regulation of DNA replication in the nuclei of the slime mold *Physarum polycephalum*. Transplantation of nuclei by plasmodial coalescence. *J. Cell Biol.* **37,** 761-772.

Hall, L., and Turnock, G. (1976). Synthesis of ribosomal RNA during the mitotic cycle in the slime mould *Physarum polycephalum. Eur. J. Biochem.* **62,** 471-477.

Halvorson, H. O. (1977). A review of current models on temporal gene expression in *Saccharomyces cerevisiae. In* "Cell Differentiation in Microorganisms, Plants and Animals" (L. Nover and K. Mothes, eds.), pp. 361-376. VEB Gustav Fischer Verlag, Jena.

Hara, K., Tydeman, P., and Kirschner, M. (1980). Cytoplasmic clock with same period as division cycle in *Xenopus* eggs. *Proc. Natl. Acad. Sci. U.S.A.* **77,** 462-466.

Hardie, D. G., Matthews, H. R., and Bradbury, E. M. (1976). Cell-cycle dependence of two nuclear histone kinase enzyme activities. *Eur. J. Biochem.* **66,** 37-42.

Hartwell, L. H., and Unger, M. W. (1977). Unequal division in *Saccharomyces cerevisiae* and its implications for the control of cell division. *J. Cell Biol.* **75,** 422-435.

Hatano, S. (1970). Specific effect of calcium on movement of plasmodial fragment obtained by caffeine treatment. *Exp. Cell Res.* **61,** 199-203.

Hildebrandt, A., and Sauer, H. W. (1973). Thymidine phosphorylation in the cell cycle of *Physarum polycephalum* and the effect of 5-fluoro-2-deoxyuridine and hydroxyurea. *Biochem. Biophys. Acta* **294,** 8-14.

Holt, C. E. (1980). The nuclear replication cycle in *Physarum polycephalum. In* "Growth and Differentiation in *Physarum polycephalum*" (W. F. Dove and H. P. Rusch, eds.), pp. 9-63. Princeton Univ. Press, Princeton, New Jersey.

Hülsmann, N., and Wohlfarth-Botterman, K. E. (1978). Spatio-temporal analysis of contraction dependent surface movements in *Physarum polycephalum. Cytobiologie* **17,** 23-41.

Hüttermann, A., Porter, M. T., and Rusch, H. P. (1970). Activity of some enzymes in *Physarum polycephalum.* I. In the growing plasmodium. *Arch. Mikrobiol.* **74,** 90-100.

Inglis, R. J., Langan, T. A., Matthews, H. R., Hardie, D. G., and Bradbury, E. M. (1976). Advance of mitosis by histone phosphokinase. *Exp. Cell Res.* **97,** 418-425.

3. Periodic Phenomena in *Physarum*

Jeter, J. R., Glas, P. S., and Kupetz, I. S. (1978). Changes in the nuclear activity of cyclic AMP and cyclic GMP phosphodiesterase during the cell cycle of *Physarum polycephalum*. *J. Cell Biol.* **79**, 13a.

Jeter, J. R., Cameron, I. L., Smith, N. K. R., Steffens, W. L., and Wille, J. J. (1981). Cell cycle fluctuations in concentration of various elements in cytoplasm and in nucleus/chromatin of *Physarum polycephalum*. *J. Cell Biol.* (in press).

Johnston, G. C., Pringle, J. R., and Hartwell, L. H. (1977). Coordination of growth with cell division in the yeast *Saccharomyces cerevisiae*. *Exp. Cell Res.* **105**, 79-98.

Kamiya, N. (1959). Protoplasmic streaming. *Protoplasmatologia* **8**(3a), 1-199.

Kamiya, N. (1970). Contractile properties of the plasmodial strand. *Proc. Jpn. Acad.* **46**, 1026-1031.

Kauffman, S., and Wille, J. J. (1975). The mitotic oscillator in *Physarum polycephalum*. *J. Theor. Biol.* **55**, 47-93.

Killander, D., and Zetterberg, A. (1965). Determination of DNA, RNA and mass content of age determined mouse fibroblasts *in vitro* and of intercellular variation in generation time. *Exp. Cell Res.* **38**, 272-281.

Kincaid, R. L., and Mansour, T. E. (1979). Cyclic-3',5'-AMP phosphodiesterase in *Physarum polycephalum*. II. Kinetic properties. *Biochim. Biophys. Acta* **588**, 342-350.

Klevecz, R. R. (1969). Temporal order in mammalian cells. I. The periodic synthesis of lactate dehydrogenase in the cell cycle. *J. Cell Biol.* **43**, 207-219.

Komnick, H., Stockem, W., and Wohlfarth-Bottermann, K. E. (1973). Cell motility: Mechanisms in protoplasmic streaming and ameboid movement. *Int. Rev. Cytol.* **34**, 169-249.

Krüger, J., and Wohlfarth-Bottermann, K. E. (1978). Oscillating contractions in protoplasmic strands of *Physarum*: Stretch induced phase shifts and their synchronization. *J. Interdiscipl. Cycle Res.* **9**, 61-71.

Kuehn, G. D. (1972). Cell cycle variation in cyclic adenosine 3',5'-monophosphate-dependent inhibition of a protein kinase from *Physarum polycephalum*. *Biochem. Biophys. Res. Commun.* **49**, 414-419.

Kuroiwa, T., Hizume, M., and Kawano, S. (1978). Studies on mitochondrial structure and function in *Physarum polycephalum*. IV. Mitochondrial division cycle. *Cytologia* **43**, 119-136.

Loewy, A. G. (1950). Protoplasmic streaming under anaerobic conditions in a myxomycete. *J. Cell. Comp. Physiol.* **35**, 151-153.

Loidl, P. (1979). "Regulation der Synchronen Mitose und DNA-Synthese in *Physarum polycephalum*." Univ. of Innsbruck, Innsbruck, Austria.

Loidl, P., and Gröbner, P. (1981). Acceleration of nuclear division by extracts of *Physarum polycephalum*. *In* "Biology of Physarum: Current Problems" (L. Rakoczy, ed.), pp. 84-89. Jagiellonian Univ. Press, Cracow, Poland.

Loidl, P., Linortner, C., and Sachsenmaier, W. (1979). Timing of mitosis and DNA replication in mixed heterophasic plasmodia of *Physarum polycephalum*. *In* "Current Research on Physarum" (W. Sachsenmaier, ed.), pp. 157-162. Univ. Innsbruck, Innsbruck, Austria.

Lovely, J. R., and Threlfall, R. J. (1976). Fluctuations in cyclic adenosine 3',5'-monophosphate and cyclic guanosine 3',5'-monophosphate during the mitotic cycle of *Physarum polycephalum*. *Biochem. Biophys. Res. Commun.* **71**, 789-795.

Lovely J., and Threlfall, R. J. (1978). Adenylate cyclase and cyclic AMP phosphodiesterase activity during the mitotic cycle of *Physarum polycephalum*. *Biochem. Biophys. Res. Commun.* **85**, 579-584.

Lovely, J. R., and Threlfall, R. J. (1979). The activity of guanylate cyclase and cyclic GMP phosphodiesterase during synchronous growth of the acellular slime mould *Physarum polycephalum*. *Biochem. Biophy. Res. Commun.* **86**, 365-370.

Ludlow, C. T., and Durham, A. C. H. (1977). Calcium ion fluxes across the external surface of *Physarum polycephalum*. *Protoplasma* **91**, 107-113.

Lutkenhaus, J. F., Moore, B. A., Masters, M., and Donachie, W. D. (1979). Individual proteins are synthesized continuously throughout the *Escherichia coli* cell cycle. *J. Bacteriol.* **138**, 352–360.

Magun, B. (1979). Changes in cytoplasmic DNA-binding phosphoproteins during the cell cycle of *Physarum polycephalum*. *Cell Differ.* **8**, 157–172.

Matthews, L. M., Jr. (1977). Ca^{2+} regulation in caffeine-derived microplasmodia of *Physarum polycephalum*. *J. Cell Biol.* **72**, 502–505.

Matthews, H. R. (1978). *In* "4th European Cell Cycle Workshop. Abstracts and Addresses," p. 7.

Melera, P. W., and Rusch, H. P. (1973). Aminoacylation of transfer ribonucleic acid *in vitro* during the mitotic cycle of *Physarum polycephalum*. *Biochemistry* **12**, 1307–1311.

Meyer, R., and Stockem, W. (1979). Studies on microplasmodia of *Physarum polycephalum*.V. Electrical activity of different types of microplasmodia and macroplasmodia. *Cell Biol. Int. Rep.* **3**, 321–330.

Minor, P. D., and Smith, J. A. (1974). Explanation of degree of correlation of sibling generation times in animal cells. *Nature (London)* **248**, 241–243.

Mitchell, J. L. A., and Carter, D. D. (1977). Physical and kinetic distinction of two ornithine decarboxylase forms in *Physarum*. *Biochim. Biophys. Acta* **483**, 425–434.

Mitchell, J. L. A., and Kottas, G. E. (1979). Osmotically-induced modification of ornithine decarboxylase in *Physarum*. *FEBS Lett.* **102**, 265–268.

Mitchell, J. L. A., and Rusch, H. P. (1973). Regulation of polyamine synthesis in *Physarum polycephalum* during growth and differentiation. *Biochim. Biophys. Acta* **297**, 503–516.

Mitchell, J. L. A., and Sedory, M. J. (1974). Cycloheximide induced *in vivo* modification of ornithine decarboxylase in *Physarum polycephalum*. *FEBS Lett.* **49**, 120–124.

Mitchell, J. L. A., Campbell, H. A., and Carter, D. D. (1976). Multiple ornithine decarboxylase forms in *Physarum polycephalum:* Interconversion induced by cycloheximide. *FEBS Lett.* **62**, 33–37.

Mitchell, J. L. A., Carter, D. D., and Rybski, J. A. (1978). Control of ornithine decarboxylase activity in *Physarum* by polyamines. *Eur. J. Biochem.* **92**, 325–331.

Mitchelson, K., Chambers, T., Bradbury, E. M., and Matthews, H. R. (1978). Activation of histone kinase in G2 phase of the cell cycle in *Physarum polycephalum*. *FEBS Lett.* **92**, 339–342.

Mitchison, J. M. (1969). Enzyme synthesis in synchronous cultures. *Science* **165**, 657–663.

Mitchison, J. M. (1971). "The Biology of the Cell Cycle." Cambridge Univ. Press, London and New York.

Mitchison, J. M. (1977). Enzyme synthesis during the cell cycle. *In* "Cell Differentiation in Microorganisms, Plants and Animals" (L. Nover and K. Mothes, eds.), pp. 377–401. VEB Gustav Fischer Verlag, Jena.

Mittermayer, C., Braun, R., and Rusch, H. P. (1964). RNA synthesis in the mitotic cycle of *Physarum polycephalum*. *Biochim. Biophys. Acta* **91**, 399–405.

Mittermayer, C., Braun, R., Chayka, T. G., and Rusch, H. P. (1966a). Polysome patterns and protein synthesis during the mitotic cycle of *Physarum polycephalum*. *Nature (London)* **210**, 1133–1137.

Mittermayer, C., Braun, R., and Rusch, H. P. (1966b). Ribonucleic acid synthesis *in vitro* in nuclei isolated from the synchronously dividing *Physarum polycephalum*. *Biochim. Biophys. Acta* **114**, 536–546.

Mohberg, J., and Rusch, H. P. (1969). Growth of large plasmodia of the myxomycete *Physarum polycephalum*. *J. Bacteriol.* **97**, 1411–1418.

Mohberg, J., Babcock, K. L., Haugli, F. B., and Rusch, H. P. (1973). Nuclear DNA content and chromosome numbers in the myxomycete *Physarum polycephalum*. *Dev. Biol.* **34**, 228–245.

Mohberg, J., Dworzak, E., Sachsenmaier, W., and Haugli, F. B. (1980). Thymidine kinase-deficient mutants of *Physarum polycephalum;* relationships between enzyme activity levels and ploidy. *Cell Biol. Int. Rep.* **4**, 137-148.

Nagai, R., Yoshimoto, Y., and Kamiya, N. (1978). Cyclic production of tension force in the plasmodial strand of *Physarum polycephalum* and its relation to microfilament morphology. *J. Cell Sci.* **33**, 205-225.

Nasmyth, K. A. (1979). A control acting over the initiation of DNA replication in the yeast *Schizosaccharomyces pombe. J. Cell Sci.* **36**, 155-168.

Nurse, P. (1975). Genetic control of cell size at cell division in yeast. *Nature (London)* **256**, 547-551.

Nurse, P., and Thuriaux, P. (1977). Controls over the timing of DNA replication during the cell cycle of fission yeast. *Exp. Cell Res.* **107**, 365-375.

Oleinick, N. L. (1972). The radiation-sensitivity of mitosis and the synthesis of thymidine kinase in *Physarum polycephalum. Radiat. Res.* **51**, 638-653.

Oleinick, N. L., Daniel, J. W., and Brewer, E. N. (1981a). Absence of a correlation between cyclic nucleotide fluctuations and cell cycle progression. *Exp. Cell Res.* **131**, 373-377.

Oleinick, N. L., Brewer, E. N., and Blank, D. J. (1981b). Histone kinase activity during radiation-induced mitotic delay in *Physarum polycephalum* (preprint).

Oppenheim, A., and Katzir, N. (1971). Advancing the onset of mitosis by cell free preparations of *Physarum polycephalum. Exp. Cell Res.* **68**, 224-226.

Parisi, E., Filosa, S., DePetrocellis, B., and Morvy, A. (1978). The pattern of cell division in the early development of the sea urchin, *Paracentrotus lividus. Dev. Biol.* **65**, 38-49.

Pierron, G., and Sauer, H. W. (1980a). More evidence for replication-transcription-coupling in *Physarum polycephalum. J. Cell Sci.* **41**, 105-113.

Pierron, G., and Sauer, H. W. (1980b). RNA Polymerase B levels during the cell cycle of *Physarum polycephalum. Wilhelm Roux's Arch. Dev. Biol.* **189**, 165-169.

Plaut, B. S., and Turnock, G. (1975). Coordination of macromolecular synthesis in the slime mould *Physarum polycephalum. Mol. Gen. Genet.* **137**, 211-225.

Prescott, D. M. (1956). Relation between cell growth and cell division. II. The effect of cell size on cell growth rate and generation time in *Amoeba proteus. Exp. Cell Res.* **11**, 86-98.

Rao, P. N., and Johnson, R. T. (1970). Mammalian cell fusion: Studies on the regulation of DNA synthesis and mitosis. *Nature (London)* **225**, 159-164.

Rhea, R. P. (1966). Macrocinematographic, electron microscopic, and electrophysiological studies on shuttle streaming in the slime mold *Physarum polycephalum. In* "Dynamics of Fluids and Plasmas" (S. I. Pai, A. J. Faller, T. L. Lincoln, D. A. Tidman, G. N. Trytten, and T. D. Wilkerson, eds.), pp. 35-58. Academic Press, New York.

Ridgway, E. B., and Durham, A. C. H. (1976). Oscillations of calcium ion concentrations in *Physarum polycephalum. J. Cell Biol.* **69**, 223-226.

Rossow, P. W., Riddle, V. G. H., and Pardee, A. B. (1979). Synthesis of labile, serum-dependent protein in early G1 controls animal cell growth. *Proc. Natl. Acad. Sci. U.S.A.* **76**, 4446-4450.

Rusch, H. P., Sachsenmaier, W., Behrens, K., and Gruter, V. (1966). Synchronization of mitosis by the fusion of the plasmodia of *Physarum polycephalum. J. Cell Biol.* **31**, 204-209.

Sachsenmaier, W. (1976). Control of synchronous nuclear mitosis in *Physarum polycephalum. In* "The Molecular Basis of Circadian Rhythms" (J. W. Hastings and H. G. Schweiger, eds.), pp. 409-420. Dahlem Konf., Berlin.

Sachsenmaier, W., and Ives, D. H. (1965). Periodische Aenderungen der Thymidinkinase-Aktivitaet im Synchronen Mitosecyclus von *Physarum polycephalum. Biochem. Z.* **343**, 399-406.

Sachsenmaier, W., von Fournier, D., and Gürtler, K. F. (1967). Periodic thymidine kinase production in synchronous plasmodia of *Physarum polycephalum:* Inhibition by actinomycin and actidion. *Biochem. Biophys. Res. Commun.* **27**, 655-660.

Sachsenmaier, W., Immich, H., Grunst, J., Scholz, R., and Bücher, T. (1969). Free ribonucleotides of *Physarum polycephalum*. *Eur. J. Biochem.* **8**, 557-561.

Sachsenmaier, W., Dönges, K. H., Rupff, H., and Czihak, G. (1970). Advanced initiation of synchronous mitosis in *Physarum polycephalum* following UV-irradiation. *Z. Naturforsch. Anorg. Chem. Org.* **25**, 866-871.

Sachsenmaier, W., Remy, U., and Plattner-Schobel, R. (1972). Initiation of synchronous mitosis in *Physarum polycephalum*. A model of the control of cell division in Eukarioes. *Exp. Cell Res.* **73**, 41-48.

Sachsenmaier, W., Blessing, J., Brauser, B., and Hansen, K. (1973). Protoplasmic streaming in *Physarum polycephalum*. Observations of the oscillatory pattern by photometric and fluorometric techniques. *Protoplasma* **77**, 381-396.

Sachsenmaier, W., Bohnert, E., Clausnizer, B., and Nygaard, O. (1970). Cycle dependent variation of X-ray effects on synchronous mitosis and thymidine kinase induction in *Physarum polycephalum*. *FEBS Lett.* **10**, 185-189.

Scheffey, C., and Wille, J. J. (1978). Cycloheximide-induced mitotic delay in *Physarum polycephalum*. *Exp. Cell Res.* **113**, 259-262.

Schel, J. H. N., Steenbergen, L. C. A., Bekers, A. G. M., and Wanka, F. (1978). Change of the nuclear pore frequency during the nuclear cycle of *Physarum polycephalum*. *J. Cell Sci.* **34**, 225-232.

Schiebel, W. (1973). The cell cycle of *Physarum polycephalum*. *Ber. Dtsch. Bot. Ges.* **86**, 11-38.

Schiebel, W., and Baer. A. (1979). DNA polymerase of *Physarum polycephalum*. *In* "Current Research on Physarum" (W. Sachsenmaier, ed.), pp. 77-81. Univ. of Innsbruck, Innsbruck, Austria.

Schmidt, R. R. (1974). Transcriptional and post-transcriptional control of enzyme levels in eucaryotic microorganisms. *In* "Cell Cycle Controls" (G. M. Padilla, I. L. Cameron, and A. Zimmerman, eds.), pp. 201-233. Academic Press, New York.

Schneiderman, M. H., Dewey, W. C., and Highfield, D. P. (1971). Inhibition of DNA synthesis in synchronized Chinese hamster cells treated in G1 with cycloheximide. *Exp. Cell Res.* **67**, 147-155.

Sedory, M. J., and Mitchell, J. L. A. (1977). Regulation of ornithine decarboxylase activity during the *Physarum* mitotic cycle. *Exp. Cell Res.* **107**, 105-110.

Shields, R. (1978). Further evidence for a random transition in the cell cycle. *Nature (London)* **273**, 755-758.

Shilo, B., Shilo, V., and Simchen, G. (1976). Cell-cycle initiation in yeast follows first-order kinetics. *Nature (London)* **264**, 767-769.

Smith, D. L., and Mansour, T. E. (1978). An adenosine-3',5'-monophosphate activated adenylate cyclase in the slime mold *Physarum polycephalum*. *FEBS Lett.* **92**, 57-62.

Smith, J. A., and Martin, L. (1973). Do cells cycle? *Proc. Natl. Acad. Sci. U.S.A.* **70**, 1263-1267.

Smith, S. S., and Braun, R. (1978). A new method for the purification of RNA polymerase II from the lower eukaryote *Physarum polycephalum*. *Eur. J. Biochem.* **82**, 309-320.

Salao, P., and Shall, S. (1971). Control of DNA replication in *Physarum polycephalum*. I. Specific activity of NAD pyrophosphorylase in isolated nuclei during the cell cycle. *Exp. Cell Res.* **69**, 295-300.

Sudbery, P. E., and Grant, W. D. (1975). The control of mitosis in *Physarum polycephalum*. The effect of lowering the DNA: mass ratio by UV irradiation. *Exp. Cell Res.* **95**, 405-415.

Sudbery, P. E., Goodey, A. R., and Carter, B.L.A. (1980). Genes which control cell proliferation in the yeast *Saccharomyces cerevisiae*. *Nature (London)* **288**, 401-404.

Threlfall, R. J., and Thomas, A. J. (1979). Fluctuations in proline and other free amino acids during the mitotic cycle of the myxomycete *Physarum polycephalum*. *Eur. J. Biochem.* **93**, 129-133.

Thuriaux, P., Nurse, P., and Carter, B. (1978). Mutants altered in the control co-ordinating cell division with cell growth in the fission yeast *Schizosaccharomyces pombe*. *Mol. Gen. Genet.* **161**, 215–220.

Tyson, J. J. (1975). Classification of instabilities in chemical reaction systems. *J. Chem. Phys.* **62**, 1010–1015.

Tyson, J. J. (1979). Periodic enzyme synthesis: Reconsideration of the theory of oscillatory repression. *J. Theor. Biol.* **80**, 27–38.

Tyson, J. J. (1981). Nonlinear analysis of simple metabolic control circuits. In "Applications of Nonlinear Analysis in the Physical Sciences" (H. Amann, N. W. Bazley, and K. Kirchgässner, eds.), pp. 310–322. Pitman, London.

Tyson, J., and Sachsenmaier, W. (1978). Is nuclear division in *Physarum* controlled by a continuous limit cycle oscillator? *J. Theor. Biol.* **73**, 723–738.

Tyson, J. J., and Sachsenmaier, W. (1979). Derepression as a model for control of the DNA-division cycle in eukaryotes. *J. Theor. Biol.* **79**, 275–280.

Tyson, J. J., Alivisatos, S. G. A., Grun, F., Pavlidis, T., and Richter, O. (1976). Mathematical background: Group report. In "The Molecular Basis of Circadian Rhythms" (J. W. Hastings and H. G. Schweiger, eds.), pp. 85–108. Dahlem Konf., Berlin.

Tyson, J. J., Garcia-Herdugo, G., and Sachsenmaier, W. (1979). Control of nuclear division in *Physarum polycephalum*. Comparison of cycloheximide pulse treatment, UV irradiation, and heat shock. *Exp. Cell Res.* **119**, 87–98.

Ueda, T., and Götz von Olenhusen, K. (1978). Replacement of endoplasm with artificial media in plasmodial strands of *Physarum polycephalum*. *Exp. Cell Res.* **116**, 55–62.

Ueda, T., Maratsugu, M., Kurihara, K., and Kobatake, Y. (1976). Chemotaxis in *Physarum polycephalum:* Effects of chemicals on isometric tension in relation to chemotactic movement. *Exp. Cell Res.* **100**, 337–344.

Wendelberger-Schieweg, G., Hüttermann, A., and Haugli, F. B. (1980). Multiple sites of action of cycloheximide in addition to inhibition of protein synthesis in *Physarum polycephalum*. *Arch. Microbiol.* **126**, 109–115.

Wick, R. (1977). Poly (A)-haltige RNA im Lebenszyklus von *Physarum polycephalum*. *Verh. Dtsch. Zool. Ges.* **70A**, 22–34.

Wille, J. J. (1977). Preferential binding of S-phase proteins to temporally-characteristic units of replication in *Physarum polycephalum*. *Nucleic Acid Res.* **4**, 3143–3154.

Wille, J. J., Scheffey, C., and Kauffman, S. A. (1977). Novel behaviour of the mitotic clock in *Physarum*. *J. Cell Sci.* **27**, 91–104.

Winfree, A. T. (1974). Patterns of phase compromise in biological cycles. *J. Math. Biol.* **1**, 73–95.

Woertz, G., and Sachenmaier, W. (1979). Deoxyadenosine kinase in *Physarum polycephalum*. In "Current Research on Physarum" (W. Sachsenmaier, ed.), pp. 123–129. Univ. of Innsbruck, Innsbruck, Austria.

Wohlfarth-Bottermann, K. E. (1962). Weitreichende, Fibrilläre, Protoplasmadifferenzierungen und Ihre Bedeutung für die Protoplasmaströmung. I. Elektronmikroskopischer Nachweis und Feinstruktur. *Protoplasma* **54**, 514–539.

Wohlfarth-Bottermann, K. E. (1974). Plasmalemma invaginations as characteristic constituents of plasmodia of *Physarum polycephalum*. *J. Cell Sci.* **16**, 23–37.

Wohlfarth-Bottermann, K. E. (1975). Tensiometric demonstration of endogenous oscillating contractions in plasmodia of *Physarum polycephalum*. *Z. Pflanzenphysiol.* **76**, 14–27.

Wohlfarth-Bottermann, K. E. (1979). Oscillatory contraction activity in *Physarum*. *J. Exp. Biol.* **81**, 15–32.

Wohlfarth-Bottermann, K. E., and Götz von Olenhusen, K. (1977). Oscillating contractions in protoplasmic strands of *Physarum:* Effects of external Ca-depletion and Ca-antagonistic drugs on contraction automaticity. *Cell Biol. Int. Rep.* **1**, 239–247.

Wolf, H., Finkenstedt, G., Woertz, G., and Sachsenmaier, W. (1973). Regulation von Nucleosidphosphorylierenden Enzymen im Synchronen Mitosezyklus von *Physarum polycephalum*. *H.S. Z. Physiol. Chem.* **354,** 1260.

Wolf, R., Wick, R., and Sauer, H. (1979). Mitosis in *Physarum polycephalum:* Analysis of time-lapse films and DNA replication of normal and heat-shocked macroplasmodia. *Eur. J. Cell Biol.* **19,** 49–59.

Wright, M., and Tollon, Y. (1978). Heat sensitive factor necessary for mitosis onset in *Physarum polycephalum*. *Mol. Gen. Genet.* **163,** 91–99.

Wright, M., and Tollon, Y. (1979a). *Physarum* thymidine kinase: A step or peak enzyme depending upon temperature of growth. *Eur. J. Biochem.* **96,** 177–181.

Wright, M., and Tollon, Y. (1979b). Regulation of thymidine kinase synthesis during the cell cycle of *Physarum* by the heat-sensitive system which triggers mitosis and S phase. *Exp. Cell Res.* **122,** 273–279.

Yen, A., Fried, J., Kitahara, T., Strife, A., and Clarkson, B. D. (1975). The kinetic significance of cell size. I. Variation of cell cycle parameters with size measured at mitosis. *Exp. Cell Res.* **95,** 295–302.

Yoshimoto, Y., and Kamiya, N. (1978). Studies on contraction rhythm of the plasmodial strand. II. Effect of externally applied forces. *Protoplasma* **95,** 101–109.

Zeuthen, E., and Williams, N. E. (1969). Division-limiting morphogenetic processes in *Tetrahymena*. *In* "Nucleic Acid Metabolism, Cell Differentiation and Cancer Growth" (E. V. Cowdry and S. Seno, eds.), pp. 203–216. Pergamon, Oxford.

CHAPTER 4

Chemotaxis in Plasmodia of *Physarum polycephalum*

TETSUO UEDA and YONOSUKE KOBATAKE

I.	Introduction	112
II.	The Experiment	113
	A. Measurement of Chemotaxis at the Behavioral Level	113
	B. Double-Chamber Method	115
	C. Measurement of Tension Generation in the Plasmodial Strand .	115
	D. Determination of Electrophoretic Mobility	116
III.	Response to Carbohydrates	117
	A. Response of the Plasmodium to Sugars at the Membrane Level	117
	B. Response to Sugars at the Behavioral Level	118
	C. Metabolizability and Positive Chemotaxis	118
IV.	Hydrophobicity and Heterogeneity of Functional Membranes	119
	A. Response to n-Alcohols	119
	B. Dependence of the Threshold Concentration on Alkyl-Chain Length	120
	C. Hydrophobicity of the Membranes	120
V.	Electrostatic Interactions in Relation to Chemoreceptive Thresholds ..	122
	A. The Schulze–Hardy Rule in Chemoreception	122
	B. Water Structure Around the Surface Membrane	123
	C. Fluorescence Analysis of Anilinonaphthalene Sulfonate in Chemoreception	124
	D. Nature of the Structural Change in the Plasmodial Membrane Effected by Chemoreception	126
VI.	Correlation between Tension Generation and Tactic Movement	127
VII.	Manipulation of Intracellular Components in the Plasmodial Strands .	128
	A. Replacement of Endoplasm with Artificial Media	128
	B. Effects of Injected Chemicals on Tension Generation	129
VIII.	Intracellular Control of Chemotaxis	132
	A. Intracellular ATP as a Modulator of Contractility in Chemotaxis	132
	B. Intracellular Mediators in the Chemotactic Response	133
IX.	Rhythmicity of the Plasmodial Contractile System	134
	A. Rhythm in Tension Generation and Protoplasmic Streaming	134

X.	Effects of Temperature on Membrane Activities	136
	A. Effects of Temperature on Chemoreception	136
	B. Effects of Temperature on Pseudopod Formation and Periodicity	137
XI.	Summary	138
XII.	Glossary of Symbols	140
	References	141

I. INTRODUCTION

Even with the simplest forms of chemotaxis, we have to ask (1) how the chemical stimuli are sensed at the surface membrane, (2) how the sensed information is transduced into the motile system, and (3) how the behavior takes place.

The first problem concerns the mechanism of the primary process of chemoreception. Here we will analyze it in terms of physicochemical concepts and techniques. The general characteristics of chemoreception at the membrane level in living organisms will be emphasized by using the plasmodium as a model organism.

The transduction mechanism related to problems (2) and (3) is distinctive in chemotaxis and occurs widely among unicellular organisms. Ciliates and flagellates have definite motile organs. Unlike these organisms, amoeboid cells such as the plasmodia have no solid structures for motility; the motile apparatus is composed of microfilaments of actomyosin which are in a continuous birth-and-death cycle. Furthermore, there exists strong coupling between the membrane and motile systems in amoeboid motility. Important roles of structural change of the surface membrane both in regulating the rhythmic contractility and in recognizing stimulus chemicals in the plasmodium of *Physarum polycephalum** will be made clear.

II. THE EXPERIMENT

A. Measurement of Chemotaxis at the Behavioral Level

The plasmodium in a starved state migrates randomly in the absence of external stimuli on agar gel. Imposition of a concentration gradient of a chemical stimulus in agar gel leads to a directional movement or chemotaxis. For example,

---

*The plasmodium *P. polycephalum* generally used was a Carolina strain provided by Professor N. Kamiya, who obtained it from Professor R. D. Allen. The strain LU 647 X LU 861 was provided by Dr. M. J. Carlile, who obtained it from Dr. J. Dee. The latter strain was derived by Dr. Dee from crossing M_3C (Wisconsin) and Colonia (CL) strains. See Chapter 6, by Dee, and Chapter 8, by Mohberg and Babcock, this volume.

4. Chemotaxis in Plasmodia of *Physarum polycephalum* 113

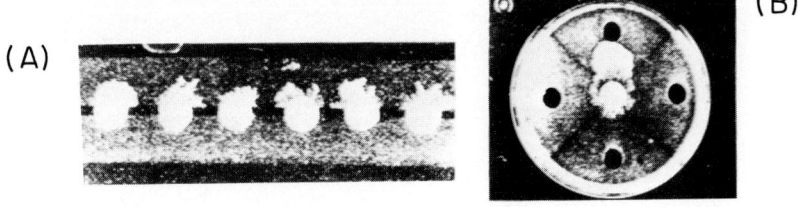

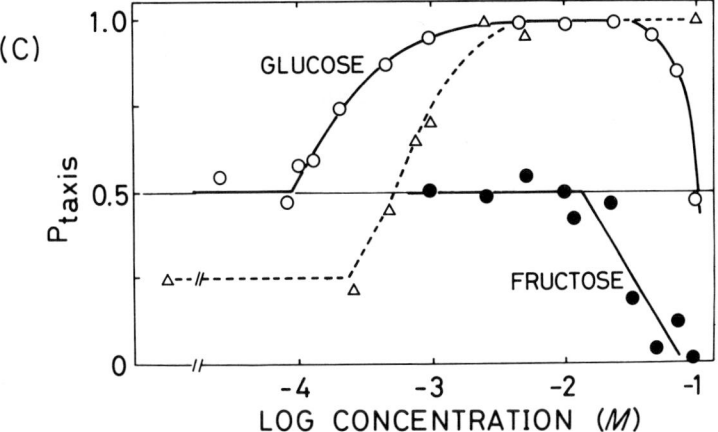

Fig. 1. Measurement of chemotaxis at the behavioral level. (A) The double-strip test showing the response of plasmodia to salt agar plus 10 mM glucose; 4 hours later after onset of the experiment. (B) The quadrant test. (C) The responses of plasmodia to sugars. - - - - - -, quadrant test (P, 0.25 = zero response). Glucose gives a positive response. ———, double-strip test (P, 0.5 = zero response). Glucose gives a positive and fructose a negative response. After Knowles and Carlile (1978). (●) Fructose; (○,△) glucose.

plasmodia are placed between a strip of agar containing the substance to be tested and a control strip lacking the test substances, as shown in Fig. 1A (Knowles and Carlile, 1978). A count is then made of how often the plasmodia are found at the testing site 4 hours later. Thus one can determine the probability P_{taxis} of finding the plasmodia at the testing site. The value of P_{taxis} depends on the concentration of stimulus chemicals. In this arrangement P_{taxis} = 0.5 for random migration, 1.0 for complete attraction, and 0.0 for complete repulsion. This method can be applied to thermotaxis (Tso and Mansour, 1975) and possibly to phototaxis.

One can easily modify the experimental setup to the quadrant test, as shown in Fig. 1B. In this experiment P_{taxis} becomes 0.25 for random selection of the well. Typical results are shown in Fig. 1C, where P_{taxis} in double-strip and quadrant

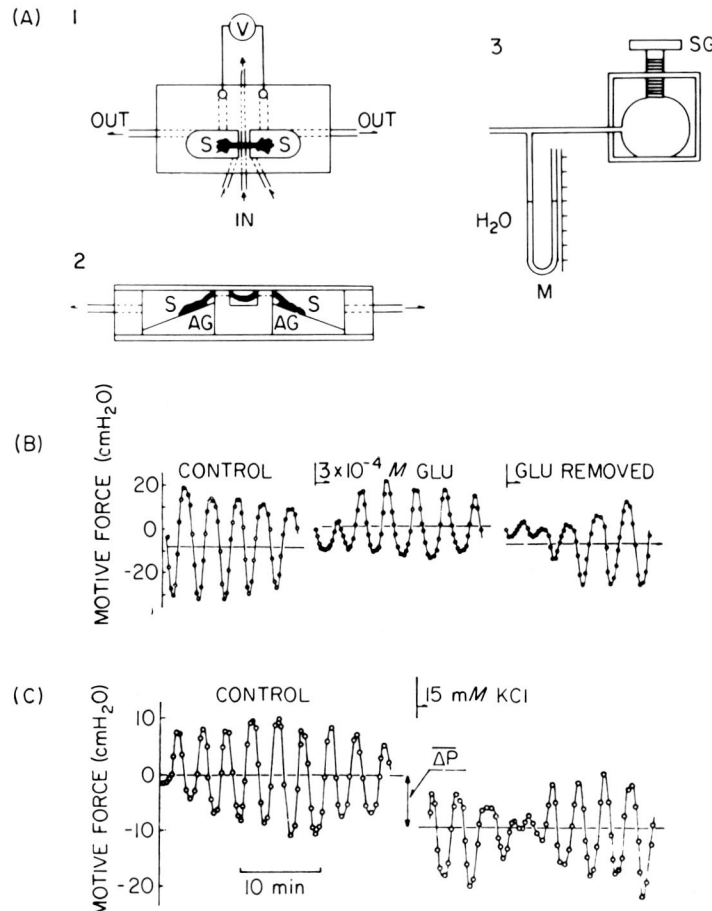

Fig. 2. Schematic diagram illustrating the experimental setup for measuring the membrane potential and the motive force of protoplasmic streaming of slime mold. (A) 1, the double chambers used. S, slime mold; V, potentiometer. Solutions in chambers are exchanged as indicated by arrows. 2, side view of the double chambers. AG, agar gels. 3, apparatus supplying the pressure difference between two compartments. M, U-tube manometer containing H_2O. The pressure difference is controlled by pressing a rubber ball with a screw, SG. (B, C) Dynamoplasmograms. (B) 3×10^{-4} M glucose was applied to one compartment and removed at the times indicated by the arrows. (C) Solution in one compartment was exchanged from water to 15 mM KCl at the arrow indicated in the figure. Motive force of taxis of the plasmodium is the difference in the area-averaged pressure, $\overline{\Delta P}$, as indicated in the figure. After Ueda et al. (1975).

tests is plotted against concentration of carbohydrates. Glucose began to attract the plasmodium at 0.1 mM, whereas the organism moved away from fructose at values higher than 10 mM (Carlile, 1970). One can also use a filter instead of an agar gel (Kincaid and Mansour, 1978a).

4. Chemotaxis in Plasmodia of *Physarum polycephalum*

The quantity P_taxis thus measured is the final output of the initial stimulation. Between the two, there must be many steps. Therefore we need other methods to measure individual steps involved in chemotaxis.

B. Double-Chamber Method

The motive force of protoplasmic streaming and the change in membrane potential can be measured by using the double-chamber method introduced by Kamiya (1942), as depicted in Fig. 2A.

A plasmodium is separated into two portions hydrostatically as well as electrically through a narrow ditch, except for the strand portion at the middle which connects the two sections. The measurement is based on the fact that the protoplasmic streaming in the middle strand can be stopped or counterbalanced by regulating a difference in hydrostatic pressure between the two compartments, thus affording a quantitative measure of the motive force of protoplasmic streaming. Application of a chemical to one compartment caused a deviation of the motive force of protoplasmic streaming toward the attractants or away from the repellents, as shown in Fig. 2B,C. The area-averaged deviation of the motive force $\overline{\Delta P}$* is called the "chemotactic motive force," which drives the protoplasm to move toward or away from the chemical stimulus applied (Ueda *et al.*, 1975).

One can measure accurately the electrical potential difference between the two compartments. For this measurement, the plasmodium was placed on 1% agar gel containing 0.5 mM KCl to provide electrical conductivity in the medium. This concentration of KCl in the external medium did not interfere with the chemoreception of other chemicals by the plasmodia. The potential difference between the two chambers was picked up by a pair of calomel electrodes with salt bridges. Application of chemicals to one compartment caused a potential change when the concentrations exceeded their respective threshold values, and removal of the chemical from the milieu restored the original level of the potential. Potential differences determined by this method agreed with those measured by a microelectrode inserted into the cell, so far as the change in the potential was concerned (Hato *et al.*, 1976).

This method can be extended to small organisms with amoeboid motility, such as *Amoeba proteus* (Kamiya, 1964) and *Acanthamoeba* (Gicquard, 1978), or possibly to leukocytes or macrophages.

C. Measurement of Tension Generation in the Plasmodial Strand

If we dissect a plasmodial strand a few centimeters long and put it onto a tension meter, we can observe periodic tension generation in the strand. The

---

*See Section XII, Glossary of Symbols, for further identification.

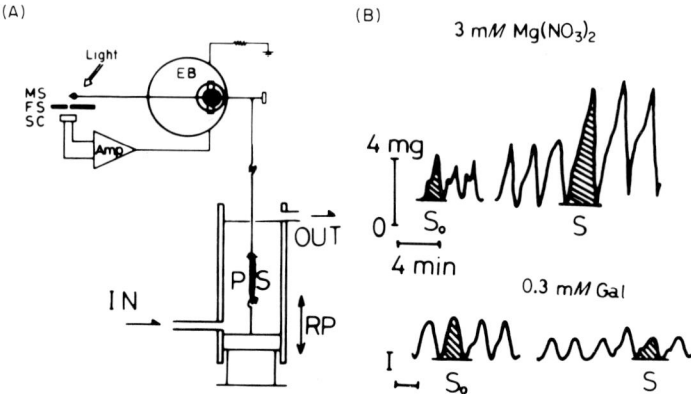

Fig. 3. (A) Schematic illustration of the apparatus used for measuring isometric tension. MS, moving shutter; FS, fixed iris; SC, solar cell; EB, electrobalance; RP, rack-pinion; PS, plasmodium. (B) Periodic changes of isometric tension, and the effects of chemicals on the tension. Quantitative evaluation of the effect is obtained by integration for one period, as shown by the shaded area for a repellent [3 mM Mg(NO$_3$)$_2$] and an attractant (0.3 mM Gal). After Ueda *et al.* (1976).

system of isometric tension may be regarded as a one-dimensional version of ΔP measurement. Thus, the plasmodial strand is a simple and useful experimental system.

Figure 3A illustrates a schematic arrangement which was devised by Kamiya (1970) for measuring isometric tension in the plasmodial strand. As shown in Fig. 3B, isometric tensions were modified by the application of chemicals. The effects may be quantified by taking the area S in the tension curve, as illustrated by the shaded area in the figure. The quantity S is proportional to the chemical energy consumed during contraction in one period. Repellents increased the S value, whereas attractants reduced it (Ueda *et al.*, 1976).

D. Determination of Electrophoretic Mobility

With the use of microplasmodia (50–100 μm in diameter), we can measure the electrophoretic mobility (μ) of a cell by applying cell electrophoresis under various external media. With the help of the Helmholtz–Smoluchowski equation, we can calculate the zeta potential ζ:

$$\zeta = (4\pi\eta/D)\mu \tag{1}$$

where η and D are the viscosity and the dielectric constant of the medium, respectively, and u is the electrophoretic mobility of a cell. The charge density σ at the surface of the membrane is calculated by use of the Gouy–Chapman equation:

4. Chemotaxis in Plasmodia of *Physarum polycephalum*

$$\sigma = \sqrt{\frac{DRT}{500\pi}} \sqrt{I} \cdot \sinh \frac{F\zeta}{2RT} \qquad (2)$$

where I denotes the ionic strength in the medium and the other symbols have their usual thermodynamic meanings.

III. RESPONSE TO CARBOHYDRATES

A. Response of the Plasmodium to Sugars at the Membrane Level

Figure 4A shows the zeta and membrane potentials as a function of log concentration of sugars. Both the membrane potential, $\Delta\psi$, and zeta potential, ζ, remain at a constant level until the concentration of the reagent reaches its threshold, and then they start to change in a positive direction as the concentration becomes higher than the threshold (Hato *et al.* 1975).

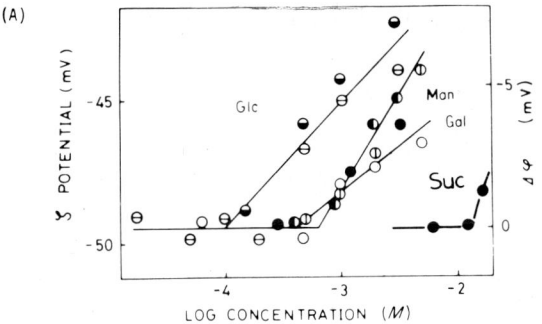

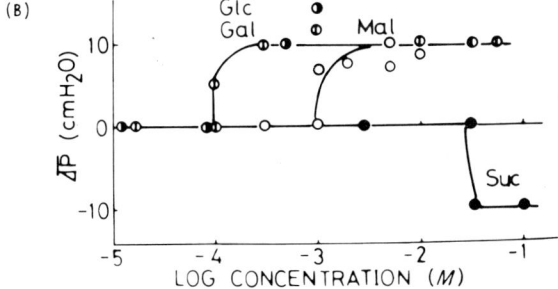

Fig. 4. (A) Zeta potential, ζ, and membrane potential, $\Delta\psi$, of a slime mold measured with the double-chamber method as a function of sugar concentrations in the external media. ζ: ⊖, glucose; ○, galactose; ⊕, mannose. $\Delta\psi$: ◓, glucose; ●, galactose; ◐, mannose; ●, sucrose. Temperature 20°C. (B) Tactic motive force, $\overline{\Delta P}$, as a function of concentration of sugars. ●, sucrose; ○, maltose; ⊕, glucose; ◐, galactose. (A) After Hato *et al.* (1975). (B) After Ueda *et al.* (1976).

These changes in the zeta potential can be attributed to changes in the surface charge density with the help of Eq. (2), because the sugars used are nonelectrolytes and hence the ionic strength in the milieu remains constant. Furthermore, this change in the surface charge density can probably be attributed to the conformational change of the receptor membrane. Thus, for example, in 1 mM glucose solution, 10–15% of the negative net charge is buried in the receptor membrane when a conformational change of the receptor membrane is induced by glucose reception. The gradual change in σ may be interpreted as a successive increase in the number of domains, the conformation of which is changed by sugar reception, with increase of sugar concentration. A similar model can be applied to the reception of hydrophobic substances, salts, nucleotides, etc.

Agreement between the changes of zeta and membrane potentials due to chemoreception, with no appreciable change in membrane resistance for all chemical stimuli applied, suggests that the change in intracellular potential stems mainly from the change in the interfacial potential at the membrane–solution interface rather than from ionic diffusion across the membrane. This concept has been supported by work on various other chemoreceptive membranes and provides an alternative mechanism to ionic diffusion for the generation of membrane potential (Kurihara et al., 1978).

B. Response to Sugars at the Behavioral Level

Figure 4B shows the concentration dependence of changes in chemotactic motive force $\overline{\Delta P}$ for various sugars. $\overline{\Delta P}$ did not change below the threshold concentration, C_{th}, and altered sharply at the C_{th}, with further increase of concentration $\overline{\Delta P}$ remaining at the constant level of $\pm$ 10 cm H_2O. This fact does not imply that the plasmodium cannot distinguish the difference in concentration of chemicals in a region higher than the C_{th}. In fact, the plasmodium generated $\overline{\Delta P} = 10$ cm H_2O when placed between 1 mM and 10 mM glucose solutions.

Comparing the data of $\overline{\Delta P}$ (Fig. 4) and P_{taxis} (Fig. 1), we notice that the thresholds agree with each other, e.g., 0.1 mM for glucose. This coincidence of different cellular activity may be explained as follows: The surface membrane plays the primary role in receiving the external chemical stimuli, and then the sensed information is transduced to motile systems such as pseudopod formation and contractility.

C. Metabolizability and Positive Chemotaxis

Sugars such as glucose, galactose, and maltose support the growth of the plasmodium and are found to be attractants, whereas sugars which are not metabolizable, such as sucrose or fructose, are no longer attractants. Thus we may conclude that there is a strong correlation between metabolizability and

4. Chemotaxis in Plasmodia of *Physarum polycephalum*

chemotactic activity. However, as already demonstrated in bacterial chemotaxis (Adler, 1975), this does not imply that metabolism is essential for chemotaxis. In fact, nonmetabolizable carbohydrates such as 2-deoxyglucose attract the plasmodium.

In addition, only a poor correlation is found between metabolizability and chemotaxis for amino acids (Chet *et al.*, 1977; Kincaid and Mansour, 1978b).

IV. HYDROPHOBICITY AND HETEROGENEITY OF FUNCTIONAL MEMBRANES

A. Response to *n*-Alcohols

Figure 5 shows the concentration dependence of changes in zeta potential of the plasmodia in response to *n*-alcohols with varying alkyl-chain lengths. The zeta potential started to change at their respective thresholds and depolarized almost linearly, with log C above C_{th}. A 10-fold increase of the concentration depolarized the membrane by about 20 mV.

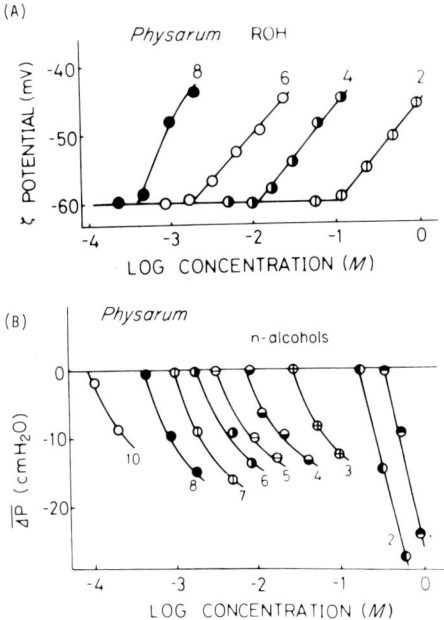

Fig. 5. Effects of *n*-alcohols on (A) membrane and zeta potentials and on (B) chemotactic motive force as a function of concentration. Numbers in the figure indicate the number of carbon atoms in *n*-alcohols. After Ueda and Kobatake (1977b).

Chemotactic behavior did not take place below the C_{th}. Above it, $\overline{\Delta P}$ decreased approximately linearly with log C. This dependence of $\overline{\Delta P}$ on concentration differs from that for sugars and salts and seems to be characteristic of hydrophobic substances (Ueda and Kobatake, 1977a and b).

The threshold in chemotaxis agreed with that of the potential response. These observations imply that the structural changes of the membrane may trigger the chemotactic response as well as changes in charge density at the surface of the membrane.

B. Dependence of the Threshold Concentration on Alkyl-Chain Length

As seen in Fig. 5, the threshold concentration for n-alcohols decreased systematically with increase in the length of the alkyl chain. Similar observations can be made for n-aldehydes and n-fatty acids. Fig. 6 shows the dependence of the threshold on alkyl-chain length when log C_{th} is plotted against the number of carbon atoms (n) in the homologous compounds. The data fall on a straight line for a given functional group at the end of the alkyl chain. Thus we have the following empirical relation:

$$\log C_{th} = -An + B \qquad (3)$$

where A and B are constants depending on the functional end groups involved.

C. Hydrophobicity of the Membranes

We can determine hydrophobicity and sensitivity of the chemoreceptive membranes from the empirical relation of Eq. (3). Assuming that the stimulus chemicals are in absorption equilibrium between membrane and solution phases, we can derive the relationship which relates parameter A in Eq. (3) with a difference in the standard chemical potential of a —CH_2— group, $\Delta\mu^o_{CH_2}$, between the membrane and solution phases:

$$A = -(\Delta\mu^o_{CH_2})/RT \qquad (4)$$

Eq. (4) indicates that parameter A in Eq. (3) is a measure of the hydrophobicity of the membrane in question.

The relationship expressed by Eq. (3) is found to be applicable to chemoreceptive membranes in a variety of organisms from bacteria to higher vertebrates. With the help of Eq. (4), we can compare the hydrophobicity of various biomembranes. The results are summarized in Table I. Hydrophobicity of a biological membrane depends on the species of the end group of the compounds applied. This fact implies that the surface membrane discriminates the end group of the chemical at different receptor domains and that each domain has a different

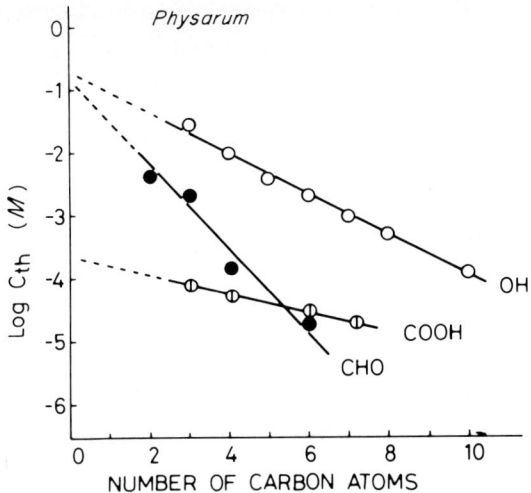

Fig. 6. Dependence of threshold concentration on the length of the alkyl chain. After Ueda and Kobatake (1977b).

hydrophobic character. In this sense, the membrane surface seems heterogeneous in nature.

Hydrophobic interaction plays a role in the reception of odorants other than homologous compounds. Comparison between recognition thresholds of human olfaction T and chemotactic response C_{th} gives the following empirical relationship:

$$\log T = \alpha \log C_{th} + \beta \tag{5}$$

where α and β are constants depending on biological species; α is the ratio of

TABLE I

Hydrophobicity of Various Biological Membranes Determined from Chemoreceptive Threshold of n-Alcohols, Aldehydes, and Fatty Acids[a]

Species	$\Delta\mu^0_{CH_2}$ (OH)	$\Delta\mu^0_{CH_2}$ (CHO)	$\Delta\mu^0_{CH_2}$ (COOH)
Man	0.95 (1–8)		0.10
Rat	0.90 (1–10)		
Blowfly	0.71 (1–10)	0.72 (3–8)	
Nitella	0.87 (3–8)	0.84 (2–9)	0.45 (4–7)
Tetrahymena	0.73 (1–10)	0.73 (2–9)	0.39 (3–10)
Paramecium	0.71 (1–5)		
Physarum	0.45 (3–10)	0.9 (2–6)	0.20 (2–7)

[a] Units in kcal/mole. Numbers in parentheses show the number of methylene groups examined.

hydrophobicity between the human olfactory epithelium and the membranes of lower organisms. Similar parallelism was found in the response to bitter substances such as quinine, strychnine, and phenylthiourea (Ataka et al., 1978).

The fact that the thresholds of olfactory reception and of bitter taste in higher animals parallel each other and the processes of chemoreception and taxis in protozoa indicates that these reception processes are not governed by specific receptor molecules, but that more general characteristics such as the hydrophobicity of the chemoreceptive membrane are responsible for the reception of odor and bitter substances. This notion can also be applied to the actions of anesthetics (Seeman, 1972).

V. ELECTROSTATIC INTERACTIONS IN RELATION TO CHEMORECEPTIVE THRESHOLDS

A. The Schulze–Hardy Rule in Chemoreception

Let us now discuss the electrostatic interactions between the plasmodial membrane and stimulus chemicals. Figure 7 shows the concentration dependence of the membrane potential $\Delta\psi$, of the chemotactic motive force $\overline{\Delta P}$, and of tension generation in the plasmodial strand S/S_0 when the plasmodia were stimulated with various electrolytes. The membrane potential did not change until the concentrations of stimulus chemicals reached their respective C_{th}. Above the C_{th}, $\Delta\psi$ depolarized linearly with log C. Changes in $\Delta\psi$ agreed with those of zeta potential, as described previously. Thresholds determined both from potential and from behavioral measurements agreed with each other.

The threshold concentration decreased systematically with increase in the valency of cation z. We obtained a linear relationship with a slope of -6 when log C_{th} was plotted against log z; hence we have (Ueda et al., 1975)

$$C_{th} = K_H z^{-6} \qquad (6)$$

This is a well-known relationship in the field of colloid science (the Schulze–Hardy rule), and K_H is referred to as the "Hamaker constant."

For the plasmodia of *P. polycephalum* used extensively in Japanese studies, the value of K_H is 3 mM for all polyvalent cations examined, whereas a white plasmodium (strain LU 647 × LU 861) has two distinct K_H values: 3 mM for Ca, La, and Th (group I cations) and 0.2 mM for Na, K, Mg, Mn, and Al (group II cations) (Ueda and Kobatake, 1979).

According to the theoretical interpretation of the Schulze–Hardy rule, the instability of lyophobic colloids appears to be a result of competition between the repulsive force due to the electrical double layer and the attractive force of the van der Waals interaction. However, the structural similarity between colloidal systems and the slime mold should not be overstressed.

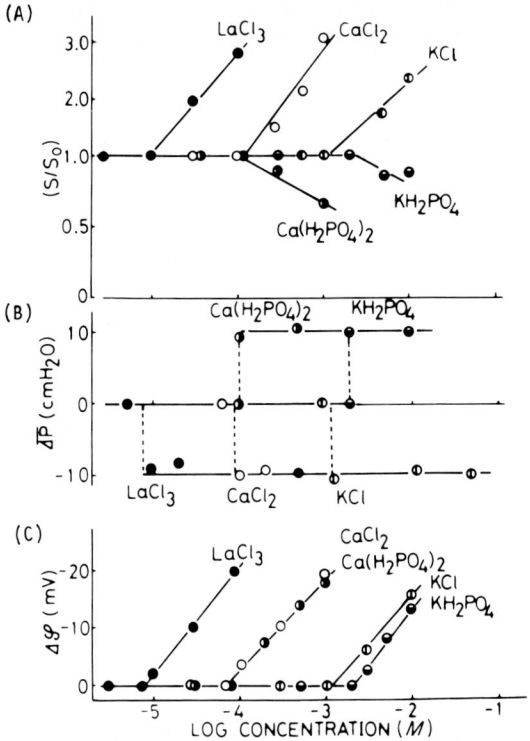

Fig. 7. (A) Tension generation, S/S_0, (B) chemotactic motive force, $\overline{\Delta P}$, and (C) the membrane potential, $\Delta \varphi$, as a function of salt concentration. ●, $LaCl_3$; ○, $CaCl_2$; ◐, $Ca(H_2PO_4)_2$; ⊕, KCl; ◔, KH_2PO_4. After Ueda et al. (1976).

B. Water Structure Around the Surface Membrane

Figure 8 shows the dependence of log C_{th} on the lyotropic number of the anions, N. The lyotropic numbers of the anions were taken from the literature (Voet, 1939). Figure 8 illustrates that for various salts containing a common cation and different anions, the plot of log C_{th} versus N of the anion is linear, that is

$$\log C_{th} = -aN + b \tag{7}$$

in which a and b are constants depending on cation species and environment (Terayama et al., 1977a). We have determined that the parameter a decreases for the following sequence of cations: $H^+ > Li^+ > K^+ > Na^+ > Rb^+ > Cs^+ > NH_4^+$. However, changes in the environment such as alterations in the chemical composition of the external solution can alter the cation sequence in various ways (Terayama et al., 1977b).

For example, the presence of sugars interferes with the recognition thresholds for various salts when the sugar concentration is increased above its threshold

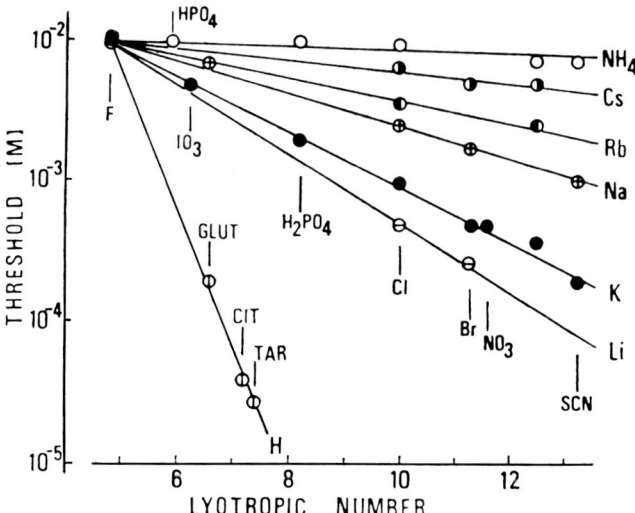

Fig. 8. Relation between the threshold concentration and the lyotropic number of anions. Cation species are indicated in the figure at the right of the respective lines. After Ueda et al. (1975). Glutamate (GLUT), citrate (CIT), tartarate (TAR), thiocyanate (SCN), hydrogen ion (H).

concentration. A double logarithmic plot of C_{th} for univalent salts and sugar concentration, [SUG], gave different straight lines for different sugars:

$$\log C_{th} = a^* \log [SUG] + \text{constant}$$
$$(\text{for } [SUG] > C_{th, \text{sugar}}) \quad (8)$$

The parameter a^* depends on the species of monovalent cations.

Comparison of the parameters a^* and a reveals a close correlation between them. Figure 9 shows the plot of a^* ($= \delta \log C_{th}/\delta \log [SUG]$) against a ($= \delta \log C_{th}/\delta N$) for various kinds of sugars and univalent cations.

The lyotropic number is a parameter which is closely related to the water structure. The correlation shown in Fig. 9 indicates that interference between salt and sugar reception takes place through the water structure around the membrane. Recent physicochemical studies indicated that sucrose and glucose tend to increase the structure of water and ribose acts to disrupt it. These results are consistent with the results shown in Fig. 9.

C. Fluorescence Analysis of Anilinonaphthalene Sulfonate in Chemoreception

We consider anilinonaphthalene sulfonate (ANS) salts as chemical stimuli which fluoresce only in hydrophobic environments such as the plasmodial membrane. With use of the white plasmodium (LU 647 × LU 861), we can apply fluorescence techniques to the study of the molecular processes of chemorecep-

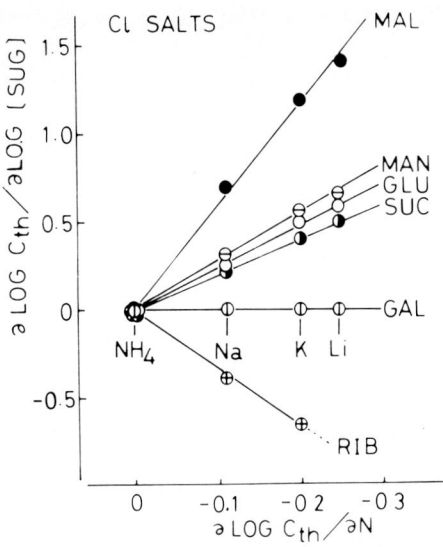

Fig. 9. Linear relations between $\delta \log C_{th}/\delta \log [\text{SUG}]$ and $\delta \log C_{th}/\delta N$ for chloride salts of NH_4^+, Na^+, K^+, and Li^+. The sugar species in media are shown in the figure. After Terayama *et al.* (1977a). ● Maltose, ⊖ mannose, ○ glucose, ◐ sucrose, ⊕ galactose, ⊕ ribose.

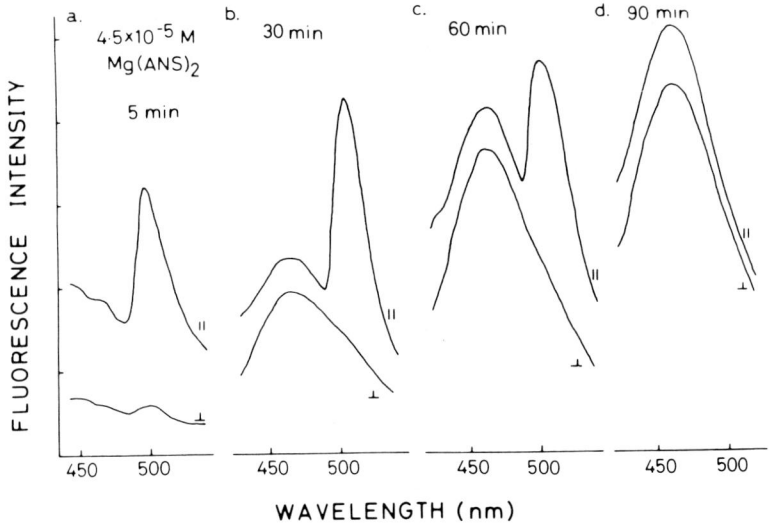

Fig. 10. Time course of the spectral pattern of ANS fluorescence after immersion of the plasmodia in $4.5 \times 10^{-5} M$ Mg(ANS)$_2$. ANS was excited by a vertically polarized light at 360 nm. ∥ and ⊥ indicate parallel and vertical components of fluorescence intensity with respect to incident light. After Ueda and Kobatake (1979).

tion on the plasmoidial membrane (Ueda and Kobatake, 1979).

The frontal region of the migrating plasmodium was cut together with agar gel, placed in a chamber made of black Lucite, and subjected to fluorometric measurements. Thus we measured the fluorescent light emitted from the surface of the plasmodium. Figure 10 shows changes in the spectral pattern of ANS fluorescence with time when 4.5×10^{-5} M Mg(ANS)$_2$ was applied to the white plasmodium. Just after the application of ANS salts, the fluorescence spectra exhibited a maximum at 500 nm, but as time went on, the fluorescence at 460 nm grew larger and larger and at the final stage became the major emission. Application of Mg(ANS)$_2$ below C_{th} ($= 1 \times 10^{-5}$ M) induced no time variation in the spectral pattern, and fluoresced at 500 nm, as shown in Fig. 10a. Similar variation in the spectral pattern with time was observed with NaANS applied to the white plasmodium. In this case, the threshold appeared at 1×10^{-4} M, which agreed with the C_{th} determined from the membrane potential response (see Fig. 11).

D. Nature of the Structural Change in the Plasmodial Membrane Effected by Chemoreception

According to physicochemical studies of ANS fluorescence, the maximum emission of ANS is located at 500–520 nm in polar environments and shifts to 460 nm in hydrophobic media. Based on these characteristics of ANS, the data shown in Fig. 10 are interpreted as follows. Below C_{th}, the surface membrane (plasma membrane) of the plasmodium is predominantly hydrophilic, and hence penetration or binding of ANS molecules at the membrane surface is prevented. On the other hand, the hydrophobic portion of the membrane having access to ANS increases discontinuously above C_{th}. This picture is consistent with data on the adhesive properties of the plasmodia, i.e., adhesion of the plasmodia to a hydrophobic substrate such as a polypropylene surface increases above C_{th} for each chemical added to the media (Ishida et al., 1977).

Data shown in Fig. 10 are also analyzed in terms of the fluorescence polarization, p, which is a parameter of membrane fluidity of the binding sites of ANS on the membrane. The parameter p is defined as follows:

$$p = (I_\| - I_\perp)/(I_\| + I_\perp) \tag{9}$$

where $I_\|$ and $I_\perp$ stand for the parallel and perpendicular components of the fluorescence intensity with respect to the incident light.

In Fig. 11, a comparison is made between changes in the membrane potential and changes in the fluorescence polarization p at 460 nm when Mg(ANS)$_2$ or NaANS was applied to the white plasmodium. The membrane potential changed similarly for chlorides and depolarized linearly with log C above the respective C_{th}. Contrary to the case of the membrane potential, p changed in an all-or-

4. Chemotaxis in Plasmodia of *Physarum polycephalum*

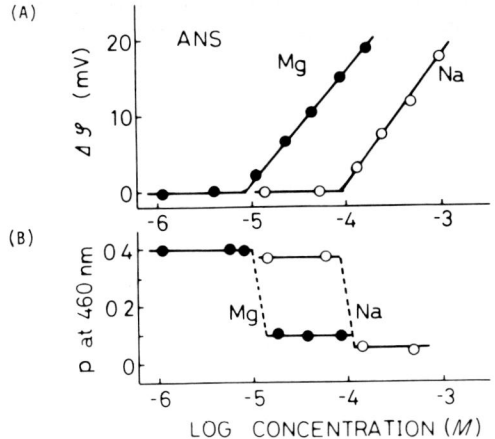

Fig. 11. Dependences of (A) membrane potential and (B) fluorescence polarization at 460 nm on the concentration of ANS salts. ○, NaANS; ●, Mg(ANS)$_2$. After Ueda and Kobatake (1979).

nothing manner at the threshold. The value of p decreased from 0.4 to 0.0–0.1 for both Mg and Na salts of ANS.

Fluorescence polarization p is related to the rotational motion of the chromophore by the well-known Perrin's equation:

$$(1/p - 1/3) = (1/p_o - 1/3)(1 + 3\tau/\rho) \qquad (10)$$

where p_o is a constant, τ the fluorescent life time, and ρ the rotational relaxation time of the solute. Here ρ is proportional to the microviscosity of the medium surrounding the chromophore. Assuming that τ stays constant, we may attribute the decrease in p to a lowering of microviscosity of the membrane. Thus, the data shown in Fig. 11 indicate that the chemoreceptive membrane becomes more fluid discontinuously on reception of Mg and Na salts.

In white plasmodia, the addition of Ca, La, or Th salts caused an increase in p, whereas the addition of NH$_4$, K, and Mg decreased p for extrinsic ANS. This classified cations into two groups: group I (Ca, La, and Th) and group II (Na, K, NH$_4$, Mg, and Al). This classification agrees with that based on the Schulze–Hardy rule, as shown in Section V,A.

VI. CORRELATION BETWEEN TENSION GENERATION AND TACTIC MOVEMENT

As seen in Fig. 7A, the relative tension of the plasmodial strand S/S_o decreased when an attractant [Ca(H$_2$PO$_4$)$_2$, NaH$_2$PO$_4$, etc.] was applied to the plasmodia, whereas S/S_o increased with the reception of a repellent [CaCl$_2$, NaCl, etc.]. With increase in the stimulus concentration, S/S_o did not change until the threshold C_{th} was reached and either decreased or increased gradually above C_{th}. On the contrary, the chemotactic motive force $\overline{\Delta P}$ changed at C_{th} in an almost all-or-nothing manner by $+10$ cm H_2O for attractants and by -10 cm H_2O for repellents. Similar changes in $\overline{\Delta P}$ and S/S_o were observed for sugars and nucleotides.

Comparison between S/S_o and $\overline{\Delta P}$ provides a possible explanation for the mechanisms of chemotactic movement in the plasmodium, even though the relation between the two is nonlinear. Local application of an attractant reduces the tension at that portion, which produces a difference in internal hydrostatic pressures, pushing the plasmasol to move toward that portion. Thus, the protoplasm accumulates there, or a positive taxis occurs. In the case of a repellent, the reverse mechanism works to push the plasmasol away from that portion. A similar mechanism is applicable to phototactic movement in *Physarum* (Hato *et al.*, 1976).

VII. MANIPULATION OF INTRACELLULAR COMPONENTS IN THE PLASMODIAL STRANDS

Making use of the fact that the endoplasm in a plasmodial vein flows passively according to the gradient of the hydrostatic pressure in an ectoplasmic tube, we can replace the endoplasm with an appropriate artificial solution without loss of contractility. Thus we can study the direct effect of chemicals in the internal fluid on the contractility of the plasmodial strand without stimulating the membrane. By this method, we can suggest the probable identity of the modulators of contractility in the natural state.

A. Replacement of Endoplasm with Artificial Media

Figure 12 shows the experimental arrangement for measuring the tension of the plasmodial strand before and after the injection of artificial solution. The crucial features are these: (1) the injection should be performed so rapidly as to be completed before the plasmasol turns to gel, and (2) the applied pressure should not be so strong as to collapse the ectoplasmic tube (Ueda *et al.*, 1978). We can replace plasmasol with an artificial solution more than 3 cm long along the vein without blowing up the strand in the radial direction because of the viscoelastic properties of the ectoplasmic wall. Figures 12B and C show the cross sections of a plasmodial strand before and after the injection of an aqueous solution. We can see that the predominant part of the endoplasm has been replaced with an artificial solution by this operation. The remaining ectoplasmic tube retains the contractile activity.

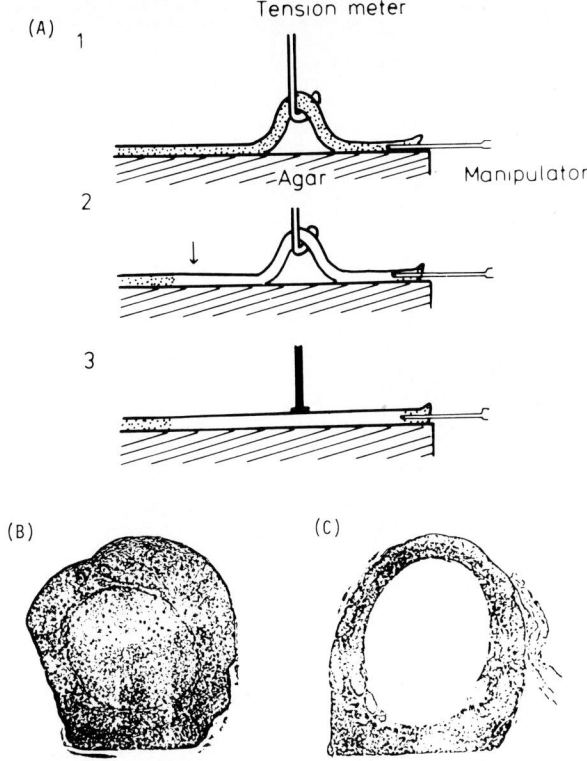

Fig. 12. (A) Measurement of the contraction behavior of a plasmodial strand (1) before and (2, 3) after the injection of chemicals. (1, 2) Longitudinal measurement of tension; (3) radial measurement of the tension. (B) Cross section (optical micrograph) of a plasmodial strand before and after (C) injection of artificial solution. After Ueda *et al.* (1978).

B. Effects of Injected Chemicals on Tension Generation

Figure 13A shows the time courses of tension generation before and after the injection of various solutions. An addition of ATP to Hinssen's solution (30 mM KCl, 3 mM MgCl$_2$, 1 mM CaCl$_2$, 10 mM Tris-HCl, pH 7.1) elicited transient contraction.

Figure 13B shows the dependence of the tension on the concentration of injected ATP. From the figure, we can see that optimal contraction was obtained at about 0.2 mM ATP. Of course, we may expect that a living cell consumes its chemical energy in the most effective manner possible. We therefore conclude that the intracellular ATP level is maintained at about the optimal concentration of 0.2 mM. In a subsequent study, the ATP concentration was fixed at this level.

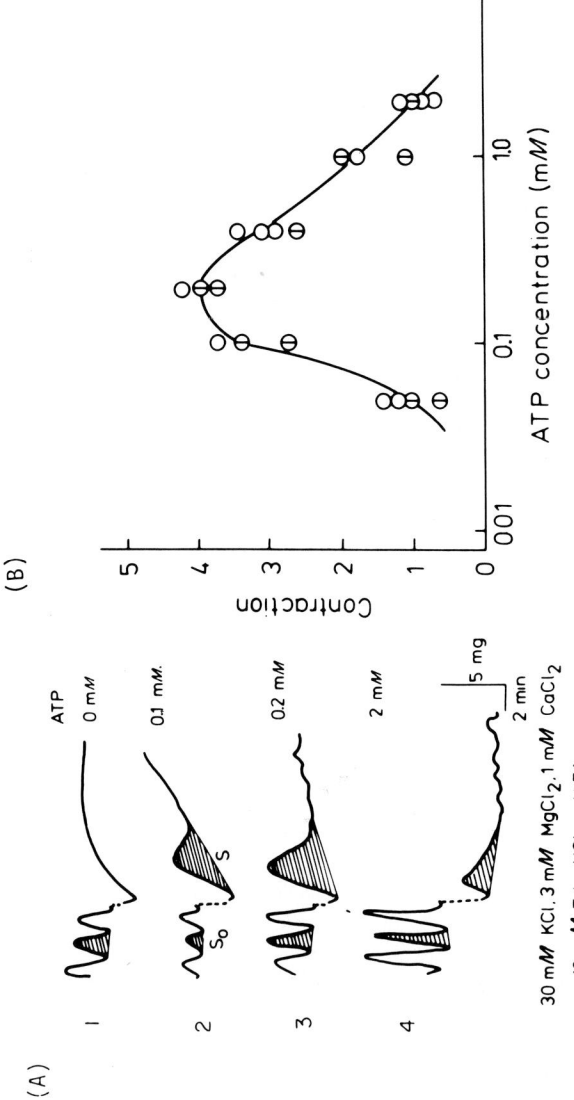

Fig. 13. (A) Time course of the isometric tension development before and after the injection of different ATP concentrations. The basal solution is indicated in the figure. ATP injection at break in curve. (B) Concentration dependence of ATP-induced contraction *in vivo*. ⊖, longitudinal; ○, radial measurements of contractions. Contraction was evaluated by a shaded area, S, which determined the area surrounded by a curve and a line connecting the nearest minimum. After Ueda *et al.* (1978).

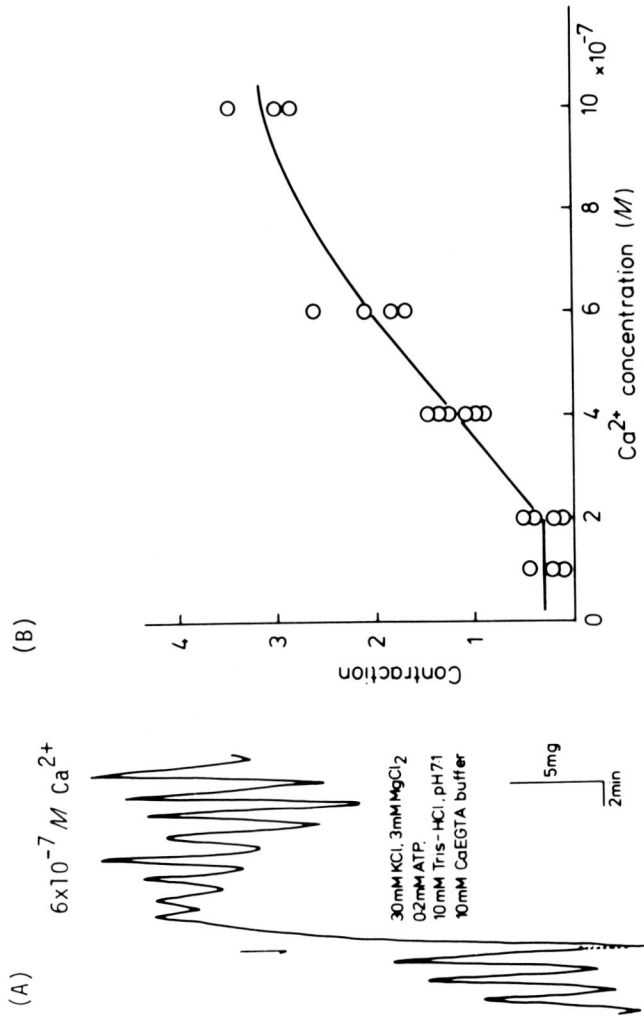

Fig. 14. (A) Effect of internally applied Ca^{2+} on the contractility of plasmodial strands. The composition of the perfused solution is indicated in the figure. (B) Dependence of contraction activity on free Ca^{2+} concentration. Contraction is relative to the amplitude of control oscillation. The free Ca^{2+} level was maintained by a Ca–EGTA buffer. After Ueda *et al.* (1978).

Lowering the Ca^{2+} level, e.g., a solution containing 30 mM KCl, 3 mM $MgCl_2$, 10 mM Tris-HCl, pH 7.1, 6×10^{-7} M Ca^{2+}, 10 mM EGTA-Ca buffer, did not disturb the contractile activity appreciably, though slight uniform contraction took place, as shown in Fig. 14A.

Figure 14B illustrates concentration dependence of injected Ca^{2+} on the contractility of the plasmodial strand. Contraction took place above 2×10^{-7} M Ca^{2+}. If we could inject an appropriate physiological solution compatible with the protoplasm into the plasmodia, tension generation should continue without any change in the contractility. By this method, we were able to estimate that the intracellular free Ca^{2+} level is approximately 10^{-6} M in the plasmodium of *Physarum*.

The effects of injected ATP and Ca^{2+} on tension generation may be compared to those of the ATPase activity of plasmodial actomyosin *in vitro*. Activity of the ATPase is known to become optimal at 0.2 mM of ATP in the presence of Ca^{2+}, and its threshold is about 1×10^{-7} M Ca^{2+}.

The effects of injected high-salt solutions paralleled those on the solubility of myosin B. Plasmodial myosin B is soluble in salt solutions of a few hundred millimolars. Solubilizing power exhibits cation specificity in the following order: K > Cs > Li > Na. Similar results are obtained for tension decrease (Ueda *et al.*, 1980).

All these results indicate that the injected chemicals act directly on the motile system of the strand. Thus this method is useful in studying *in vivo* the regulating factors of contractility.

VIII. INTRACELLULAR CONTROL OF CHEMOTAXIS

A. Intracellular ATP as a Modulator of Contractility in Chemotaxis

By use of an injection technique we can find, as shown above, what chemicals affect the contractility of plasmodia. However, this information is not sufficient. We should add analytical results: A true modulator functioning in the chemotactic response should change its concentration during the response.

Figure 15 shows the response of the intracellular concentration of ATP to stimulus chemicals after the plasmodia were dipped in various chemical solutions for 15 minutes. ATP was determined by the luciferin-luciferase method. The ATP concentration remained at a constant level and started to increase above the respective thresholds. In a strain of LU 647 × LU 861 with about 10 times the sensitivity to the same chemicals, the increase in ATP level took place at onetenth the concentrations required in the wild type. These results show a close parallel with those of membrane and zeta potentials, fluorescence, tension generation, and chemotactic behavior, as shown previously. This fact indicates that ATP works as a regulator of contractility in the chemotactic response of the plasmodium.

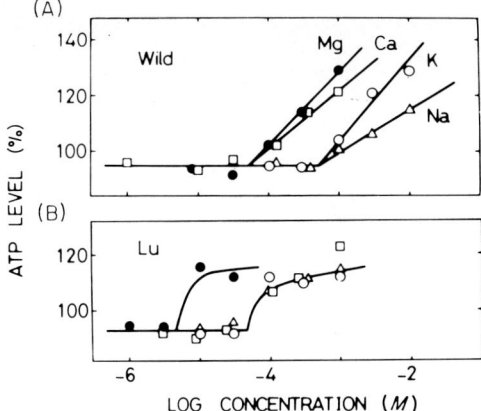

Fig. 15. Intracellular ATP level of (A) a wild-type and (B) a mutant white plasmodium as a function of chemical stimuli applied. ●, $MgCl_2$; □, $CaCl_2$; ○, KCl; △, NaCl. The ATP level of the plasmodia dipped in H_2O, about 2 mM, was taken as 100%. After T. Hirose, T. Ueda, and Y. Kobatake, unpublished data.

Table II summarizes the results of changes in the intracellular ATP level when the plasmodium was stimulated by various chemicals. In terms of behavior and effects on ATP level, stimulus chemicals are classified into three groups. Group I (repellents which increase the ATP level) contains electrolytes. Hydrophobic substances belong to group II (repellents which decrease or cause no change in ATP level). Group III (attractants) induced no change in ATP level in a short time. Metabolizable carbohydrates such as glucose increased the ATP level after 60 minutes.

The fact that the repellents are subdivided into two based on ATP change may result from different interaction of stimulus chemicals with the membrane. The sensed information may be transduced to motile systems through different pathways. One depends on ATP; the other is independent of it. This assumes that the contraction takes place with or without ATP.

B. Intracellular Mediators in the Chemotactic Response

Several inhibitors of the enzyme cyclic 3′,5′-AMP phosphodiesterase such as theophylline and IBMX were found to be attractants in the plasmodium of *P. polycephalum*, suggesting the possible involvement of cAMP in the chemotactic response (Kincaid and Mansour, 1979). In the cellular slime mold *Dictyostelium discoideum*, intracellular cGMP increased as a result of external stimulation with the attractant cAMP (Mato and Konijn, 1977). Thus, cyclic nucleotides might play a role as a second messenger inside a cell, as in many other receptor cells. However, Ca^{2+} also takes part in these cases. Positive chemotaxis in the plasmodium was converted to a negative one by Ca^{2+}-ionophore treatment (Satoh *et al.*, 1980), and Ca^{2+} outflow was enhanced in *D. discoideum* (Chi and Francis, 1971). Also, positive chemotaxis in leukocytes was correlated with the release of

TABLE II

Intracellular ATP Level in the Presence of Various Chemicals[a]

Chemicals	Concentration (mM)	Changes in ATP level (%)	Comments
KCl	20.0	154	Repellents
NaCl	20.0	132	
CaCl$_2$	3.0	151	Electrolytes
MgCl$_2$	3.0	130	
Ethanol	400.0	96	Repellents
Actealdehyde	10.0	100	
Butyric acid	1.0	113	Hydrophobic
Quinine	1.0	74	substances
Picrate	3.0	108	
Glucose	3.0	97	Attractants
2-Deoxyglucose	3.0	90	
Maltose	10.0	109	
Alanine	3.0	113	
Phenylalanine	10.0	86	
KH$_2$PO$_4$	10.0	111	
H$_2$O		100	

[a] The ATP level of the plasmodia dipped in H$_2$O, about 2 mM, it was taken as 100%. Temperature, 22°C.

Ca^{2+} (Gallin and Rosenthal, 1974; Boucek and Snyderman, 1976). Thus we have to determine the sequence of events in sensory transduction in the chemotactic response. Intracellular pH is another possible mediator, as shown in cell division of sea urchin eggs. Now that it is possible to measure intracellular pH in amoeboid cells (Heiple and Taylor, 1980), we will soon know the role of pH. We have emphasized the mechanism of chemotaxis in the plasmodium of *P. polycephalum*. General reviews of chemotactic responses in microorganisms are those of Perez-Miravete (1973), Carlile (1975), and Levandowsky and Hauser (1978).

IX. RHYTHMICITY OF THE PLASMODIAL CONTRACTILE SYSTEM

A. Rhythm in Tension Generation and Protoplasmic Streaming

Plasmodia of *Physarum* exhibit regular back-and-forth streaming of the protoplasm and generate tension periodically. This contraction rhythm depends not

4. Chemotaxis in Plasmodia of *Physarum polycephalum*

only on external factors, such as temperature and oxygen concentration, but also on internal factors, such as the amount of the protoplasm. In fact, Vouk (1910) reported that the contraction rhythm became slower as the size (volume) of the plasmodium increases.

Figure 16 shows the relationship between the period of isometric tension and the amplitude F of the oscillation. Here the amplitude was varied by using plasmodial strands of various thicknesses. For a small oscillation, the period starts from τ_0 and becomes slower as the amplitude increases until a certain value F_c is attained. The period remains constant for values larger than F_c. Thus we get the following rhythmicity–contractility relationship (Ueda and Kobatake, 1980):

$$\begin{aligned} \tau &= aF + \tau_0 \quad &\text{(for } F < F_c\text{)} \\ &= \tau_s \quad &\text{(for } F > F_c\text{)} \end{aligned} \qquad (11)$$

where a, τ_0, and τ_s are constants independent of F but dependent on environmental factors.

A similar relationship holds between the period and the amplitude ΔP of the motive force of protoplasmic streaming.

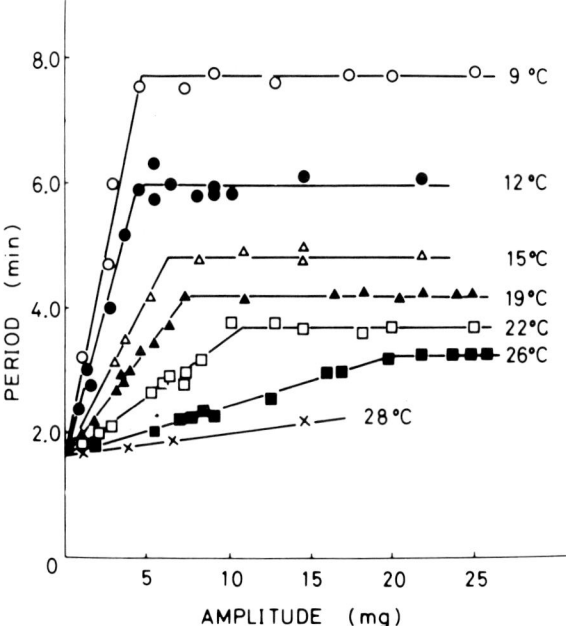

Fig. 16. Relationship between period (τ) and amplitude (F) of isometric tension of a plasmodial strand at various temperatures. Plasmodial strands with various thicknesses (0.07 ~ 1.0 mm) were immersed in distilled water at different temperatures. The period was measured after stationary oscillation was attained. After Ueda and Kobatake (1980).

$$\tau = A\Delta P + \tau_o \tag{12}$$

By comparing the size dependence in tension and the motive force generation, as expressed in Eqs. (11) and (12), we have

$$\Delta P = (a/A)F \tag{13}$$

Equation (13) states that the generation of the motive force is the result of contractility, and that the basic contractile mechanism may be similar in the plasmodial strand and in the frontal region, the morphological differences notwithstanding.

The empirical relations of Eqs. (11) and (12) indicate that the period of the contraction rhythm is composed of at least three components. As seen from Fig. 16, τ_o is independent of temperature and gives a value of 1.6 minutes in water. Measurement of the contraction rhythm in air determines the τ_o value as 1.0 minute. Thus τ_o implies a limit cycle of a series of chemical reactions which support the contraction rhythm. These chemical reactions might involve mitochondria.

External chemical stimuli differently affected the period. Ethanol slowed down the periodicity through parameter a, whereas KCl accelerated it by affecting parameter a. Heavy water slowed the rhythm while maintaining τ_s at a constant level. Chemicals such as NaCl, CaCl$_2$, MgCl$_2$, and glucose did not change the τ-F relationship.

In the presence of attractants, the period of the contraction rhythm becomes faster. This does not mean that the periodicity was a cause of tactic movement. As we know from Eqs. (11) or (12), periodicity and contractility are interrelated.

X. EFFECTS OF TEMPERATURE ON MEMBRANE ACTIVITIES

Temperature has a great influence on cellular activities. In the following subsections, we will show some effects which may result from the thermal transition of membrane structure.

A. Effects of Temperature on Chemoreception

As seen in Fig. 7, electrical responses to stimulus chemicals are expressed by

$$\Delta \phi = R \log (C/C_{th}) \quad \text{(for } C > C_{th}) \tag{14}$$

Parameters R and C_{th} vary differently with temperature. Threshold C_{th} depended on temperature systematically, as expressed by

$$\log C_{th} = \Delta H_{app}/RT + \text{constant} \tag{15}$$

where ΔH_{app} is determined to be -12 kcal/mol for all chemicals studied when distilled water is taken as the reference solution (Ueda et al., 1975).

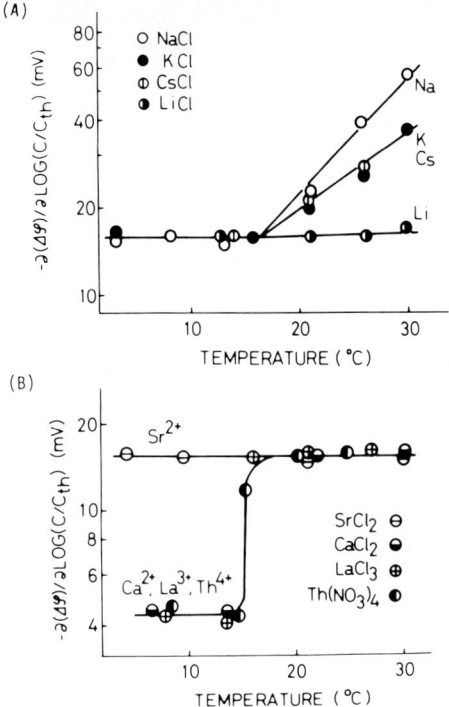

Fig. 17. (A, B) Temperature dependence of the magnitude of the potential response $R = -\delta\Delta\psi/\delta \log(C/C_{th})$. Symbols are indicated in the figure. R is calculated from similar data in Fig. 7C or Fig. 11A. Chemicals were dissolved in distilled water. After Ueda and Kobatake (1978).

On the other hand, parameter R showed a sharp and discontinuous change with temperature variation. Figure 17 shows a temperature dependence of R for various electrolytes. For monovalent salts, R remained -18 mV/decade below $T_c = 15°C$ and increased gradually above T_c. NaCl induced the largest response at higher temperatures. For polyvalent salts such as Ca and Th, lowering the temperature abruptly diminished the response of the membrane potential at $T_c = 15°C$. Similar results were obtained for the zeta potential. The fact that responses change abruptly at a certain critical temperature is ascribed to the location of the reactions at the surface membrane (Ueda and Kobatake, 1978).

B. Effects of Temperature on Pseudopod Formation and Periodicity

Fig. 18A shows the Arrhenius plot of the period of isometric tension at the stationary phase τ_s. The data are taken from Fig. 16. Data points fall on a straight line, and an inflection takes place at $T_c = 15°C$. The apparent enthalpy changes

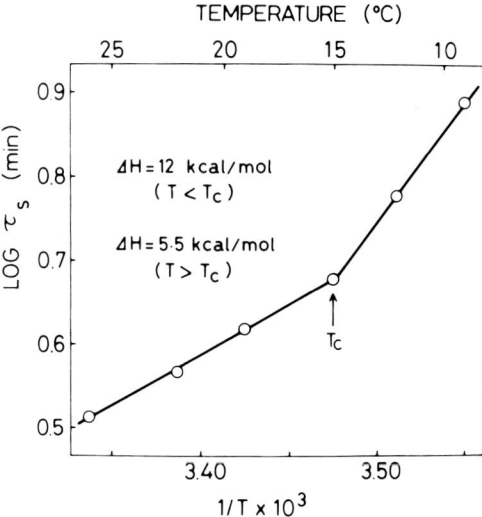

Fig. 18. The Arrhenius plot of the period of isometric tension, τ_s, at the stationary state. The values of τ_s are averages of τ which are independent of F, as shown in Fig. 16. After Ueda et al. (1980).

were found to be 12 kcal/mol below T_c and 5.5 kcal/mol above T_c (Ueda et al., 1980).

Figure 19A shows the migration velocity of the plasmodium as a function of temperature. For the sake of comparison, the temperature dependence of the amplitude of the motive force of protoplasmic streaming was plotted in the same figure. The motive force showed a slight increase as a consequence of temperature decrease. Migration velocity, on the contrary, decreased sharply at T_c = 15°C. Slow migration may thus be attributed to a reduction of pseudopod formation since the motive force did not decrease.

Figure 19B shows the temperature variation of intracellular ATP level in the plasmodia. The organism was adapted at the respective temperatures for about an hour. ATP content decreased sharply at 15°C by one-half as a result of temperature decrease (Ueda et al., 1980).

All these results show that the plasmodia of *P. polycephalum* have two distinct states depending on temperature, the critical temperature being about 15°C. These phenomena provide examples, in addition to chemotaxis, demonstrating strong coupling between membrane and cytoplasmic activities.

XI. SUMMARY

1. Methods useful for the study of chemotaxis in the plasmodium are described, such as measurements of tactic behavior, motive force of protoplasmic

streaming, zeta and membrane potentials, tension generation, and fluorescence. The results obtained are analyzed in terms of physicochemical principles.

2. Changes in membrane and zeta potentials parallel each other in response to chemical stimuli. The charge density at the membrane surface gradually decreased above recognition thresholds.

3. Hydrophobicity of various chemoreceptive membranes was determined.

4. The recognition threshold for salts is governed by the Schulze–Hardy rule. The plasmodium of a strain LU 647 × LU 861 has a different Hamaker constant, indicating cation specificity.

5. Water structure around the membrane–solution interface is involved in the recognition process that occurred at the surface membrane.

6. Conformation of the surface membrane changes when chemoreception oc-

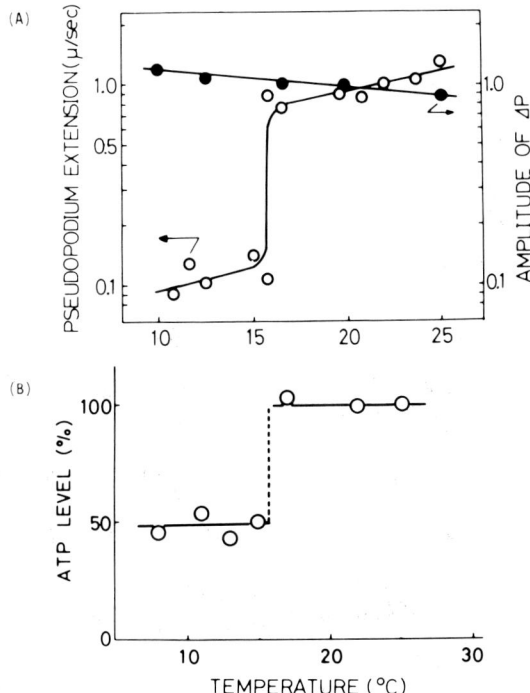

Fig. 19. (A) Temperature dependence of locomotion velocity and amplitude of the motive force of protoplasmic streaming. ○, locomotion velocity; ●, amplitude of ΔP. (B) Intracellular ATP concentration as a function of temperature. The ATP concentration at 22°C was about 2 mM and is taken as 100%. After stimulation with various chemicals, plasmodia were dipped in hot water (∼95°C) for 5 to 10 minutes. After centrifugation, extracted ATP was measured by the luciferin–luciferase method and the protein content was measured by the Lowry method. After Hirose, Ueda and Kobatake, unpublished data (B) and Ueda and Kobatake (1978) (A).

curs. Fluorescence analysis with ANS showed that the surface membrane became hydrophobic and more fluid on exposure to Na and Mg salts.

7. Parallelism between tactic behavior and contractility was demonstrated.

8. Factors modulating contractility were determined by injecting chemicals into the plasmodium and by studying their effects on tension generation.

9. The significance of ATP as a modulator in chemotaxis was demonstrated.

10. The fact that the surface membrane has a transition temperature affects various relevant membrane phenomena, such as chemoreception, locomotion, rhythmicity, and ATP content.

All these results indicate that the structural change of the surface membrane plays important roles in recognizing chemical stimuli as well as in regulating the contraction rhythm of the plasmodium of *P. polycephalum*.

XII. GLOSSARY OF SYMBOLS

$\overline{\Delta P}$, chemotactic motive force defined as the area averaged difference in the motive force of the protoplasmic streaming before and after chemical stimulation

The Helmholtz–Smoluchowski equation:
$$\zeta = (4\pi\eta/D)\, u$$
ζ, zeta potential; η, viscosity of the medium; D, dielectric constant of the medium; u, electrophoretic mobility of a cell.

The Gouy–Chapman equation:
$$\sigma = \sqrt{\frac{DRT}{500\pi}}\, \sqrt{I}\, \sinh\left(\frac{F\zeta}{2RT}\right)$$
σ, charge density at the cell surface; R, gas constant; T, absolute temperature; I, ionic strength of the medium; F, Faraday constant.

Empirical relationship in chemoreception of homologous compounds:
$$\log C_{th} = -An + B$$
C_{th}, threshold concentration; n, number of carbon atoms in homologous compounds; A, B, constants
$$A = -(\Delta\mu^{\circ}_{CH_2})/RT$$
$\Delta\mu^{\circ}_{CH_2}$ = difference in standard chemical potential of a methylene group in solution and membrane phases

Comparison of human olfactory threshold and chemotactic threshold:
$$\log T = \alpha \log C_{th} + \beta$$
T, human olfactory threshold; α, β, constants.

The Schulze–Hardy rule:
$$C_{th} = K_H z^{-6}$$
K_H, Hamaker constant; z, valency of cations.

Dependence of the threshold on the lyotropic number:
$$\log C_{th} = -aN + b$$
N, lyotropic number of anions; a, b, constants.

Perrin equation:
$$(1/p - 1/3) = (1/p_o - 1/3)(1 + 3\tau/\rho)$$
p, fluorescence polarization defined by
$$p = \frac{I_\| - I_\perp}{I_\| + I_\perp}$$

where $I_\|$ and $I_\perp$ are parallel and vertical components, respectively, of fluorescent light intensity relative to incident light;
τ, fluorescence life time; ρ, rotational relaxation time of the solute.

Empirical relationship between period and amplitude in isometric tension:
$\tau = aF + \tau_0$ (for $F < F_c$); τ, τ_s (for $F > F_c$); τ, the period of the contraction rhythm; F, the amplitude of the oscillation; τ_0 and τ_s, constants independent of F; F_c, critical amplitude.

Period–amplitude relationship in the protoplasmic streaming:
$$\tau = A\Delta P + \tau_0$$
τ, period; ΔP, amplitude of the protoplasmic streaming; A, τ_o, constants.

Empirical relationship between electrical response and chemical stimulation:
$$\Delta\phi = R \log(C/C_{th}) \text{ (for } C > C_{th})$$
$\Delta\phi$, difference in membrane potential induced by chemical stimulation; C, concentration of chemical stimuli; C_{th}, threshold concentration; R, parameter indicating magnitude of response.

Arrhenius plots for sensitivity:
$$\log C_{th} = \Delta H_{app}/RT + \text{constant}$$
C_{th}, threshold concentration; ΔH_{app}, apparent enthalpy change; R, gas constant; T, absolute temperature;

ACKNOWLEDGMENT

The authors thank Dr. M. J. Carlile at the Imperial College, London, for reading the chapter.

REFERENCES

Adler, J. (1975). Chemotaxis in bacteria. In "Primitive Sensory and Communication Systems, Taxes and Tropisms of Micro-Organisms and Cells" (M. Carlile, ed.), pp. 91–100. Academic Press, New York.

Ataka, M., Tsuchii, A., Ueda, T., Kurihara, K., and Kobatake, Y. (1978). Comparative studies on the reception of bitter stimuli in the frog, *Tetrahymena*, slime mould and *Nitella*. *Comp. Biochem. Physiol.* **61A,** 109–115.

Boucek, M. N., and Snyderman, R. (1976). Calcium influx requirement for human neutrophil chemotaxis: Inhibition by lanthanum chloride. *Science* **193,** 905-907.

Carlile, M. J. (1970). Nutrition and chemotaxis in the myxomycete *Physarum polycephalum:* The effect of carbohydrates on the plasmodium. *J. Gen. Microbiol.* **63,** 221-226.

Carlile, M. J. (1975). "Primitive Sensory and Communication Systems, Taxes and Tropisms of micro-organisms and cells." Academic Press, New York.

Chet, I., Naveh, A., and Henis, Y. (1977). Chemotaxis of *Physarum polycephalum* towards carbohydrates, amino acids and nucleotides. *J. Gen. Microbiol.* **102,** 145-148.

Chi, Y.-Y., and Francis, D. (1971). Cyclic GMP and calcium exchange in a cellular slime mould. *J. Cell. Physiol.* **77,** 169-174.

Gallin, J. I., and Rosenthal, A. S. (1974). The regulatory role of divalent cations in human granulocyte chemotaxis: Evidence for an association between calcium exchange and microtubule assembly. *J. Cell Biol.* **62,** 594-609.

Gicquaud, C. R. (1978). Technique de mesure de la force motrice de tres petites cellules amiboides. *Rev. Can. Biol.* **37,** 201-208.

Hato, M., Ueda, T., Kurihara, K., and Kobatake, Y. (1975). Changes in zeta potential and membrane potential of slime mold *Physarum polycephalum* in response to chemical stimuli. *Biochim. Biophys. Acta* **426,** 73-80.

Hato, M., Ueda, T., Kurihara, K., and Kobatake, Y. (1976). Phototaxis in true slime mold *Physarum polycephalum. Cell Struct. Funct.* **1,** 269-278.

Heiple, J. M., and Taylor, D. L. (1980). Intracellular pH in single motile cells. *J. Cell Biol.* **86,** 885-890.

Ishida, N., Kurihara, K., Kobatake, Y. (1977). Changes in adhesive properties of slime mold *Physarum polycephalum* accompanied with chemoreception. *Cytobiologie* **15,** 269-274.

Kamiya, N. (1942). Physical aspects of protoplasmic streaming. *In* "The Structure of Protoplasm" (W. Seifriz, ed.), pp. 199-244. Iowa State Univ. Press, Ames, Iowa.

Kamiya, N. (1964). The motive force of endoplasmic streaming in the ameba. *In* "Primitive Motile Systems in Cell Biology" (R. D. Allen and N. Kamiya, eds.), pp. 257-277. Academic Press, New York.

Kamiya, N. (1970). Contractile properties of the plasmodial strand. *Proc. Jpn. Acad.,* **46,** 1026-1031.

Kincaid, R. L., and Mansour, T. E. (1978a). Measurement of chemotaxis in the slime mold *Physarum polycephalum. Exp. Cell Res.* **116,** 365-375.

Kincaid, R. L., and Mansour, T. E. (1978b). Chemotaxis toward carbohydrates and amino acids in *Physarum polycephalum. Exp. Cell Res.* **116,** 377-385.

Kincaid, R. L., and Mansour, T. E. (1979). Cyclic 3',5' AMP phosphodiesterase in *Physarum polycephalum.* I. Chemotaxis toward inhibitors and cyclic nucleotides. *Biochim. Biophys. Acta* **588,** 333-341.

Knowles, D. J. C., and Carlile, M. J. (1978). The chemotactic response of plasmodia of the myxomycete *Physarum polycephalum* to sugars and related compounds. *J. Gen. Microbiol.* **108,** 17-25.

Kurihara, K., Kamo, N., and Kobatake, Y. (1978). Transduction mechanism in chemoreception. *Adv. Biophys.* **10,** 27-95.

Levandowsky, M., and Hauser, D. C. R. (1978). Chemosensory responses of swimming algae and protozoa. *Int. Rev. Cytol.* **53,** 145-210.

Mato, J. M., and Konijn, T. M. (1977). Chemotactic signal and cyclic GMP accumulation in *Dictyostelium. In* "Development and Differentiation in the Cellular Slime Moulds" (P. Capuccinelli and J. M. Ashworth, eds.), pp. 93-103. Elsevier, Amsterdam.

Perez-Miravete, A., ed. (1973). "Behaviour of Microorganisms." Plenum, New York.

Satoh, H., Mito, Y., Ueda, T., Kurihara, K., and Kobatake, Y. (1980). Conversion of positive chemotaxis to negative one by Ca-ionophore treatment in the plasmodium of *Physarum polycephalum*. *Biochim. Biophys. Acta* **633**, 436-443.
Seeman, P. (1972). The membrane actions of anaesthetics and tranquilizers. *Pharmocol. Rev.* **24**, 583.
Terayama, K., Ueda, T., Kurihara, K., and Kobatake, Y. (1977a). Effect of sugars on salt reception in true slime mould *Physarum polycephalum:* Physicochemical interpretation of interaction between salt and sugar reception. *J. Membr. Biol.* **34**, 369-381.
Terayama, K., Kurihara, K., and Kobatake, Y. (1977b). Variation in selectivity of univalent cations in slime mould *Physarum polycephalum* caused by reception of polyvalent cations. *J. Membr. Biol.* **37**, 1-12.
Tso, W., and Mansour, T. E. (1975). Thermotaxis in a slime mould *Physarum polycephalum. J. Behav. Biol.* **14**, 499-504.
Ueda, T., and Götz von Olenhusen, K. (1978). Replacement of endoplasm with artificial media in plasmodial strands of *Physarum polycephalum:* Effects on contractility and morphology. *Exp. Cell Res.* **116**, 55-62.
Ueda, T., and Kobatake, Y. (1977a). Hydrophobicity of biosurfaces as shown by chemoreceptive thresholds in *Tetrahymena, Physarum* and *Nitella. J. Membr. Biol.* **34**, 351-368.
Ueda, T., and Kobatake, Y. (1977b). Changes in membrane potential, zeta potential, and chemotaxis of *Physarum polycephalum* in response to n-alcohols, n-aldehydes and n-fatty acids. *Cytobiologie* **16**, 16-26.
Ueda, T., and Kobatake, Y. (1978). Discontinuous change in membrane activities of plasmodium of *Physarum polycephalum* caused by temperature variation: Effects on chemoreception and amoeboid motility. *Cell Struct. Funct.* **3**, 129-139.
Ueda, T., and Kobatake, Y. (1979). Spectral analysis of fluorescence of 8-anilino-1-naphthalenesulfonate in chemoreception with a white plasmodium of *Physarum polycephalum:* Evidence for conformational change in chemoreceptive membrane. *Biochim. Biophys. Acta* **557**, 199-207.
Ueda, T., and Kobatake, Y. (1980). Contraction rhythm in the plasmodium of *Physarum polycephalum:* Effects of size, temperature, and chemicals on the periodicity. *Eur. J. Cell Biol.* **23**, 37-42.
Ueda, T., Terayama, K., Kurihara, K., and Kobatake, Y. (1975). Threshold phenomena in chemoreception and taxis in slime mould *Physarum polycephalum. J. Gen. Physiol.* **65**, 233-234.
Ueda, T., Muratsugu, M., Kurihara, K., and Kobatake, Y. (1976). Chemotaxis in *Physarum polycephalum:* Effects of chemicals on isometric tension of the plasmodial strand in relation to chemotactic movement. *Exp. Cell. Res.* **100**, 337-344.
Ueda, T., Götz von Olenhusen, K., and Wohlfarth-Bottermann, K. E. (1978). Reaction of the contractile apparatus in *Physarum* to injected Ca^{2+}, ATP, ADP and 5' AMP. *Cytobiologie* **18**, 76-94.
Ueda, T., Hirose, T., and Kobatake, Y. (1980). Membrane biophysics of chemoreception and taxis in the plasmodium of *Physarum polycephalum. Biophys. J.* **11**, 461-473.
Voet, A. (1939). Quantitative lyotropy. *Chem. Rev.* **20**, 169-179.
Vouk, V. (1910). Untersuchungen über die Bewegung der Plasmodien: Die Rhythmik der Protoplasmaströmung. *Sitzungsber. Kais. Akad. Wiss. Wien* **119**, 853-876.

CHAPTER 5

Plasmodial Structure and Motility

DIETRICH KESSLER

I. Introduction 145
   A. The Plasmodium as a Motile System 145
   B. Shuttle Streaming in the Plasmodium of *Physarum* 147
   C. Protoplasmic Streaming: Conclusions from Descriptive Data ... 149
II. Plasmodial Morphology 151
   A. Light Microscopic Observations 151
   B. Ultrastructural Studies 155
III. Mechanism of Protoplasmic Streaming in the Plasmodium 169
   A. Contraction–Hydraulic Mechanism: Actomyosin Contraction in the Ectoplasm Causes Endoplasmic Flow 169
   B. Mechanism of Actomyosin Contraction: A Sliding Filament Model 172
   C. State of Actin Polymerization 175
   D. Calcium Regulation of the Contractile System 179
IV. An Oscillating Regulatory System Controls Shuttle Streaming 185
   A. The Whole Plasmodium Is a Monorhythmic System 185
   B. Local Ectoplasmic Contraction Rhythms Are Coordinated by Endoplasmic Flow 186
   C. Hypothesis: Coordination of Local Contraction Rhythms by Stretch Entrainment 188
   D. Molecular Nature of the Local Oscillator: A Calcium–Cyclic AMP Control Loop? 190
V. A Preliminary Model for the Molecular Mechanism of Shuttle Streaming in the Plasmodium 192
   References 196

I. INTRODUCTION

A. The Plasmodium as a Motile System

The Myxomycetes are a group of free-living protists in which the active vegetative stage of the life cycle, the plasmodium, is a large multinucleate mass of actively moving protoplasm possessing an external slime layer but no cell wall. The size and motile characteristics of the plasmodia vary somewhat among the group (Gray and Alexopoulos, 1968; Alexopoulos, this volume, Chapter 1). The plasmodium of both *Physarum* and *Didymium* belongs to the most common type, the phaneroplasmodium, characterized by its large size (up to a square foot or more!) and the anterioposterior polarity of the migrating plasmodium into a network of interconnected strands or veins at the rear which merge into a fanlike sheet at the front (Fig. 1). Within the strands at the rear, and in channels throughout the fanlike front, the protoplasm is differentiated into a tubelike cortical gel layer, the ectoplasm, and a more fluid inner sol, the endoplasm, which is engaged in rapid shuttle streaming characterized by a rhythmic reversal of direction of flow with a period of about 1.5–3 minutes (Kamiya, 1959; Wohlfarth-Bottermann and Isenberg, 1976). A large, starving plasmodium may migrate at a rate of about 2–3 cm/hr (up to 5 cm/hr for short periods), pausing during the synchronous mitosis (Miller and Anderson, 1966, 1971). In smaller, well-synchronized plasmodia (1.5–2 cm or less in diameter), the velocity of streaming slows markedly or actually stops in all channels throughout the plasmodium in late metaphase (Sachsenmaier *et al.*, 1973; Kessler and Lathwell, 1979). In larger plasmodia, net movement of the plasmodium ceases, although streaming does not completely stop in mitosis (Guttes and Guttes, 1963). These changes of motility may be related to the increase of calcium efflux during metaphase observed by Holmes and Stewart (1977). Plasmodia growing in pure

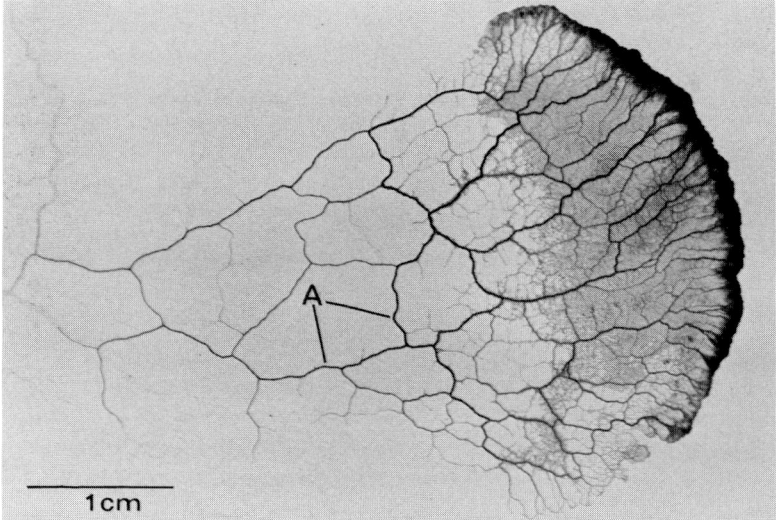

Fig. 1. Plasmodium of *P. confertum*. The plasmodium is organized into a fanlike front with a network of strands or veins in the rear. From Stiemerling (1970).

5. Plasmodial Structure and Motility

culture on the surface of filter paper over semidefined nutrient medium in petri dishes do not migrate with an anterioposterior orientation. They expand with growth over the filter paper equally in all directions, making a broad, flat, circular plasmodial sheet in which the channels of active endoplasmic flow are embedded.

The ease of culture and large size of the surface-grown plasmodium of *P. polycephalum* have made it a convenient subject for biochemical studies of the contractile proteins. Furthermore, observations of the microscopic structure and biophysical analysis of the contractile force may conveniently be performed on the strands, either in the intact plasmodium or by excising individual strands, which are relatively unharmed by this treatment and recover to exhibit the characteristic shuttle-streaming pattern after a short time.

The immense size of a large plasmodium, with its complex array of interconnecting strands, tends to emphasize the uniqueness of the organism. In fact, as Lewis (1942) and Gawlitta *et al.* (1980a) have clearly observed, the movements of the protoplasm in the plasmodium are very similar to, if not identical with, those seen during the movement of large amoebae such as *Amoeba proteus*, with the exception of the characteristic rhythmic reversal of the direction of endoplasmic flow seen in the slime mold. The plasmodium of the myxomycete, then, is a superb example of an organism migrating by amoeboid motion.

B. Shuttle Streaming in the Plasmodium of *Physarum*

In very small plasmodia of *P. polycephalum*, which originated from 1-mm-diameter pieces teased from a larger plasmodium, Lewis (1942) examined the anterioposterior organization of the migrating plasmodium and made some of the following observations. The middle portion of each tiny plasmodium is a single strand or vein differentiated into the outer gel-like ectoplasm and the inner more fluid endoplasm in which the rapid shuttle streaming occurs. The posterior of the plasmodium is merely the termination of the vein by ectoplasm somewhat thicker than that in the walls of the rest of the strand. The anterior portion of the strand opens into a broad, fan-shaped transition zone in which several endoplasmic channels of decreasing diameter radiate out into the lobulated, crescentic anterior end, terminating in numerous rather blunt pseudopodia. At the start of endoplasmic streaming in the anterior direction, contraction of the ectoplasm in the posterior end and along the vein occurs, resulting in a general shortening of the posterior part of the vein which may pull the posterior end forward rather rapidly, often undergoing various contortions in movement. The vein at the posterior end and in fact along its entire length appears to constrict radially and longitudinally as the endoplasm flows toward the front. Partial solation occurs in the ectoplasm of the posterior end and along the vein as part of the ectoplasm is converted to endoplasm and joins the anterior flow. Occasionally small regions of the vein may contract locally and constrict the streaming of the endoplasm, but normal endoplasmic flow resumes after a short time. At the anterior end, the endoplasm is forced into the branching channels of the fan-shaped sheet, slowing in velocity and ultimately extending forward into the pseudopodia, many of which exhibit a granule-free hyaloplasm or hyaline cap where at least some of the endoplasm is converted into ectoplasm. As endoplasmic flow decreases in the anterior direction, the ectoplasm of the anterior portion of the plasmodium begins to exert pressure, thus reversing the endoplasmic flow, which now proceeds in the posterior direction down the vein toward the rear. A net forward gain of endoplasm at the anterior end occurs with each cycle, moving the plasmodium forward. As in the large amoebae, the slime mold moves forward by tearing down the posterior end and a portion of the tube and building it up at the anterior end, making the posterior portion of the strand the oldest part of the plasmodium.

This streaming may be observed in time-lapse cinematographic films now available, demonstrating motility in myxomycetes at different stages of the life cycle (Kerr, 1970; Haskins, 1974). Moderately large macroscopic plasmodia (Fig. 1) with five or six anastomosing veins in the posterior region and one large, multilobed, fanlike front appear to move in a very similar manner to Lewis's tiny plasmodia. Recent studies have shown that the streaming pattern in these plasmodia is also monorhythmic. That is, the entire system of major veins in the posterior region of the plasmodium contracts in unison, coinciding with the advancement of the front. This is followed by a reciprocal contraction of the front and expansion of the veins at the rear (see Section IV,A).

Kamiya (1959) has recorded values up to 1350 μm/sec for the velocity of protoplasmic flow in major veins of *P. polycephalum,* which is extremely fast when compared with 27.5 μm/sec for the velocity of endoplasmic flow in *Chaos carolinensis* (Allen, 1973), 10 μm/sec for the velocity of cyclosis in *Elodea* leaves, and 4 μm/sec for granule movement in the cells of the inner epidermis of *Allium cepa.* In the typical cycle of protoplasmic streaming in a plasmodial vein, the rate of flow slowly increases in a particular direction, reaching a peak velocity followed by a slow decline to zero before the subsequent reversal of flow, which follows immediately. Flow in one direction typically lasts 30 to 90 seconds. Kamiya (1959) analyzed the velocity of flow in an optical cross section of a vein during peak velocity aand found that the maximum velocity occurs in the middle of the endoplasm, whereas flow tapers off to zero at the interface between endoplasm and ectoplasm. Artificially induced streaming by external air pressure applied to one end of a plasmodium has the same pattern. Kamiya concluded that the force producing streaming does not reside in the endoplasm and is not created by shear forces in the interface between the endoplasm and ectoplasm (as in *Nitella*), but is passively transmitted to the endoplasm by constriction of the ectoplasmic wall.

If two ends of a small plasmodium are placed in two different airtight com-

5. Plasmodial Structure and Motility

partments but are joined by a strand of protoplasm penetrating the central septum between the chambers, the contractile force (or "motive force") that produces streaming may be studied by recording the amount of air pressure applied to one compartment which is just sufficient to prevent the protoplasm from flowing either forward or backward between the two chambers (Kamiya, 1942, 1959). With this "double-chamber" method, Kamiya found that the counterpressure needed to prevent protoplasmic flow varied continuously, with about the same frequency as that expected for the streaming cycle. In other words, the contractile force, which normally produces endoplasmic flow, continues to oscillate normally despite the enforced cessation of streaming. This experiment reveals the presence of an oscillating control system in the plasmodium for the regulation of force production in the ectoplasm which appears to be independent of the actual movement it controls.

C. Protoplasmic Streaming: Conclusions from Descriptive Data

Before turning to more recent evidence on the mechanism of protoplasmic streaming in the plasmodium, the following conclusions may be made on the basis of the observational evidence and earlier experiments discussed thus far.

1. The migrating plasmodium is organized into a posterior portion composed of a network of veins or strands which converge into an anterior, fan-shaped sheet of protoplasm. The strands all along their length are capable of various contractions, most importantly radial and longitudinal, which are associated with endoplasmic streaming to the anterior. The fanlike front is also capable of producing force to create endoplasmic flow to the posterior.

2. The tubelike ectoplasmic layer possesses the contractile mechanism to move the more fluid endoplasm in the interior.

3. The constriction of the ectoplasm is applied with steadily increasing force, rising to a peak followed by a steady decrease over a total period of time of approximately 30–90 seconds.

4. The ectoplasm and endoplasm are interconvertible at various times in the shuttle-streaming cycle.

5. Since ectoplasm and endoplasm are interconvertible, the endoplasm must contain the contractile mechanism in an inactive form.

6. Local regulatory systems must exist for controlling endoplasmic flow all along each vein since localized asynchronous constrictions or relaxations of the ectoplasm may occasionally interrupt the overall pattern of endoplasmic flow in a vein.

7. A monorhythmic, oscillating regulatory mechanism exists in all except the very large plasmodia which ordinarily coordinates local ectoplasmic constrictions so that all the veins contract simultaneously to produce anterior flow, and

then later relax in unison while the ectoplasm at the front exerts pressure for posterior flow.

8. The anterior and posterior portions of the plasmodium are morphologically and functionally different since the overall migration of the plasmodium proceeds in the anterior direction.

Many of these descriptive phenomena of shuttle streaming in the plasmodium may now be understood in terms of a molecular model for protoplasmic movement in which force is produced by a calcium-sensitive actomyosin constriction system in the ectoplasm. Further elaborations of this model will be discussed in subsequent sections of this chapter. The reader is referred to reviews for further information about the molecular basis of protoplasmic streaming in amoeboid cells (Komnick *et al.*, 1973; Pollard, 1973; Pollard and Weihing, 1974; Condeelis and Taylor, 1977; Taylor and Condeelis, 1979; Taylor *et al.*, 1979a,b; Stebbings and Hyams, 1979; Britz, 1979; Stossel *et al.*, 1980).

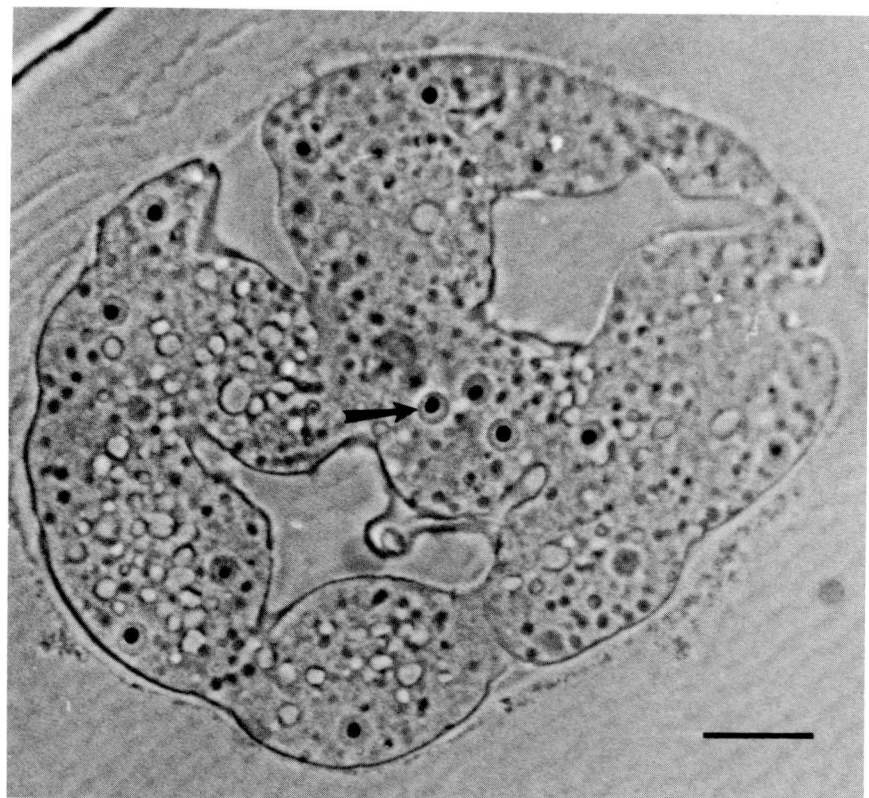

Fig. 2. Microplasmodium of *P. polycephalum*. Section photographed under phase contrast microscopy. There are a number of nuclei in this section, each with a large, darkly stained central nucleolus (arrow). Bar = 10 μm. Zegeye and Kessler, from Ringertz and Savage (1976).

II. PLASMODIAL MORPHOLOGY

A. Light Microscopic Observations

1. INTRODUCTION

The most recent general review of cytological studies of myxomycete plasmodia with the light microscope may be found in Gray and Alexopoulos (1968).

If microplasmodia growing in shake flasks on a semidefined medium (Daniel and Baldwin, 1964) are fixed with osmium tetroxide–glutaraldehyde solution and embedded in plastic, sections about 1 μm thick may be examined with phase contrast microscopy (Fig. 2). Nuclei with a single large central nucleolus are visible throughout the cytoplasm, as well as numerous rounded vacuoles of various sizes. Any refractile material contained in the vacuoles has been dissolved during specimen preparation. Small, dark mitochondria-like bodies are scattered in the cytoplasm. Large invaginations from the plasmalemma protrude deeply into the cytoplasm, often appearing as large vesicles of irregular shape in cross section, but no systematic organization of these invaginations occurs in shake flask microplasmodia. A thin layer of slime, the glycocalyx, may be seen just outside the plasmalemma. No clear differentiation of ectoplasm and endoplasm may be seen, and no cytoplasmic fibrils are visible. This correlates with the observation that living microplasmodia taken directly from shake flasks exhibit no obvious cytoplasmic movement for several minutes. More detailed observations of structures visible with the light microscope are summarized below.

2. THE CYTOPLASM AND ITS INCLUSIONS

a. Nuclei. In living plasmodia, the numerous round interphase nuclei with a diameter of 6 μm each contain a single rather large nucleolus about 3 μm in diameter, and about 30–50 small Feulgen-staining heterochromatic granules 0.5 μm in diameter scattered throughout the nucleoplasm (Andresen and Pollock, 1952; Wolf *et al.*, 1979b). Nuclei are difficult to observe in live plasmodia without phase optics (Lewis, 1942). The diameter of nuclei in fixed preparations is about 25% smaller than in living material (Guttes *et al.*, 1961). The nuclei may be seen in smears with phase optics after fixation in 95% ethanol or in 50% glycerol–ethanol. Alternatively, they may be stained with the Feulgen reaction, toluidine blue, azocarmine, acetoorcein, or with the fluorescent stain, acridine orange. Each cubic millimeter of plasmodium contains about 800,000 nuclei, corresponding to about 7% of the total plasmodial volume (Andresen and Pollock, 1952). Further cytological studies of nuclei and their changes in mitosis are discussed in Chapter 9, by Lafontaine and Cadrin, this volume.

b. Mitochondria. The mitochondria stain with Janus Green *in vivo* (Lewis, 1942) and with acid fuchsin in fixed preparations (Guttes *et al.*, 1961). Andresen and Pollock (1952) have calculated that mitochondria take up 1–2% of the total volume of the plasmodium. Mitochondrial size and shape vary systematically throughout the mitochondrial division cycle, from spheres of about 1.5 μm in diameter just after mitochondrial division to an oval and then a dumbbell shape about 3 μm in length just before synchronous mitochondrial fission (Kuroiwa *et al.*, 1978). DNA in the nucleoid within each mitochondrion may be visualized with the light microscope by the Feulgen (Kuroiwa, 1973) or, even more clearly, by the fluorescent-Feulgen staining techniques (Schuster, 1967; Grunfeld and Kessler, 1967). The mitochondrial nucleoid and its changes in the mitochondrial division cycle will be discussed in more detail in Section II,B. Intracellular bacteria or DNA-containing endosymbionts have not been observed using these staining methods in plasmodia of *P. polycephalum* growing in pure culture.

c. Granules. In plasmodia of *P. polycephalum,* the numerous highly refractile yellow pigment granules are 0.2–4.0 μm in diameter, enclosed in vacuoles, and vary somewhat in abundance and intensity of color depending on growth conditions. On the average, they occupy about 5% of the volume of the plasmodium (Andresen and Pollock, 1952; Guttes *et al.*, 1962; Gray and Alexopoulos, 1968).

Small cytoplasmic granules staining metachromatically with toluidine blue, presumably containing polyphosphate, have been observed in *Physarum* plasmodia (Guttes *et al.*, 1961; Guttes and Guttes, 1964). They are less than 0.2–0.5 μm in diameter.

d. Vacuoles. The contractile vacuole activity in plasmodia has been studied by Camp (1937) and Andresen and Pollock (1952). The numerous contractile vacuoles are moved passively with the flow of the protoplasm, growing in size up to a maximum of 10 μm. Sooner or later they lie near the plasmalemma, where they contract, forcing their content through the membrane. When systole is complete, a small vacuole remains which is flattened parallel to the outer surface of the plasmodium. It quickly becomes spherical, about 2 μm in diameter, and begins to increase in size, often fusing with other small vacuoles before contraction occurs again, as little as 2 minutes later. In the larger amoebae, the contractile vacuoles are surrounded by a layer of mitochondria. This has not been observed in *Physarum* plasmodia.

Ingestion of food particles was observed by Camp (1937). A flask-shaped depression is formed at the surface of the plasmodium containing the food particle. The neck constricts, isolating the particle in a food vacuole which is released into the cytoplasm. Egestion of solid particles occurs when dense older food vacuoles come to lie very close to the surface of the plasmodium. The number and appearance of food vacuoles in the cytoplasm vary greatly with the food source.

5. Plasmodial Structure and Motility

Pinocytosis occurs in the plasmodium of *P. polycephalum,* beginning in the hyaline portion of the ectoplasm as a beaker-shaped ectoplasmic protrusion which is pulled in, becoming shaped like a sphere with a narrow channel to the outside. The channel elongates, and the attached vacuole may reach the granular portion of the ectoplasm. Finally, the channel is obliterated and the vacuole is released into the streaming protoplasm (Guttes and Guttes, 1960). Pinocytotic-like channels and vacuoles have been observed by Rhea (1966a) with the electron microscope.

In addition to food and contractile vacuoles, Andresen and Pollock (1952) observed many empty vacuoles with a diameter up to 50 μm in living plasmodia. They are noncontractile. A few other vacuoles were seen which contained some scattered granules of debris together with a structureless, rather fluffy material. Since they were also present in starved cultures, they appeared to be distinct from food vacuoles. No lipid-containing vacuoles were observed after staining plasmodia growing on rolled oats with Sudan Black; other culture conditions may promote accumulation of lipid inclusions.

The total vacuole content in starved plasmodia without food vacuoles, but including contractile vacuoles, empty but noncontractile vacuoles, and vacuoles with bits of debris is about 18% of the volume of the plasmodium.

e. The Glycocalyx. The extracellular slime coat of the plasmodium, or glycocalyx, has been shown in biochemical studies to contain an acidic polysaccharide containing galactose, sulfate, and trace amounts of rhamnose (McCormick *et al.,* 1970). Histochemical observations of plasmodia reveal that this slime layer stains strongly with the colloidal iron reaction for acid mucopolysaccharides (Sheen *et al.,* 1969). This procedure also demonstrates similar staining material in the ectoplasm, in agreement with the observation that slime-covered invaginations of the plasma membrane extend into the ectoplasm. No staining is seen in the endoplasm, where these invaginations from the surface are absent. The staining reaction for slime is heaviest in the periphery of the posterior veins and least in the areas of advancing front where the slime layer is rather thin. During active migration, the slime layer is deposited on the surface of the substrate as an empty sheath behind the moving plasmodium. The thickness of the slime layer varies with different culture conditions and increases with the age of the culture. When the plasmodium migrates on the surface of a liquid, the slime layer expands in that area of the plasmodium in contact with the liquid. In the myxomycetes, this unusually profuse glycocalyx probably helps prevent desiccation and provides a relatively constant external ionic environment for the actively motile plasmodium, which is capable of migrating for long distances exposed to air on a glass surface, for example, if the atmosphere is reasonably moist.

3. FIBRILS IN THE ECTOPLASM AND INVAGINATIONS OF THE PLASMALEMMA

Sections of veins or strands from plasmodia which had been actively moving before fixation reveal numerous cytoplasmic fibrils, often associated with elongated vesicles in the cytoplasm of the outer cortex of the vein, whereas in the inner portion of the vein no fibrils or elongated vesicles may be seen (Fig. 3). This rather complex arrangement of elongated vesicles and fibrils has been carefully studied by Wohlfarth-Bottermann (1964b, 1974, 1975b), who has shown that the elongated vesicles are actually invaginations of the plasma membrane. The associated fibrils may be oriented longitudinally, radially, or spirally along the long axis of the vein. Since birefringent fibrils may be seen in the veins of living plasmodia with the same location and orientation as those observed in fixed specimens, they represent actual structures in the ectoplasm of living plasmodia (Nakajima, 1964; Nakajima and Allen, 1965). The outer area of the vein containing the fibrils and the invaginations of the plasmalemma corresponds to the ectoplasm, and the inner portion, which is free of these structures, is the endoplasm of the vein (Fig. 4). Since the number of invaginations along a strand changes with the availability of nutrients in the medium (40% of the plasmalemma is invaginated in nonnutrient conditions, whereas 80% is invaginated

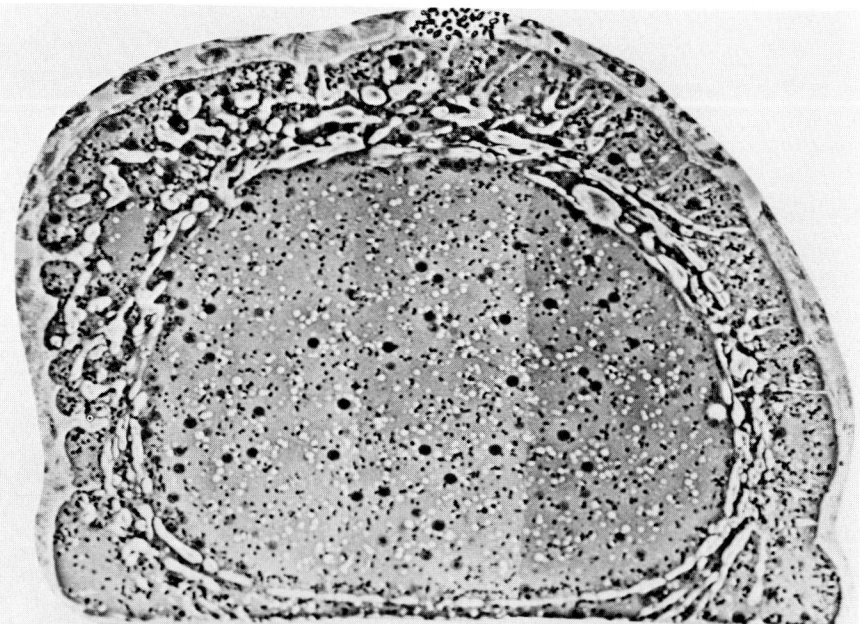

Fig. 3. Cross section of a protoplasmic vein from *P. polycephalum* growing on agar containing semidefined medium. Extensive invaginations of the plasmalemma extend throughout the ectoplasm. Photomicrograph was trimmed before mounted; border around plasmalemma represents the extent of trimming, not part of structure of vein. From Achenbach *et al.* (1979b).

5. Plasmodial Structure and Motility

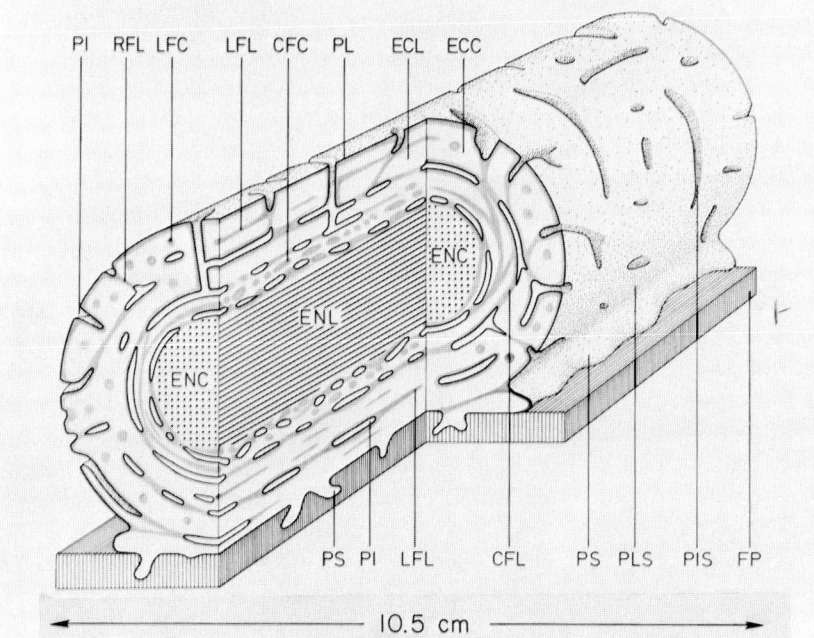

Fig. 4. The arrangement of actomyosin fibrils in the ectoplasm of plasmodial veins. No fibrils are found in the endoplasm. CFC, circular fibrils, cross section; CFL, circular fibrils, longitudinal section; ECC, ectoplasm, cross section; ECL, ectoplasm, longitudinal section; ENC, endoplasm, cross section; ENL, endoplasm, longitudinal section; FP, filter paper; LFC, longitudinal fibrils, cross section; LFL, longitudinal fibrils, longitudinal section; PI, plasmalemma invaginations; PIS, plasmalemma invaginations, surface view; PL, plasmalemma; PLS, plasmalemma surface; PS, pseudopodium; RFL, radial fibrils, longitudinal section. From Fleischer and Wohlfarth-Bottermann (1975).

when nutrients are present), the invaginations appear to be important for absorption as well as for protoplasmic streaming (Achenbach *et al.*, 1979b). In protoplasmic drops formed by puncturing a rapidly streaming vein, *de novo* formation of plasmalemmal invaginations occurs first by indentation of the plasma membrane, forming more or less unbranched channels reaching into the ectoplasm. This is followed by fusion of intracellular vacuoles with these indentations to produce the final labyrinth of interconnecting channels in the ectoplasm which communicates with the slime layer surrounding the plasmodium (Achenbach *et al.*, 1979a).

B. Ultrastructural Studies

1. INTRODUCTION

For observations of the general ultrastructure of the plasmodial cytoplasm and its organelles, consult Charvat *et al.* (1973a) on *Perichaena vermicularis* and

Daniel and Järlfors (1972a) and Rhea (1966a) on *P. polycephalum*. Earlier work suffered in varying degrees from inadequate fixation of the plasmodia or other artifacts associated with electron microscopic technique, e.g., Sponsler and Bath (1953), Stewart and Stewart (1959), Dugas and Bath (1962), and Terada (1962). Of this latter group, the observations by McManus (1965) are the most useful. Hoffmann and Stockem (1979) reported that the fewest ultrastructural artifacts of membrane structure are produced after fixing microplasmodia in buffered 2.5% glutaraldehyde for 1 hour at room temperature, thoroughly washing in buffer, followed by postfixation in 1% osmium tetroxide for 30 minutes, dehydration, embedding in plastic, and staining the thin sections in uranyl acetate followed by lead citrate. Adequate penetration of fixative or embedding medium may be a problem when working with large pieces of plasmodium.

Even under the best circumstances, the cytoplasm of myxomycete plasmodia can be difficult to interpret at the ultrastructural level, presenting a confusing array of numerous vacuoles of different sizes and shapes intermingled with the more readily recognized outlines of mitochondria and nuclei in a groundplasm packed with small granules and occasional microfilaments and fibrils. The ultrastructural morphology of the plasmodium seems much closer to that of the large, free-living amoebae (Flickinger, 1973; Daniels, 1973) than to that of most cells of higher plants and animals.

2. THE GLYCOCALYX

Outside the plasmalemma the slime layer or glycocalyx surrounding the entire plasmodium, often much more than 2 μm thick, is composed of a fibrillar network of curved and branched filaments 30–75 Å in diameter and of indefinite length (Rhea, 1966a). This material also fills the lumen of the numerous elongated invaginations of the plasma membrane in the ectoplasm (Daniel and Järlfors, 1972a). Identical filaments make up the empty slime sheaths found in the tracks left by migrating plasmodia (Rhea, 1966a). In a cytochemical study, Stiemerling (1970) found that the acid mucopolysaccharide of the slime layer could also be identified as staining material in cytoplasmic vacuoles as small as 1 μm in diameter. The Golgi body vesicles contain branched filaments typical of the slime layer but do not stain for acid mucopolysaccharide. This suggests that the mucopolysaccharide is collected in the Golgi body, packaged in vacuoles derived from the Golgi body where it is sulfated (McCormick *et al.*, 1970), and then transported to the plasmalemma for deposition at the surface of the plasmodium.

3. THE NUCLEUS

Rounded interphase nuclei about 4–5 μm in diameter, enclosed in a nuclear envelope and containing one large central nucleolus with numerous heterochromatic areas scattered in the nucleoplasm, are present throughout the ecto-

5. Plasmodial Structure and Motility

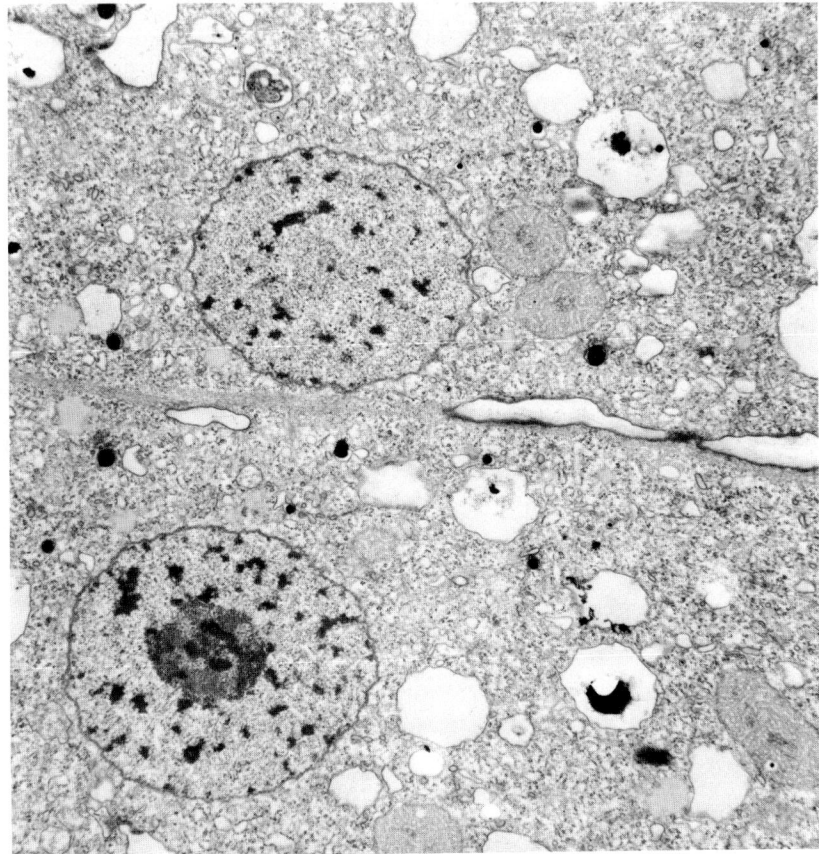

Fig. 5. Electron micrograph of a cross section of a plasmodium showing nuclei, mitochondria, and invaginations of the plasmalemma. × 15,000. From Kessler (1972b).

plasm and endoplasm (Fig. 5). Nuclear pores have been described by several investigators (Rhea, 1966a; Ryser, 1970; Daniel and Järlfors, 1972a; Charvat *et al.*, 1973a; Schel *et al.*, 1978). The ultrastructure of nuclei throughout the mitotic cycle is considered in Chapter 9 by Lafontaine and Cadrin, this volume.

4. THE CYTOPLASM

a. Ribosomes and Glycogen Granules. Numerous ribosomes about 300 Å in diameter are scattered throughout the cytoplasm, often aggregated in small groups suggestive of polyribosomes. Most ribosomes in the plasmodium are free in the groundplasm and are not associated with the surface of membrane systems. Glycogen particles 500–900 Å in diameter are present in large numbers in the groundplasm, often obscuring the ribosomes and microfilaments or fibrous network also present in the groundplasm (McManus, 1965; Goodman and Rusch, 1970; Daniel and Järlfors, 1972a; Charvat et al., 1973a). Glycogen particles decrease in number during starvation (Goodman and Rusch, 1970).

The staining procedure of Ploton and Gontcharoff (1979), enhancing the contrast of RNA-containing structures, demonstrates the presence of ribosomes in large numbers in the groundplasm. Hoffmann and Stockem (1979) found that the contrast of the glycogen granules could be decreased by staining thin sections with uranyl acetate in alcoholic rather than aqueous solution, or by the addition to the glutaraldehyde fixative of agents which selectively extract low-molecular-weight material from the groundplasm, such as spermidine phosphate (Hauser, 1978) or osmium tetroxide (Daniel and Järlfors, 1972a), revealing ribosomes and microfilaments to better advantage. Wohlfarth-Bottermann and his colleagues enhance cytoplasmic microfilament contrast in the same manner after brief prefixation in buffered osmium tetroxide containing potassium dichromate (Wohlfarth-Bottermann, 1957; Gotz von Olenhusen et al., 1979).

b. Mitochondria. Mitochondria in the plasmodium of *P. polycephalum* are 1.5–3.0 μm long and 1.5 μm wide. They contain numerous tubular cristae, with the moderately electron-dense matrix between the cristae containing occasional fine particles about 140 Å in diameter having the appearance of ribosomes (Kuroiwa, 1973). Larger granules of variable size containing calcium have also been observed in the mitochondrial matrix under conditions enhancing calcium accumulation in cytoplasmic vacuoles (Daniel and Järlfors, 1972b).

The mitochondrial nucleoid, first noted by Schuster (1965) in the mitochondria of myxamoebae and swarm cells of *Didymium nigripes,* has also been observed in the mitochondria of plasmodia of *P. polycephalum* (Guttes et al., 1966). In longitudinal section, the nucleoid is a rod-shaped structure up to 1.5 by 0.3 μm occupying the central matrix of the mitochondrion (Fig. 6). The central region of the nucleoid along the long axis is slightly less dense than the peripheral region and seems to contain fibrous material which cytochemical studies have identified as DNA (Seavey et al., 1967; Stockem, 1968; Kuroiwa, 1973), verifying earlier studies with the light microscope using autoradiography (Guttes and Guttes, 1964), and DNA-specific stains (see Section II,A). Note that the quantity of DNA per mitochondrion in *P. polycephalum* is 10 times more than the DNA in mitochondria of most other organisms (although not as much as the kinetoplast of trypanosomes, which contains 10^3 times more DNA than most other mitochondria), explaining why it can be readily visualized by light and electron microscopic studies. Biochemical characterization of plasmodial mitochondrial DNA is summarized in by T. E. Evans in Chapter 11, this volume. RNA is present in the peripheral regions of the mitochondrial nucleoid and in the ribosomes of the mitochondrial matrix. Kuroiwa and Hizume (1974), using a cytochemical stain,

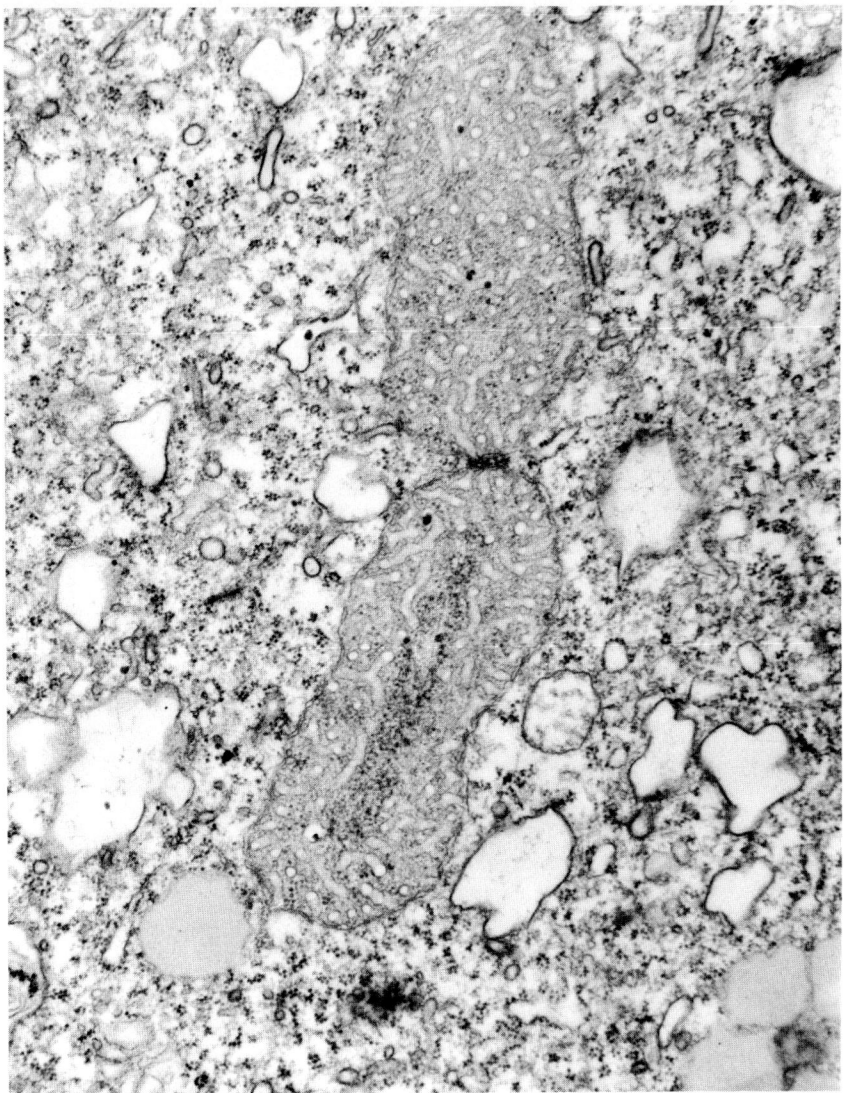

Fig. 6. Mitochondria in a section of the plasmodium showing the central DNA-containing nucleoid. × 30,000.

demonstrated the presence of a basic protein similar to histone in the mitochondrial nucleoid, suggesting that at least some of the DNA may be complexed with protein.

The mitochondria and the nucleoids they contain undergo a synchronous division cycle in the plasmodium of *P. polycephalum* which does not coincide with the synchronous nuclear division cycle; however, both nuclear and mitochondrial cycles are the same length (see the review by Goodman, 1980). Although Guttes *et al.* (1969) saw division stages of mitochondria throughout the mitotic cycle in plasmodia of *P. polycephalum*, Kuroiwa *et al.* (1977, 1978) found after a more thorough study that dumbbell-shaped mitochondria, i.e., elongated mitochondria 3 μm long by 1.5 μm wide with a very pronounced constriction in the center of the long axis, occur almost exclusively at one time in the mitotic cycle (about 1 hour after the synchronous nuclear mitosis in a nuclear division cycle of 14 hours), followed immediately by a rapid increase in the number of mitochondria now reduced in size to spheres 1.5 μm in diameter. DNA synthesis in mitochondria is also more or less synchronized. Although at least some mitochondria may be found synthesizing DNA at any time in the division cycle, a peak of 5 times more mitochondria are labeled with a [^{3}H]thymidine pulse during the nuclear G_2 period than during the nuclear S period, when mitochondrial DNA synthesis is at its lowest level and mitochondrial fission is occurring. The nucleoid also changes in shape during the mitochondrial division cycle. As the shape of the mitochondrion changes from spherical to oval to an elongated dumbbell, the nucleoid lengthens from about 0.4 μm to about 1.5 μm long. The nucleoid also becomes dumbbell-shaped with the mitochondrion but does not divide in two until the mitochondrion completely constricts into two new spherical mitochondria. Since the nucleoid appears to be attached to the side of the mitochondrial membrane at the constriction site in newly divided mitochondria, moving to the center again a short time later, Kuroiwa *et al.* (1977) hypothesize a role for the constricting mitochondrial membrane in separating the nucleoid into two parts, as in bacteria.

  c. Endoplasmic Reticulum. As in the large, free-living amoebae, the endoplasmic reticulum in myxomycete plasmodia is seen in thin sections as short lengths of tubules about 0.25 μm long and 0.04 μm wide, and as small spherical vesicles 0.04–0.3 μm in diameter sparsely distributed in the cytoplasm (Terada, 1962; McManus, 1965; Rhea, 1966a; Ryser, 1970; Scheetz, 1972; Daniel and Järlfors, 1972b; Charvat *et al.*, 1973a). The lumen often appears empty but occasionally contains dense material. The endoplasmic reticulum is best identified by first locating membrane-bound ribosomes, i.e., the rough endoplasmic reticulum. Smooth vesicles and short tubules about the same size but lacking ribosomes on their outer surface, presumably the smooth endoplasmic reticulum, may now be observed. This dispersed membrane system, rough or smooth, is not elaborated into a complex labyrinth of interconnected lamellae throughout the cytoplasm and so, strictly speaking, does not have the appearance of a reticulum. Biochemical studies on the nature of these membranes and their enzymatic properties have not yet been performed.

  Goodman and Rusch (1970) found much more elaboration of rough endoplasmic reticulum in plasmodia treated with cycloheximide. They observed more endoplasmic reticulum in the cytoplasm of starving plasmodia which were differentiating into spherules than in growing plasmodia. During sporulation, the plasmodial endoplasmic reticulum also becomes more extensive (Charvat et al., 1973b).

d. Golgi Body. Membranes with the appearance of the Golgi body are not numerous in the plasmodia of myxomycetes. They were first identified by McManus (1965), and several other investigators have reported their existence (Guttes et al., 1968; Goodman and Rusch, 1970; Stiemerling, 1970; Charvat et al., 1973a). They are seen most often as a group of two to four flattened discs or cisternae, about 0.4 μm long by about 0.02 μm wide, often with bulblike enlargements about 0.08 μm in diameter at the ends of the discs. The Golgi bodies increase in number during starvation just before spherulation (Goodman and Rusch, 1970). Stiemerling (1970) found that they may become distended during slime production, presumably collecting the nascent mucopolysaccharide, each flattened disc becoming an oblong vacuole 0.3 μm by 0.17 μm, and finally a sphere about 0.25 μm in diameter.

e. Food Vacuoles and Lysosomes. Food vacuoles are readily observed in the plasmodium, their size varying with culture conditions. Charvat et al. (1973a) distinguished between heterophagic vacuoles containing bacteria or other particles phagocytosed from the external environment and autophagic vacuoles containing degenerating nuclei and mitochondria. The formation of autophagic vacuoles is greatly increased during sporulation as large portions of the plasmodium degenerate during differentiation as part of sporangium formation (Charvat et al., 1973b). Autophagic and heterophagic vacuoles are secondary lysosomes, as indicated by their acid phosphatase activity, identified cytochemically. Certain smooth membranes, perhaps derived from stacks of the Golgi cisternae, also give a positive test for acid phosphatase. Wolf et al. (1979a) present a model for intracellular digestion in microplasmodia of *P. polycephalum* in which primary lysosomes derived from the Golgi body fuse with endocytotic vacuoles or with autolysosomes to form secondary lysosomes which may further fuse with each other as digestion proceeds. They later form defecation vacuoles which are transported to the cell surface.

f. Other Vacuoles. The plasmodium contains a great many vacuoles of various sizes which have not been clearly characterized at the ultrastructural level.

Lipid droplets or osmiophilic granules about 0.3–0.5 μm in diameter have been reported (Rhea, 1966a; Stockem, 1968; Stiemerling, 1970; Daniel and Järlfors, 1972a; Sakai and Shigenaga, 1972). Their contents are often not well preserved, so that their lumen either scarcely has greater contrast than background or occasionally is filled with a moderately dense, uniformly homogeneous substance, very rarely appearing extremely dense. Rhea (1966a) states that they are more numerous than mitochondria. Charvat *et al.* (1973a) distinguish them from other types of vacuoles because they lack a tripartite membrane.

Coated vesicles, i.e., spherical vesicles about 140 nm in diameter with a fuzzy coat on the outer membrane surface, have been observed by Charvat *et al.* (1973a). Ryser (1970) shows a coated pit about 100 nm long with bristles on the cytoplasmic surface located in a depression of the plasma membrane.

Pigment granules such as those seen in abundance in the living yellow plasmodia of *P. polycephalum* (see Section II,A) have rarely been reported in ultrastructural studies. Evidently they are extracted by most fixation and embedding procedures, leaving behind empty or nearly empty vacuoles. McManus (1965) examined an unidentified yellow phaneroplasmodium similar in appearance to *P. polycephalum* and observed many large cytoplasmic vacuoles, about 1.2 μm in diameter, each containing one very dense granule about 0.5 μm in diameter.

Contractile vacuoles have not been positively identified in electron microscopic studies of plasmodia, although they have been observed and carefully studied *in vivo* using the light microscope. On the ultrastructural level, they may be confused with the numerous large, empty, cytoplasmic, noncontractile vacuoles, or with vacuoles resulting from the extraction of pigment granules, as discussed above.

Calcium-containing vacuoles have been identified with the electron microscope in the plasmodium, especially under certain culture conditions, and are discussed in Section II,B,4,*h*. They may be identical with the polyphosphate granules observed with the light microscope (Section II,A,2,*c*).

g. Microtubules. A controversy developing from electron microscopic studies of plasmodial ultrastructure concerns the relative importance of microtubules or microfilaments in the production of force necessary for protoplasmic streaming.

Conditions for the adequate fixation of microfilaments and fibrils were not adequate in the earlier ultrastructural investigations of cytoplasmic morphology. Stewart and Stewart (1959), for example, could not find filaments in the cytoplasm of *P. polycephalum* plasmodia and thus discounted the idea of a sliding filament mechanism for force production. McManus and Roth (1965, 1967), after seeing microtubules in the cytoplasm of *P. melleum* and *Fuligo septica,* preferred a microtubule-based streaming mechanism. In the meantime, Wohlfarth-Bottermann (1962, 1964b) discovered procedures for the visualization of cytoplasmic microfilaments in plasmodia. Shortly thereafter, only cytoplasmic microfilaments and fibrils, and not cytoplasmic microtubules, were observed in studies of *P. polycephalum* plasmodia (Porter *et al.,* 1965; Rhea, 1966a). Of course, microtubules in the nucleus have always been seen making up the intra-nuclear spindle in mitosis (Goodman, 1980; Hinchee and Haskins, 1980; Chapter 9).

With better fixation procedures, occasional single microtubules about 250 Å in diameter and indefinite length have now been observed in the cytoplasm of plasmodia from several species of myxomycetes, e.g., *Perichaena vermicularis* (Charvat *et al.*, 1973a) and *Ceratiomyxa fruticulosa* (Scheetz, 1972). Nevertheless, cytoplasmic microfilaments and fibrils derived from their aggregation are much more numerous than microtubules in all the plasmodia studied. These microfilaments, rather than microtubules, are believed to be the structural elements important in shuttle streaming (see Section III,A,B).

h. Invaginations, Fibrils, Microfilaments, and Calcium-Containing Vacuoles. The invaginations of the plasmalemma extending into the ectoplasm in the veins of migrating plasmodia visualized so well in sectioned material with the light microscope have been studied extensively at the ultrastructural level (see especially Rhea, 1966a; Daniel and Järlfors, 1972a; Achenbach *et al.*, 1979b). These large, elongated structures are easily distinguished from the rounded cytoplasmic vacuoles on the basis of size and shape, and because their lumen contains filamentous material, probably a continuation of the mucopolysaccharide slime layer which surrounds the plasmodium. Fibrillar material composed of numerous filaments about 50–70 Å in diameter (i.e., the diameter of F-actin) is often closely associated with the cytoplasmic side of the invagination membrane, often parallel to the long axis of the invagination. Fibrils composed of the same microfilaments are observed throughout the ectoplasm, often joining two invaginations (Fig. 7). These fibrils were first described by Wohlfarth-Bottermann (1962, 1964a,b) and have been observed subsequently by other investigators (Nagai and Kamiya, 1966). Cytoplasmic fibrils composed of filaments are sometimes seen associated with nuclei in which the shape of the nuclear envelope is distorted to suggest the application of force by the fibril (Kessler, 1972b). In addition, amorphous material containing occasional filaments or nonparallel arrays and networks of filaments of the same density as the fibrils are often present in the ectoplasm. The cytoplasmic side of the entire plasmalemma is usually composed of a layer of amorphous material or filamentous network about 1 μm or less thick which excludes organelles and which appears to be continuous with the fibrils and microfilaments of the invaginations (Daniel and Järlfors, 1972a). Although the differentiation between ectoplasm and endoplasm is less definite in the fanlike front of the plasmodium (in contrast to the strands in the rear), fibrils are also found in the cytoplasm there, often associated with the plasmalemma (Rhea, 1966a; Usui, 1971). The only ultrastructural difference between endoplasm and ectoplasm in the plasmodial veins is the lack of microfilaments, fibrils, and plasmalemma invaginations in the endoplasm.

Daniel and Järlfors (1972b) found that light (required for sporulation) induces

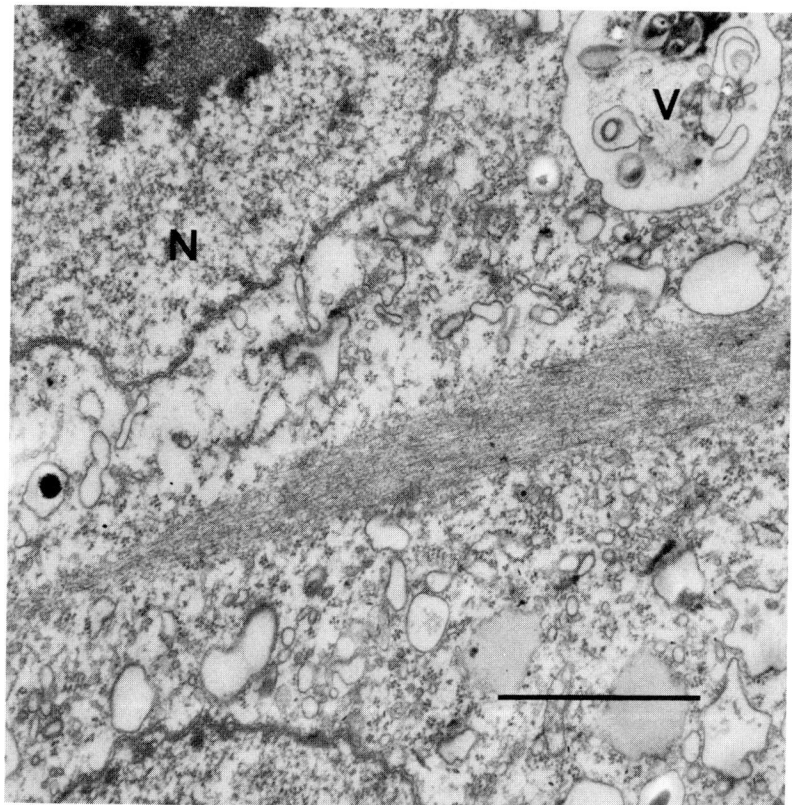

Fig. 7. Large cytoplasmic fibril seen in a thin section from a starved microplasmodium. The fibril is composed of numerous microfilaments. Bar = 1 μm; N = nucleus; V = vacuole. From Kessler (1972b).

the formation of striking electron-dense deposits in a great many cytoplasmic vacuoles (Fig. 8). These were shown to contain calcium by their removal after treating sections with EGTA, and by X-ray spectroscopic analysis. These rounded, calcium-containing vacuoles are variable in size, but a majority appear to be about 1-1.5 μm in diameter. These authors also observed smaller calcium-containing granules in the intermembrane spaces of mitochondria after illuminating plasmodia. In a histochemical study, Braatz and Komnick (1970) demonstrated that vacuoles in the cytoplasm of briefly fixed or glycerinated plasmodia of *P. confertum* and *P. polycephalum* are capable of accumulating calcium in the presence of ATP. Ettienne (1972) also identified calcium-containing vacuoles in the cytoplasm of *P. polycephalum*. It seems likely that these investigators have been studying the same vacuoles, induced to accumulate calcium by different methods. These findings are very relevant to the problem of regulating the calcium concentration of the cytoplasm (see Section III,D).

5. Plasmodial Structure and Motility

5. MORPHOLOGICAL AND BIOCHEMICAL IDENTIFICATION OF PLASMODIAL ACTIN AND MYOSIN

Observations of shuttle streaming in the plasmodium strongly indicate (Section I,B and C) that some kind of contraction mechanism in the ectoplasm produces pressure on the endoplasm to cause protoplasmic streaming. Light microscopic and electron microscopic examination of the plasmodial ectoplasm reveals fibrils and networks composed of filaments about 50–70 Å in diameter which are absent from the endoplasm (Section II,A,3 and B,4,h). It is logical to conclude that the filaments must somehow be responsible for the contractile properties of the endoplasm.

The mechanism of contraction in skeletal muscle fibers also involves filaments: Thin, actin-containing filaments about 50–70 Å in diameter slide past thick, myosin-containing filaments with a diameter of about 100–120 Å in

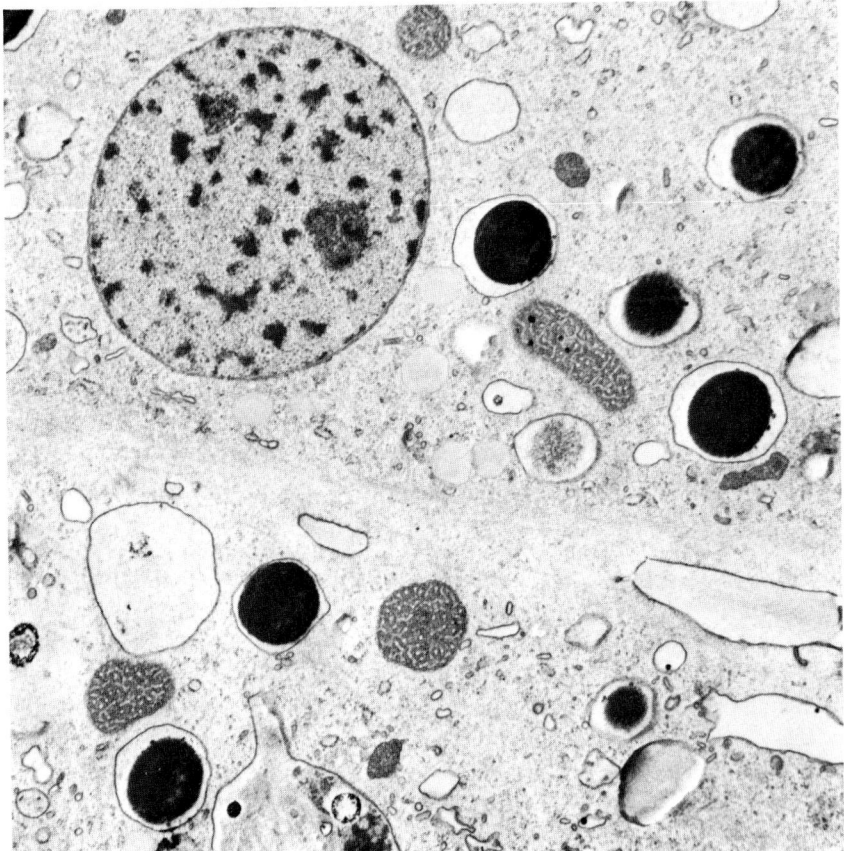

Fig. 8. Plasmodial vein (cross section) after illumination. Electron-dense, calcium-containing deposits are seen in vacuoles and mitochondria. From Daniel and Järlfors (1972b).

skeletal muscle. ATP is hydrolyzed to ADP during this interaction. Hence, actomyosin is an ATPase. The conditions for polymerization and the properties of monomeric actin and myosin from muscle are known (Huxley, 1972). A close examination of proteins from the plasmodium of *Physarum* which have actomyosin-like properties is the initial step in establishing the similarity of the contractile mechanism in the plasmodium with that of skeletal muscle.

Loewy (1952) obtained the first biochemical evidence for a contractile system in *Physarum* by observing the viscosity decrease followed by a rise which occurs when ATP is added to plasmodial extracts at high ionic strength. Since muscle actomyosin also behaves in this manner, he correctly concluded that an actomyosin ATPase was present in the plasmodium. (On the addition of ATP at high ionic strength, the rigor bonds joining the myosin monomers with the actin filaments are broken, decreasing the viscosity of the solution; myosin permanently reattaches to the actin, causing a rise in viscosity again after the ATP is hydrolyzed by the actomyosin ATPase.) This ATPase seemed to be a very large molecule with a diameter of about 70 Å (Ts'o *et al.*, 1957). Glycerinated plasmodia could be made to contract on the addition of ATP (Kamiya and Kuroda, 1965), providing additional evidence that an ATPase is associated with contraction in *Physarum*. Subsequently, Hatano and Oosawa (1966a,b) isolated a protein from the plasmodium which had properties very similar to those of muscle actin: ability to associate with muscle myosin, drop in viscosity when ATP is added to the complex followed by a viscosity rise, and conversion of the plasmodial G-actin monomer to the F-actin filament by addition of salts. This protein has been biochemically characterized in subsequent studies (Adelman and Taylor, 1969a; Hinssen, 1972; Adelman, 1977; Hatano and Owaribe, 1977). In the electron microscope (Fig. 9), *Physarum* F-actin has the same diameter, the same axial periodicity, and interacts with muscle heavy meromyosin to form the same arrowhead configuration as F-actin from muscle (Nachmias *et al.*, 1970). In fact, *Physarum* actin is very conservative evolutionarily: It differs in only 8% of its amino acids from rabbit skeletal muscle actin and in only 4% from mammalian cytoplasmic actin (Vandekerckhove and Weber, 1978). By treating glycerinated plasmodia with muscle heavy meromyosin, Alléra *et al.*, (1971) proved that the filaments 50–70 Å in diameter seen in the ectoplasm of the plasmodium were F-actin: They became decorated in an arrowhead pattern.

Information about *Physarum* myosin has been summarized by Jacobson *et al.* (1976) and Nachmias (1979b). Adelman and Taylor (1969b) and Hatano and Ohnuma (1970) first studied purified myosin from *Physarum* plasmodia. Its ATPase activity is very similar to that of muscle myosin, with a few minor differences. The molecular weight of the slime mold myosin monomer is only slightly larger than that of rabbit muscle myosin, and it can form actomyosin complexes with rabbit F-actin. Hatano and Takahashi (1971) found that *Physarum* myosin monomers look almost exactly the same as muscle myosin

5. Plasmodial Structure and Motility

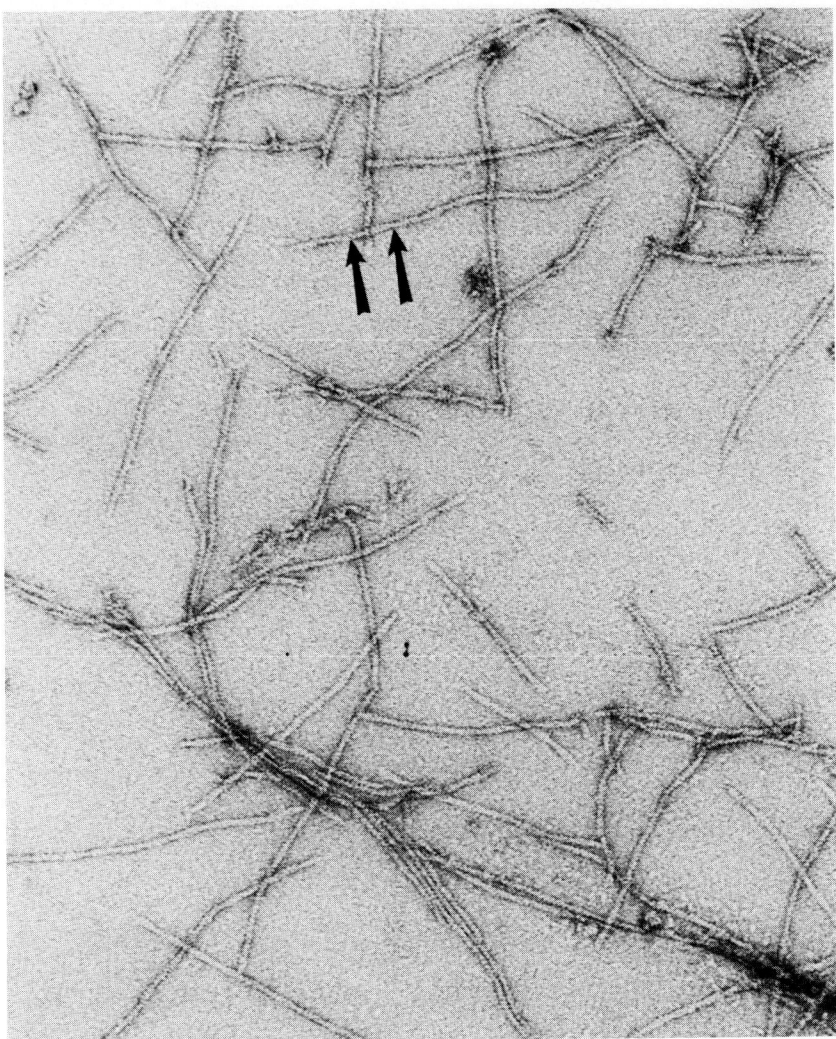

Fig. 9. *Physarum* actin from plasmodial actomyosin treated with a solution of ATP, MgCl$_2$, and EGTA to dissociate the myosin from the actin. Arrows point to crossover points on the actin. × 110,000. From Nachmias (1972a).

monomers after shadow casting and examination with the electron microscope (Fig. 10). The *Physarum* myosin monomer of about 1210 Å is composed of a long, thin rod with a globular region on one end where two heads can often be seen. The rodlike tail of myosin has also been seen in *Physarum* actomyosin preparations (Nachmias and Ingram, 1970; Nachmias, 1972a). *Physarum*

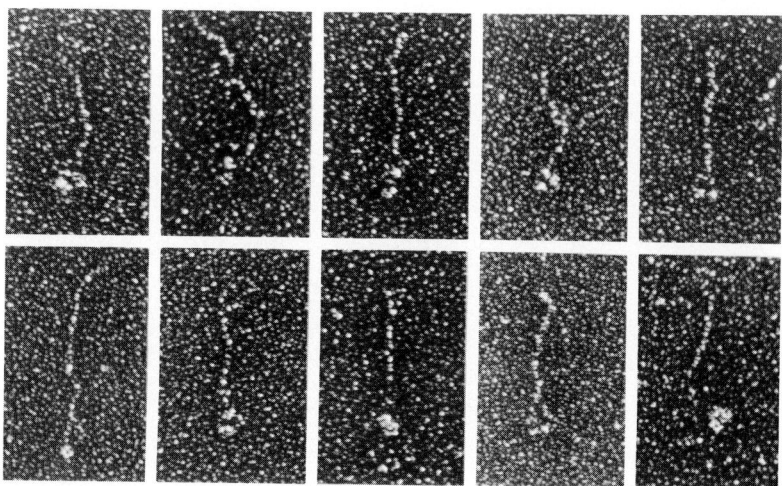

Fig. 10. A composite of selected rotary-shadowed *Physarum* myosin monomers. The head of the myosin consists of two globular subunits. Average length of the monomer is 1210 Å. From Hatano and Takahashi (1971).

myosin has a curious property different from skeletal muscle myosins: The purified monomers in KCl at low ionic strength do not aggregate into thick filaments. It has been subsequently found that cytoplasmic myosins, in general, polymerize into filaments with more difficulty than muscle myosins (Pollard, 1979). Hinssen (1970) was the first to observe myosin thick filaments in extracts of plasmodial actomyosin. After a careful investigation, Nachmias (1972b,c, 1974) found that purified *Physarum* myosin at low ionic strength and physiological pH requires the presence of Mg^{2+} or Ca^{2+} in order to aggregate into thick filaments. Thick filaments composed of myosin are not normally seen in the ectoplasm of plasmodia. However, the presence of myosin among the F-actin filaments of the ectoplasmic fibrils is revealed after glycerinating the plasmodium at low ionic strength in the presence of divalent cations: Tapered thick filaments 100–120 Å in diameter now appear, with dimensions similar to those produced by the aggregation of muscle myosin molecules (Alléra and Wohlfarth-Bottermann, 1972; Kessler, 1972b).

These studies on the biochemical and ultrastructural properties of *Physarum* contractile proteins demonstrate that actin and myosin with many properties similar to actin and myosin from muscle are indeed present in the plasmodium. The fibrils in the ectoplasm seen in numerous microscopic investigations contain *Physarum* actin filaments, probably interspersed with myosin oligomers containing fewer monomers than the thick filaments, which may be produced *in vitro*. Estimates of the amount of actin and myosin in the cytoplasm of the plasmodium (e.g., Kessler *et al.*, 1976) have shown a striking difference from their concentrations in skeletal muscle. Myosin makes up about 0.8% of the total protein in the plasmodium, with a concentration of only 0.6 mg myosin per gram of plasmodium (wet weight). In contrast, in skeletal muscle, myosin makes up 38% of the total protein at a concentration of 72 mg myosin per gram of muscle (wet weight). The situation for actin is somewhat different. The amount of actin in the plasmodium has been estimated at 15–25% of the total protein, or at least 3 mg actin per gram of fresh plasmodium. Actin comprises 23% of muscle protein at 43 mg actin per gram of fresh muscle. On a molar basis, the actin:myosin ratio in muscle is about 7:1, but in the plasmodium the actin:myosin ratio is 200:1 or higher, making the plasmodium a very actin-rich contractile system. Cytoplasmic actomyosins from other cells have also been found to be actin-rich (Pollard and Weihing, 1974).

Actin and myosin have not only been identified in the cytoplasm of *Physarum*. Both proteins have been found in nuclei and nucleoli from the plasmodium (Jockusch *et al.*, 1971, 1974; Jockusch, 1973, Hauser *et al.*, 1975; LeStourgeon *et al.*, 1975; Schwärzler *et al.*, 1977; LeStourgeon, 1977). After glycerination of the plasmodia, myosin-like filaments are seen in postmitotic nuclei (Jockusch *et al.*, 1973). It is not known if these two proteins have a role in chromosome movement in anaphase or in the elongation of the late anaphase nucleus, separating the daughter nuclei and terminating the intranuclear mitosis in this organism.

III. MECHANISM OF PROTOPLASMIC STREAMING IN THE PLASMODIUM

A. Contraction–Hydraulic Mechanism: Actomyosin Contraction in the Ectoplasm Causes Endoplasmic Flow

1. INTRODUCTION

A number of investigators who have examined endoplasmic flow in *Physarum* have come to the conclusion, on the basis of microscopic observations alone, that the ectoplasm somehow possesses contractile properties which exert pressure on a more or less passive endoplasm (Lewis, 1942; Stewart, 1964; but see Siefriz, 1942; Loewy, 1949). Kamiya's study (1959) of the velocity profile of streaming in a cross section of the endoplasm, in which he found that maximum speed is obtained in the middle of the endoplasmic channel (see Section I,B), further established the fact that endoplasmic flow has almost the same velocity profile as water flowing in a pipe in passive response to hydraulic pressure. The microscopic evidence (Section II,A,3 and B,4,*h*) demonstrates the presence of fibrils made up of filaments in the ectoplasm of the plasmodium, and further work has established that these fibrils are composed of actin and myosin (Section II,B,5).

The filamentous network associated with the cytoplasmic side of the plasmalemma is also probably composed of actomyosin. Furthermore, actin and myosin are known to interact in skeletal muscle to produce contraction (Huxley, 1972). The logical path from this flow of reasoning ends with the conclusion that actomyosin contraction in the ectoplasm of the plasmodium produces the force which is responsible for the protoplasmic streaming seen in the endoplasm (Jahn and Bovee, 1967; Komnick *et al.*, 1973). Current theories on the molecular basis of the movement of amoebae also employ a version of this mechanism with some elaborations (Taylor *et al.*, 1979a).

Actomyosin-based systems of protoplasmic movement are very common in the biological world. Differences exist among them, so that the contraction–hydraulic mechanism proposed here does not explain all instances of motion produced by actomyosin (Allen and Allen, 1978; Stebbings and Hyams, 1979). Even in *Physarum*, mechanisms controlling calcium homeostasis in the cytoplasm may be disrupted, producing fountain streaming or cyclosis (Hatano, 1970; Kuroda, 1979), protoplasmic movements which cannot be easily explained by a hydraulic pressure mechanism per se, although the system is still based on the interaction of actin and myosin. Therefore, the physical arrangement of actomyosin in the ectoplasm and the nature of the mechanism controlling free cytoplasmic calcium concentrations are crucial problems in considering the evidence for the contraction–hydraulic mechanism proposed here for shuttle streaming in the plasmodium.

2. CORRELATING THE ORIENTATION AND QUANTITY OF ECTOPLASMIC FIBRILS WITH ENDOPLASMIC FLOW AND FORCE PRODUCTION

A number of experiments and observations have been made supporting a model for shuttle streaming in the plasmodium of *Physarum* in which fibrils in the ectoplasm contract to produce pressure on the endoplasm, resulting in endoplasmic streaming. See the review by Komnick *et al.* (1973) for particularly persuasive evidence.

a. Observations of Birefringent Fibrils in Vivo. Nakajima and Allen (1965) examined small, motile plasmodia of *P. polycephalum* with the polarizing microscope and found birefringent fibrils in the ectoplasm oriented parallel, circumferentially, and spirally along the long axis of the veins. The fibrils are positioned so that contractions within them could bring about a decrease in girth and a shortening of the veins in which they are embedded. Some fibrils remain unchanged in birefringence for a number of minutes, but others, particularly the circumferential fibrils, undergo rhythmic changes in birefringence during the shuttle-streaming cycle. These observations support the idea that the fibrils contract to cause endoplasmic flow. Similar changes of birefringence in fibrils at the

5. Plasmodial Structure and Motility

anterior end of the plasmodium have been reported (Yoshimoto and Kamiya, 1978c).

b. Fibril Orientation along Veins Is Correlated with Tension Production. Wohlfarth-Bottermann (1975b) observed parallel, circularly, and spirally oriented ectoplasmic fibrils in sections of plasmodial strands in the same location and orientation as the birefringent fibrils seen *in vivo* by Nakajima and Allen (1965). These are the actomyosin-containing fibrils identified in ultrastructural studies. Measurements of tension activity at certain points along actively streaming veins with tension transducers reveal both longitudinal and radial contractions correlated with the shuttle-streaming cycle (Wohlfarth-Bottermann, 1975a). Hence, the correlation of force production along a vein with fibril orientation is compatible with a mechanism of fibril contraction.

c. De Novo Appearance of Fibrils in Protoplasmic Droplets Is Correlated with Contraction. Large plasmodial veins may be punctured, so that a drop of fluid endoplasm is extruded quickly under pressure. The droplet remains attached to the vein but is temporarily stationary. Examination of the cytoplasm reveals that fibrils are absent throughout the drop just after it has emerged. During the subsequent 10-minute period, the drop begins to differentiate into an ectoplasmic portion in which fibrils begin to appear and an endoplasmic portion which lacks them. At about 10 minutes after formation, the endoplasm starts to exhibit shuttle streaming and begins to flow back into the endoplasm of the vein (Wohlfarth-Bottermann, 1964b). Motion starts several minutes after fibrils first become visible, never before fibrils are seen (Isenberg and Wohlfarth-Bottermann, 1976).

3. THE FILAMENTOUS NETWORK ALONG THE INSIDE OF THE PLASMALEMMA ALSO PRODUCES TENSION

In active macroplasmodia, the filamentous network on the inner surface of the plasmalemma is continuous with the fibrils containing parallel arrays of F-actin filaments attached to the inner side of the membrane of the ectoplasmic invaginations (Section II,B,4,*h*). The force created by the contraction of the fibrils in the ectoplasm can therefore be communicated to this plasmalemma-associated network, directing the contractile force inward to the endoplasm and preventing herniation of the plasmalemma.

Gawlitta *et al.* (1980a) studied the motile behavior of microplasmodia of different shapes recently removed from shake flasks. In bell-shaped microplasmodia, the well-organized shuttle streaming is correlated with the presence of fibrils in the ectoplasm. However, in amoeboid-shaped microplasmodia, the motive force for the irregular protoplasmic streaming activity must be produced in the filamentous network just inside the plasmalemma, since neither ectoplasmic fibrils nor invaginations of the plasmalemma are present. Hence, it is likely that the plasmalemma-associated filamentous network is composed of actomyosin. Whether the filaments in this network are bound together by an actin-binding protein or gelation factor, as described in other amoeboid cells (Taylor *et al.*, 1979a; Stossel *et al.*, 1979), is not yet known. Likewise, the presence of a spectrin–actin complex at the inner surface of the membrane has not been demonstrated in *Physarum* (Marchesi, 1979).

B. Mechanism of Actomyosin Contraction: A Sliding Filament Model

Indirect evidence supports a sliding filament model for fibril contraction in the plasmodium. This mechanism in muscle involves the sliding together of oppositely polarized F-actin (thin) filaments on either side of bipolar myosin (thick) filaments, shortening the fiber of which they are a part by the increasing overlap of the filaments (Huxley, 1963, 1969, 1972).

1. ACTIN POLARITY

Each ectoplasmic fibril seen in the plasmodium contains a great many F-actin filaments, often in parallel array. The polarity of F-actin filaments may be determined by the direction of the arrowhead pattern resulting from combination with heavy meromyosin. Examination of the polarity of neighboring F-actin filaments within a particular fibril in glycerinated plasmodia (Alléra *et al.*, 1971) yields equivocal results on this point, but probably each fibril is composed of a mixture of F-actin filaments polarized in either direction. This arrangement is a necessary requirement for a sliding filament mechanism.

2. MYOSIN POLARITY

Bipolar myosin filaments must also be present in the fibrils if contraction occurs by filaments sliding together, as in striated muscle. The state of *Physarum* myosin aggregation *in vivo* is unknown, since myosin thick filaments are not observed in the fibrils unless the plasmodia have been glycerinated or convulsed by slow-acting fixatives (Kessler, 1972b; Alléra and Wohlfarth-Bottermann, 1972). Filaments slightly thicker in diameter than the 50–70-Å actin filaments are occasionally observed in the fibrils of normal plasmodia. They may represent myosin oligomers in which polarity is too difficult to ascertain. Evidence concerning *Physarum* myosin filament polarity is therefore limited to *in vitro* studies of myosin polymerization.

Purified *Physarum* myosin aggregates to form large, thick filaments at low ionic strength in the presence of divalent cations (Fig. 11) or upon isoelectric precipitation (Nachmias, 1972b,c, 1974; Hinssen and D'Haese, 1974). Polarity is clearly bipolar in smaller filaments. More recent studies of *Physarum* myosin aggregation demonstrate that the large filaments which have no clear bare zone

5. Plasmodial Structure and Motility

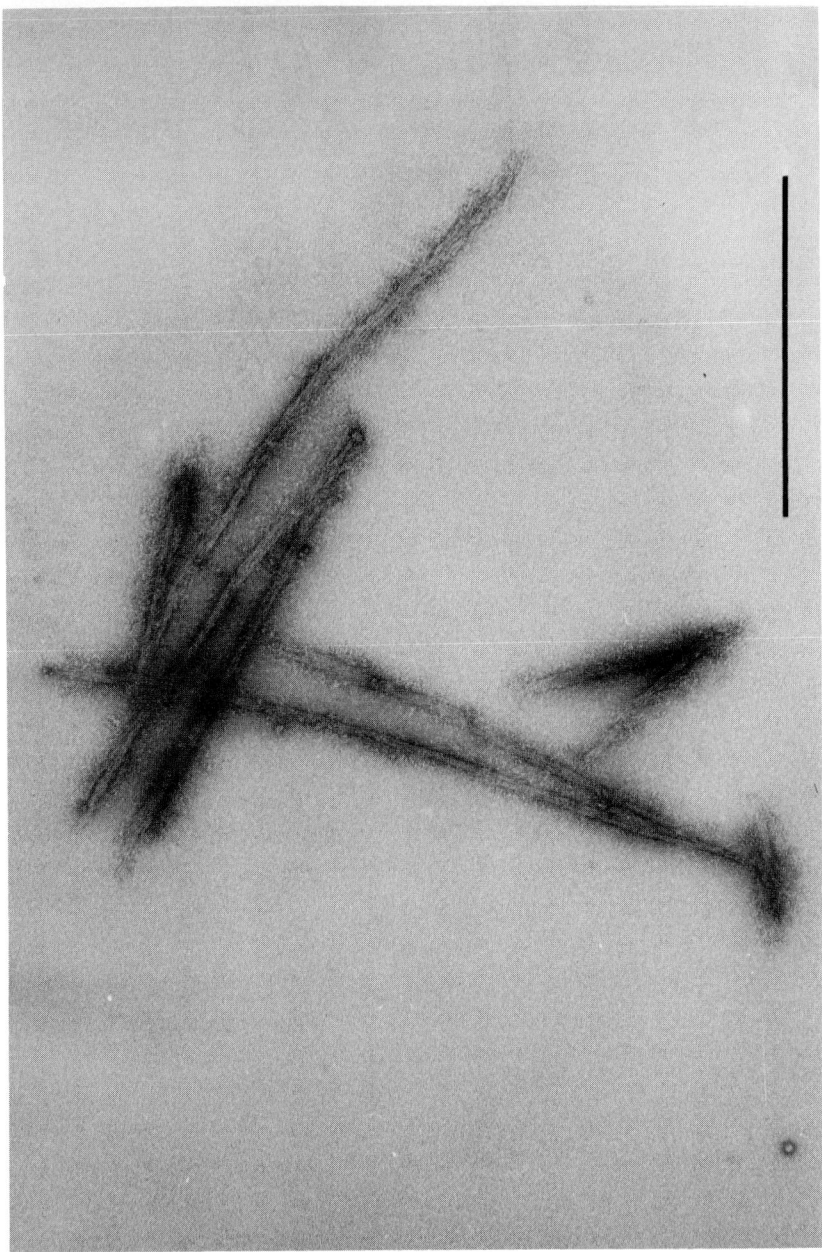

Fig. 11. Long filaments of purified *Physarum* myosin. The monomers aggregate into filaments at low ionic strength in the presence of Mg^{2+} or Ca^{2+}. × 46,000. From Nachmias (1974).

separating the two poles nevertheless are obliquely bipolar, i.e., polarized in opposite directions on each side of the filament, and comparable to smooth muscle myosin filaments (D'Haese and Hinssen, 1979). This work on *Physarum* myosin polymerization *in vitro* supports the notion that myosin oligomers are bipolar in the functioning ectoplasmic fibril.

3. ACTOMYOSIN THREAD STUDIES AS MODELS FOR FIBRIL CONTRACTION

Actomyosin in the ectoplasm of plasmodia, both in the fibrils and in the filamentous network just inside the plasmalemma, probably contracts in a manner similar to the actomyosin threads which have been studied *in vitro* (see Komnick *et al.*, 1973; Hinssen and D'Haese, 1976). A suspension of unoriented filaments composed of F-actin and bipolar myosin filaments may contract at low ionic strength in the presence of ATP by a sliding filament mechanism. This is due to the properties of the bipolar myosin filaments: Not only can they bind two (or more) oppositely polarized F-actin filaments, but they tend to draw these filaments together rather than to push them apart, due to the opposing position and movement of the myosin heads in the two regions of polarity (Huxley, 1969). In such a suspension, actin filament overlap greatly increases upon addition of ATP, leading to superprecipitation. Monomeric myosin cannot produce opposing filament overlap. When myosin is in the monomer form, for example, by dissolving the actomyosin at high ionic strength, skeletal muscle actomyosin does not superprecipitate. *Physarum* actomyosin can be made to superprecipitate *in vitro* upon the addition of ATP at low ionic strength (Hatano and Tazawa, 1968; Matsumura and Hatano, 1978; Hatano *et al.*, 1979). Therefore, if the mechanism of superprecipitation is the same for *Physarum* and muscle actomyosin, *Physarum* myosin must be in a bipolar state.

The systematic studies of actomyosin threads made by extruding actomyosin isolated from different organisms into a solution of low ionic strength demonstrate that threads made from *Physarum* actomyosin have the same contractile properties as those from muscle actomyosins (Beck *et al.*, 1969, 1970a, 1970b; Komnick *et al.*, 1973; D'Haese and Hinssen, 1978). When the actin filaments are oriented parallel to the longitudinal axis of the thread, either by the application of tension or by parallel aggregation of the actin during polymerization in the case of artificial fibrils (Hinssen and D'Haese, 1976), then subsequent contraction on the addition of ATP is primarily along the longitudinal axis of the system, much as the plasmodial actomyosin fibrils appear to contract in the living organism. During isometric contraction of ectoplasmic fibrils *in vivo*, the actin filaments may be observed to be aligned in a parallel array (Nagai *et al.*, 1975, 1978; Fleischer and Wohlfarth-Bottermann, 1975), supporting the idea of a sliding filament mechanism and ruling out fibril contraction by a change in actin polymerization from a linear F-actin form to a shortened, nonlinear polymer (Hatano, 1972).

5. Plasmodial Structure and Motility

In summary, the most likely mechanism for the contraction of actomyosin in the ectoplasm of the living plasmodium is the increased overlap of F-actin filaments by sliding against oligomeric bipolar myosin.

C. State of Actin Polymerization

1. STRUCTURAL EVANESCENCE OF THE CONTRACTILE APPARATUS

Careful microscopic observation of the plasmodium during active streaming leads to the conclusion that the ectoplasm and the endoplasm are interchangeable forms of the same substance (Lewis, 1942). Grebecki and Cieślawska (1978a) estimate that 29% of the ectoplasm moves into the endoplasmic stream at the rear of the plasmodium and along the wall of the vein during forward contraction, whereas endoplasm is transformed to a stationary ectoplasm at the front. It seems probable that only a portion of the contractile apparatus in the ectoplasm along a vein is disassembled after each ectoplasmic contraction in the shuttle-streaming cycle. Observations of motile plasmodia with the polarizing microscope indicate that the duration of at least some of the birefringent ectoplasmic fibrils is longer than one cycle (Nakajima and Allen, 1965; Taylor and Wang, 1978). In isotonic contraction, when the plasmodial strand is allowed to change in length, fibrils are composed of packed filaments aligned parallel to the long axis of the fibril in the beginning of the contracting phase, while the strand is still at maximum length, and become transformed into feltlike networks of filaments at minimal strand length (Nagai *et al.*, 1975, 1978; Kamiya, 1979). During the initial stages of contraction, the filaments probably become aligned along the long axis of the fibril by the developing tension exerted as the two ends of the contracting fibril meet resistance from their attachment to the filamentous network along the inner surface of the plasmalemma. When the whole strand shortens and the surface of the plasmalemma changes greatly in configuration, the parallel alignment of the filaments may become distorted as the amount of resistance at the two ends of each fibril changes unpredictably.

In summary, complete disassembly followed by reassembly occurs for only a portion of the contractile apparatus in the ectoplasm of a vein during each shuttle-streaming cycle. However, during active migration of the plasmodium, a net disassembly of the contractile apparatus must occur in the ectoplasm of the veins at the rear for easy transportation in the endoplasm to the front, where a net reassembly occurs.

3. EFFECT OF DRUGS INTERFERING WITH ACTIN RECYCLING: CYTOCHALASIN AND PHALLOIDIN

The depolymerization and repolymerization of F-actin would be an effective system for the disassembly and reassembly of the contractile apparatus occurring during migration. Such a mechanism requires control of the net polymerization and depolymerization at opposite ends of the F-actin filament during steady-state recycling between G-actin monomer and F-actin polymer (Wegner, 1976). Inhibitors of actin polymerization and depolymerization produce profound effects on protoplasmic streaming in *Physarum*, indicating that actin recycling is an important mechanism in the streaming cycle *in vivo*. For example, cytochalasin is known to inhibit actin polymerization under conditions of active recycling (Brenner and Korn, 1979, 1980; Flanagan and Lin, 1980). In *Physarum* external application of cytochalasin appears to induce localized conversion of ectoplasmic wall to fluid endoplasm, resulting in herniation and evagination of the endoplasmic stream under pressure from ectoplasmic contraction elsewhere along the vein (Kessler, 1972a; Mante *et al.*, 1978). Phalloidin, a toxin from mushrooms which preserves the F-actin state, upon injection into a plasmodial strand causes cessation of contraction and the formation of an abnormally large amount of F-actin in the cytoplasm (Götz von Olenhusen and Wohlfarth-Bottermann, 1979b).

3. EVIDENCE THAT DEPOLYMERIZED ACTIN EXISTS IN THE ENDOPLASM

In nonmuscle cells, the pool of actin in nonfilamentous form is rather high (Bray and Thomas, 1976), making it likely that cytoplasmic actin normally recycles between the G and F states. Both Hinssen (1972) and Adelman (1977) extracted actin from the plasmodium of *Physarum* and found that at least half of it is in monomer or nonfilamentous form even though the total cytoplasmic actin concentration is 3–10 mg/ml (Hinssen, 1979), well above the critical concentration for polymerization under physiological conditions (Hatano and Oosawa, 1979).

If living strands of the plasmodium are suddenly stretched by 50% of their original length and held there, endoplasmic streaming ceases. Within 5 seconds after stretching, plaques of homogeneous material devoid of organelles appear throughout the endoplasm, and by 35 seconds after stretching, fibrils composed of F-actin filaments aligned in parallel have appeared in the plaques. The strands begin to exert a longitudinal contraction at this time (Fleischer and Wohlfarth-Bottermann, 1975; Wohlfarth-Bottermann and Fleischer, 1976; Wohlfarth-Bottermann and Isenberg, 1976). If the strands are released and allowed to return to resting length, the endoplasmic fibrils disappear within a short time and normal streaming gradually resumes.

De novo formation of fibrils also occurs in endoplasmic droplets which are formed by the protrusion of endoplasm under pressure after puncturing a vein during streaming. Drops at 0 minutes after formation do not contain fibrils, nor are they differentiated into ectoplasm and endoplasm. However, 5 minutes later, fibrils begin to appear in plaques throughout the newly forming ectoplasmic layer of the drop, and in 10-minute drops they are very abundant in the ectoplasm as the endoplasm of the drop begins to flow back into the veins. Portions of the cytoplasm from these drops are placed on a spreading solution containing buffer prior to negative staining and are then examined in the electron microscope. F-actin filaments are not seen in the spread cytoplasm of 0-minute drops, but they are more and more abundant in preparations from older drops. Additions of polylysine (a nucleating agent for actin polymerization) to the spreading solution produces large numbers of F-actin filaments even in the 0-minute drops. Therefore, the endoplasm must contain actin in a nonfilamentous form which can be converted to F-actin under the appropriate experimental conditions (Isenberg and Wohlfarth-Bottermann, 1976; Wohlfarth-Bottermann and Isenberg, 1976).

The biochemical mechanisms which regulate the polymerization of endoplasmic actin are currently the subject of intensive research. When high-speed supernatants of plasmodial extracts are examined in the electron microscope, no F-actin is seen (Nachmias, 1979b). However, the addition of 5'-AMP stimulates the production of F-actin, together with a rise in viscosity of the supernatant. Other mononucleotides, ATP, and cyclic AMP are not as effective in inducing the appearance of F-actin. When the filaments are collected and resuspended in dialyzed high-speed supernatant, they disappear! These experiments point to the existence of a factor in the cytoplasm of the slime mold which regulates actin polymerization.

4. FRAGMIN

A possible candidate for such a regulatory factor influencing actin polymerization is fragmin (Maruyama et al., 1976; Hatano and Owaribe, 1976, 1979; Hatano et al., 1979; Hasegawa et al., 1980), a protein discovered as a contaminant in *Physarum* plasmodial actin preparations and originally named "*Physarum* actinin." Hinssen (1979) has also isolated this protein from the plasmodium. Fragmin has the same molecular weight as G-actin but has quite a different amino acid composition. Unlike actin, fragmin lacks cysteine residues. Fragmin co-migrates with actin after electrophoresis on SDS–polyacrylamide gels, but they migrate slightly differently on urea–SDS gels. The molecular weight and amino acid composition indicate that *Physarum* fragmin is not closely related to muscle α and β actinin, or to profilin, a protein from platelets and spleen which binds to G-actin and prevents its polymerization (Carlsson et al., 1976; Markey et al., 1978; Harris and Weeds, 1978).

When fragmin is added to a solution of F-actin *in vitro*, the viscosity of the solution decreases, as though fragmin cuts the actin filaments at random positions into smaller F-actin fragments (Hasegawa et al., 1980). When actin is polymerized in the presence of various amounts of fragmin (Fig. 12), shorter filaments are formed with increasing fragmin concentration. Fragmin appears to bind to the initiation end of nascent F-actin filaments during polymerization, resulting in the availability of more nuclei for polymerization and therefore an

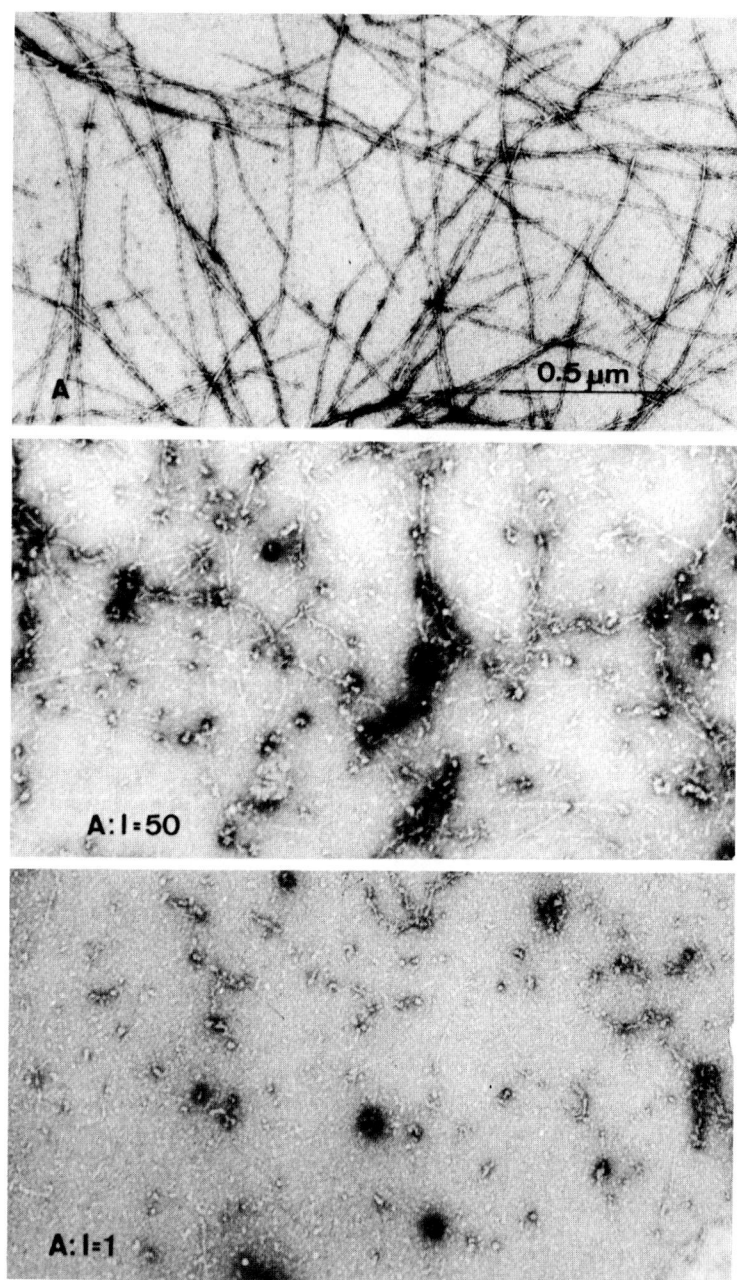

Fig. 12. Muscle actin polymerized in the presence of *Physarum* fragmin, which can inhibit actin polymerization. (A) Actin alone polymerized with 0.1 M KCl. (A:I = 50) Same conditions as in (A), but inhibitor was added at a weight ratio of 1:50. (A:I = 1) Polymerization inhibitor was added at a weight ratio of 1:1. From Hinssen (1979).

increased number of shorter filaments. The interaction of fragmin with actin requires free Ca^{2+} at a concentration higher than 10^{-7} M (Hasegawa et al., 1980). This requirement for calcium is puzzling since it suggests that fragmin is most active in that region of the plasmodium in which actomyosin contraction is occurring, i.e., assuming the interaction of actin and myosin in the plasmodium is also calcium sensitive (Section III,D). This problem would be resolved if fragmin were bound to the inner surface of the plasmalemma and its major physiological role in the presence of calcium were the creation of nuclei for F-actin polymerization. The localization and state of fragmin in living plasmodia is the subject of continuing investigation (Hasegawa et al., 1980).

D. Calcium Regulation of the Contractile System

1. EVIDENCE FROM *IN VIVO* EXPERIMENTS

A number of studies strongly support the hypothesis that calcium is required for protoplasmic movement in *Physarum*.

a. Aequorin is a protein which, if injected into cells, emits light in response to small increases in free cytoplasmic calcium levels. By this method, Ridgway and Durham (1976) observed a periodic rise in light emission at each contraction in the end of a small plasmodial strand during active shuttle streaming. They concluded that the free calcium concentration rises in the cytoplasm during contraction. However, these fluctuations in light emission may have been caused merely by the periodic increase and decrease of strand thickness which occurs during shuttle streaming (Sachsenmaier et al., 1973), rather than by oscillations in free cytoplasmic calcium concentration. This important study should be repeated to clarify the interpretation of the data. The work discussed below suggests that the idea that calcium is required for ectoplasmic contraction is correct.

b. Calcium–EGTA buffers of known free calcium concentration may be injected into the endoplasm of plasmodia (Ueda et al., 1978; and Ueda and Kobatake, this volume, Chapter 4). Strong radial and longitudinal contractions occur only in veins injected with solutions containing more than 10^{-7} M Ca^{2+}. The half-maximal response lies in the area of 4×10^{-7} M Ca^{2+}. Below 10^{-7} M the veins do not contract, but they may recover a normal contraction rhythm after a period of time. In experiments involving injection of nucleotides, maximum contraction response for ATP occurs at 0.2 mM, whereas higher concentrations are required for maximal contraction with ADP (0.5 mM) and AMP (2.5 mM). The effects of ADP and AMP may be indirect, i.e., on the ATP-regenerating system or on the system controlling actin polymerization.

The levels of Ca^{2+} and ATP required for maximal contraction of plasmodial veins are very similar to those in contracting muscle (Huxley, 1972), emphasizing the similarity of the two contractile systems.

c. The plasmodium may be treated with 5–10 mM caffeine, which causes the ectoplasm to be weakened so that the endoplasm gushes out through the broken places to form separate small, spherical droplets (Hatano and Oosawa, 1971). Adenosine or adenine treatment of plasmodia produces similar droplets (Nachmias, 1979a; Nachmias and Meyers, 1980). These droplets consist of protoplasmic membrane, hyaloplasm, and a granular cytoplasm which has a movement more like fountain streaming than the usual shuttle streaming. Similar movements are seen in membrane-free droplets of cytoplasm from plasmodia (Kuroda, 1979). Hatano (1970) found that these droplets are much more sensitive to external calcium concentrations than normal microplasmodia. If the droplets are suspended in a solution containing the calcium chelator EGTA, active movements of the cytoplasm gradually cease. Movement is activated again in the presence of calcium at a concentration no lower than 10^{-7}–10^{-5} M, about the same as that required for skeletal muscle contraction. Magnesium is not effective in restoring movement. Matthews (1977) resuspended the droplets in caffeine-free medium containing EGTA and applied test solutions by micropipette to the surface of individual inactive droplets. Local application of caffeine stimulated movement again for a short time. After repeated application, the droplets gradually lost the ability to respond, but they revived on the addition of $CaCl_2$. The ionophore A23187 was even more effective than caffeine in reviving movement. These experiments clearly demonstrate that calcium is necessary for streaming and that it can be mobilized from internal storage sites.

2. CYTOPLASMIC VACUOLES CAPABLE OF SEQUESTERING AND RELEASING CALCIUM

Calcium-sequestering vacuoles have been induced to appear in large numbers in the cytoplasm of plasmodia by two methods: in presporulating plasmodia by illumination (Daniel and Järlfors, 1972b) and in briefly fixed plasmodia by ATP-induced accumulation of calcium into vacuoles (Braatz and Komnick, 1970, 1973). Braatz (1975) extended these studies by demonstrating with histological techniques a change in the cytoplasmic location of calcium correlated with the shuttle-streaming cycle. Relaxed regions of plasmodial veins had more calcium in vacuoles, less calcium free in the cytoplasm. In contrast, contracted areas had fewer vacuoles containing calcium but much more calcium in the groundplasm. Mitochondria did not contain large amounts of observable calcium. These findings support the hypothesis that the presence of calcium in the groundplasm is required for contraction and that cyclic release and uptake of calcium by vacuoles in the cytoplasm is an important part of the regulatory system for shuttle streaming.

In recent biochemical experiments, vesicles isolated from plasmodia of *Physarum* have been shown to accumulate Ca^{2+} in the presence of ATP. The

5. Plasmodial Structure and Motility

half-maximal calcium concentration for activation of the Ca^{2+}, Mg^{2+} membrane-bound ATPase is about 10^{-6} M, analogous to that of the sarcoplasmic reticulum of skeletal muscle (Kato and Tonomura, 1977; Kato, 1979). Antibodies directed against the Ca^{2+}, Mg^{2+}-ATPase from muscle sarcoplasmic reticulum cross-react with protein isolated from plasmodial vesicles. Both the muscle and the plasmodial membrane-bound ATPases have a molecular weight of 105,000 by SDS–polyacrylamide gel electrophoresis (Zubruzycka-Gaarn et al., 1979). The mechanism for controlling cytoplasmic calcium levels in contracting cells is evidently very conservative evolutionarily.

3. CALCIUM REGULATION OF ACTOMYOSIN CONTRACTION

a. Calcium Regulation of Actomyosin in Other Contractile Systems. In muscle and several nonmuscle contractile systems, calcium has been found to regulate actomyosin ATPase and contraction. At least three mechanisms for this regulation have been discovered, and others (Ebashi et al., 1979) will probably be found as work continues on this problem.

In mammalian striated muscle, calcium regulation occurs by thin filament control. That is, the calcium-binding protein, troponin C, is part of a complex on the actin filaments with tropomyosin and two other subunits of troponin (Gergely, 1976). Upon stimulation, calcium released from the sarcoplasmic reticulum and increases in the sarcoplasm to about 10^{-5} M, at which concentration troponin C is easily capable of binding calcium. This results in a series of conformation changes permitting actin and myosin to interact, splitting ATP, and producing the force necessary for contraction.

Not all actomyosin contractile systems exhibit thin filament control of actomyosin ATPase regulation. Two examples of myosin control are known. In mollusk muscles and in certain other invertebrates, calcium binds directly to a subunit of the myosin, the regulatory light chain, resulting in an enhanced capability of actomyosin interaction, thus triggering muscle contraction (Szent-Györgyi et al., 1973). In at least some smooth muscles and in platelets (Sherry et al., 1978; Hathaway and Adelstein, 1979; Adelstein and Eisenberg, 1980; Small and Sobieszek, 1980), a myosin light chain kinase possessing the calcium-binding protein calmodulin as a subunit catalyzes the phosphorylation of one of the myosin light chains in the presence of calcium. At this point, the phosphorylated myosin is capable of interacting with actin, resulting in contraction.

Troponin C, myosin light chains, and calmodulin, the polypeptides capable of binding calcium at low cytoplasmic concentrations in these systems of contraction, are evolutionarily related. For further information about their structure and their roles as calcium-modulated proteins, see Kretsinger (1979, 1980). Myosin light chains have been discussed by Weeds and McLachlan (1974),

Collins (1976), and Weeds *et al.* (1977). Detailed reviews of the wide-ranging functions of calmodulin may be found in Wolff and Brostrom (1979), Dedman *et al.* (1979), Wang and Waisman (1979), Cheung (1980), and Klee *et al.* (1980).

b. Physarum *Actomyosin ATPase.* The properties of purified *Physarum* myosin have been summarized by Nachmias (1979b). Compared with skeletal muscle myosin, its properties are strikingly similar, including the presence of ATPase activity at high ionic strength. The relatively minor differences between *Physarum* myosin and mammalian skeletal muscle myosin are the low sulfhydryl content of slime mold myosin, its slightly higher molecular weight, and the absence of a K^+, EDTA-activated ATPase in the slime mold protein. The polymerization properties of *Physarum* myosin *in vivo* are quite different from those of skeletal muscle myosin, although both can form thick filaments *in vitro* (see Section III,B,2).

Physarum actomyosin has ATPase activity at low ionic strength in the presence of magnesium. This is the physiologically significant ATPase activity of myosin, and therefore its properties are very relevant in any discussion of the regulation of contraction. *Physarum* actomyosin does not require calcium for superprecipitation. In other words, it is not calcium sensitive (Hatano and Tazawa, 1968). When purified *Physarum* myosin and purified actin from rabbit muscle or *Physarum* are combined, the ATPase activity at low ionic strength is increased 4–10 times, indicating actin activation of the myosin ATPase, albeit somewhat lower than that observed with skeletal muscle myosin (Hatano and Ohnuma, 1970; Nachmias, 1974).

Since purified *Physarum* actomyosin does not require calcium for ATPase activity, one mechanism for the regulation of contraction *in vivo* may be ruled out. Activation of the actomyosin ATPase by direct calcium binding to the myosin, as in mollusk muscle, cannot be the calcium-sensitive regulatory system in *Physarum*.

Experiments with crude preparations of *Physarum* actomyosin initially indicated the presence of a calcium-sensitive troponin-like regulatory factor on the actin similar to the calcium control system of skeletal muscle actomyosin (Nakajima, 1964; Tanaka and Hatano, 1972; Nachmias and Asch, 1974, 1976; Kato and Tonomura, 1975; Nachmias, 1975). However, the calcium requirement in these crude preparations for ATPase activity is due to a contaminating calcium-activated ATP pyrophosphohydrolase (Jacobson *et al.,* 1976; Nachmias, 1979b). Therefore, a thin-filament calcium-control system for regulating actomyosin ATPase in *Physarum* seems unlikely.

c. Probable Calcium Control Mechanism in Physarum. The mechanism of actomyosin regulation by calcium in *Physarum* is assumed to be reasonably similar to known mechanisms in other contractile systems. If so, the most promising model to account for the calcium requirement for protoplasmic streaming in

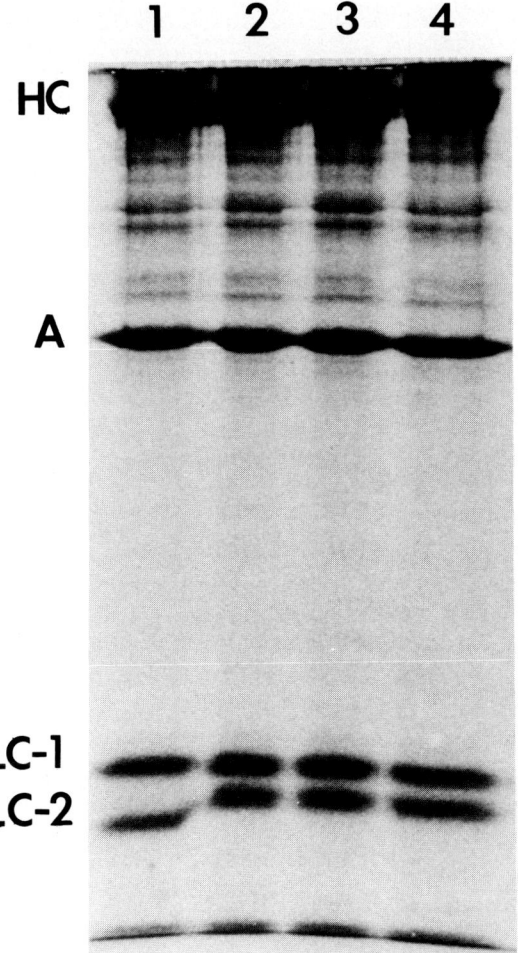

Fig. 13. Change in relative mobility of *Physarum* myosin LC-2 caused by the addition of Ca^{2+}, but not Mg^{2+}. *Physarum* actomyosin in 1 mM EGTA was subjected to electrophoresis on an SDS polyacrylamide gel after adjusting the sample to the indicated concentration of $CaCl_2$ or $MgCl_2$. Well 1, 5 mM $CaCl_2$; well 2, EGTA alone; well 3, 5 mM $MgCl_2$; well 4, 100 mM $MgCl_2$. HC, myosin heavy chain; A, actin; LC-1, myosin light chain-1; LC-2, myosin light chain-2. From Kessler *et al.* (1980).

Physarum is myosin light chain phosphorylation, resulting in activation of the actomyosin ATPase, as in smooth muscle. Calcium control in this model is indirect, required by the calmodulin subunit of a myosin light chain kinase. The model requires a phosphatase for inactivating the system by removing the phosphate from the myosin light chain. Evidence for this mechanism is now being accumulated.

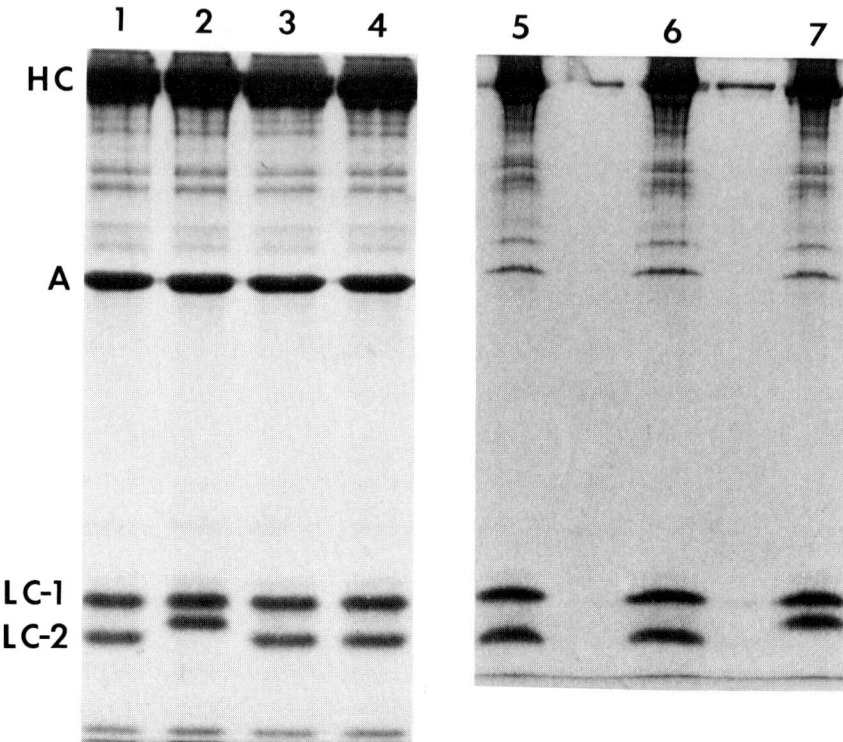

Fig. 14. Addition of Ca^{2+}, Cd^{2+}, or La^{3+} induces a change in relative mobility of *Physarum* myosin LC-2. SDS–polyacrylamide gel electrophoresis of *Physarum* actomyosin in 1 mM EGTA was performed after adding to the sample buffered cation stock solution to the concentration indicated. Well 1, 5 mM $CaCl_2$; well 2, EGTA alone; well 3, 5 mM Cd $(NO_3)_2$ (Baker, Reagent Grade); well 4, 5 mM Cd $(NO_3)_2$ (Puratronic, Grade I); well 5, 2.6 mM $LaCl_3$; well 6, 5 mM $CaCl_2$; well 7, EGTA alone; A, actin; Other abbreviations as indicated in Fig. 13. From Kessler *et al.* (1980).

Calmodulin has been isolated from the plasmodium (Kuźnicki and Drabikowski, 1979; Kuźnicki *et al.*, 1979), and experiments are underway to determine if it is required for myosin light chain phosphorylation. *Physarum* myosin has two classes of light chains, LC-1 and LC-2, with molecular weights of 17,700 and 16,900, respectively, after electrophoresis in SDS–polyacrylamide gels (Kessler *et al.*, 1980). Experiments are now in progress, which indicate that *Physarum* myosin LC-1 can be phosphorylated, and tests are being conducted to determine if the phosphorylated myosin differs in actomyosin ATPase activity from the unphosphorylated form (V. Nachmias, personal communication).

Physarum myosin LC-2 has an interesting property seen after electrophoresis in SDS–polyacrylamide gels (Kessler *et al.*, 1980). In the presence of the metal ion chelator EGTA, it migrates with an apparent molecular weight of 16,900, but if calcium ions are added to the sample prior to electrophoresis, the apparent molecular weight decreases to 16,100 (Fig. 13). We assume that LC-2 is capable of binding Ca^{2+} even after boiling in buffer containing SDS and 2-mercaptoethanol, resulting in a more compact polypeptide that migrates more quickly during electrophoresis in the presence of Ca^{2+}. The presence of Mg^{2+} at a concentration of 5–100 mM in the sample does not produce the same relative change in migration of LC-2 that is noted with Ca^{2+}, indicating the binding is more specific for Ca^{2+} (Fig. 14). Lanthanum and cadmium ions, but not magnesium, can substitute for calcium. Because the ionic radii of Ca^{2+}, La^{3+}, and Cd^{2+} are almost identical, *Physarum* myosin LC-2 must possess a very size-specific binding site for calcium (Kessler *et al.*, 1980). Troponin C and calmodulin, both known to be important in actomyosin regulation by calcium, share with *Physarum* myosin LC-2 the property of changing their electrophoretic mobility in SDS–PAGE depending on the calcium concentration of the sample (Klee *et al.*, 1979; Burgess *et al.*, 1980). We are uncertain if native *Physarum* myosin binds calcium under physiological conditions, i.e., in the presence of millimolar concentrations of magnesium ions. If so, then calcium binding by the light chain may be required before a subsequent regulatory step, such as phosphorylation, can take place.

IV. AN OSCILLATING REGULATORY SYSTEM CONTROLS SHUTTLE STREAMING

A. The Whole Plasmodium Is a Monorhythmic System

Recent cinematographic analysis of the pulsation rhythms in the veins of moderately small plasmodia (8 cm or less in diameter) during streaming in the forward direction lead to the rather startling conclusion that the entire system of major veins in one plasmodium contracts simultaneously while the frontal end expands (Grebecki and Cieślawska, 1978b; Grebecki, 1979; Cieślawska and Grebecki, 1979). Upon streaming in the reverse direction, the front contracts and the veins expand. The duration of the entire cycle of shuttle streaming is 1.5–3 minutes. Occasionally, less than a 1-minute difference in phase may develop among the major veins at any time in the shuttle-streaming cycle. The pulsation usually returns to the common rhythm in the course of one or two subsequent cycles. The small lateral veins are well synchronized in their contraction cycle with the major veins, but the shuttle-streaming rhythm within them may be irregular due to hydrostatic pressure differences between the larger veins they connect (Grebecki and Moczon, 1978). Coordination of plasmodial endoplasmic flow among the major veins is monorhythmic rather than peristaltic or polyrhythmic. Exceptions to monorhythmic contraction occur when well-separated veins are observed in a large plasmodium where several fronts exist (Loewy, 1949; Stewart, 1964).

B. Local Ectoplasmic Contraction Rhythms Are Coordinated by Endoplasmic Flow

1. AN OSCILLATING CONTROL SYSTEM

In Kamiya's double-chamber experiment (Section I,B), the amount of air pressure just required to prevent endoplasmic flow in a strand joining two parts of a plasmodium varies rhythmically with the same period as that observed previously for the shuttle-streaming cycle (Kamiya, 1942, 1959). This demonstrates the existence of an oscillating control system which regulates the amount of actomyosin contraction. Ordinarily coupled with the frequency of shuttle streaming, the control system is revealed when endoplasmic flow is prevented but ectoplasmic contraction continues in a rhythmic manner. Small changes in the motive force occur with time, indicating that some local control of the oscillating frequency is possible despite overall coordination in the plasmodium.

2. COORDINATION OF LOCAL ISOTONIC CONTRACTION RHYTHMS ALONG A NEWLY ISOLATED VEIN

A strand with active protoplasmic flow may be excised from the plasmodium. Streaming ceases for a while, but rhythmic contractions in the strand resume again in 15–20 minutes. Isotonic length changes may be measured in subsegments of the strand by measuring the changes in the distances separating small resin particles attached along the vein. Between the time of strand excision and the start of coordinated rhythmic contraction of the whole strand, small, irregular local contraction rhythms may be observed. The contractions of individual subsegments are diverse at first, and then become coordinated as the ectoplasm along the whole strand begins to contract simultaneously (Yoshimoto and Kamiya, 1978a; Kamiya, 1979). These experiments lead to the conclusion that the regulatory system which controls actomyosin contraction must be composed of local oscillators which begin to operate independently, but soon become coordinated along the strand so that they all adopt the same rhythm.

3. COORDINATION OF CONTRACTION RHYTHM BETWEEN TWO VEINS BY ENDOPLASMIC FLOW

The following experiments indicate that endoplasmic flow is necessary to maintain synchrony of contraction rhythm between two strands. Two unconnected strands may be excised from a plasmodium and hung on an apparatus in which the longitudinal isotonic or isometric contraction of each of the two sepa-

5. Plasmodial Structure and Motility

rate strands may be measured simultaneously. The contraction rhythm in each strand is different, indicating that each strand has its own system coordinating the rhythm of contraction along the strand. However, if the two veins are joined by a strand of plasmodium between them, then after about 40 minutes the two strands develop synchronous contraction rhythms. If the strand connecting them is severed, the two strands gradually lose synchrony after about 10 minutes.

The two strands may be put in separate airtight chambers and a connecting vein allowed to penetrate the septum between the chambers. If endoplasmic flow is allowed to shuttle normally in the connecting vein between the strands, the strands maintain a synchronous rhythm of contraction. If the flow of endoplasm through the connecting piece is stopped by balanced air pressure in one chamber, within about 10 minutes the two strands no longer have synchronous contraction. When the flow is allowed to proceed between them again, resynchrony occurs within 5 minutes (Yoshimoto and Kamiya, 1978b; Kamiya, 1979).

4. COORDINATION OF CONTRACTION RHYTHM IS NOT DUE TO AN ACTION POTENTIAL SPREADING ALONG THE STRANDS

The transmembrane potential along a strand of the plasmodium is difficult to measure for an extended period using internal microelectrodes because formation of new membrane quickly seals the end of the electrode, causing the potential difference to drop to zero. Values of -90 mV have been obtained (Hato *et al.*, 1976), but reported average values for the membrane potential are often somewhat less, ranging from -45 to -55 mV (Rhea, 1966b; Miller *et al.*, 1968; Meyer and Stockem, 1979). With either intracellular or surface electrodes, slow, rhythmic oscillations of potential with amplitudes of 1–6 mV and a peak-to-peak distance of about 90 seconds can be correlated with the shuttle-streaming cycle. They are also seen when endoplasmic flow is stopped artificially by air pressure. Change in membrane potential in the depolarization direction occurs during contraction rather than relaxation. Action potentials having a large amplitude with a complete reversal of sign are not observed during shuttle streaming, nor can they be induced by electrical stimulation (Miller *et al.*, 1968; Kamiya, 1979). Some investigators have recorded rapid changes in surface potential having much smaller amplitudes than action potentials. They are not transmitted along the strand. For example, Miller *et al.* (1968) found that sudden local depolarizations of small amplitude (2 mV) and rapid duration (1 second) may be superimposed on the depolarizing phase of the slower rhythmic change in potential or may be induced by mechanical stimulation.

The ion fluxes responsible for the changes in membrane potential have not been identified with certainty. Ridgway and Durham (1976) found a correlation between the change in surface potential between two ends of an actively streaming vein and the increase of free calcium in the cytoplasm during contraction measured by aequorin light emission. Mechanical stimulation also induces an increase in free cytoplasmic calcium. Several sources for this increase in cytoplasmic calcium are possible, one of which may be an influx from the external medium, as in *Paramecium* (Eckert, 1972). Calcium containing vacuoles in the cytoplasm may also release a large amount of free calcium into the cytoplasm at this time (see Section IV,D). Potassium fluxes across the plasmalemma have been studied by Anderson (1964) and Miller et al. (1968). Migrating plasmodia constantly lose K^+ to the substrate. Plasmodia which are forced to migrate toward the cathode by the application of direct current on their substrate lose much more potassium than controls in the absence of current. Holmes and Stewart (1977) incubated streaming plasmodia in medium containing ^{45}Ca and measured the change in radioactive calcium content of plasmodial aliquots at 1-minute intervals thereafter. They observed regular fluctuations with about the same period as shuttle streaming, suggesting oscillating rates of calcium efflux and/or influx during streaming.

In conclusion, no evidence supports the idea that action potentials traveling along the length of the plasmodium coordinate the systems controlling local rhythms of ectoplasmic contraction. However, small, nonpropagated, rhythmic changes in membrane potential occur along the length of the plasmodium and are well correlated with the shuttle-streaming cycle. The available evidence implicates K^+ and Ca^{2+} fluxes across the plasmalemma as the ions associated with these small, oscillating changes in membrane potential. No evidence links these small changes in membrane potential directly with the control of actomyosin contraction in the ectoplasm.

5. THE ENDOPLASMIC FLOW ITSELF MUST COORDINATE CONTRACTION RHYTHMS

The evidence presented so far in this section suggests that production of a monorhythmic plasmodial shuttle-streaming cycle must involve coordination of local contraction rhythms along the veins by the endoplasmic flow itself, not by a regenerative change of membrane potential propagated along the vein. Since exchange of cytoplasmic material between endoplasm and ectoplasm is incomplete during endoplasmic flow in a particular direction in the shuttle-streaming cycle (Section III,C,1), it seems unlikely that the spread of a chemical substance in the endoplasm is the mechanism linking endoplasmic flow with coordination of local contraction in the ectoplasm, where the force is produced.

C. Hypothesis: Coordination of Local Contraction Rhythms by Stretch Entrainment

Some types of smooth muscle which contract rhythmically without neural stimulation can be entrained by stretching. That is, a slightly different internal oscillation of contraction can be induced by periodic mechanical stretching if the

5. Plasmodial Structure and Motility

driving frequency is close to the autonomous frequency (Rapp and Berridge, 1977).

A stretch entrainment model could explain the coordination of local contraction rhythms along a vein of the plasmodium. A common contraction rhythm would develop by autoentrainment as each section of a strand became influenced by the endoplasmic flow coming from neighboring sections. Stretch entrainment would also be responsible for the coordination of the contraction rhythm between two plasmodial strands connected by a vein.

One experiment (Achenbach and Wohlfarth-Bottermann, 1980) demonstrates that stretch can produce local changes in contraction rhythm along a plasmodial vein. These authors measured simultaneously the isometric contraction of both ends of one strand and observed synchrony in their rhythm during normal conditions. When one end of the strand is stretched by 50% of its length, a phase change occurs between the contractions of the ends of the strand, which spontaneously resynchronize in 20–25 minutes. If the ends are separated by amputation just after stretching, they never resynchronize. Temporary phase shifting of contractions between the ends of a strand can also be induced by short, unilateral temperature decreases at one end.

Fleischer and Wohlfarth-Bottermann (1975) observed a spontaneous increase in tension produced in plasmodial strands by mechanical stretching. The tension

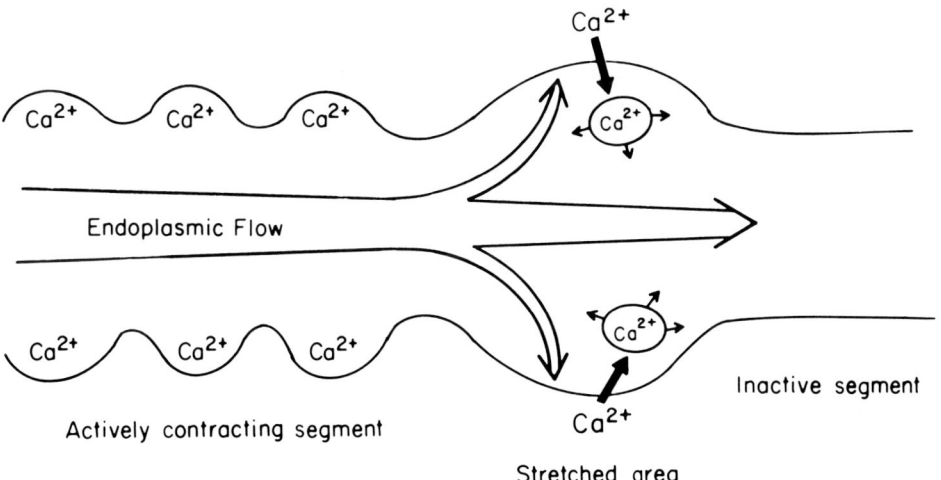

Fig. 15. Stretch entrainment hypothesis explaining the coordination of local contraction rhythms along a newly isolated vein. Endoplasmic flow from an actively contracting segment of the vein stretches the plasmalemma in an adjoining inactive segment, causing an increase in ion influx, perhaps Ca^{2+}, stimulating a much larger release of Ca^{2+} from cytoplasmic vacuoles. The increase in free cytoplasmic Ca^{2+} activates contraction in the ectoplasm of the previously inactive segment, resulting in endoplasmic flow.

increase begins in the strand about 35 seconds after the stretch stimulus is applied and lasts for about 44 seconds. Since the contraction phase in a strand during normal streaming is about 30–90 seconds in duration, the experimental values of Fleischer and Wohlfarth-Bottermann suggest that plasmodial strands are capable of responding with a contraction of the duration found in normal shuttle streaming when stimulated by stretch.

More specifically, the stretch entrainment hypothesis proposed here suggests that the increase in plasmodial strand volume caused by a rhythmic rise in endoplasmic flow from neighboring areas of a newly excised strand is a type of stretch stimulus (Fig. 15). This stimulus, perhaps by causing a small increase in calcium influx across the plasmalemma, stimulates further release of calcium into the cytoplasm from the localized internal vacuoles along the strand which are part of a system regulating rhythmic changes in local free cytoplasmic calcium concentration (see Section V). The actomyosin in the ectoplasm then responds to the large increase in free cytoplasmic calcium by contraction and to the subsequent decrease in free calcium by relaxation as the cytoplasmic vesicles sequester the calcium again.

D. Molecular Nature of the Local Oscillator: A Calcium—Cyclic AMP Control Loop?

1. CALCIUM MOVEMENT AND STORAGE IN CELLS

In relaxed cells the concentration of free calcium in the cytoplasm is very low, around 10^{-7} M (Carafoli and Crompton, 1978; Dedman *et al.*, 1979). During activation, calcium may enter the cytosol from three different sites of high concentration: the external medium, the mitochondria, and the smooth endoplasmic reticulum or its equivalent in nonmuscle cells, the calcium-containing vacuoles (Fig. 16). Different cells vary in the relative importance of each of the three calcium storage sites in contributing to the rise of free cytoplasmic calcium during activation (Carafoli and Crompton, 1978; Bolton, 1979).

Since the external calcium concentration is usually more than 1000 times the internal concentration, some leakage of calcium normally occurs into the cytoplasm across the plasmalemma. In all cells, an active transport system at the plasma membrane pumps calcium out of the cell into the external medium. ATP and magnesium are required. Another Ca^{2+}, Mg^{2+}-ATPase is located in the membrane of the sarcoplasmic reticulum and its equivalent in nonmuscle cells (Tada *et al.*, 1978). This calcium pump transports free calcium from the cytoplasm into sarcoplasmic reticulum vesicles, where it is stored. Mitochondria accumulate calcium by a respiration-driven electrophoretic influx.

Not much is known about the molecular mechanism of calcium release from these storage sites in nonmuscle cells during activation. I have proposed (Section

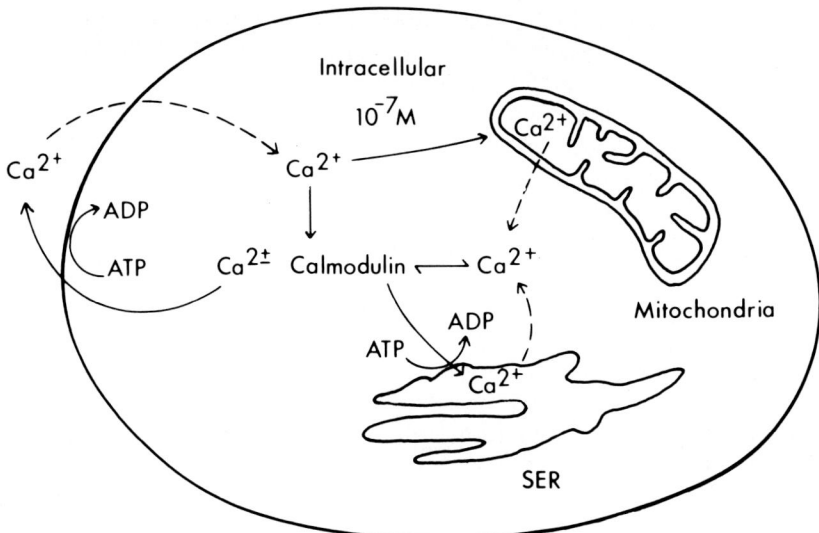

Fig. 16. Regulation of cellular calcium. Involvement of the three Ca^{2+} transport systems and calmodulin in the regulation of intracellular Ca^{2+} concentrations. From Dedman *et al.* (1979).

IV,C) that the stretch stimulus caused by endoplasmic flow in *Physarum* causes a small increase in calcium influx across the plasmalemma which coordinates local free calcium oscillations in the ectoplasm along the vein (Fig. 15), probably involving periodic release and storage of calcium in cytoplasmic vesicles. In various cell types, small increases in calcium levels in the cytosol stimulate an additional sudden release of calcium stored in the vacuoles or reticulum, resulting in high enough free calcium concentrations in the cytoplasm to activate the cell (Fabiato and Fabiato, 1979; Wier, 1980). The flux of mitochondrial calcium may also account for some of the sudden change in free calcium levels which occur in the shuttle-streaming cycle (Holmes and Stewart, 1979a).

2. THE OSCILLATOR: A CALCIUM–CYCLIC AMP CONTROL LOOP?

Data are lacking to give a complete account of the control of oscillating cytoplasmic calcium levels in *Physarum*. A hypothesis has been offered by Rapp and Berridge (1977) to account for oscillating calcium concentrations in various cell types which may be applicable to *Physarum*. They have proposed a calcium–cyclic AMP (cAMP) control loop in which oscillations in internal calcium and cAMP concentrations may be started and maintained by feedback of these two substances on the reactions which determine their cytoplasmic concentrations. The hypothesis has been given credence by recent discoveries about the relationship of cAMP and calcium in various biochemical reactions.

The two enzymes controlling the synthesis and degradation of cAMP are both activated by calcium, which binds to their calmodulin subunits (Cheung, 1980). Calmodulin is also required by the Ca^{2+}, Mg^{2+}-ATPase transport enzymes in both the plasmalemma and sarcoplasmic reticulum in various cells (Fig. 16). The calcium transport enzyme of the sarcoplasmic reticulum in cardiac muscle and various other cell types is activated by phosphorylation catalyzed by a kinase which requires cAMP. Therefore, cAMP helps to regulate the cytoplasmic calcium concentration in these cells (Tada *et al.*, 1978; Dedman *et al.*, 1979). Adelstein *et al.* (1978) have discovered another effect of cAMP on contractile systems. The phosphorylation of myosin light chain kinase from smooth muscle requires cAMP. The resulting kinase is less active in catalyzing the phosphorylation of myosin light chains. Thus, a high level of cAMP results in the inactivation of actomyosin in smooth muscle. Detailed kinetic data regarding feedback mechanisms such as this proposed cAMP–Ca^{2+} control loop in myxomycete plasmodia must be obtained.

V. A PRELIMINARY MODEL FOR THE MOLECULAR MECHANISM OF SHUTTLE STREAMING IN THE PLASMODIUM

The available information about the molecular nature of shuttle streaming in myxomycete plasmodia has been organized in a tentative model presented in Fig. 17. As discussed in Section III, the force required for endoplasmic flow is produced by the interaction of actin and myosin in the ectoplasm, probably by a sliding mechanism similar to contraction in skeletal muscle fibers. The force-producing mechanism in plasmodia differs in detail from that in skeletal muscle in the molecular nature of actomyosin regulation by calcium; in the properties of myxomycete myosin aggregation into small, bipolar oligomers rather than large, thick filaments; and in the location and control of actin polymerization. During normal shuttle streaming, interaction of actin and myosin in force production promotes the formation of fibrils composed of parallel actomyosin filaments in the ectoplasm, often seen to be associated with invaginations of the plasmalemma (Section II,A,B). Presumably these fibrils originate from the less organized, ubiquitous filamentous network seen on the cytoplasmic side of the plasmalemma, where the actin is anchored. A change in the state of actin polymerization may also occur at this time. The role of actin-binding gelation factors in plasmodial streaming is not known at the present time.

The model in Fig. 17 assumes that actomyosin interaction occurs when the calcium concentration in the ectoplasm increases from 10^{-7} M or lower in the relaxed state to 10^{-6} or 10^{-5} M in the contracting state (Section III,D). The cytoplasmic ATP concentration is believed to remain at a relatively high level during shuttle streaming, and therefore is not responsible for fluctuations in actomyosin interaction (Wohlfarth-Bottermann, 1979). During the mitotic cycle the plasmodial ATP concentration varies between 0.5 and 1.0 mM, assuming a range of 6–15 μmole ATP/mg protein during the mitotic cycle (Fink, 1975) and assuming that the living plasmodium contains 7.8% protein by weight (Kessler *et al.*, 1976). This variation in ATP concentration is above the level required for actomyosin contraction. However, a detailed study of ATP fluctuations within the plasmodium during the shuttle-streaming cycle has not been made.

As indicated in Fig. 17, oscillating free calcium concentrations in the ectoplasm determine which regions of the plasmodium will undergo active contraction or relaxation. During ectoplasmic contraction in the veins in the posterior portion of an actively migrating plasmodium (Fig. 17A), endoplasmic flow proceeds in the forward direction. Conversely, during flow to the rear, contraction occurs in the ectoplasm of the fanlike front (Fig. 17B). Since shuttle streaming caused by periodic ectoplasmic contraction reverses direction rhythmically, the free calcium concentration must undergo rhythmic changes in the ectoplasm. As discussed in Section III,D, the immediate source for this calcium is believed to be release from cytoplasmic vacuoles (or perhaps mitochondria). Although calcium could periodically enter the ectoplasm from the external medium by oscillations in calcium permeability across the plasmalemma, several investigators have noted that the rhythmic oscillations in contraction and streaming direction continue for 30–60 minutes despite incubation of the plasmodium in EGTA or EDTA buffers sufficient to lower the calcium concentration of the external medium to a level below that required for streaming (Wohlfarth-Bottermann and Götz von Olenhusen, 1977; Ludlow and Durham, 1977; Ishida *et al.*, 1979). Normal shuttle streaming also occurs for at least 40 minutes in the presence of 5 mM La^{3+}, which is believed to block Ca^{2+} influx across the plasmalemma (Achenbach and Achenbach, 1979). Therefore, an increase of calcium influx across the plasmalemma is not required to explain the sudden increase in free cytoplasmic calcium which activates each contraction in the shuttle-streaming cycle. This suggests the presence of a cytoplasmic oscillator for the regulation of ectoplasmic calcium levels, in contrast to an oscillator controlling calcium flux at the plasmalemma (Wohlfarth-Bottermann, 1979).

Although quite speculative, the best available model for such a cytoplasmic oscillator is the cAMP–Ca^{2+} control loop (Rapp and Berridge, 1977) discussed in Section IV,D. Components of such an oscillator include the reservoirs of sequestered calcium in the cytoplasm, i.e., the calcium-containing vacuoles and perhaps the mitochondria (see Fig. 16), and some kind of feedback system involving the effect of changes of cAMP concentration on the calcium flux across the vacuoles. The resulting change in the concentration of the free cytoplasmic calcium would subsequently influence the reactions controlling the cytoplasmic cAMP levels. These cytoplasmic calcium oscillators are distributed locally

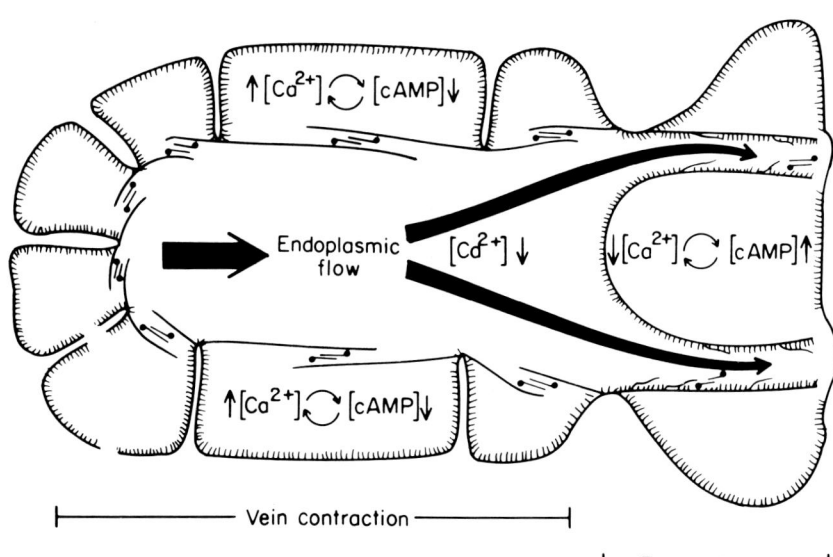

A

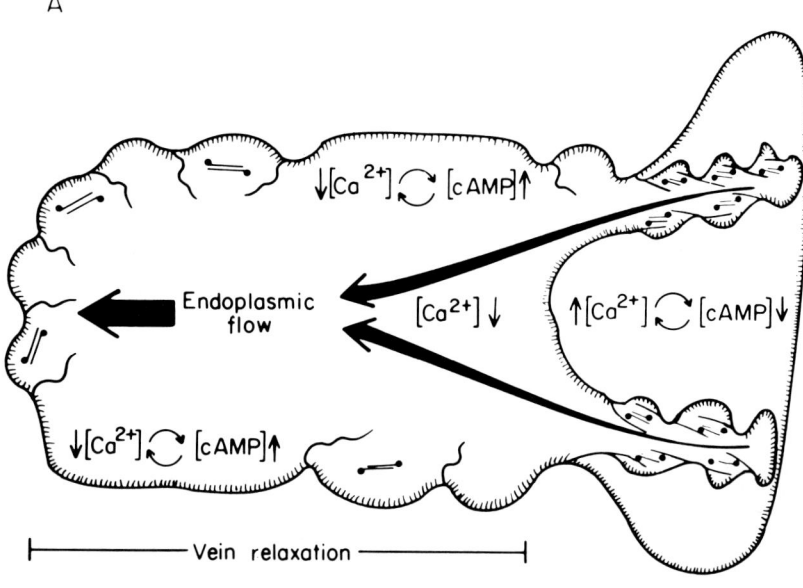

B

5. Plasmodial Structure and Motility

throughout the cytoplasm along the plasmodial veins and in the fanlike front. Thus, these local oscillators must be capable of coordination in some manner so that the different contraction rhythms initiated in segments along a recently excised vein, or those in two recently joined veins, may become synchronized throughout the whole plasmodium. I have suggested (Section IV,C) that this coordination is initially created by stretch entrainment by endoplasmic flow influencing the calcium release from cytoplasmic vacuoles in neighboring segments. After the local calcium oscillators are coordinated, stretch entrainment need be necessary only as segments drift out of phase after relatively long periods of coordinated contraction.

Calcium efflux and influx across the plasmalemma have been detected in normal plasmodia (Ludlow and Durham, 1977; Holmes and Stewart, 1977; Holmes and Stewart, 1979b). Aside from ectoplasmic fibrils and the filamentous network on the cytoplasmic surface of the plasmalemma, the ectoplasm and endoplasm appear to differ ultrastructurally only in their relative distance from the plasmalemma (Sections II,A,3 and II,B,4,h). The most likely explanation for this preferred distribution of aggregated cytoplasmic filaments near the plasmalemma is that the constant rate of calcium influx from the external medium across the plasmalemma is normally rather high, providing a higher free calcium concentration along the inner surface of the plasmalemma compared with the calcium content of the endoplasm. This gradient of calcium may allow the cyclic fluctuation of free calcium regulated by the local oscillators to reach activating levels in the ectoplasm, but not in the endoplasm. Preferential polymerization of actin on sites along the inside of the plasmalemma may also be influenced by the free cytoplasmic calcium concentration, as indicated by experiments with caffeine (Götz von Olenhusen et al., 1979; Götz von Olenhusen and Wohlfarth-Bottermann, 1979a).

The plasmalemma must be involved in the modulation of the fluctuations in calcium concentration controlled by the local cytoplasmic calcium oscillators. This is implied in the stretch entrainment model (Fig. 15), in which stretching the plasmalemma changes its permeability to ions, perhaps increasing calcium influx.

Fig. 17. Model for shuttle streaming in *Physarum* plasmodia. Local cytoplasmic oscillators, here suggested to involve a Ca^{2+}-cyclic AMP control loop, produce rhythmic fluctuations of free cytoplasmic Ca^{2+} concentrations, resulting in the interaction of bipolar myosin oligomers with F-actin filaments in different regions of the ectoplasm during shuttle streaming. A filamentous network is present along the inner side of the plasmalemma at all times in the shuttle-streaming cycle. The calcium concentration in the endoplasm is believed to remain permanently below the threshold for actomyosin activation. (A) Forward flow. The ectoplasm throughout the veins contracts, forcing the endoplasm to stream forward to the relaxed front. Actomyosin fibrils are particularly evident along the invaginations of the plasmalemma in the veins. (B) Backward flow. Interaction of actomyosin in the ectoplasm of the fanlike front forces endoplasmic flow back into the relaxed posterior veins.

In living *Amoeba proteus* examined after the injection of chlorotetracycline, which allows the visualization of membrane bound calcium, calcium has been detected on the cytoplasmic side of the plasmalemma, as well as in the membranes of some cytoplasmic vesicles and in the glycocalyx surrounding the organism (Gawlitta et al., 1980a). Substances inducing pinocytosis appear to act as local stimuli on the outer surface of the plasmalemma, resulting in the loss of calcium from the glycocalyx and in small changes in the calcium content of the inner surface of the plasma membrane, perhaps by local changes in calcium flux across the membrane. Studies of chemotaxis in *P. polycephalum* plasmodia also implicate changes in the properties of the plasmalemma as one part of a system which ultimately modulates the level of free calcium in the ectoplasm. Attractants promote ectoplasmic relaxation and repellents cause contraction (Satoh et al., 1980). Both attractants and repellents produce a graded depolarization of the membrane potential, starting at a threshold concentration unique for each substance (Hato et al., 1976; Ueda and Kobatake, 1977 and this volume). Although the chemotactic motive force is not graded, i.e., it is all or none, its threshold is identical to the threshold for membrane depolarization. These facts suggest that an intermediate step is required upon binding of the chemotactic agent with membrane receptor which is not coupled with the absolute value of membrane depolarization, but which results ultimately in changes in the concentration of free calcium in the ectoplasm. The molecular nature of the intermediate step is not clear, but calcium may somehow be involved at this level, at least in the action of attractants, since a decrease in external calcium using EDTA buffers can eliminate the response of the plasmodium to attractants while leaving the normal magnitude and frequency of shuttle streaming unchanged for at least 20 minutes (Ishida et al., 1979).

ACKNOWLEDGMENTS

The work of the author described in this chapter was supported by research grants GM 22924 and RR 07141 from NIH, PCM-8016620 from NSF, and a Cottrell College Science Grant from the Research Corporation. I am grateful to the investigators who contributed figures to this chapter, as noted in the figure legends. My thinking about protoplasmic streaming in the Myxomycetes has been greatly influenced by recent discussions with Vivianne T. Nachmias, Ariel G. Loewy, and K. E. Wohlfarth-Bottermann, whom I thank.

REFERENCES

Achenbach, F., and Achenbach, U. (1979). Oscillating contractions in protoplasmic strands of *Physarum:* Effects of externally applied ouabain, sodium-, potassium-, and calcium-ions. *Cell Biol. Int. Rep.* **3**, 141–149.

5. Plasmodial Structure and Motility

Achenbach, F., Achenbach, U., and Wohlfarth-Bottermann, K. E. (1979a). Plasmalemma invaginations, contraction and locomotion in normal and caffeine-treated protoplasmic drops of *Physarum*. *Eur. J. Cell Biol.* **20,** 12-23.
Achenbach, F., Naib-Majani, W., and Wohlfarth-Bottermann, K. E. (1979b). Plasmalemma invaginations of *Physarum* dependent on the nutritional content of the plasmodial environment. *J. Cell Sci.* **36,** 355-359.
Achenbach, U., and Wohlfarth-Bottermann, K. E. (1980). Oscillating contractions in protoplasmic strands of *Physarum*. *J. Exp. Biol.* **85,** 21-31.
Adelman, M. R. (1977). *Physarum polycephalum* actin: Observations on its presence, stability, and assembly in plasmodial extracts and development of an improved purification procedure. *Biochemistry* **16,** 4862-4871.
Adelman, M. R., and Taylor, E. W. (1969a). Isolation of an actomyosin-like protein complex from slime mold plasmodium and the separation of the complex into actin- and myosin-like fractions. *Biochemistry* **8,** 4964-4975.
Adelman, M. R., and Taylor, E. W. (1969b). Further purification and characterization of slime mold myosin and slime mold actin. *Biochemistry* **8,** 4976-4988.
Adelstein, R. S., and Eisenberg, E. (1980). Regulation and kinetics of the actin-myosin-ATP interaction. *Annu. Rev. Biochem.* **49,** 921-957.
Adelstein, R. S., Conti, M. A., Hathaway, D. R., and Klee, C. B. (1978). Phosphorylation of smooth muscle myosin light chain kinase by the catalytic subunit of adenosine 3':5'-monophosphate-dependent protein kinase. *J. Biol. Chem.* **253,** 8347-8350.
Allen, R. D. (1973). Biophysical aspects of pseudopodium formation and retraction. *In* "The Biology of Amoeba" (K. W. Jeon, ed.), pp. 201-247. Academic Press, New York.
Allen, R. D., and Allen, N. S. (1978). Cytoplasmic streaming in amoeboid movement. *Annu. Rev. Biophys. Bioeng.* **7,** 469-495.
Alléra, A., and Wohlfarth-Bottermann, K. E. (1972). Weitreichende fibrilläre protoplasmadifferenzierungen und ihre bedeutung für die protoplasmaströmung IX. Aggregationszüstande des myosins und bedingungen zur entstehung von myosin filamenten in den plasmodien von *Physarum polycephalum*. *Cytobiologie* **6,** 261-286.
Alléra, A., Beck, R., and Wohlfarth-Bottermann, K. E. (1971). Weitreichende fibrilläre protoplasmadifferenzierungen und ihre bedeutung für die protoplasmaströmung VIII. Identifizierung der plasmafilamente von *Physarum polycephalum* als F-actin durch anlagerung von heavy meromyosin *in situ*. *Cytobiologie* **4,** 437-449.
Anderson, J. D. (1964). Regional differences in ion concentration in migrating plasmodia. *In* "Primitive Motile Systems in Cell Biology" (R. D. Allen and N. Kamiya, eds.), pp. 125-136. Academic Press, New York.
Andresen, N., and Pollock, B. M. (1952). A comparison between the cytoplasmic components in the myxomycete, *Physarum polycephalum,* and in the amoeba, *Chaos chaos. C. R. Trav. Lab. Carlsberg, Ser. Chim.* **28,** 247-264.
Beck, R., Komnick, H., Stockem, W., and Wohlfarth-Bottermann, K. E. (1969). Weitreichende, fibrilläre protoplasmadifferenzierungen und ihre bedeutung für die protoplasmaströmung. IV. Vergleichende unter suchungen an actomyosin-fadën unt glyceunierten zellen. *Cytobiologie* **1,** 99-114.
Beck, R., Hinssen, H., Komnick, H., Stockem, W., and Wohlfarth-Bottermann, K. E. (1970a). Weitreichende, fibrilläre protoplasmadifferenzierungen und ihre bedeutung für die protoplasmaströmung V. Kontraktion, ATPase-aktivität und feinstruktur isolierter actomyosin-fäden von *Physarum polycephalum. Cytobiologie* **2,** 259-274.
Beck, R., Komnick, H., Stockem, W., and Wohlfarth-Bottermann, K. E. (1970b). Weitreichende, fibrilläre protoplasmadifferenzierungen und ihre bedeutung für die protoplasmaströmung VI. Vergleichende unter suchungen an isolierton actomyosin-fadën schräggestreifter und glatter muskeln. *Cytobiologie* **2,** 413-428.

Bolton, T. B. (1979). Mechanisms of action of transmitters and other substances on smooth muscle. *Physiol. Rev.* **59**, 606-718.
Braatz, R. (1975). Differential histochemical localization of calcium and its relation to shuttle streaming in *Physarum*. *Cytobiologie* **12**, 74-78.
Braatz, R., and Komnick, H. (1970). Histochemischer nachweis eines calcium-pumpenden systems in plasmodien von schleimpilzen. *Cytobiologie* **2**, 457-463.
Braatz, R., and Komnick, H. (1973). Vacuolar calcium segregation in relaxed myxomycete protoplasm as revealed by combined electrolyte histochemistry and energy dispersive analysis of X-rays. *Cytobiologie* **8**, 158-163.
Bray, D., and Thomas, C. (1976). Unpolymerized actin in fibroblasts and brain. *J. Mol. Biol.* **105**, 527-544.
Brenner, S. L., and Korn, E. D. (1979). Substoichiometric concentrations of cytochalasin D inhibit actin polymerization. *J. Biol. Chem.* **254**, 9982-9985.
Brenner, S. L., and Korn, E. D. (1980). The effects of cytochalasins on actin polymerization and actin ATPase provide insights into the mechanism of polymerization. *J. Biol. Chem.* **255**, 841-844.
Britz, S. J. (1979). Cytoplasmic streaming in *Physarum*. *In* "Encyclopedia of Plant Physiology." (W. Haupt and M. E. Feinleib, eds.), Vol. 7, pp. 127-149. Springer-Verlag, Berlin and New York.
Burgess, W. H., Jemiolo, D. K., and Kretsinger, R. H. (1980). Interaction of calcium and calmodulin in the presence of sodium dodecyl sulfate. *Biochim. Biophys. Acta* **623**, 257-270.
Camp, W. G. (1937). The structure and activities of myxomycete plasmodia. *Bull. Torrey Bot. Club* **64**, 307-335.
Carafoli, E., and Crompton, M. (1978). The regulation of intracellular calcium. *Curr. Top. Membr. Transp.* **10**, 151-216.
Carlsson, L., Nyström, L.-E., Lindberg, U., Kannan, K. K., Cid-Dresdner, H., Lövgren, S., and Jörnvall, H. (1976). Crystallization of a non-muscle actin. *J. Mol. Biol.* **105**, 353-366.
Charvat, I., Ross, I. K., and Cronshaw, J. (1973a). Ultrastructure of the plasmodial slime mold *Perichaena vermicularis* I. Plasmodium. *Protoplasma* **76**, 333-351.
Charvat, I., Ross, I. K., and Cronshaw, J. (1973b). Ultrastructure of the plasmodial slime mold *Perichaena vermicularis* II. Formation of the peridium. *Protoplasma* **78**, 1-19.
Cheung, W. Y. (1980). Calmodulin plays a pivotal role in cellular regulation. *Science* **207**, 19-27.
Cieślawska, M., and Grebecki, A. (1979). Synchronal pulsation in plasmodia of *Physarum polycephalum*. *In* "Current Research on *Physarum*" (W. Sachsenmaier, ed.), Vol. 120, pp. 167-170. Univ. of Innsbruck, Austria.
Collins, J. H. (1976). Structure and evolution of troponin C and related proteins. *Soc. Exp. Biol. Symp.* **30**, 303-334.
Condeelis, J. S., and Taylor, D. L. (1977). The contractile basis of amoeboid movement V. The control of gelation, solation, and contraction in extracts from *Dictyostelium discoideum*. *J. Cell Biol.* **74**, 901-927.
Daniel, J. W., and Baldwin, H. H. (1964). Methods of culture for plasmodial myxomycetes. *Methods Cell Physiol.* **1**, 9-41.
Daniel, J. W., and Järlfors, U. (1972a). Plasmodial ultrastructure of the myxomycete *Physarum polycephalum*. *Tissue Cell* **4**, 15-36.
Daniel, J. W., and Järlfors, U. (1972b). Light-induced changes in the ultrastructure of a plasmodial myxomycete. *Tissue Cell* **4**, 405-426.
Daniels, E. W. (1973). Ultrastructure. *In* "The Biology of Amoeba" (K. W. Jeon, ed.), Chap. 5, pp. 125-169. Academic Press, New York.
Dedman, J. R., Brinkley, B. R., and Means, A. R. (1979). Regulation of microfilaments and microtubules by calcium and cyclic AMP. *Adv. Cyclic Nucleotide Res.* **11**, 131-174.

5. Plasmodial Structure and Motility

D'Haese, J., and Hinssen, H. (1978). Contraction properties of isolated slime mold actomyosin I. Comparison of thread models made of natural, recombined, and hybridized actomyosins from slime mould and muscle. *Protoplasma* **95,** 273-295.

D'Haese, J., and Hinssen, H. (1979). Aggregation properties of non-muscle myosins. *In* "Cell Motility: Molecules and Organization" (S. Hatano, H. Ishikawa, and H. Sato, eds.), pp. 105-118. Univ. Park Press, Baltimore, Maryland.

Dugas, D. J., and Bath, J. D. (1962). Electron microscopy of the slime mold *Physarum polycephalum*. *Protoplasma* **54,** 421-431.

Ebashi, S., Nonomura, Y., Mikawa, T., Hirata, M., and Saida, K. (1979). Regulatory mechanisms of muscle contraction. *In* "Cell Motility: Molecules and Organization" (S. Hatano, H. Ishikawa, and H. Sato, eds.), pp. 225-237. Univ. Park Press, Baltimore, Maryland.

Eckert, R. (1972). Bioelectrical control of ciliary activity. *Science* **176,** 473-481.

Ettienne, E. (1972). Subcellular localization of calcium repositories in plasmodia of the acellular slime mold *Physarum polycephalum*. *J. Cell Biol.* **54,** 179-184.

Fabiato, A., and Fabiato, F. (1979). Calcium and cardiac excitation-contraction coupling. *Annu. Rev. Physiol.* **41,** 473-484.

Fink, K. (1975). Fluctuations in deoxyribo- and ribonucleoside triphosphate pools during the mitotic cycle of *Physarum polycephalum*. *Biochim. Biophys. Acta* **414,** 85-89.

Flanagan, M. D., and Lin, S. (1980). Cytochalasins block actin filament formation elongation by binding to high affinity sites associated with F-actin. *J. Biol. Chem.* **255,** 835-838.

Fleischer, M., and Wohlfarth-Bottermann, K. E. (1975). Correlation between tension force generation, fibrillogenesis, and ultrastructure of cytoplasmic actomyosin during isometric and isotonic contractions of protoplasmic strands. *Cytobiologie* **10,** 339-365.

Flickinger, C. J. (1973). Cellular membranes of amoebae. *In* "The Biology of Amoeba" (K. W. Jeon, ed.), Chap. 6, pp. 171-199. Academic Press, New York.

Gawlitta, W., Stockem, W., Wehland, J., and Weber, K. (1980a). Pinocytosis and locomotion of amoeba XV. Visualization of Ca^{2+}-dynamics by chlorotetracycline (CTC) fluorescence during induced pinocytosis in living *Amoeba proteus*. *Cell Tissue Res.* **213,** 9-20.

Gawlitta, W., Wolf, K. V., Hoffmann, H.-U., and Stockem, W. (1980b). Studies on microplasmodia of *Physarum polycephalum* I. Classification and locomotion behavior. *Cell Tissue Res.* **209,** 71-86.

Gergely, J. (1976). Troponin-tropomyosin-dependent regulation of muscle contraction by calcium. *In* "Cell Motility, Book A" (R. Goldman, T. Pollard, and J. Rosenbaum, eds.), pp. 137-149. Cold Spring Harbor Lab., Cold Spring Harbor, New York.

Goodman, E. M. (1980). *Physarum polycephalum:* A review of a model system using a structure-function approach. *Int. Rev. Cytol.* **63,** 1-58.

Goodman, E. M., and Rusch, H. P. (1970). Ultrastructural changes during spherule formation in *Physarum polycephalum*. *J. Ultrastruct. Res.* **30,** 172-183.

Götz von Olenhusen, K., and Wohlfarth-Bottermann, K. E. (1979a). Effects of caffeine and D_2O on persistence and *de novo* generation of intrinsic oscillatory contraction automaticity in *Physarum*. *Cell Tissue Res.* **197,** 479-499.

Götz von Olenhusen, K., and Wohlfarth-Bottermann, K. E. (1979b). Evidence for actin transformations during the contraction-relaxation cycle of cytoplasmic actomyosin: Cycle blockade by phalloidin-injection. *In* "Cell Motility: Molecules and Organization" (S. Hatano, H. Ishikawa, and H. Sato, eds.), pp. 379-397. Univ. Park Press, Baltimore, Maryland.

Götz von Olenhusen, K., Jücker, H., and Wohlfarth-Bottermann, K. E. (1979). Induction of a plasmodial stage of *Physarum* without plasmalemma invaginations. *Cell Tissue Res.* **197,** 463-477.

Gray, W. D., and Alexopoulos, C. J. (1968). "Biology of the Myxomycetes." Ronald Press, New York.

Grebecki, A. (1979). Interrelations of motory phenomena in junctions of plasmodial veins of *Physarum polycephalum. In* "Current Research on *Physarum*" (W. Sachsenmaier, ed.), Vol. 120, pp. 171-175. Univ. of Innsbruck, Austria.
Grebecki, A., and Cieślawska, M. (1978a). Dynamics of the ectoplasmic walls during pulsation of plasmodial veins of *Physarum polycephalum. Protoplasma* **97**, 365-371.
Grebecki, A., and Cieślawska, M. (1978b). Plasmodium of *Physarum polycephalum* as a synchronous contractile system. *Cytobiologie* **17**, 335-342.
Grebecki, A., and Moczón, M. (1978). Correlation of contractile activity and of streaming direction between branching veins of *Physarum polycephalum* plasmodium. *Protoplasma* **97**, 153-164.
Grunfeld, C., and Kessler, D. (1967). Detection of mitochondrial DNA in the slime mold, *Physarum polycephalum. J. Cell Biol.* **35**, 168a.
Guttes, E., and Guttes, S. (1960). Pinocytosis in the myxomycete *Physarum polycephalum. Exp. Cell Res.* **20**, 239-241.
Guttes, E., and Guttes, S. (1963). Arrest of plasmodial motility during mitosis in *Physarum polycephalum. Exp. Cell Res.* **30**, 242-244.
Guttes, E., and Guttes, S. (1964). Thymidine incorporation by mitochondria in *Physarum polycephalum. Science* **145**, 1057-1058.
Guttes, E., Guttes, S., and Rusch, H. P. (1961). Morphological observations on growth and differentiation of *Physarum polycephalum* grown in pure culture. *Dev. Biol.* **3**, 588-614.
Guttes, S., Guttes, E., and Hadek, K. (1966). Occurrence and morphology of a fibrous body in the mitochondria of the slime mold *Physarum polycephalum. Experientia* **22**, 452-454.
Guttes, E., Guttes, S., and Ellis, R. A. (1968). Electron microscope study of mitosis in *Physarum polycephalum. J. Ultrastruct. Res.* **22**, 508-529.
Guttes, E., Guttes, S., and Devi, R. V. (1969). Division stages of the mitochondria in normal and actinomycin-treated plasmodia of *Physarum polycephalum. Experientia* **25**, 66-68.
Harris, H. E., and Weeds, A. G. (1978). Platelet actin: Subcellular distribution and association with profilin. *FEBS Lett.* **90**, 84-88.
Hasegawa, T., Takahashi, S., Hayashi, H., and Hatano, S. (1980). Fragmin: A calcium ion sensitive regulatory factor on the formation of actin filaments. *Biochemistry* **19**, 2677-2683.
Haskins, E. F. (1974). *Stemonitis flavogenita* (myxomycetes)—Plasmodial Phase (Aphanoplasmodium) Film No. E-2000. Instit. Wissenschaftlichen Film, Göttingen, Germany.
Hatano, S. (1970). Specific effect of Ca^{2+} on movement of plasmodial fragment obtained by caffeine treatment. *Exp. Cell Res.* **61**, 199-203.
Hatano, S. (1972). Conformational changes of plasmodium actin polymers formed in the presence of Mg^{2+}. *J. Mechanochem. Cell Motil.* **1**, 75-80.
Hatano, S., and Ohnuma, J. (1970). Purification and characterization of myosin A from the myxomycete plasmodium. *Biochim. Biophys. Acta* **205**, 110-120.
Hatano, S., and Oosawa, F. (1966a). Extraction of an actin-like protein from the plasmodium of a myxomycete and its interaction with myosin A from rabbit striated muscle. *J. Cell. Physiol.* **68**, 197-202.
Hatano, S., and Oosawa, F. (1966b). Isolation and characterization of plasmodium actin. *Biochim. Biophys. Acta* **127**, 488-498.
Hatano, S., and Oosawa, F. (1971). Movement of cytoplasm in plasmodial fragment obtained by caffeine treatment I. Its Ca^{++} sensitivity. *J. Physiol. Soc. Jpn.* **33**, 589-590.
Hatano, S., and Oosawa, F. (1979). Cell motility and the organization of actin and myosin in non-muscle cells. *In* "Cell Motility: Molecules and Organization" (S. Hatano, H. Ishikawa, and H. Sato, eds.), pp. 675-683. Univ. Park Press, Baltimore, Maryland.
Hatano, S., and Owaribe, K. (1976). Actin and actinin from myxomycete plasmodia. *In* "Cell Motility, Book B" (R. Goldman, T. Pollard, and J. Rosenbaum, eds.), pp. 499-511. Cold Spring Harbor Lab., Cold Spring Harbor, New York.

5. Plasmodial Structure and Motility

Hatano, S., and Owaribe, K. (1977). A simple method for the isolation of actin from myxomycete plasmodia. *J. Biochem. (Tokyo)* **82**, 201-206.
Hatano, S., and Owaribe, K. (1979). Some properties of *Physarum* actinin, a regulatory protein of actin polymerization. *Biochim. Biophys. Acta* **579**, 200-215.
Hatano, S., and Takahashi, K. (1971). Structure of myosin A from the myxomycete plasmodium and its aggregation at low salt concentrations. *J. Mechanochem. Cell Motil.* **1**, 7-14.
Hatano, S., and Tazawa, M. (1968). Isolation, purification, and characterization of myosin B from myxomycete plasmodium. *Biochim. Biophys. Acta* **154**, 507-519.
Hatano, S., Matsumura, F., Hasegawa, T., Takahashi, S., Sato, H., and Ishikawa, H. (1979). Assembly and disassembly of F-actin filaments in *Physarum* plasmodium and *Physarum* actinin. *In* "Cell Motility: Molecules and Organization" (S. Hatano, H. Ishikawa, and H. Sato, eds.), pp. 87-104. Univ. Park Press, Baltimore, Maryland.
Hathaway, D. R., and Adelstein, R. S. (1979). Human platelet myosin light chain kinase requires the calcium-binding protein calmodulin for activity. *Proc. Natl. Acad. Sci. U.S.A.* **76**, 1653-1657.
Hato, M., Ueda, T., Kurihara, K., and Kobatake, Y. (1976). Change in zeta potential and membrane potential of slime mold *Physarum polycephalum* in response to chemical stimuli. *Biochim. Biophys. Acta* **426**, 73-80.
Hauser, H., Beinbrech, G., Gröschel-Stewart, U., and Jockusch, B. M. (1975). Localization by immunological techniques of myosin in nuclei of lower eukaryotes. *Exp. Cell Res.* **95**, 127-135.
Hauser, M. (1978). Demonstration of membrane-associated and oriented microfilaments in *Amoeba proteus* by means of a Schiff base/glutaraldehyde fixative. *Cytobiologie* **18**, 95-106.
Hinchee, A. A., and Haskins, E. F. (1980). Closed spindle nuclear division in the plasmodial phase of the acellular slime mold *Eschinostelium minutum*. *Protoplasma* **102**, 235-252.
Hinssen, H. (1970). Synthetische myosin-filamente von schleimpilzplasmodien. *Cytobiologie* **2**, 326-331.
Hinssen, H. (1972). Actin in isoliertem grundplasma von *Physarum polycephalum*. *Cytobiologie* **5**, 146-164.
Hinssen, H. (1979). Studies on the polymer state of actin in *Physarum polycephalum*. *In* "Cell Motility: Molecules and Organization" (S. Hatano, H. Ishikawa, and H. Sato, eds.), pp. 59-85. Univ. Park Press, Baltimore, Maryland.
Hinssen, H., and D'Haese, J. (1974). Filament formation by slime mold myosin isolated at low ionic strength. *J. Cell Sci.* **15**, 113-129.
Hinssen, H., and D'Haese, J. (1976). Synthetic fibrils from *Physarum* actomyosin: Self-assembly, organization and contraction. *Cytobiologie* **13**, 132-157.
Hoffman, H.-U., and Stockem, W. (1979). Comparative fine structure in microplasmodia of the acellular slime mold *Physarum polycephalum*. *In* "Current Research on *Physarum*" (W. Sachsenmaier, ed.), Vol. 120, pp. 181-186. Univ. of Innsbruck, Austria.
Holmes, R. P., and Stewart, P. R. (1977). Calcium uptake during mitosis in the myxomycete *Physarum polycephalum*. *Nature (London)* **269**, 592-594.
Holmes, R. P., and Stewart, P. R. (1979a). The isolation of coupled mitochondria from *Physarum polycephalum* and their response to Ca^{2+}. *Biochim. Biophys. Acta* **545**, 94-105.
Holmes, R. P., and Stewart, P. R. (1979b). The response of *Physarum polycephalum* to extracellular Ca^{2+}: Studies on Ca^{2+} nutrition, Ca^{2+} fluxes and Ca^{2+} compartmentation. *J. Gen. Microbiol.* **113**, 275-285.
Huxley, H. E. (1963). Electron microscope studies on the structure of natural and synthetic protein filaments from striated muscle. *J. Mol. Biol.* **7**, 281-308.
Huxley, H. E. (1969). The mechanism of muscular contraction. *Science* **164**, 1356-1366.
Huxley, H. E. (1972). Molecular basis of contraction in cross-striated muscles. *In* "The Structure

and Function of Muscle" (G. H. Bourne, ed.), Vol. 1, 2nd ed., pp. 301-387. Academic Press, New York.

Isenberg, G., and Wohlfarth-Bottermann, K. E. (1976). Transformation of cytoplasmic actin: Importance for the organization of the contractile gel reticulum and the contraction-relaxation cycle of cytoplasmic actomyosin. *Cell Tissue Res.* **173,** 495-528.

Ishida, N., Kurihara, K., and Kobatake, Y. (1979). Selective modification of positive chemotaxis in the true slime mold *Physarum polycephalum* by ethylendiaminetetraacetic acid treatment. *Biochim. Biophys. Acta* **587,** 89-98.

Jacobson, D. N., Johnke, R. M., and Adelman, M. R. (1976). Studies on motility in *Physarum polycephalum*. *In* "Cell Motility, Book B" (R. Goldman, T. Pollard, and J. Rosenbaum, eds.), pp. 749-770. Cold Spring Harbor Lab., Cold Spring Harbor, New York.

Jahn, T. L., and Bovee, E. C. (1967). Motile behavior of protozoa. *In* "Research in Protozoology" (T. T. Chen, ed.), Vol. I, pp. 41-200. Pergamon, Oxford.

Jockusch, B. M. (1973). Nuclear proteins in *Physarum polycephalum*. *Ber. Dtsch. Bot. Ges.* **86,** 39-54.

Jockusch, B. M., Brown, D. F., and Rusch, H. P. (1971). Synthesis and some properties of an actin-like nuclear protein in the slime mold *Physarum polycephalum*. *J. Bacteriol.* **108,** 705-714.

Jockusch, B. M., Ryser, U., and Behnke, O. (1973). Myosin-like protein in *Physarum* nuclei. *Exp. Cell Res.* **76,** 464-466.

Jockusch, B. M., Becker, M., Hindennach, I., and Jockusch, J. (1974). Slime mould actin: Homology to vertebrate actin and presence in the nucleus. *Exp. Cell Res.* **89,** 241-246.

Kamiya, N. (1942). Physical aspects of protoplasmic streaming. *In* "The Structure of Protoplasm" (W. Seifriz, ed.), pp. 199-244. Iowa State College Press, Ames, Iowa.

Kamiya, N. (1959). Protoplasmic streaming. *Protoplasmatologia* **8** (3a), 1-199.

Kamiya, N. (1979). Dynamic aspects of movement in the myxomycete plasmodium. *In* "Cell Motility: Molecules and Organization" (S. Hatano, H. Ishikawa, and H. Sato, eds.), pp. 399-414. Univ. Park Press, Baltimore, Maryland.

Kamiya, N., and Kuroda, K. (1965). Movement of the myxomycete plasmodium I. A study of glycerinated models. *Proc. Jpn. Acad.* **41,** 837-841.

Kato, T. (1979). Ca^{2+} uptake of *Physarum* microsomal vesicles. *In* "Cell Motility: Molecules and Organization" (S. Hatano, H. Ishikawa, and H. Sato, eds.), pp. 211-223. Univ. Park Press, Baltimore, Maryland.

Kato, T., and Tonomura, Y. (1975). *Physarum polycephalum* tropomyosin-troponin complex: Isolation and properties. *J. Biochem. (Tokyo)* **78,** 583-588.

Kato, T., and Tonomura, Y. (1977). Uptake of calcium ions into microsomes isolated from *Physarum polycephalum*. *J. Biochem.* **81,** 207-213.

Kerr, N. S. (1970). *Didymium nigripes* (Myxomycetes)—Plasmodium Phase (Phaneroplasmodium), Film No. E-1569. Instit. Wissenschaftlichen Film, Göttingen, Germany.

Kessler, D. (1972a). Effect of cytochalasin B on cytoplasmic streaming in the slime mold, *Physarum polycephalum*. *J. Cell Biol.* **55,** 134a.

Kessler, D. (1972b). On the location of myosin in the myxomycete *Physarum polycephalum* and its possible function in cytoplasmic streaming. *J. Mechanochem. Cell Motil.* **1,** 125-137.

Kessler, D., and Lathwell, M. J. (1979). Cessation of protoplasmic streaming during mitosis in plasmodia of *Physarum polycephalum*. *In* "Motility in Cell Function" (F. A. Pepe, J. W. Sanger, and V. T. Nachmias, eds.), pp. 463-465. Academic Press, New York.

Kessler, D., Nachmias, V. T., and Loewy, A. G. (1976). Actomyosin content of *Physarum polycephalum* plasmodia and detection of immunological crossreactions with myosins from related species. *J. Cell Biol.* **69,** 393-406.

Kessler, D., Eisenlohr, L. C., Lathwell, M. J., Huang, J., Taylor, H. C., Godfrey, S. D., and Spady, M. L. (1980). *Physarum* myosin light chain binds calcium. *Cell Motil.* **1,** 63-71.

Klee, C. B., Crouch, T. H., and Krinks, M. H. (1979). Calcineurin: A calcium- and calmodulin-binding protein of the nervous system. *Proc. Natl. Acad. Sci. U.S.A.* **76,** 6270-6273.

Klee, C. B., Crouch, T. H., and Richman, P. G. (1980). Calmodulin. *Annu. Rev. Biochem.* **49,** 489-515.

Komnick, H., Stockem, W., and Wohlfarth-Bottermann, K. E. (1973). Cell motility: Mechanisms in protoplasmic streaming and ameboid movement. *Int. Rev. Cytol.* **34,** 169-249.

Kretsinger, R. H. (1979). The informational role of calcium in the cytosol. *Adv. Cyclic Nucleotide Res.* **11,** 1-26.

Kretsinger, R. H. (1980). Structure and evolution of calcium modulated proteins. *CRC Crit. Rev. Biochem.* **8,** 119-174.

Kuroda, K. (1979). Movement of cytoplasm in a membrane-free system. *In* "Cell Motility: Molecules and Organization" (S. Hatano, H. Ishikawa, and H. Sato, eds.), pp. 347-361. Univ. Park Press, Baltimore, Maryland.

Kuroiwa, T. (1973). Studies on mitochondrial structure and function in *Physarum polycephalum*. I. Fine structure, cytochemistry and ^{3}H-uridine autoradiography of a central body in mitochondria. *Exp. Cell Res.* **78,** 351-359.

Kuroiwa, T., and Hizume, M. (1974). Mitochondrial nucleoid staining with ammoniacal silver. *Exp. Cell Res.* **87,** 406-409.

Kuroiwa, T., Kawano, S., and Hizume, M. (1977). Studies on mitochondrial structure and function in *Physarum polycephalum* V. Behavior of mitochondrial nucleoids throughout mitochondrial division cycle. *J. Cell Biol.* **72,** 687-694.

Kuroiwa, T., Hizume, M., and Kawano, S. (1978). Studies on mitochondrial structure and function in *Physarum polycephalum*. IV. Mitochondrial division cycle. *Cytologia* **43,** 119-136.

Kuźnicki, J., and Drabikowski, W. (1979). Purification and properties of the Ca^{2+}-binding modulator protein from *Physarum polycephalum*. *In* "Current Research on *Physarum*" (W. Sachsenmaier, ed.), Vol. 120, pp. 99-105. Univ. of Innsbruck, Austria.

Kuźnicki, J., Kuźnicki, L., and Drabikowski, W. (1979). Ca^{2+}-binding modulator protein in protozoa and myxomycete. *Cell Biol. Int. Rep.* **3,** 17-23.

LeStourgeon, W. M. (1977). Identification of contractile proteins in nuclear protein fractions. *Methods Cell Biol.* **16,** 269-281.

LeStourgeon, W. M., Forer, A., Yang, Y.-Z., Bertram, J. S., and Rusch, H. P. (1975). Contractile proteins: Major components of nuclear and chromosomal non-histone proteins. *Biochim. Biophys. Acta* **379,** 529-553.

Lewis, W. H. (1942). The relation of the viscosity changes of protoplasm to ameboid locomotion and cell division. *In* "The Structure of Protoplasm," (W. Seifriz, ed.), pp. 163-197. Iowa State College Press, Ames, Iowa.

Loewy, A. G. (1949). A theory of protoplasmic streaming. *Proc. Am. Phil. Soc.* **93,** 326-329.

Loewy, A. G. (1952). An actomyosin-like substance from the plasmodium of a myxomycete. *J. Cell Comp. Physiol.* **40,** 127-156.

Ludlow, C. T., and Durham, A. C. H. (1977). Calcium ion fluxes across the external surface of *Physarum polycephalum*. *Protoplasma* **91,** 107-113.

McCormick, J. J., Blomquist, J. C., and Rusch, H. P. (1970). Isolation and characterization of an extracellular polysaccharide from *Physarum polycephalum*. *J. Bacteriol.* **104,** 1110-1118.

McManus, Sister M. A. (1965). Ultrastructure of myxomycete plasmodia of various types. *Am. J. Bot.* **52,** 15-25.

McManus, Sister M. A., and Roth, L. E. (1965). Fibrillar differentiation in myxomycete plasmodia. *J. Cell Biol.* **25,** 305-318.

McManus, Sister M. A., and Roth, L. E. (1967). Microtubular structure in myxomycete plasmodia. *J. Ultrastruct. Res.* **20**, 260-266.

Mante, S. D., Flashner, M., and Tanenbaum, S. W. (1978). Effects of cytochalasin A on the morphology of plasmodia and sclerotia of *Physarum polycephalum. Cytobiologie* **17**, 10-22.

Marchesi, V. T. (1979). Spectrin: Present status of a putative cytoskeletal protein of the red cell membrane. *J. Membr. Biol.* **51**, 101-131.

Markey, F., Lindberg, U., and Ericksson, L. (1978). Human platelets contain profilin, a potential regulator of actin polymerisability. *FEBS Lett.* **88**, 75-79.

Maruyama, K., Kamiya, R., Kimura, S., and Hatano, S. (1976). Beta actinin-like protein from plasmodium. *J. Biochem. (Tokyo)* **79**, 709-715.

Matsumura, F., and Hatano, S. (1978). Reversible superprecipitation and bundle formation of plasmodium actomyosin. *Biochim. Biophys. Acta* **533**, 511-523.

Matthews, L. M., Jr. (1977). Calcium ion regulation in caffeine derived microplasmodia of *Physarum polycephalum. J. Cell Biol.* **72**, 502-506.

Meyer, R., and Stockem, W. (1979). Studies on microplasmodia of *Physarum polycephalum*. V: Electrical activity of different types of microplasmodia and macroplasmodia. *Cell Biol. Int. Rep.* **3**, 321-330.

Miller, D. M., and Anderson, J. D. (1966). The morphology, migration and pressure development of oriented plasmodia of the slime mold. *Trans. Ill. State Acad. Sci.* **59**, 352-357.

Miller, D. M., and Anderson, J. D. (1971). Migration and biopotentials in slime mold plasmodia. *In* "Experiments in Physiology and Biochemistry" (G. A. Kerkut, ed.), Vol. 4, pp. 183-202. Academic Press, New York.

Miller, D. M., Anderson, J. D., and Abbott, B. C. (1968). Potentials and ionic exchange in slime mold plasmodia. *Comp. Biochem. Physiol.* **27**, 633-646.

Nachmias, V. T. (1972a). Electron microscope observations on myosin from *Physarum polycephalum. J. Cell Biol.* **52**, 648-663.

Nachmias, V. T. (1972b). Filament formation by purified *Physarum* myosin. *Proc. Natl. Acad. Sci. U.S.A.* **69**, 2011-2014.

Nachmias, V. T. (1972c). *Physarum* myosin: Two new properties. *Cold Spring Harbor Symp. Quant. Biol.* **37**, 607-612.

Nachmias, V. T. (1974). Properties of *Physarum* myosin purified by a potassium iodide procedure. *J. Cell Biol.* **62**, 54-65.

Nachmias, V. T. (1975). Calcium sensitivity of hybrid complexes of muscle myosin and *Physarum polycephalum* proteins. *Biochim. Biophys. Acta* **400**, 208-221.

Nachmias, V. T. (1979a). From ameba to muscle: On some work by and with John M. Marshall. *In* "Motility in Cell Function" (F. A. Pepe, J. W. Sanger, and V. T. Nachmias, eds.), pp. 9-26. Academic Press, New York.

Nachmias, V. T. (1979b). The contractile proteins of *Physarum polycephalum* and actin polymerization in plasmodial extracts. *In* "Cell Motility: Molecules and Organization" (S. Hatano, H. Ishikawa, and H. Sato, eds.), pp. 33-57. Univ. Park Press, Baltimore, Maryland.

Nachmias, V. T., and Asch, A. (1974). Actin mediated calcium dependency of actomyosin in a myxomycete. *Biochem. Biophys. Res. Commun.* **60**, 656-664.

Nachmias, V. T., and Asch, A. (1976). Regulation and polarity: Results with myxomycete plasmodium and with human platelets. *In* "Cell Motility, Book B" (R. Goldman, T, Pollard, and J. Rosenbaum, eds.), pp. 771-783. Cold Spring Harbor Lab., Cold Spring Harbor, New York.

Nachmias, V. T., and Ingram, W. C. (1970). Actomyosin from *Physarum polycephalum:* Electron microscopy of myosin-enriched preparations. *Science* **170**, 743-745.

Nachmias, V. T., and Meyers, C. H. (1980). Cytoplasmic droplets produced by the effect of adenine on *Physarum* plasmodia. *Exp. Cell Res.* **128**, 121-126.

5. Plasmodial Structure and Motility

Nachmias, V. T., Huxley, H. E., and Kessler, D. (1970). Electron microscope observations on actomyosin and actin preparations from *Physarum polycephalum*, and on their interaction with heavy meromyosin subfragment I from muscle myosin. *J. Mol. Biol.* **50**, 83-90.

Nagai, R., and Kamiya, N. (1966). Movement of the myxomycete plasmodium. II. Electron microscopic studies on fibrillar structures in the plasmodium. *Proc. Jpn. Acad.* **42**, 934-939.

Nagai, R., Ishima, Y., Kukita, F., and Takenaka, T. (1975). Calcium and magnesium contents of ectoplasm and endoplasm of *Physarum polycephalum* plasmodium. *Protoplasma* **86**, 169-174.

Nagai, R., Yoshimoto, Y., and Kamiya, N. (1978). Cyclic production of tension force in the plasmodial strand of *Physarum polycephalum* and its relation to microfilament morphology. *J. Cell Sci.* **33**, 205-226.

Nakajima, H. (1964). The mechanochemical system behind streaming in *Physarum*. In "Primitive Motile Systems in Cell Biology" (R. D. Allen and N. Kamiya, eds.), pp. 111-123. Academic Press, New York.

Nakajima, H., and Allen, R. D. (1965). The changing pattern of birefringence in plasmodia of the slime mold, *Physarum polycephalum*. *J. Cell Biol.* **25**, 361-374.

Ploton, T. D., and Gontcharoff, M. (1979). Ultrastructural study of the nucleolar cycle in *Physarum polycephalum* using EDTA preferential stain for RNP. *Exp. Cell Res.* **118**, 418-423.

Pollard, T. D. (1973). Progress in understanding amoeboid movement at the molecular level. In "The Biology of Amoeba" (K. W. Jeon, ed.), pp. 291-317. Academic Press, New York.

Pollard, T. D. (1979). Cytoplasmic myosin filaments. In "Motility in Cell Function" (F. A. Pepe, J. W. Santer, and V. T. Nachmias, eds.), pp. 117-125. Academic Press, New York.

Pollard, T. D., and Weihing, R. R. (1974). Actin and myosin and cell movement. *CRC Crit. Rev. Biochem.* **2**, 1-65.

Porter, K. R., Kawakami, N., and Ledbetter, M. C. (1965). Structural basis of streaming in *Physarum polycephalum*. *J. Cell Biol.* **27**, 78a.

Rapp, P. E., and Berridge, M. J. (1977). Oscillations in calcium-cyclic AMP control loops form the basis of pacemaker activity and other high frequency biological rhythms. *J. Theor. Biol.* **66**, 497-526.

Rhea, R. P. (1966a). Electron microscopic observations on the slime mold *Physarum polycephalum* with specific reference to fibrillar structures. *J. Ultrastruct. Res.* **15**, 349-379.

Rhea, R. P. (1966b). Microcinematographic, electron microscopic, and electrophysiological studies on shuttle streaming in the slime mold *Physarum polycephalum*. In "Dynamics of Fluids and Plasmas" (S. I. Pai, ed.), pp. 35-58. Academic Press, New York.

Ridgway, E. B., and Durham, A. C. H. (1976). Oscillations of calcium ion concentrations in *Physarum polycephalum*. *J. Cell Biol.* **69**, 223-226.

Ringertz, N. R., and Savage, R. E. (1976). "Cell Hybrids." Academic Press, New York.

Ryser, U. (1970). Die ultrastruktur der mitosekerne in den plasmodien von *Physarum polycephalum*. *Z. Zellforsch. Mikrosk. Ana.* **110**, 108-130.

Sachsenmaier, W., Blessing, J., Brauser, B., and Hansen, K. (1973). Protoplasmic streaming in *Physarum polycephalum:* Observation of spontaneous and induced changes of the oscillatory pattern by photometric and fluorometric techniques. *Protoplasma* **77**, 381-396.

Sakai, A., and Shigenaga, M. (1972). Electron microscopy of dividing cells. IV. Behavior of spindle microtubules during nuclear division in the plasmodium of the myxomycete, *Physarum polycephalum*. *Chromosoma* **37**, 101-116.

Satoh, H., Mito, Y., Ueda, T., Kurihara, K., and Kobatake, Y. (1980). Conversion of positive to negative chemotaxis by Ca^{2+} ionophore treatment in plasmodium of *Physarum polycephalum*. *Biochim. Biophys. Acta* **633**, 436-443.

Scheetz, R. W. (1972). The ultrastructure of *Ceratiomyxa fruticulosa*. *Mycologia* **69**, 38-54.

Schel, J. H. N., Steenbergen, L., Beckers, A., and Wanka, F. (1978). Change of the nuclear pore frequency during the nuclear cycle of *Physarum polycephalum*. *J. Cell Sci.* **34**, 225–232.

Schuster, F. L. (1965). A deoxyribose nucleic acid component in mitochondria of *Didymium nigripes*, a slime mold. *Exp. Cell Res.* **39**, 329–345.

Schwärzler, M., Jockusch, B. M., Hall, L., and Braun, R. (1977). Synthesis and transport of myosin in *Physarum polycephalum*. *Eur. J. Biochem.* **80**, 43–50.

Seavey, D., Goldmark, P., and Kessler, D. (1967). Mitochondrial DNA synthesis in the slime mold, *Physarum polycephalum*. *J. Cell Biol.* **35**, 187a.

Seifriz, W. (1942). Some physical properties of protoplasm and their bearing on structure. *In* "The Structure of Protoplasm" (W. Seifriz, ed.), pp. 245–264. Iowa State College Press, Ames, Iowa.

Sheen, S. J., Gailey, F. B., Miller, D. M., Anderson, J. D., Bargmann, T. J., and Carter, D. A. (1969). Sol-gel differences in plasmodia of the acellular slime mold, *Physarum polycephalum*. *BioScience* **19**, 1003–1005.

Sherry, J. M. F., Górecka, A., Aksoy, M. O., Dabrowska, R., and Hartshorne, D. J. (1978). Roles of calcium and phosphorylation in the regulation of the activity of gizzard myosin. *Biochemistry* **17**, 4411–4418.

Small, J. V., and Sobieszek, A. (1980). The contractile apparatus of smooth muscle. *Int. Rev. Cytol.* **64**, 241–306.

Sponsler, O. L., and Bath, J. D. (1953). A view of submicroscopic components of protoplasm as revealed by the electron microscope. *Protoplasma* **42**, 69–76.

Stebbings, H., and Hyams, J. S. (1979). "Cell Motility." Longman Group Limited, London.

Stewart, P. A. (1964). The organization of movement in slime mold plasmodia. *In* "Primitive Motile Systems in Cell Biology" (R. D. Allen and N. Kamiya, eds.), pp. 69–78. Academic Press, New York.

Stewart, P. A., and Stewart, B. T. (1959). Protoplasmic streaming and the fine structure of slime mold plasmodia. *Exp. Cell Res.* **18**, 374–377.

Stiemerling, R. (1970). Produktion und ausscheidung des schleimes von *Physarum polycephalum*. *Cytobiologie* **1**, 273–282.

Stockem, W. (1968). Über den DNS- und RNS- gehalt der mitochondrien von *Physarum polycephalum*. *Histochemie* **15**, 160–183.

Stossel, T. P., Hartwig, J. H., Yin, H. L., and Davies, W. A. (1979). Actin-binding protein. *In* "Cell Motility: Molecules and Organization" (S. Hatano, H. Ishikawa, and H. Sato, eds.), pp. 189–209. Univ. Park Press, Baltimore, Maryland.

Stossel, T. P., Hartwig, J. H., Yin, H. L., and Stendahl, O. I. (1980). The motor of ameboid leukocytes. *In* "The Cell Surface: Mediator of Developmental Processes" (S. Subtelny and N. K. Wessells, eds.), pp. 9–21. Academic Press, New York.

Szent-Györgyi, A. G., Szentkiralyi, E. M., and Kendrick-Jones, J. (1973). The light chains of scallop myosin as regulatory subunits. *J. Mol. Biol.* **74**, 179–203.

Tada, M., Yamamoto, T., and Tonomura, Y. (1978). Molecular mechanism of active calcium transport by sarcoplasmic reticulum. *Physiol. Rev.* **58**, 1–79.

Tanaka, H., and Hatano, S. (1972). Extraction of native tropomyosin-like substances from myxomycete plasmodium and the cross-reaction between plasmodium F-actin and muscle native tropomyosin. *Biochim. Biophys. Acta* **257**, 445–451.

Taylor, D. L., and Condeelis, J. S. (1979). Cytoplasmic structure and contractility in amoeboid cells. *Int. Rev. Cytol.* **56**, 57–144.

Taylor, D. L., and Wang, Y.-L. (1978). Molecular cytochemistry incorporation of fluorescently labeled actin into living cells. *Proc. Natl. Acad. Sci. U.S.A.* **75**, 857–861.

Taylor, D. L., Hellewell, S. B., Virgin, H. W., and Heiple, J. (1979a). The solation-contraction coupling hypothesis of cell movements. *In* "Cell Motility: Molecules and Organization" (S.

Hatano, H. Ishikawa, and H. Sato, eds.), pp. 363-377. Univ. Park Press, Baltimore, Maryland.
Taylor, D. L., Wang, Y.-L., Heiple, J., and Hellewell, S. (1979b). The contractile cytoskeleton of amoeboid cells *in vitro* and *in vivo*. *In* "Motility in Cell Function" (F. A. Pepe, J. W. Sanger, and V. T. Nachmias, eds.), pp. 263-283. Academic Press, New York.
Terada, T. (1962). Electron microscope studies on the slime mold *Physarum polycephalum*. Faculty of Science, Osaka University, Osaka, Japan 1962, pp. 47-58.
Ts'o, P. O. P., Eggman, L., and Vinograd, J. (1957). Physical and chemical studies of myxomyosin, an ATP-sensitive protein in cytoplasm. *Biochim. Biophys. Acta* **25**, 532-554.
Ueda, T., and Kobatake, Y. (1977). Changes in membrane potential, zeta potential, and chemotaxis of *Physarum polycephalum* in response to n-alcohols, n-aldehydes and n-fatty acids. *Cytobiologie* **16**, 16-26.
Ueda, T., Goetz von Olenhusen, K., and Wohlfarth-Bottermann, K. E. (1978). Reaction of the contractile apparatus in *Physarum* to injected calcium, ATP, ADP, and $5'$-AMP. *Cytobiologie* **18**, 76-94.
Usui, N. (1971). Fibrillar differentiation in a macroplasmodium of the slime mold *Physarum polycephalum*. *Dev. Growth Differ.* **13**, 241-255.
Vandekerckhove, J., and Weber, K. (1978). The amino acid sequence of *Physarum polycephalum* actin. *Nature (London)* **276**, 720-721.
Wang, J. H., and Waisman, D. M. (1979). Calmodulin and its role in the second messenger system. *Curr. Top. Cell. Regul.* **15**, 47-107.
Weeds, A. G., and McLachlan, A. D. (1974). Structural homology of myosin alkali light chains, troponin C and carp calcium binding protein. *Nature (London)* **252**, 646-649.
Weeds, A., Wagner, P., Jakes, R., and Kendrick-Jones, J. (1977). Structure and function of myosin light chains. *In* "Calcium-Binding Proteins and Calcium Function" (R. H. Wasserman, R. A. Corradino, E. Carafoli, R. H. Kretsinger, D. H. MacLennon, and F. L. Siegel, eds.), pp. 222-231. Elsevier North-Holland, New York.
Wegner, A. (1976). Head to tail polymerization of actin. *J. Mol. Biol.* **108**, 139-150.
Wier, W. G. (1980). Calcium transients during excitation-contraction coupling in mammalian heart: Aequorin signals of canine purkinje fibers. *Science* **207**, 1085-1087.
Wohlfarth-Bottermann, K. E. (1957). Die kontrastierung tierischer zellen und gewebe im rahmen ihrer elektronenmikroskopischen untersuchung an ultradünnen schnitten. *Naturwissenschaften* **44**, 287-288.
Wohlfarth-Bottermann, K. E. (1962). Weitreichende, fibrilläre protoplasmadifferenzierungen und ihre bedeutung für die protoplasmaströmung. I. Elektronenmikroskopischer nachweis und feinstruktur. *Protoplasma* **54**, 514-539.
Wohlfarth-Bottermann, K. E. (1964a). Cell structures and their significance for ameboid movement. *Int. Rev. Cytol.* **16**, 61-132.
Wohlfarth-Bottermann, K. E. (1964b). Differentiations of the ground cytoplasm and their significance for the generation of the motive force of ameboid movement. *In* "Primitive Motile Systems in Cell Biology" (R. D. Allen and N. Kamiya, eds.), pp. 79-109. Academic Press, New York.
Wohlfarth-Bottermann, K. E. (1974). Plasmalemma invaginations as characteristic constituents of plasmodia of *Physarum polycephalum*. *J. Cell Sci.* **16**, 23-37.
Wohlfarth-Bottermann, K. E. (1975a). Tensiometric demonstration of endogenous, oscillating contractions in plasmodia of *Physarum polycephalum*. *Z. Pflanzenphysiol.* **76**, 14-27.
Wohlfarth-Bottermann, K. E. (1975b). Weitreichende fibrilläre protoplasmadifferenzierungen und ihre bedeutung für die protoplasmastroemung .X. Die anordnung der actomyosin-fibrillen in experimentell unbeeinflussten protoplasmaadern von *Physarum in situ*. *Protistologica* **11**, 19-30.

Wohlfarth-Bottermann, K. E. (1979). Oscillatory contraction activity in *Physarum*. *J. Exp. Biol.* **81**, 15-32.
Wohlfarth-Bottermann, K. E., and Fleischer, M. (1976). Cycling aggregation patterns of cytoplasmic F-actin coordinated with oscillating tension force generation. *Cell Tiss. Res.* **165**, 327-344.
Wohlfarth-Bottermann, K. E., and Götz von Olenhusen, K. (1977). Oscillating contractions in protoplasmic strands of *Physarum:* Effects of external Ca^{++}-depletion and Ca^{++}-antagonistic drugs on intrinsic contraction automaticity. *Cell Biol. Int. Rep.* **1**, 239-247.
Wohlfarth-Bottermann, K. E., and Isenberg, G. (1976). Dynamics and molecular basis of the contractile system of *Physarum*. *In* "Contractile Systems in Non-Muscle Tissues" (S. V. Perry, *et al.*, eds.), pp. 297-308. Elsevier/North Holland Biomedical Press, New York.
Wolf, K. V., Hoffmann, H. U., and Stockem, W. (1979a). Fine structure of the mucous layer and endocytotic activity of microplasmodia of the acellular slime mold *Physarum polycephalum*. *In* "Current Research on *Physarum*" (W. Sachsenmaier, ed.), Vol. 120, pp. 187-190. Univ. of Innsbruck, Austria.
Wolf, R., Wick, R., and Sauer, H. (1979b). Mitosis in *Physarum polycephalum:* Analysis of time-lapse films and DNA replication of normal and heat-shocked macroplasmodia. *Eur. J. Cell Biol.* **19**, 49-59.
Wolff, D. J., and Brostrom, C. O. (1979). Properties and functions of the calcium-dependent regulator protein. *Adv. Cyclic Nucleotide Res.* **11**, 27-88.
Yoshimoto, Y., and Kamiya, N. (1978a). Studies on contraction rhythm of the plasmodial strand I. Synchronization of local rhythms. *Protoplasma* **95**, 89-97.
Yoshimoto, Y., and Kamiya, N. (1978b). Studies on contraction rhythm of the plasmodial strand III. Role of endoplasmic streaming in synchronization of local rhythms. *Protoplasma* **95**, 111-121.
Yoshimoto, Y., and Kamiya, N. (1978c). Studies on contraction rhythm of the plasmodial strand IV. Site of active oscillation in an advancing plasmodium. *Protoplasma* **95**, 123-133.
Zubruzycka-Gaarn, E., Korczak, B., and Osińska, H. E. (1979). Identification of sarcoplasmic reticulum-like system in *Physarum polycephalum*. *FEBS Lett.* **107**, 335-339.

PART III

Genetics

CHAPTER 6

Genetics of *Physarum polycephalum*

JENNIFER DEE

I. Genetic Control of the Life Cycle and of Plasmodium Development . 212
   A. Heterothallic Isolates 212
   B. The "Colonia Isolate" 214
   C. Construction of "Colonia Background" Strains for Mutant Isolation and Analysis 217
   D. Mutations Affecting Plasmodium Formation 220
   E. Some Attempts at Interpretation 227
II. Genes Affecting Processes Other Than Development 235
   A. Plasmodial Compatibility 235
   B. Nutritional Requirements 237
   C. Drug Resistance and Sensitivity 238
   D. Temperature-Sensitive Growth and Cell Cycle Events 239
   E. Motility, Morphology, and Color 242
III. General Comments on Methods 244
   A. Mutant Isolation 244
   B. Mutant Analysis 245
   C. Storage and Maintenance of Strains 246
   D. Genetic Nomenclature 246
   References 247

There are at least two good reasons for applying genetic analysis to *Physarum polycephalum*. Perhaps the most obvious reason is that biochemical analysis has been used so successfully to study certain basic biological processes, in particular those associated with the "cell cycle" or "nuclear cycle" in this species, that the application of genetic analysis can clearly add a valuable new approach to a fertile field. If biochemical and genetic techniques are combined, it should be possible to identify mutants with specific lesions and to investigate the genetic control of these processes.

A second good reason for *P. polycephalum* genetics has become apparent, however, from the study of a process which has not yet been analyzed biochemically. Genetic analysis has shown that the development of plasmodia from amoebae (the "amoebal–plasmodial transition") is a promising experimental system in which to investigate gene activity during differentiation. Like sporulation and spherulation, plasmodium development resembles cell differentiation in higher organisms because it apparently involves changes in gene expression, which can occur in single cells in the absence of growth and without change in gene content. All three processes can be induced in homogeneous cultures with good synchrony and therefore offer favorable experimental systems. However, although sporulation and spherulation have been studied biochemically, they are not readily accessible to genetic analysis because they occur in multinucleate plasmodia. Plasmodium development, on the other hand, occurs initially in uninucleate cells, and mutants can easily be isolated and analyzed.

For any genetic analysis of *P. polycephalum*, it is obviously essential to understand the genetic control of the life cycle so that crosses can be carried out and haploid or diploid stages isolated when they are required. A number of genes have now been identified, both by analysis of natural isolates and by detection of mutants in the laboratory, which control the life cycle by determining the ability of amoebae to form plasmodia. Thus, genetic analysis of plasmodium development, as a model of cell differentiation, and the discovery of new techniques for the genetic manipulation of *P. polycephalum* have progressed concurrently. These topics are therefore considered together in the first and major section of this chapter. A historical approach is adopted so that the strains and techniques now available will be introduced gradually and because it is hoped that the accidents, both fortunate and unfortunate, of the past may act as lessons for the future. In the second section, progress in the genetic analysis of other processes is summarized, and in the final section, some practical matters are briefly discussed.

I. GENETIC CONTROL OF THE LIFE CYCLE AND OF PLASMODIUM DEVELOPMENT

A. Heterothallic Isolates

Physarum polycephalum genetics began with analysis of heterothallic isolates of various origin. The plasmodial strain which had been cultured in the McArdle Laboratory in Madison for some years and which had been isolated by Professor M. P. Backus (the "Wisconsin 1" isolate) was sporulated, and two "mating types" were identified among the amoebal progeny (Dee, 1960; Haugli, 1971; Dee, 1973). Amoebae were cloned on bacterial lawns on agar, and clones of a single mating type could apparently be subcultured indefinitely in these

6. Genetics of *Physarum polycephalum*

conditions. Plasmodia were normally formed only when amoebae of different types were mixed, and progeny amoebae carrying two mating types were recovered in a 1:1 ratio when plasmodia sporulated. Thus the mating types behaved as expected for alleles of a single gene; this gene was designated *mt*, with alleles *mt-1* and *mt-2*.

Isolates of *P. polycephalum* from different localities in the United States revealed the presence of additional mating types; each isolate contained two new types, which were compatible with each other and with all the rest, and at least 14 different mating types have now been identified (Collins, 1975; Collins and Tang, 1977). Analysis of crosses indicated that all the mating types were inherited as alleles of the same gene (*mt*). Four of the mating types have been used extensively in later work: *mt-1* and *mt-2* from the Wisconsin 1 isolate and *mt-3* and *mt-4* from the "Indiana" isolate (Dee, 1966a).

Using a mutant drug-resistant strain of amoebae, it was shown that when amoebae of different mating types were mixed, they gave rise to plasmodia in which diploid, heterozygous nuclei underwent meiosis at the time of sporulation; spores contained amoebae which were the haploid products of meiosis and among which parental and "recombinant" types were recovered in equal proportions (Dee, 1962). Thus it appeared that *P. polycephalum* underwent a conventional sexual cycle with an alternation of nuclear fusion and meiosis. Amoebae behaved as haploid gametes which could be crossed; as an added advantage, they could be cloned and cultured and thus subjected to techniques of mutagenesis and mutant isolation similar to those used on other uninucleate microorganisms. Plasmodia presumably provided a diploid phase in which tests of dominance and complementation could be done. Conclusive evidence that amoebae were haploid and plasmodia were diploid was obtained later from measurements of nuclear DNA content (Mohberg and Rusch, 1971). Aldrich (1967) and Laane and Haugli (1976) demonstrated that meiosis normally occurred in spores but was not strictly coupled with spore cleavage. It appeared from both genetic and cytological evidence that a spore sometimes contained more than one product of meiosis; thus amoebae obtained from spores were recloned before analysis. Using this method, clones regularly gave the segregation and recombination frequencies expected for the haploid progeny of a cross.

It therefore appeared that the *mt* gene was the major determinant of the sexual cycle in *P. polycephalum* and that manipulation of amoebal strains of appropriate *mt* would allow geneticists to isolate mutants and analyze crosses. However, certain results suggested that other genes modified the rate or efficiency of crossing; in crosses between Wisconsin 1 and Indiana strains (Dee, 1966a), the results suggested that more distantly related strains crossed more rapidly and implied that the isolates may differ in many unidentified genetic factors; this was also suggested by later studies of plasmodial fusion genes in these isolates (Poulter and Dee, 1968). To reduce the genetic differences between laboratory stocks, so that analysis could be carried out in a uniform genetic background, representative amoebal strains were chosen from each isolate and care was then taken to record the origin and pedigree of all strains used. Analysis of mutants was at first confined to strains of Wisconsin 1 origin, and interisolate crosses were done only when elucidation of the differences between isolates was the aim. At Leicester University, strains *a* and *i* represented the Wisconsin 1 isolate and strains B173 and B174 the Indiana isolate. These four strains, together with the "Colonia" isolate (Section I,B), gave rise to many of the strains now used for genetic analysis. Some additional strains of Wisconsin 1 origin have also been used for genetics, in particular, the RSD series (Haugli, 1971) and the DJ series (Jacobson and Dove, 1975).

B. The "Colonia Isolate"

1. INHERITANCE OF SELFING ABILITY

The Colonia isolate was first described by Dr. H. A. Von Stosch of the University of Marburg, Germany, who obtained it from the Botanical Institute of Cologne (personal communication). Its origin in nature is unknown, but it is unlikely to have been collected locally since *P. polycephalum* has never been reported in Germany and rarely in Europe (Martin and Alexopoulos, 1969).

Unlike the heterothallic strains, the Colonia strain was shown by Von Stosch *et al.* (1964) and later by Wheals (1970) to give rise to plasmodia in amoebal clones; this behavior will be referred to as "selfing" throughout this chapter to distinguish it from "crossing," which describes the formation of diploid, hybrid plasmodia in mixtures of amoebae of different genotypes. "Selfing," unlike "homothallism" or "apogamy," does not imply any particular mechanism of plasmodium formation.

Selfing was shown by Wheals (1970) to be inherited through successive generations derived from the Colonia isolate. In each generation, clones were isolated by recloning amoebae after spore germination, and all of them gave rise to plasmodia. Spores obtained from these plasmodia again gave rise to selfing clones of amoebae. Formation of plasmodia in amoebal colonies had occasionally been observed during analysis of crosses with heterothallic strains, but in no case investigated was selfing behavior inherited by the next generation of amoebae; nor were there any "consistently self-fertile" strains found among the natural isolates studied by Collins (1975).

A strain capable of producing plasmodia in clones of haploid amoebae had long been recognized as desirable for *P. polycephalum* genetics since it should allow the isolation of homozygous mutant plasmodia after mutagenesis of amoebae. Although some types of mutant could be isolated by screening amoebae, this method was not suitable for identifying mutations affecting processes studied in plasmodia, such as the synchronous nuclear division cycle or sporulation. The Colonia isolate was the most promising candidate yet discovered for the task of isolating plasmodial mutants and was therefore greeted with some excitement. However, for such an approach to work, it was essential that Colonia amoebae should be haploid so that they could be effectively mutagenized. Since plasmodia were normally diploid, it seemed all too likely that the Colonia strain might be diploid or aneuploid throughout its life cycle.

Some encouraging genetic evidence was obtained by Wheals (1970), who showed that Colonia amoebae could be crossed with amoebae of a heterothallic (*mt-4*) strain; plasmodia heterozygous for plasmodial fusion genes (Section II,A) were obtained from the cross, and recombination of genetic markers was demonstrated among the progeny clones; so, it appeared that Colonia amoebae could behave as gametes in crosses in the same way as heterothallic amoebae. The progeny consisted of selfing clones and heterothallic (*mt-4*) clones in equal ratio, indicating that selfing ability was due to an allele of *mt* (designated *mt-h*), though the possibility that a closely linked gene was responsible was not excluded. Wheals (1970) suggested that the Colonia strain was homothallic, that is, that haploid amoebae of identical genotype mated to produce diploid homozygous plasmodia. Initial measurements of nuclear DNA content and chromosome counts (Mohberg *et al.*, 1973) did not confirm that Colonia strains were homothallic; the plasmodia of two Colonia clones (C50 and C5.1) studied at this time had mean nuclear DNA contents and chromosome numbers intermediate between those of amoebae and plasmodia of heterothallic strains, suggesting that they might be aneuploid. However, mutant isolation was attempted.

2. VARIATION AMONG COLONIA STRAINS

Although all the Colonia clones investigated by Wheals (1970) selfed eventually, there was variation in the time taken for plasmodium formation and plasmodia were not normally formed unless the amoebae were growing densely in thick streaks of bacteria or as a confluent lawn. From some spores formed by one of the clones (C50), however, a strain of amoebae (C5.1) was isolated which produced plasmodia rapidly and synchronously in small colonies of amoebae after only a few days' growth on a bacterial lawn (Wheals, 1973). Strain C5.1 was used for the first attempts at plasmodial mutant isolation since large numbers of clonally formed plasmodia could be isolated from colonies derived from mutagenized amoebae. From several thousand clones screened, several potentially mutant plasmodia were isolated; however, attempts to analyze them met with difficulties (J. Dee and C. E. Holt, unpublished). It was found that strain C5.1, unlike strain C50, failed to produce spores and apparently would not cross with heterothallic strains.

In an attempt to find a Colonia strain that combined the favorable characteristics of C50 and C5.1, Cooke and Dee (1975) screened clones derived from C50 spores. Although all clones selfed, some again gave plasmodia unable to sporulate. A progeny clone was selected which selfed rapidly in small colonies and produced plasmodia capable of sporulation. The next generation of amoebae derived from these spores behaved in the same way, and these amoebae were chosen as the strain (CL, Colonia Leicester) to be used for mutant isolation in future. Since it was clearly unsafe to assume that all clones derived by selfing from the Colonia isolate were genetically identical and since changes might also occur during subculturing, care was taken to maintain the original stocks of CL and to return to them frequently for the inoculation of new cultures. Using strain CL, however, the problem of crossing a rapidly selfing strain with heterothallic amoebae still remained (Section I,C).

3. APOGAMIC DEVELOPMENT IN *mt-h* CLONES

In strain CL and in a number of *mt-h* strains derived from it, the nuclear DNA content of amoebae and of clonally formed plasmodia was found to be close to that of heterothallic amoebae (Cooke and Dee, 1974; Mohberg, 1977). Thus homothallism, which implies nuclear fusion in zygotes formed by identical gametes, was clearly not occurring in these strains; it appeared that the nuclei remained haploid throughout the life cycle. Cinematographic analysis of CL cultures then revealed that plasmodium development involved the formation of binucleate cells from single uninucleate cells by nuclear division unaccompanied by cell division; these uninucleate cells were significantly larger than normal amoebae (Anderson *et al.*, 1976). The binucleate cells grew, underwent synchronous mitoses, fused with one another, and gradually developed the distinctive characteristics of plasmodia. Plasmodium development was concluded to be apogamic, occurring in single amoebae without cell or nuclear fusion, as previously reported for a mutant strain of *Didymium nigripes* (Kerr, 1967).

Cultures of CL amoebae, plated on lawns of bacteria, could be maintained without any visible plasmodium formation if the colonies were subcultured before they reached the size at which plasmodia normally appeared (Cooke and Dee, 1974; Anderson *et al.*, 1976). Thus it appeared that events occurring during the growth of a colony triggered plasmodium formation. Colonies harvested and replated after plasmodia had begun to appear were shown to contain, in addition to amoebae, cells irreversibly committed to plasmodium development; the amoebae proliferated, giving rise to new colonies, whereas the committed cells gave rise directly to plasmodia. Youngman *et al.* (1977) described the kinetics of plasmodium formation by assaying the numbers of committed cells at frequent intervals after inoculation of CL amoebal cultures. They found that a large proportion of the amoebae became committed and that this happened with a high degree of synchrony within a culture before macroscopic plasmodia appeared. Burland *et al.* (1981) provided evidence that cells became irreversibly committed while still uninucleate. Youngman *et al.* (1977) also showed that depletion of the food supply and a diffusible inducer produced by amoebae were involved in initiating plasmodium formation.

C. Construction of "Colonia Background" Strains for Mutant Isolation and Analysis

For the isolation of plasmodial mutants in *P. polycephalum,* it was essential that CL amoebae be haploid, though it was immaterial whether plasmodium development was homothallic, as originally supposed (Wheals, 1970; Dee, 1973), or apogamic; in either case, recessive mutations induced in amoebae would be expressed in clonally formed plasmodia. Thus plans for the isolation and analysis of plasmodial mutants of various kinds in CL proceeded as soon as the nuclear DNA content had been determined (Cooke and Dee, 1974). For the analysis of mutants, crosses between *mt-h* amoebae and heterothallic amoebae would be necessary. As a result of discussion between the *Physarum* genetics groups at Leicester and at the Massachusetts Institute of Technology, it was decided that attempts should be made to construct heterothallic strains isogenic with CL in order to avoid the difficulties that might arise in interpreting the results of crosses between strains of different isolates. The Wisconsin 1 isolate was initially chosen to provide the heterothallic alleles for strain construction, since it already carried some of the same alleles as the Colonia isolate and therefore appeared to be less distantly related to it than the Indiana isolate (Wheals, 1970; Cooke and Dee, 1974). However, all attempts to isolate crossed plasmodia from mixtures of CL and Wisconsin 1 amoebae, using a range of experimental techniques, failed (Adler and Holt, 1974a; Cooke and Dee, 1975); only selfed plasmodia could be isolated.

1. COLONIA × WISCONSIN 1 STRAINS

Crosses between Colonia and Wisconsin 1 strains were eventually achieved, at Leicester University, by isolating a derivative of CL with "delayed" clonal plasmodium formation, strain CLd (Cooke and Dee, 1975). CLd was obtained by repeatedly replating CL amoebae after allowing plasmodium formation. This effectively enriched the cell population for amoebae which had failed to form plasmodia, since plasmodia tended to fuse and often did not survive replating.

CLd amoebae selfed less readily than CL and were successfully crossed with the Wisconsin 1 strain *a* (*mt-1*), giving plasmodia with diploid DNA content and progeny in which genetic markers had recombined. A recombinant *mt-1* progeny clone was back-crossed to CLd and progeny analysis repeated. After three further back-crosses of this type, the *mt-1* strain, LU648, was isolated; it was estimated to have derived 97% of its genes from CLd (Cooke and Dee, 1975). CLd amoebae were also crossed with an *mt-2* strain derived from *i*; however, attempts to back-cross *mt-2* progeny with CLd were unsuccessful. The strain LU688, estimated to have derived 94% of its genes from CLd, was constructed by crosses between *mt-1* and *mt-2* strains. The "Colonia background" strains, LU648 (*mt-1*) and LU688 (*mt-2*), carried different alleles from CL and CLd at a locus *fusA* controlling plasmodial fusion. It was essential to maintain this difference throughout the back-crossing series so that crossed and selfed plasmodia from mixtures of *mt-h* and heterothallic amoebae could be distinguished (see Section II,A). A plasmodial nutritional mutant isolated in CLd was successfully analyzed by crossing amoebae of the mutant clone with LU648 and scoring haploid plasmodia derived from *mt-h* progeny clones for mutant and wild-type growth (Cooke and Dee, 1975).

2. COLONIA × INDIANA STRAINS

At MIT, *mt-3* and *mt-4* strains derived from the Indiana isolate were successfully crossed with CL. A series of back-crosses was carried out to construct Colonia background *mt-3* and *mt-4* strains, and these were successfully used for analysis of mutants isolated in CL (Adler and Holt, 1974a). Since the original Indiana strains (B173 and B174) were suspected to be rather distantly related to the Colonia isolate, a number of tests were carried out to determine whether the strains constructed by back-crossing were genetically more similar to CL. The results indicated that the differences between the isolates had been substantially reduced. Heterogeneity for one pair of plasmodial fusion alleles at a locus *fusC* was intentionally retained so that crossed plasmodia could be distinguished from selfed plasmodial (see Section II,A).

In successive generations of the back-crossing series, there was a gradual reduction in the efficiency of crossing between heterothallic (*mt-3* and *mt-4*) amoebae and CL; however, no clear segregation of mating ability was detected (Adler and Holt, 1974a). The ability of CL to cross with *mt-3* and *mt-4* amoebae and not with *mt-1* and *mt-2* was not understood.

3. DISCOVERY OF HETEROGENEITY BETWEEN COLONIA BACKGROUND STRAINS

Following the construction of Colonia background strains and the isolation and analysis of mutants, a range of genetically marked strains became available which were exchanged between laboratories. As a result of such exchanges between Leicester University and MIT it became clear that the Colonia background strains of Wisconsin 1 and Indiana origin differed in genes which had a major effect on the efficiency of crossing. These genetic differences were discovered and analyzed in both laboratories independently and led to the identification of loci, unlinked to *mt*, designated *matB* at MIT (Youngman *et al.*, 1979a) and *rac* at Leicester (Dee, 1978); the loci are almost certainly identical (Kirouac-Brunet *et al.*, 1981). Alleles of *matB* (*rac*) segregated clearly

6. Genetics of *Physarum polycephalum*

in the progeny of crosses between strains derived from the two heterothallic isolates.

Heterothallic strains differing in *mt* cross more rapidly and efficiently if they also carry different *matB* (*rac*) alleles. Strains differing in *matB* (*rac*) but not in *mt* fail to cross. Thus the "new" locus appeared to have only a secondary effect on crossing between heterothallic strains. However, in crosses involving *mt-h* amoebae, *matB* (*rac*) heteroallelism proved crucial. It was shown that a rapidly selfing *mt-h* strain (resembling CL) could be crossed with heterothallic amoebae (*mt-1* or *mt-3*) only if it differed from them in *matB* (*rac*) (Dee, 1978).

Further analysis (Youngman *et al.*, 1979a; Kirouac-Brunet *et al.*, 1981) has indicated that *matB* has multiple alleles, each of which promotes efficient crossing with all the others. The Wisconsin 1 isolate originally carried *matB1* (*rac-1*) and *matB4* (Kirouac-Brunet *et al.*, 1981), whereas the Indiana isolate carried *matB2* and *matB3* (*rac-2*) (Youngman *et al.*, 1979a). It is interesting that at least four alleles of this kind are necessary to account for the differences in rate of crossing observed between Wisconsin 1 and Indiana strains during early investigations of the *mt* locus (Dee, 1966a).

Colonia strains resembled the Wisconsin 1 isolate in carrying *matB1* (*rac-1*). This probably accounts for the failure of attempts to cross Wisconsin 1 strains with CL; *matB4* could easily have been absent from the few strains chosen to represent the Wisconsin 1 isolate (Cooke and Dee, 1975). Certainly the *mt-1* and *mt-2* Colonia background strains, LU648 and LU688, are both *matB1* (*rac-1*).

It is clear that one reason for the successful crossing of the Indiana strains with CL was that they carried different alleles of *matB*. These differences were apparently retained throughout the back-crossing series; the Colonia background *mt-3* and *mt-4* strains still differ from CL (*matB1*) in carrying *matB2* and *matB3*. However, since a gradual decrease in crossing ability, rather than a clear segregation, was observed during analysis of successive generations (Adler and Holt, 1974a), it seems likely that differences in other loci also promoted crossing between the two isolates and that these were gradually removed by back-crossing until only *matB* differences remained. It is known that at least one additional locus (*imz*) affecting the efficiency of crossing was heterogeneous in the Indiana isolate and that the allele (*imz-2*) which stimulates crossing in certain conditions has been lost from the most highly inbred Colonia background strains (Shinnick *et al.*, 1978).

Heteroallelism at the *matB* locus has now proved to be a valuable tool for genetic analysis in *Physarum* and essential for the analysis of certain classes of mutants (Anderson, 1979; Anderson and Holt, 1981). The fact that such valuable genetic heterogeneity was maintained inadvertently during a research program designed to remove it may be regarded as a salutary experience. Although it is probably necessary to reduce the natural diversity of a species to some extent in order to produce strains in which repeatable and understandable analysis can be done, it is clear that such efforts can be carried too far. Although many of the mutations present in natural populations will appear again in the laboratory, alleles affecting the life cycle may not be so easily replaced; in *Physarum,* mutations from one incompatibility allele to another are rarely observed. Thus some of the genes by which the life cycle can be manipulated might have been irreparably lost.

D. Mutations Affecting Plasmodium Formation

Apogamic development of plasmodia in clones of amoebae is an attractive system in which to investigate the genetic control of differentiation (Dee, 1973, 1975). Since development occurs in single cells, without change in DNA content (Section I,B,3), and since amoebae and plasmodia differ in many ways (Dee, 1973; Gorman and Wilkins, 1980), it is generally assumed that control of gene activity is involved, as in cell differentiation of higher organisms. Since development can be induced synchronously in large populations of cells, there are good prospects for biochemical analysis of the changes occurring. Also, since amoebae are uninucleate microorganisms, giving rise to plasmodia in haploid clones, the prospects for genetic analysis are excellent. Two genes involved in the control of plasmodium formation (*mt* and *matB*) were identified by analyzing strains from different isolates. Further genetic analysis was concentrated on mutants isolated in the laboratory.

1. MUTANTS OF *mt-h* STRAINS WITH DECREASED SELFING ABILITY

If a strain of *mt-h* amoebae capable of producing plasmodia in small colonies is plated on lawns of bacteria, mutant clones which fail to produce plasmodia can easily be identified. Such mutants were first isolated and analyzed by Wheals (1973); similar projects were later carried out in several laboratories. In most studies, the amoebae were subjected to mutagenesis before plating. In all studies except that of Wheals, an "enrichment" procedure designed to favor the survival of amoebae unable to form selfed plasmodia was applied before plating-out. This procedure varied in detail, but in all cases it was similar to that used by Cooke and Dee (1975) to isolate CLd from CL (see Section I,B). After mutagenesis or enrichment, amoebae were plated out and colonies which failed to give rise to plasmodia were retested. In most cases, further analysis was restricted to mutants which selfed rarely or not at all; mutants in which selfing was only delayed were generally regarded as unsuitable for analysis.

The mutants were crossed with heterothallic strains and the progeny scored for *mt* and for selfing ability. Recombinants carrying *mt-h* and the wild-type allele of the "non-selfing" mutation could be recognized by their ability to self rapidly in small colonies. Thus the mutations could be mapped in relation to *mt*. The mutants were also tested for their ability to cross with one another. If they crossed, forming heterozygous, diploid plasmodia, they were assigned to different "complementation groups."

The most striking outcome of the three later studies (Anderson and Dee, 1977; Davidow and Holt, 1977; Honey *et al.*, 1979) was that the majority of mutants isolated fell into only two complementation groups, both very closely linked to *mt*. As a result of tests to correlate the mutants isolated in different laboratories (Anderson, 1979; Honey *et al.*, 1979), these groups can now be designated *npfB* (= *aptB* = *difB*) and *npfC* (= *aptC* = *difA*) (see Table I). In crosses between representative mutants and heterothallic strains, no recombination between either of these two genes and *mt* was conclusively demonstrated, although approximately 1.6×10^4 progeny were analyzed in the three studies. Some possible selfing recombinants were investigated, but in most cases they were shown not to be haploid progeny of the cross. While 46 mutants of separate origin fell into groups *npfB* and *npfC*, only one mutant was assigned to a third complementation group, *npfA*, which was unlinked to *mt*.

The four mutants isolated by Wheals (1973), however, apparently all fell into different complementation groups, and the only mutant analyzed in detail (*aptA1*) proved to be unlinked to *mt*. It has now been shown that *aptA*, *npfA*, *npfB*, and *npfC* mutants represent four separate complementation groups, crossing in all pairwise combinations to give diploid heterozygous plasmodia (Anderson, 1979); for these tests, *matB* heteroallelism was essential. The three other *apt* mutants isolated by Wheals were unfortunately no longer available when this later analysis was done, so it is not known whether they represented additional groups.

The methods used by Wheals (1973) differed from those of the three later studies in several ways. His mutants were isolated from C5.1, whereas later studies used CL (see Section I,B); the mutagenic agent was different, and no enrichment procedure was employed (Table I). The enrichment procedure would presumably favor the survival of mutants which continued to multiply as amoebae while other cells around them were becoming irreversibly committed to plasmodium formation; mutants in which plasmodium development was initiated would tend to be lost. The enrichment method cannot wholly account for the absence of *aptA* mutants in the three later studies, however, since *aptA1* amoebae were shown to survive enrichment in a reconstruction experiment (Anderson and Dee, 1977; Davidow and Holt, 1977). It would be interesting to know whether attempts to isolate mutants from CL using a different mutagenic technique and without enrichment would produce representatives of *aptA* and other complementation groups not found in the later studies.

Another method of identifying mutations in additional genes required for plasmodium formation might be to analyze some of the less extreme mutants, which have so far been discarded. As pointed out by Anderson (1979), if strains

TABLE I

Non-Selfing Mutants Isolated from *mt-h* Strains

Complementation group[a]	No. of mutants	Recombination with *mt* (%)	Parent strain	Mutagen	Enrichment	Reference
aptA	1	50	C5.1	UV/caffeine	No	Wheals (1973)
npfA	1	50	CL	NMG[b]	Yes	Anderson and Dee (1977)
npfB	8	< 0.25[c]	CL	NMG	Yes	Anderson and Dee (1977)
aptB	7	< 0.06[c]	CH357[e]	—	Yes	Davidow and Holt (1977)
difB	10	< 0.15[c]	CL	NMG	Yes	Honey *et al.* (1979)
npfC	4	< 0.50[c]	CL	NMG	Yes	Anderson and Dee (1977)
aptC	6	< 0.06[c]	CH357[e]	—	Yes	Davidow and Holt (1977)
difA	11	~ 0.03[d]	CL	NMG	Yes	Honey *et al.* (1979)

[a] The symbols for the complementation groups are as follows: *apt*, amoebal-plasmodial transition; *npf*, no plasmodium formation; *dif*, differentiation. Complementation tests have now demonstrated that *npfB* = *aptB* = *difB* and *npfC* = *aptC* = *difA* (Anderson, 1979; Honey *et al.*, 1979).
[b] *N*-Methyl-*N*-nitro-*N*-nitrosoguanidine.
[c] The figure shown is the recombination frequency which would have been estimated if *one* recombinant clone had been detected in each set of progeny analyzed. The figures for *aptB* and *aptC* are averaged, since only the total numbers of progeny for all crosses were available in Davidow and Holt (1977).
[d] The figure is based on one possible recombinant which might not have been a normal haploid progeny clone.
[e] CH357 is a clonal derivative of CL.

differing in *matB* were used, it would now be simple to test new mutants against representatives of the identified complementation groups; only mutants which complemented all known groups need be analyzed further. Mutants with a temperature-sensitive failure to self are particularly amenable to analysis. For example, *npfA1* amoebae cannot self at 28.5°C or above, although they show only slightly reduced selfing ability at lower temperatures. Thus crossing and classification of progeny from crosses can easily be carried out at the nonpermissive temperature (however, see Section III,B,1).

2. REVERSION OF MUTANTS WITH DECREASED SELFING ABILITY

Even the most extreme non-selfing mutants give rise to plasmodia occasionally if cultures of amoebae are incubated for a sufficient time. Davidow and Holt (1977) found that it was possible to characterize their mutants by plating the amoebae on bacterial lawns and measuring the mean colony diameter at the time of plasmodium formation. Whereas the parent strains, such as CL, gave rise to plasmodia when colonies were 1–3 mm in diameter, the most extreme mutants showed no plasmodium formation until colony diameter was at least 20 mm. The genetic analysis described in the previous section was for the most part restricted to these extreme mutants.

Davidow and Holt (1977) isolated plasmodia from mutants showing various degrees of expression, derived progeny clones from them, and tested these clones for expression of the mutant characteristic. If the plasmodium had been formed as a result of a revertant mutation, the progeny were expected to resemble the original *mt-h* strain in selfing ability, producing plasmodia in small colonies. If the plasmodium had been formed as a result of incomplete suppression of selfing ("leakiness") in the mutant, the progeny were expected to resemble the mutant. The results indicated that in the more extreme mutants, plasmodia had usually arisen as a result of reversion, whereas in the less extreme mutants, plasmodium formation was usually due to leakiness. Unless such tests are done, it may be difficult to distinguish between a leaky mutant and an extreme mutant with a high rate of reversion. It is now clear that CLd, originally described as a delayed selfer, can best be interpreted as an extreme non-selfing mutant giving rise to plasmodia mainly by reversion; all progeny amoebae derived from plasmodia formed in CLd clones self rapidly, like CL (Cooke and Dee, 1975; see also Section I,E,1).

Anderson (1976) obtained plasmodia from clones of all the *npf* mutants that he analyzed genetically (Anderson and Dee, 1977). The temperature-sensitive mutant that was the only representative of group *npfA* showed leaky plasmodium formation at permissive temperatures, and its mutant characteristics were inherited by amoebae derived from these plasmodia. Mutants representing groups *npfB* and *npfC* showed slight leakiness or reversion; both types of plasmodium formation were found in both groups. Anderson (1976) measured the nuclear DNA content of amoebae and clonally formed plasmodia from a large number of different mutant strains and demonstrated that all the plasmodia had been formed from haploid amoebae without change in DNA content.

3. ISOLATION OF SELFING MUTANTS FROM HETEROTHALLIC STRAINS

Heterothallic amoebae very rarely form plasmodia in clones, and the majority of such plasmodia give rise only to progeny clones carrying the original heterothallic allele; thus they have apparently arisen through leakiness. These "illegitimate" plasmodia are formed from haploid amoebae without change in nuclear DNA content (Adler and Holt, 1975, 1977). A small proportion of plasmodia, however, give progeny clones capable of selfing, and these have presumably arisen as a result of mutation. Adler and Holt (1977) designated such mutants *gad* ("greater asexual differentiation"), and in this and a subsequent study (T. M. Shinnick, R. W. Anderson, and C. E. Holt, in preparation) 12 *gad* mutants of independent origin were analyzed. The mutants were isolated without mutagenesis or enrichment from strains of four different mating types (*mt-1* to *mt-4*). These heterothallic strains were found to differ in the frequency with which they gave both illegitimate and mutant plasmodia.

Most of the *gad* mutants selfed more slowly at 30°C than at lower temperatures, and this allowed them to be crossed with heterothallic strains and with one another; T. M. Shinnick, R. W. Anderson, and C. E. Holt (in preparation) also utilized *matB* heteroallelism to promote crossing. In crosses between *gad* mutants, or between a *gad* mutant and CL, large numbers of progeny could be easily screened for recombinants unable to form plasmodia in clones; thus high-resolution mapping could be accomplished. Of 12 *gad* mutants assigned to map positions, only one (*gad-12*) was unlinked to *mt*. Eight mutants were very closely linked to *mt* and to one another, giving no recombinants among several thousand progeny; the remaining three mutants were within 12 map units of *mt*. Two of the mutants which showed no recombination with *mt* were crossed with CL, and no recombinants were found among more than 6×10^3 progeny screened.

The *gad* mutations which recombined with *mt* could be tested for their expression with different *mt* alleles; for example, *gad-11*, originally isolated in an *mt-3* strain, was shown to cause selfing also when combined with *mt-1*, *mt-2*, or *mt-4* (Shinnick and Holt, 1977; T. M. Shinnick, R. W. Anderson, and C. E. Holt, in preparation). Similar results were found for *gad-6* and *gad-12*. Such tests also showed that the *gad* mutants did not carry mutations at the *mt* locus. For the *gad* mutations inseparable from *mt*, it was not possible to demonstrate rigorously that the *mt* allele was unchanged, but attempts to cross *gad* mutants with one another suggested that they retained the mating specificity of the heterothallic strain from which they had been isolated (Adler and Holt, 1977; T. M. Shinnick, R. W. Anderson, and C. E. Holt, in preparation).

For complementation tests of *gad* mutants, it was necessary to test the selfing ability of diploid cells carrying different *gad* mutations. It had previously been discovered (see Section I,E,2) that mixtures of amoebae carrying the same *mt* allele but different *matB* alleles gave rise to diploid cells heterozygous for *matB* which were unable to develop into plasmodia (Youngman *et al.*, 1979b, 1981). It was now reasoned that if similar mixtures were made in which both strains of amoebae carried mutations in the same *gad* gene, the diploid cells would be homozygous for *gad* and might therefore be capable of developing into plasmodia. Thus complementation tests could be carried out with different combinations of *gad* mutations. Plans to test complementation in this way were foiled, however, by the very interesting discovery that *gad* mutations were dominant. Control mixtures of amoebae, in which only one strain carried a *gad* mutation and neither selfing nor crossing was expected, gave rise to numerous hybrid plasmodia (Table II; T. M. Shinnick, R. W. Anderson, and C. E. Holt, in preparation). Similar results were found for 11 of the 12 *gad* mutants. Thus a diploid cell homozygous for a heterothallic *mt* allele and heterozygous for a *gad* mutation is capable of developing into a plasmodium. The tests were not carried out quantitatively, and it is possible that the frequency of plasmodia formed by *gad* heterozygotes is lower than that formed by *gad* homozygotes (or haploids; if so, the mutations should be described as "semidominant" rather than "dominant"; this may be important in attempts to construct models of *gad* gene action (see Section I,E).

Selfing mutants similar to *gad* mutants have been isolated from heterothallic strains by Honey (1979; Honey *et al.*, 1981a) and Gorman *et al.* (1979). Honey's mutants, designated *het*$^-$, were isolated from *mt-1* and *mt-2* strains after mutagenesis by NMG (*N*-methyl-*N'*-nitro-*N*-nitrosoguanidine). In a range of strains tested, the *mt-2* allele was associated with a higher frequency of *het*$^-$

TABLE II

Evidence for Dominance of *gad* Mutations[a,b]

Strain 1: *mt-x matBx gad-x aptA1*
Strain 2: *mt-x matBy gad$^+$ apt$^+$*

[a] Plasmodia were isolated from mixtures of amoebae of the genotypes listed and incubated at 26°C. *x* and *y* represent numerical suffixes given to different alleles.

[b] Strain 1 is not expected to self, even though it carries a *gad* mutation, because *aptA1* prevents development. Strain 2 is not expected to self because it is heterothallic. Crossed plasmodia are not expected because the strains carry identical *mt* alleles. Progeny analysis showed that the plasmodia isolated were diploid heterozygotes. It is concluded that *gad-x* is dominant and *aptA1* recessive.

mutations than *mt-1*. Like many of the *gad* mutants, *het*⁻ mutants showed reduced selfing at 30°C, and crosses were attempted at this temperature. Three of the mutants apparently retained the mating specificity of the original heterothallic strains, but one mutant, derived from an *mt-2* strain, crossed with some *mt-2* tester strains as well as with *mt-1*; Honey (1979) suggested that a mutation had occurred in *mt*. Honey did not attempt to map the *het*⁻ mutations directly but based further analysis on crosses with revertant (*dif*⁻) strains (see the next section). From these, he concluded that the *het*⁻ mutations were closely linked to *mt*.

Gorman *et al.* (1979) isolated selfing mutants from several heterothallic strains and demonstrated that mutagenesis by both NMG and EMS (ethylmethane sulfonate) increased the frequency of mutants recovered. Like the previous authors, they noted differences in the frequency of mutants obtained from different strains. Crosses between 17 of the mutants, designated *cat* (clonal amoebal transition), and heterothallic strains indicated that the mutants retained the mating specificity of the strains from which they were isolated. None of the mutations recombined with *mt*. Gorman *et al.* (1979) demonstrated that *cat* mutants were haploid and formed plasmodia without change in nuclear DNA content.

In addition to mutants of the *gad* type, a class of mutants has been isolated from heterothallic strains in which plasmodia arise directly from spores; these were designated *alc* (amoebaless life cycle) by Adler *et al.* (1975). Further analysis was reported by Adler and Holt (1977), and mutants of a similar type were observed by Gorman *et al.* (1979). Full genetic analysis of one such mutant has shown that a single mutation (*alc-8*) is responsible, unlinked to *mt*, *matB*, *npfA*, *aptA*, and several other markers (Truitt and Holt, 1979). The mutant phenotype due to *alc-8* was suppressed by *aptA1* but not by *npfA1*.

4. "REVERTANTS" OF SELFING MUTANTS

The nature of selfing mutants derived from heterothallic strains has been further investigated by the isolation of "revertant" non-selfing mutants from them. Anderson and Holt (1981) extended the study of Davidow and Holt (1977) to include a total of 47 revertants from four different *gad* mutants. The revertants were isolated without mutagenesis by the enrichment method used to isolate non-selfing mutants from CL. The revertants selfed rarely, but most of them produced plasmodia at a slightly higher frequency than the original heterothallic strains. It was therefore suggested that they carried suppressors of the *gad* mutations rather than true back-mutations (Anderson and Holt, 1981).

The revertants were tested for their ability to cross with heterothallic strains (*mt-1, 2, 3, 4*) and with representatives of the four complementation groups identified among non-selfing mutants of CL. The results apparently depended upon the mating type of the heterothallic strain from which the *gad* mutant had been derived. Two *gad* mutants derived from *mt-2* strains gave rise to 18 revertants, 16 of which could be allocated to complementation group *npfB* or *npfC*.

As previously found for the mutants isolated from CL, the *npfB* mutants crossed with all heterothallic mating types except *mt-2*, whereas *npfC* mutants crossed with all four mating types. The other two revertants showed anomalous behavior but did not reveal any further complementation groups.

Two *gad* mutants derived from *mt-3* strains gave rise to 29 revertants, none of which fell into group *npfB* or *npfC*; instead they revealed two new groups, designated *npfD* and *npfE*, which complemented the four known groups. The *npfD* mutants crossed with all mating types except *mt-3*, whereas *npfE*, like *npfC*, crossed with all four mating types.

Three of the four *gad* mutants from which revertants were derived were very closely linked to *mt*; the fourth was *gad-11*, 12 map units from *mt*. Attempts to map *npfD* and *npfE* revealed no recombination between the *gad* mutation and the revertant mutation in each case. It was concluded that the revertant mutations were closely linked to the *gad* mutations even where *gad* was separable from *mt*; however, no *npfE* revertants had been isolated from the *gad-11* strain.

Honey (1979; Honey *et al.*, 1981b) used NMG mutagenesis and enrichment to isolate non-selfing (dif^-) revertants from his selfing (het^-) mutants. From six het^- mutants derived from *mt-2* strains, he identified dif^- revertants similar to the dif^- mutants isolated from CL; 12 revertants fell into the *difA* (*npfC*) complementation group, and 17 were *difB* (*npfB*) and failed to cross with *mt-2* strains. In addition, he found a third class of 10 revertants with "partial *mt-2* specificity" which crossed rarely with *mt-2* strains and with *difB* and crossed readily with other mating types and with *difA*. This class had not been found by Anderson and Holt (1981) and was not among the dif^- mutants isolated from CL by Honey *et al.* (1979). These mutants with partial *mt-2* specificity were interpreted as being due to mutations in the *mt* locus. The het^- and dif^- mutations in all these strains were all very closely linked to *mt*; no recombinants were detected.

From a het^- mutant derived from an *mt-1* strain, Honey (1979) found a class of dif^- revertants resembling *mt-1* in mating specificity which crossed with *mt-2*, *difB*, and *difA* strains; the majority of the 19 revertants were of this type. No *difA* (*npfC*) mutants were found but four revertants fell into a new class, crossing readily with *difA*, *difB*, and *mt-2* and less readily with *mt-1*. Analysis of a cross between one of these mutants and an *mt-1* strain indicated that the dif^- mutation was not linked to *mt*; one-quarter of the progeny selfed. The original het^- mutation was presumably still present and closely linked to *mt*.

E. Some Attempts at Interpretation

1. ORIGIN AND HISTORY OF COLONIA STRAINS

   a. Mutations. It has been suggested by Anderson (1979) that CL might usefully be regarded as a *gad* mutant originally derived from an *mt-2* strain; its genotype would then be symbolized by *mt-2 gad-h* instead of by *mt-h*. The basis for this suggestion is that the same classes of non-selfing mutants were obtained from CL as from *gad* mutants isolated from *mt-2* strains and that one of these (*npfB*) was identical in behavior with *mt-2*. In addition, it had often been observed that Colonia strains mated less readily with *mt-2* amoebae than with other heterothallic strains (Wheals, 1970; Cooke and Dee, 1975; Anderson and Dee, 1977; Honey *et al.*, 1979), leading to the idea that *mt-h* strains might carry an allele similar to *mt-2* and therefore relatively incompatible with it.

Since *mt-2* is characteristic of the Wisconsin 1 isolate, Honey *et al.* (1979) have proposed that Colonia was originally derived from the Wisconsin 1 isolate. Supporting evidence for this suggestion is the fact that Colonia strains carry a number of other alleles characteristic of Wisconsin 1, namely, the plasmodial fusion alleles, *fusA2* and *fusB1* (Wheals, 1970); alleles of genes controlling plasmodial lethal interactions (Honey *et al.*, 1979); and the *matB1* (*rac-1*) allele (Kirouac-Brunet *et al.*, 1981).

Since Wisconsin 1 strains were cultured at the McArdle Laboratory for some years prior to the first report on Colonia (Von Stosch *et al.*, 1964), it is certainly possible that the Botanical Institute of Cologne, from which Von Stosch obtained Colonia, had in its possession a Wisconsin 1 culture from Madison and that Von Stosch studied a mutant derivative of this culture. The Colonia strain received from Von Stosch and studied by Wheals (1970), however, did not show the rapid selfing characteristic of CL, and it appears that a second *gad* mutation must have occurred in a culture of C50, giving rise to the plasmodium from which C5.1 and CL were derived (see Section I,B,2). Prior to this, a mutation causing delayed selfing probably occurred, since the Colonia amoebae received from Von Stosch could not initially be induced to self at all. The strains studied by Wheals were derived from a single plasmodium which arose from these amoebae after several years, presumably as a result of yet another *gad* mutation. However, all these events could have involved "forward" or "reverse" mutations in the same gene.

CLd has been shown to carry a mutation belonging to the *npfC* complementation group (Anderson, 1979). It apparently has a high rate of reversion, since plasmodia are usually obtained from CLd cultures without difficulty and all amoebal progeny of these plasmodia show nondelayed selfing. The progeny of crosses between CLd and heterothallic strains are either delayed selfers or heterothallic because *npfC* is closely linked to *mt*. The isolation of CLd by repeated subculturing of CL amoebae after plasmodium formation demonstrated how selection for delayed or non-selfing mutants could occur unintentionally during ordinary laboratory maintenance of an *mt-h* strain. This has probably occurred several times since Colonia strains were first cultured, but precautions can be taken to avoid it (Cooke and Dee, 1974).

    b. Ploidy and Sporulation. Poulter and Honey (1977), using a selective technique, isolated hybrid plasmodia from a mixture of two genetically marked *npfC* amoebal strains derived from CLd. They deduced by progeny analysis that the plasmodia were diploid and had been formed by the mating of *npfC* and *npf*$^+$ revertant amoebae; progeny clones segregated for rapid and delayed selfing and showed recombination of genetic markers. They suggested that similar matings may also occur infrequently in clones. No entirely diploid plasmodium has yet been isolated from a clone of haploid amoebae; thus there is no direct evidence of homothallic development in *P. polycephalum*. However, if the majority of plasmodia formed in an *mt-h* clone are haploid, rare diploid zygotes would presumably fuse with them and would pass unnoticed. Poulter and Honey (1977) used plasmodial fusion alleles to prevent such fusion in their mixtures.

Plasmodia derived from many different haploid *mt-h* and mutant strains have been found to contain a small proportion of nuclei of diploid or higher DNA content (Anderson, 1976; Mohberg, 1977). These nuclei could have been formed by amoebal matings or at a later stage. Laffler and Dove (1977) showed that the viability of spores produced by CL plasmodia could be significantly raised (from < 1 to 8%) by increasing the proportion of diploid nuclei (to 80%) by means of heat shocks. They suggested that the only viable spores produced by a predominantly haploid plasmodium might be those that result from meiosis in the few diploid nuclei present. If this is true, the high (> 50%) spore viability observed for all the Colonia strains studied by Wheals (1970) would imply that these plasmodia were predominantly diploid; unfortunately, no measurements of DNA content are available. A C50 plasmodium isolated and studied by Mohberg (1977) had only slightly higher mean DNA content than CL and a spore viability of 2%. Laane *et al.* (1976) obtained evidence that suppression of the second meiotic division may also result in the production of some viable spores from CL plasmodia; it is possible that such a mechanism was operating efficiently in the strains studied by Wheals and has been lost during the derivation of CL.

Several clones isolated from C50 spores produced plasmodia that failed to sporulate (see Section I,B,2). If this was due to a mutation, it may still be present in the strain APT1 isolated from C5.1 (see Section I,D,1). Since APT1 amoebae fail to self, only rare revertant or crossed plasmodia have been directly derived from them, and a recessive defect of sporulation may have passed unnoticed; there is some evidence, however, that such a mutation may be detectable among progeny derived from crosses (C. E. Holt, personal communication).

2. GENETIC CONTROL OF MATING

Alleles of *mt* were originally identified because they controlled the formation of crossed plasmodia by heterothallic amoebae. However, the question of whether they acted at the level of cell fusion, nuclear fusion, or zygote development was not immediately resolved. Since amoebae of different mating types were morphologically indistinguishable, and since plasmodium formation occurred with low frequency in a mixed population of amoebae, direct cytological approaches proved impractical and the question remained open (Dee, 1966a, 1973). Control of cell fusion seemed the most likely mechanism, however, since it would avoid wastage of gametes. It was also attractive to envisage amoebal mating as a process analogous to fertilization and responsible for the initiation of plasmodium development.

With the discovery of apogamic development in Colonia strains, it became clear that neither cell nor nuclear fusion was necessary for plasmodium development. The problem also arose that the *mt* gene appeared to have a dual role; it had been supposed to control mating specificity by means of factors in the cell membrane, but it could apparently also mutate to allow development without cell fusion.

The first evidence that *mt* might control crossing by regulating the differentiation of zygotes was presented by Adler and Holt (1975). They isolated diploid amoebae heterozygous for heterothallic *mt* alleles and showed that these were capable of forming selfed plasmodia in amoebal clones without change in nuclear DNA content; thus they behaved like *mt-h* amoebae. Adler and Holt (1975) suggested that in mixtures of heterothallic amoebae, cell fusion might occur randomly and that only cells heterozygous for *mt* were capable of development; the allele *mt-h* could then be regarded as a mutant allele of a regulatory gene. However, no evidence of cell fusion in clones of amoebae had ever been obtained; even when illegitimate plasmodia were formed in heterothallic strains, they were invariably found to be haploid.

The discovery of a second locus controlling crossing has led to an elucidation of the roles of both genes. Diploid cells can be isolated from mixtures of amoebae differing in *matB* at a much higher frequency than from mixtures of amoebae carrying the same *matB* allele, regardless of the *mt* alleles present; thus *matB* clearly stimulates cell fusion even between cells of the same *mt* (Youngman *et al.*, 1979b, 1981). Diploid amoebae heterozygous for *mt* develop into plasmodia regardless of their *matB* alleles. Diploids heterozygous only for *matB*, however, fail to develop. It is concluded that *matB* controls cell fusion and *mt* controls zygote development (Youngman *et al.*, 1979b, 1981). It is presumed that when amoebae carry the same *matB* alleles, random, infrequent fusions occur; if the fusing cells differ in *mt*, they proceed to develop and a low frequency of plasmodium formation is observed. When amoebae differing in both *matB* and *mt* are mixed, fusion occurs very frequently, development is successful, and frequent plasmodia are observed.

On the basis of cytological observations, Holt *et al.* (1979) have suggested that *mt* has an effect on nuclear fusion. They reported that only in mixtures of amoebae of different *mt* was cell fusion followed by nuclear fusion during interphase. When amoebae of the same *mt* fused, a dikaryotic cell was formed which sometimes was seen to undergo nuclear fusion and cytokinesis at the first mitosis, presumably thereby giving rise to diploid amoebae.

If amoebae carrying the same *mt* can fuse, one would expect amoebal fusions to occur occasionally in *mt-h* (*mt-2 gad-h*) clones, although this may usually be precluded by the rapid development of single cells. Cooke (1974) and Anderson (1976) mixed *mt-h fusA1* and *mt-h fusA2* amoebae (carrying the same *matB* alleles) and detected a small proportion of plasmodia with a hybrid (*fusA1*/*fusA2*) fusion type among a large number of plasmodia isolated from the mixtures. Measurements of DNA content (Cooke, 1974) and progeny analysis (Anderson, 1976) indicated that these plasmodia were haploid heterokaryons, suggesting that fusion of amoebae had occurred without nuclear fusion. However, it is also possible that "illegitimate heterokaryons" were formed by fusion of incipient plasmodia (see Section II,A).

Since most natural isolates are heterogeneous for *matB* as well as for *mt* (Kirouac-Brunet *et al.*, 1981), control of cell fusion apparently does occur in natural populations. Since *matB* and *mt* are unlinked, there may still be some wastage of gametes; however, this would be avoided if cells carrying the same *mt* usually fused only transiently, separating again without nuclear fusion, as suggested by Honey *et al.* (1979). Other genes may also be involved in controlling mating in natural populations, for example, the gene *imz* (Shinnick *et al.*, 1978). Pallotta *et al.* (1979), in an investigation of the kinetics of mating, showed that diffusible factors produced by amoebae influenced the time of onset of mating.

3. GENETIC CONTROL OF PLASMODIUM DEVELOPMENT

a. mt *Alleles.* For purposes of interpreting *mt* gene action in plasmodium formation, the elucidation of *matB* action is extremely helpful. If we can release *mt* entirely from the task of controlling cell fusion, we can concentrate on understanding its role in controlling development of cells of various genotypes, whether haploid or diploid, homozygous or heterozygous, without being concerned about how such cells are formed.

Since the *mt* gene is multiallelic, and since any two *mt* alleles from the 14 so far identified combine to allow development of a diploid, hybrid plasmodium, it seems more likely that this compatibility system involves inhibition of development by single or identical *mt* alleles than stimulation by specific interaction between different alleles. Two models to account for such inhibition have been proposed: one involving a repressor–operator interaction (Honey *et al.*, 1979) and the other the formation of inhibitory oligomers composed of identical subunits (Anderson and Holt, 1981). Both models assume that fusion of cells has occurred regardless of their *mt* genotype and that development of the fusion product is prevented at an early stage in *mt* homozygotes, perhaps even prior to nuclear fusion. Both models satisfactorily explain how inhibition would be released in cells (dikaryotic or diploid) containing two different *mt* alleles and also account for the authors' own analysis of mutants isolated from CL and from heterothallic strains.

Since the mutant studies on which these models are based have involved different strains and have not yet been fully correlated, it seems premature to argue their respective merits. It would also be desirable to have some biochemical evidence of the gene products involved in plasmodium development before considering specific models of gene action. Therefore, I wish primarily to consider how far the genetic evidence can be interpreted in genetic terms.

b. gad Mutations. In natural populations of *P. polycephalum*, the heterothallic system apparently predominates; thus heterothallic alleles may be regarded as wild type. If heterothallic *mt* alleles cause inhibition of selfed plasmodium development, *gad* (het^-) mutations cause the inhibitory mechanism to fail. Since some of these mutations are widely separated from *mt* on the genetic map, it is likely that they have occurred in several different genes; thus several (gad^+) genes may be involved in effecting inhibition in heterothallic strains. The fact that many of the *gad* mutations are too close to *mt* to be randomly distributed in the genome but are probably not close enough to be part of the same functional unit (T. M. Shinnick, R. W. Anderson, and C. E. Holt, in preparation) is one of the most puzzling aspects of the genetic results.

For some of the *gad* mutations genetically separable from *mt*, there is clear evidence that they cause selfing in the presence of an unchanged *mt* allele. For a few of the *gad* mutations inseparable from *mt*, there is also some evidence that the heterothallic allele is still present; however, many of these mutations could be in *mt* itself. Since there is no conclusive evidence on this point, it is probably more useful for the present to refer to the mutant strains as *mt-x gad-y*, thus indicating the heterothallic strain from which the mutant was derived. In addition, CL may be more usefully represented as *mt-2 gad-h* than as *mt-h*, since this draws attention not only to its relationship with *mt-2* but also to its similarities with other *gad* mutants.

Since *mt-x gad-y* cells develop into plasmodia, diploid cells homozygous for these alleles would be expected to develop. The fact that heterozygous *mt-x gad-y*/*mt-x gad$^+$* cells are also capable of development (i.e., *gad-y* is dominant) shows that the *gad* mutation interferes in some way with the inhibitory mechanism even in the presence of the wild-type gene. This could be by alteration of a gene product or by dilution of an inhibitor below a critical level. The fact that many *gad* mutants (including *gad-h*) are heat sensitive may suggest that the mutant alleles produce protein products which interfere with inhibition more effectively at low temperatures than at high temperatures.

c. npf Mutations. From *mt-x gad-y* strains, mutants (*npf, apt*, or dif^-) can

6. Genetics of *Physarum polycephalum*

be isolated in which plasmodium formation is absent or greatly reduced. Such mutants could arise in several different ways.

Firstly, revertant mutations in the *gad* (*het⁻*) gene may restore a functioning inhibitory system. The mutants from *mt-2 gad-y* strains which behave like *mt-2* (group *npfB*) are presumably of this type, as are *npfD* mutants from *mt-3 gad-y* strains, which behave like *mt-3*. As expected, the *npf* mutations in these strains map very close to the *gad* genes. Since complete *gad⁺* function is apparently often not restored, some mutations may be intragenic suppressors rather than true back-mutations (Anderson and Holt, 1981).

Secondly, mutations may occur in genes which are not involved in the inhibitory mechanism but which are required for the further development of a plasmodium. Since many genes are presumably involved in plasmodial functions which are not required in amoebae, many different mutations of this type were expected when non-selfing mutants were isolated from CL. However, their isolation may have been precluded by the enrichment technique (see Section I,D,1). At present, *aptA1* appears to be the sole representative of this type. Since *aptA* is unlinked to *mt* and *gad-h*, clearly it is not a revertant, and it has been shown to prevent development of most *gad* mutants (T. M. Shinnick, R. W. Anderson, and C. E. Holt, in preparation). In a haploid strain containing a *gad* mutation and *aptA1*, development may be initiated but unable to continue. In a doubly heterozygous *gad-y/gad⁺ aptA1/apt⁺* cell, however, development occurs because *gad-y* is dominant and *aptA1* recessive (T. M. Shinnick, R. W. Anderson, and C. E. Holt, in preparation). Thus *aptA1* can simply be regarded as a loss of function. This function is apparently also required for development of plasmodia heterozygous for heterothallic *mt* alleles; no plasmodia develop if both strains carry *aptA1* (Wheals, 1973).

A third type of mutation seems to be represented by *npfA1*. This mutation is unlinked to *mt* and clearly is not a reversion of *gad-h*, but it cannot be interpreted in the same terms as *aptA1* because, even when homozygous, it does not prevent the development of plasmodia formed as the result of a heterothallic cross (Anderson and Dee, 1977). It prevents the development of most *gad* mutants, however, even (though to a lesser extent) in the presence of its wild-type (*npfA⁺*) allele (T. M. Shinnick, R. W. Anderson, and C. E. Holt, in preparation). Thus it appears to be an unlinked, semidominant suppressor of a large number of different *gad* mutations.

The mutation unlinked to *mt* recovered by Honey (1979) from *mt-1 gad-y* strains, which represented a new complementation group, is also clearly not a revertant of the *gad* mutation, but it is not known whether it resembles *aptA1* or *npfA1* in its action. The mutations giving rise to groups *npfC* and *npfE*, however, closely linked to *mt* and to the *gad* mutations from which they were derived, are more difficult to interpret. Some conclusions may be drawn from consideration of the results of complementation tests.

d. Complementation Tests of npf *Mutants.* The *npfA1* and *aptA1* mutants isolated from CL presumably still carry *mt-2* and *gad-h*. If *gad-h* is dominant, these mutants will cross with each other and with heterothallic *mt-2* strains so long as the crosses are attempted at a temperature at which *gad-h* causes plasmodium development. Anderson (1979) found that such crosses were successful at 26°C but were markedly inhibited at 30°C, the temperature at which plasmodium formation is also inhibited in CL. The *npfB* mutants behaved exactly like *mt-2* heterothallic strains in these tests and failed to cross with *mt-2* at any temperature, agreeing with the assumption that they were gad^+ revertants. All the mutants derived from CL, however, crossed well with *mt-1*, *mt-3*, and *mt-4* strains at both temperatures; presumably these crosses did not depend on *gad-h*. The fourth group of mutants (*npfC*) derived from CL crossed with all the other mutants and with *mt-2*, but again, these crosses were inhibited at 30°C, suggesting that they also depended on *gad-h*. Since *gad-h* was certainly not present in the *npfB* or *mt-2* strains, it was presumably carried by *npfC* mutants. Since crossing of *npfC* mutants with *mt-1*, *mt-3*, or *mt-4* was not reduced at 30°C, it seems likely that the *npfC* mutants still also carried *mt-2*. Thus it seems simplest to conclude that *npfC* mutants carry mutations in another gene functionally separate from *mt* and *gad* though closely linked to them; this would be the *difA* gene proposed by Honey *et al.* (1979).

Although Anderson and Holt (1981) found essentially the same results for *npf* mutants isolated from *mt-2 gad-y* strains as for those isolated from CL, Honey (1979) found a new class which crossed with *npfB* and *npfC* mutants and which he interpreted as due to mutations in *mt*. Different strains and conditions of crossing were used in the two laboratories, and no information is available from which crossing due to heterothallism can be distinguished from crossing due to *gad* dominance.

There is no published evidence from which it can be deduced whether the *gad* allele is still present in the *npfE* mutants isolated from *mt-3 gad-y* strains. However, as each apparently new "complementation group" inseparable from *mt* emerges, one feels increasingly uneasy about interpreting it as representing a new gene with a separate function. A "complementation group" does not have the same significance in this situation as, for example, in the analysis of mutants affecting a biochemical pathway. Each pair of heterothallic mating types "complements" in the sense that *npfC* and *npfE* mutants do, yet no one is likely to propose that the 14 mating types represent 14 different genes. It seems attractive, therefore, to suppose that *npfC* and *npfE* mutants might be new heterothallic mating types which have arisen as a result of mutations in *mt* (Anderson and Holt, 1981). However, if *npfC* is an allele of *mt*, it is necessary to explain why this suppresses the *gad* mutation when there is evidence that at least some *gad* mutations cause plasmodium development in combination with any *mt* allele.

6. Genetics of *Physarum polycephalum*

It is also necessary to provide a new explanation for the heat sensitivity of crossing between *npfC* mutants and *mt-2* strains.

II. GENES AFFECTING PROCESSES OTHER THAN DEVELOPMENT

Mutant strains of *P. polycephalum* have been sought for two main reasons: because they may carry mutations affecting processes of general biological interest and because they may provide genetic markers which will indirectly aid genetic analysis of these processes. Most of the mutants so far analyzed, apart from those already discussed, fulfill the second purpose rather than the first. Although the isolation of mutant haploid plasmodia is now possible, progress in isolating cell cycle mutants has been disappointingly slow, and no mutants have been isolated with specific lesions of several other processes which have been thoroughly studied in plasmodia, for example, sporulation, spherulation, protoplasmic streaming, and chemotaxis. The mutants which are available are briefly described below. They are grouped according to their phenotypic effects, with comments on the methods by which they have been obtained and the aims which originally inspired their isolation.

A. Plasmodial Compatibility

All known genes affecting plasmodial compatibility have been identified by analysis of natural isolates; no mutants have been isolated in the laboratory. Genetically identical plasmodia fuse readily. This is a normal process during the growth of plasmodia developing in an amoebal culture, though there is no reason to suppose that it is an obligatory stage in the life cycle; an isolated plasmodium would presumably eventually grow large enough to sporulate. Plasmodia which carry different alleles at the "plasmodial fusion" loci fail to fuse; those which are identical at the fusion loci but which differ at "killing" loci undergo a lethal interaction after fusion. Each natural isolate investigated has proved to be heteroallelic at a number of fusion and killing loci (Collins and Haskins, 1972; Carlile, 1973), but the significance of this heterogeneity for the *Physarum* population is unknown.

For purposes of genetic analysis, much of the heterogeneity of the natural isolates has been intentionally removed. The Wisconsin 1 strains *a* and *i* differed at two loci, *fusA* and *fusB*, though there is evidence that the original isolate was heteroallelic at additional loci (Dee, 1973). All selfed plasmodia from the Colonia isolate were *fusA2 fusB1* (Wheals, 1970). Strains derived from *i*, carrying *fusB2*, are still used for genetics (Poulter and Honey, 1977; Honey *et al.*, 1979), but all the strains derived from Wisconsin 1 × Colonia back-crosses carry *fusB1* and differ only in one pair of alleles, *fusA1* and *fusA2*, which are co-dominant.

Analysis of plasmodial fusion in crosses between Wisconsin 1 and Indiana strains gave complex results, suggesting the segregation of alleles at *fusA* and at additional loci (Poulter and Dee, 1968; Dee, 1973). The initial crosses between Indiana strains and Colonia indicated differences at many loci, but these were removed by back-crossing until only one pair of alleles was segregating: *fusC1*, carried by Colonia, and *fusC2*, derived from the Indiana isolate (Adler and Holt, 1974a); *fusC2* is dominant to *fusC1*. Further analysis (Anderson, 1976) indicated that *fusC* was a separate locus from *fusA* and *fusB*. Thus, Colonia strains (e.g., CL) are *fusA2 fusB1 fusC1*. Wisconsin 1 × Colonia strains are all *fusB1 fusC1*, and Indiana × Colonia strains are all *fusA2 fusB1*.

The plasmodial fusion genes have been used primarily as genetic markers and are essential for the isolation of hybrid plasmodia from crosses involving apogamic strains. The co-dominant *fusA* alleles are the most useful in this respect, since they prevent fusion between heterozygous (*fusA1/fusA2*) plasmodia and both selfed types and so allow the hybrids to be unambiguously identified. They were essential for the complementation studies described in Sections I, D and E. All three *fus* genes have also been used as markers for progeny analysis in many crosses. Two examples of linkage involving *fus* genes have been detected; *fusA* and *sax* (Poulter, 1969; Dee, 1973) showed about 10% recombination and *fusC* and *aptA* about 1% recombination (Pallotta et al., 1979).

Although the majority of plasmodia with a heterozygous (*fusA1/fusA2*) fusion type isolated from crosses of *fusA1* and *fusA2* amoebae are diploid, heterokaryons have also been found occasionally (Cooke, 1974; Adler, 1975; Anderson, 1976). These may have resulted from fusion of amoebae without nuclear fusion (see Section I,E,2) or from fusion of plasmodia at a stage of development before the *fus* gene became effective (illegitimate heterokaryons). Since the plasmodial fusion genes apparently have no effect on amoebal mating, it is possible that they are "switched on" early in plasmodial development. However, no successful biochemical analysis of the factors involved in controlling plasmodial fusion has yet been reported.

Genetic differences causing lethal interactions between plasmodia (Carlile and Dee, 1967; Carlile, 1972) were mainly eliminated at early stages in the construction of Colonia background strains, and only one pair of alleles, *kilA1* and *kilA2*, has been retained. *kilA1* is carried by the great majority of strains used for genetics and *kilA2* by a few strains derived from the Indiana isolate (Adler and Holt, 1974a; T. M. Shinnick, R. W. Anderson, and C. E. Holt, in preparation). Using strains derived from the original Wisconsin 1 isolate, however, Carlile (1976) analyzed the genetic basis of lethal interactions and obtained evidence that allelic differences at three loci (designated *let*) were responsible. These strains have also been used for detailed cytological and biochemical investigations of the processes involved in lethal interactions (Lane and Carlile, 1979; Schrauwen, 1981).

B. Nutritional Requirements

Since *P. polycephalum* plasmodia grow well on simple defined media, it should be possible to isolate mutants with a range of different specific nutritional requirements. Cooke and Dee (1975) isolated one lysine-requiring mutant by screening 5×10^3 plasmodia derived from clones of CL amoebae and one leucine-requiring mutant from 7.5×10^3 CLd clones. The amoebae were mutagenized with uv or NMG. These mutants were shown to carry single gene mutations, denoted by *lys-1* and *leu-1*, respectively, which have served as useful genetic markers in many crosses. A second mutation causing lysine requirement was identified later in another clone isolated from mutagenized CLd amoebae; the mutant allele (*lys-2*) complements *lys-1* (J. Dee and J. L. Foxon, unpublished). Although additional markers of this kind would be valuable and their isolation should be straightforward, no further identification of nutritional mutants by these methods has been reported.

Strains with a specific requirement for valine were identified among the progeny of a cross between the Wisconsin 1 strain *a* and the mutant APT1, derived from C5.1. However, this requirement was either due to mutations in two genes or to a chromosomal translocation (Dee *et al.*, 1973; Dee, 1973). It did not reappear among progeny of *a* × CLd (D. J. Cooke and J. Dee, unpublished) and has not been transferred to Colonia background strains. Thus it has not been used in any subsequent analysis.

Attempts to grow *P. polycephalum* amoebae on synthetic agar media have repeatedly failed, and amoebae have been cultured on lawns of living or formalin-killed bacteria for all routine genetic analysis. Several strains of amoebae, including CLd, were eventually grown in a range of defined and complex liquid synthetic media (Goodman, 1972; McCullough and Dee, 1976), including the semidefined medium used for plasmodia. The only absolute requirements demonstrated in these strains were for the same three vitamins and one amino acid that are required by plasmodia (McCullough and Dee, 1976). However, further analysis revealed that these strains, which had been subcultured many times in liquid axenic media, had become genetically different from the strains from which they had been derived (McCullough *et al.*, 1978). Analysis of crosses indicated that mutations in two genes (*axeA* and *axeB*) were responsible for growth in axenic media of the strain CLd–AXE (the mutant derivative of CLd) and that additional mutations had probably occurred in other strains. Mutations allowing growth of amoebae in axenic media had previously been demonstrated in the cellular slime mold *Dictyostelium discoideum* (Williams *et al.*, 1974); their mode of action is unknown.

Although these results suggested that it would be possible to construct various genetically marked strains capable of growth in axenic media, further analysis has indicated that the genetic control of axenic growth may be more complex than initially supposed (A. M. Chainey and J. Dee, unpublished results), and no new strains are yet available. In addition, amoebal growth in most synthetic media is prevented by the presence of agar, probably due to the inhibition by agar of an extracellular aminopeptidase (C. H. R. McCullough, J. Dee, A. Haars, and A. Hüttermann, in preparation). Although agar inhibition does not occur in fully defined media, these media support only very slow growth. Thus there are no immediate prospects of culturing amoebae on agar-based synthetic media or of isolating specific nutritional mutants in this stage of the life cycle.

Culture of amoebae in liquid axenic media has allowed some useful studies, however. First, sufficient quantities of amoebae can now be cultured in the absence of bacteria to allow comparative biochemical studies of amoebae and plasmodia (Hall *et al.*, 1975; Hall and Braun, 1977). Second, the expression of some mutations (nutritional and temperature-sensitive) initially identified in plasmodia has been tested in amoebae grown in axenic media; the results suggest that when amoebae and plasmodia are cultured in similar conditions, the differences in gene expression between them may be fewer than originally supposed (McCullough *et al.*, 1978; Burland and Dee, 1979; see Section II,D).

C. Drug Resistance and Sensitivity

Unlike nutritional mutants, drug-resistant mutants can easily be isolated in *P. polycephalum* from amoebae plated on bacterial lawns. They provide useful genetic markers not only for progeny analysis but also, for example, in tests of the effectiveness of mutagenic and enrichment techniques for the isolation of other types of mutants (Haugli and Dove, 1972; Gorman and Dove, 1974; Gorman *et al.*, 1977). The potential value of certain types of drug resistance for investigating macromolecular synthesis and function in the cell cycle has also motivated many of the attempts to isolate such mutants. The progress and prospects of this approach have been reviewed by Haugli *et al.* (1980).

Several mutants resistant to cycloheximide (actidione) have been isolated and analyzed, and most were shown to be expressed in plasmodia as well as in amoebae (Dee, 1966b; Dee and Poulter, 1970; Haugli *et al.*, 1972; Dee, 1973; Evans and Evans, 1980). Haugli *et al.* (1972) identified two loci, *actA* and *actB*, and showed that the former defined a component of the ribosomes. Strains carrying the "strong" mutation *actA169* show a high level of resistance and have been particularly useful, for example, in studies of the effects of cycloheximide on DNA synthesis (Funderud and Haugli, 1977). All these mutants were isolated in Wisconsin 1 strains prior to the construction of Colonia background strains. However, another mutant resistant to cycloheximide as a result of a change in ribosomes was isolated from the Colonia background strain LU648 by Evans and Evans (1980) and has been used to investigate further the effects of cycloheximide on DNA synthesis.

Emetine-resistant amoebal mutants were isolated from a Wisconsin 1 strain (Dee, 1962) and from CL amoebae mutagenized with EMS (Adler and Holt, 1974a) and with NMG (Laffler *et al.*, 1979). The mutant *eme-4* (Adler and Holt, 1974a) has proved to be a useful marker, is available in a range of Colonia background strains, and is linked to a gene *imz* which affects the pH sensitivity of mating (Shinnick *et al.*, 1978).

Mutant amoebae resistant to BUdR were isolated by Haugli and Dove (1972) in Wisconsin 1 strains and by Lunn *et al.* (1977) in Wisconsin 1 and *mt-h* strains. Lunn *et al.* (1977) analyzed 24 mutants and found that all were due to single gene mutations which were also expressed in plasmodia, resulting in reduced rates of thymidine incorporation. Complementation tests in heterozygous plasmodia indicated that all but one of the mutations were in the same gene, *burA*, which apparently controls thymidine kinase activity; the remaining mutant (*burB*) had normal thymidine kinase activity. Gene dosage studies (Mohberg *et al.*, 1980) suggested that the *burA* mutants carried mutations of the structural gene for thymidine kinase.

Since many studies on *Physarum* are impeded by its natural resistance to drugs which are useful inhibitors of specific processes in other organisms, the value of isolating drug-sensitive mutants has been discussed by Gorman *et al.* (1977). They successfully used an enrichment technique, based on differential survival of encysted and active amoebae, to isolate anisomycin-sensitive mutants.

D. Temperature-Sensitive Growth and Cell Cycle Events

1. ISOLATION OF TEMPERATURE-SENSITIVE GROWTH MUTANTS IN AMOEBAE AND PLASMODIA

Mutants showing temperature-sensitive growth have usually been isolated with the aim of identifying mutants with conditional defects of the cell or nuclear division cycles. Such mutants are primarily required for studies of the synchronous nuclear cycle in plasmodia, but mutant isolation can clearly be accomplished more readily in the uninucleate amoebae; thus, much attention has been given to the question of whether the mutations are expressed in only one or in both of these growth phases. This question was examined by subjecting amoebae of strain CLd to NMG mutagenesis and screening for temperature sensitivity among clones of amoebae (Wheals *et al.*, 1976) and among plasmodia derived from clones (Gingold *et al.*, 1976). Temperature-sensitive growth mutants were isolated by both methods, but when all the mutant clones were tested for sensitivity in amoebae and plasmodia, the results suggested that few mutations were expressed in both growth phases. The authors concluded that if mutants expressed in plasmodia were required, they must be isolated in plasmodia, although this method will necessarily be more laborious. This conclusion was supported by Laffler *et al.* (1979); among a total of 50 mutants isolated in amoebae in this and the previous study, only eight were temperature-sensitive as plasmodia. Del Castillo *et al.* (1978), however, using very similar methods, tested 89 mutants isolated from CL as amoebal clones and found that 65 of them were temperature-sensitive as plasmodia. For the isolation of mutants in amoebae, they successfully used a replica-plating technique devised by Wheals *et al.* (1976).

The results of Sudbery *et al.* (1978) suggested that, although many mutants appeared to be temperature-sensitive in only one phase when subjected to growth tests, enrichment and prescreening techniques which would select for temperature-sensitive amoebae considerably boosted the yield of temperature-sensitive plasmodial mutants. They concluded that many of the mutations clearly expressed only in plasmodia may be also weakly expressed in amoebae. Burland and Dee (1979) investigated the effects of different growth conditions on the expression of temperature-sensitive mutants since, in the previous studies, different media and sometimes different temperatures had been used for testing amoebae and plasmodia. They found that when amoebae and plasmodia were tested in identical conditions, the proportion of mutants sensitive in both phases was greatly increased. They concluded that temperature-sensitive plasmodial mutants could be isolated most efficiently by screening amoebae in conditions resembling as closely as possible those that will be used for testing plasmodia.

2. ENRICHMENT METHODS

Haugli and Dove (1972), the first authors to describe effective mutagenic techniques for *P. polycephalum,* also showed that an enrichment technique, involving light sensitization of growing amoebae by allowing them to incorporate BUdR, increased the yield of amoebal temperature-sensitive mutants. The effectiveness of this technique in also increasing the yield of temperature-sensitive mutants expressed in plasmodia was demonstrated by Sudbery *et al.* (1978). However, the technique apparently did not achieve the aim for which it was originally devised, that of enriching specifically for mutants with heat-sensitive DNA synthesis. An enrichment technique using the antibiotic netropsin was shown to increase the yield of temperature-sensitive amoebal mutants by Gorman and Dove (1974). However, later experiments suggested that this method would not enrich for mutants defective in DNA synthesis (Laffler *et al.,* 1979). Doubts have also been expressed about the value of applying enrichment techniques for the isolation of nuclear division cycle mutants in *Physarum* on the grounds that they might exclude certain classes of mutant, for example, those that continue

6. Genetics of *Physarum polycephalum*

DNA synthesis in the absence of nuclear division and those that continue growth for a few cycles before arresting (Burland, 1978; Haugli *et al.*, 1980). An enrichment method favoring survival of encysted cells, which might allow selection of temperature-sensitive mutants in which growth is not rapidly arrested, has been described by Gorman *et al.* (1977). However, there seems to be no compelling evidence at present that any of these enrichment techniques facilitate the isolation of nuclear cycle mutants in *Physarum*.

3. SCREENING TEMPERATURE-SENSITIVE GROWTH MUTANTS FOR DEFECTS OF THE CELL OR NUCLEAR CYCLE

Several methods have been devised for screening temperature-sensitive growth mutants in order to identify those with conditional defects of the cell or nuclear cycle. Gingold *et al.* (1976) measured the "nucleolar reconstruction index" of nuclei isolated from microplasmodia with the aim of detecting the "lining up" of nuclei that would be expected in an asynchronous culture under nonpermissive conditions if progression through the cycle was blocked at a particular point. Del Castillo *et al.* (1978) and Laffler *et al.* (1979) used similar methods to screen microplasmodial cultures. On the same principle, Wheals *et al.* (1976) screened amoebal cultures by measuring the mitotic index. Burland (1978; Burland and Dee, 1980) used a modification of this method to screen 55 temperature-sensitive mutants isolated in CLd amoebae and identified two in which the division indices were significantly different from those of the wild type. In both mutants, nuclear cycle defects were also expressed in plasmodia. It was suggested that screening amoebae might be a more rapid and efficient method of identifying interesting mutants than screening plasmodia (Burland and Dee, 1980).

Sudbery *et al.* (1978), Del Castillo *et al.* (1978), and Laffler *et al.* (1979) characterized their temperature-sensitive mutants by measuring the relative rates of DNA, RNA, and protein synthesis in microplasmodia under nonpermissive conditions. All these authors identified a few promising candidates for classification as nuclear cycle mutants, usually on the basis that DNA synthesis was arrested without immediate effects on RNA or protein synthesis. For one mutant, Laffler *et al.* (1979) then defined an "execution point" (15 minutes before mitotic metaphase) by temperature shifts on synchronous macroplasmodial cultures. Burland (1978), though not using this method for initial screening, measured macromolecular synthesis and nuclear DNA content of his potential nuclear cycle mutants. He found that in one mutant at least, DNA synthesis apparently continued in the absence of nuclear division, resulting in increased nuclear DNA content (Burland and Dee, 1980); such a mutant might be missed if only synthesis were measured.

Cytological examination of amoebae, which can easily be combined with measurement of the mitotic index, has also proved productive; Burland (1978) identified a mutant in which many multinucleate and giant uninucleate cells were present in amoebal cultures under nonpermissive conditions. A large proportion of nuclei isolated from these amoebae had increased DNA content. Analysis of cell pedigrees by time-lapse cinematography revealed that the primary defect was failure of cell division and that the other abnormalities resulted from this defect (Burland *et al.*, 1981). This interesting mutant (*hts-23*), which shows leaky, temperature-sensitive growth, would probably have been discarded if screening had been carried out on plasmodia instead of amoebae. Del Castillo *et al.* (1978) used fluorescence microscopy to screen plasmodial nuclei for cytological abnormalities and analyzed interesting mutants more thoroughly with the electron microscope.

In view of the characteristics of the mutants identified so far, it therefore seems likely that a range of screening methods should be used if all the potential cell and nuclear cycle mutants are to be identified.

4. GENETIC ANALYSIS

Most of the temperature-sensitive growth mutants in these studies were isolated from CL or CLd after mutagenesis by NMG, EMS, or uv light (followed by caffeine treatment). In several studies, some of the mutants were crossed with heterothallic strains to determine whether single gene mutations were involved and were tested for complementation in heterokaryons (Gingold *et al.*, 1976; Laffler *et al.*, 1979; Burland and Dee, 1979, 1980). A few mutants have been analyzed more thoroughly, and mutations which show clearly defined lethal or morphological effects at certain temperatures have proved to be useful genetic markers, for example, *hts-1* (Dee, 1978) and *hts-23* (Burland *et al.*, 1981).

E. Motility, Morphology, and Color

Jacobson and Dove (1975) investigated factors determining the morphology of colonies of amoebae growing on bacterial lawns and isolated several variants with altered colony morphology. Attempts at genetic analysis suggested that some of the variants were probably aneuploids or polyploids. These strains also had the increased cell size and "fuzzy" colony form, which Adler and Holt (1974b) had found to be associated with increased ploidy. Other variants, however, segregated clearly in crosses and could be ascribed to gene mutations. Detailed analysis of the behavior of one mutant, *mov-1*, by Jacobson (1979, 1980) has shown that the altered colony form is due to an increased rate of amoebal movement under certain conditions and that the doubling time of the mutant is not significantly different from that of the wild type. The mutant has apparently lost the ability to regulate its movement in response to certain environmental factors, for example, pH and the presence of bacteria, which affect the movement of wild-type amoebae. Jacobson's mutants have been isolated in a set of heterothallic strains (denoted DJ) constructed by back-crossing clones derived from the original Wisconsin 1 isolate (Jacobson and Dove, 1975).

Movement in *P. polycephalum* amoebae can also be accomplished by the development of flagella, and this interesting morphogenetic process is apparently amenable to genetic analysis. Among 76 temperature-sensitive growth mutants of independent origin isolated by replica plating of mutagenized CL amoebae (Del Castillo *et al.*, 1978), Mir *et al.* (1979) identified 16 mutants with defects of flagellation. This was a surprisingly high proportion considering that no intentional enrichment for such mutants had been carried out. In at least 11 of the strains, it seemed unlikely that the inability to flagellate was a nonspecific effect of the mutation causing temperature-sensitive growth, and in several there was evidence that separate mutations were involved. However, since in the majority of the mutants the flagellation defect was temperature-sensitive, it seems possible that the isolation technique somehow favored the survival of mutants unable to flagellate. In any case, since *P. polycephalum* amoebae can easily be induced to flagellate (Wakasugi and Ohta, 1973; Mir *et al.*, 1979) and since enrichment methods for the isolation of mutants defective in flagellation have been successfully employed in other amoeboflagellates (Fulton, 1970), it should be simple to collect a range of interesting mutants. Some detailed comparisons of flagellation in wild-type and mutant strains of *P. polycephalum* have already been carried out (Mir *et al.*, 1979).

A nonmotile plasmodium is one of the most useful types of mutant that a *Physarum* geneticist can imagine, allowing replica-plating techniques to be applied to plasmodia and immense savings of financial resources and time. However, there have been no reports of any such mutant yet. On the contrary, where abnormal plasmodial morphology has been noted in association with various mutations, it has usually involved a tendency to be more migratory than wild-type; for example, in actidione-resistant mutants (Dee and Poulter, 1970) and in plasmodia carrying the *hts-23* mutation (Burland *et al.*, 1981). It is not yet clear whether *hts-23* plasmodial morphology is due to the same mutation that causes defective cell division in amoebae (Section II,D), but there is a possibility that the mutation affects a structural component involved in the completion of cleavage and in plasmodial morphology (Burland *et al.*, 1981). A third abnormality in *hts-23* plasmodia, possibly due to the same mutation, is failure to sporulate.

Since mutations affecting plasmodial morphology or functions such as protoplasmic streaming may often have effects in amoebae, it would certainly be worthwhile to examine the effects on plasmodia of all known mutations affecting amoebal motility or structure. Further screening programs to identify such mutants in *mt-h* amoebae might also be productive.

The most useful visible marker now available, the mutant allele *whi-1* which causes white plasmodia, apparently originated in a heterothallic (*mt-2*) amoeba and was discovered only because the mutant amoeba then gave rise to an illegitimate plasmodium (Anderson, 1977). The white plasmodium gave only *mt-2* progeny, but by a series of crosses, the mutant allele was transferred to *mt-h* strains and *whi-1* was shown to be a single recessive mutation, unlinked to several other genetic markers and having a clear-cut effect on plasmodial pigmentation (Anderson, 1977). It has since proved extremely valuable in genetics; it has been used as an easily scored marker for detecting recombination, as a means of distinguishing selfed and crossed plasmodia in crosses and complementation tests, for detecting plasmodial fusion, analyzing heterokaryons, and in many other ways. It may also be useful for certain physiological and cytological investigations in which the yellow pigment is obstructive and for determining the role of pigment in sporulation.

III. GENERAL COMMENTS ON METHODS

A. Mutant Isolation

Various mutagenic, enrichment, and screening procedures for the isolation of mutants in *P. polycephalum* have been described. The choice of method will depend largely on the type of mutant required, and the details will be found in the papers cited in the appropriate section of this chapter. The most difficult question to resolve will probably be whether to screen amoebae or plasmodia, since different results have been obtained by investigators in pursuit of the same type of mutant (Section II,D). Since it will usually be interesting to examine the effects of mutations on both amoebae and plasmodia, there are obvious advantages in choosing an apogamic strain for mutant isolation regardless of which stage is to be screened.

If mutants are to be isolated in amoebae, plasmodium formation in amoebal cultures of rapidly selfing strains such as CL can be reduced by incubation at high temperatures (above 29.5°C) or by increase of the food supply (Adler and Holt, 1974a; Youngman *et al.*, 1977). However, there may still be a danger of inadvertent enrichment for delayed-selfing mutants under these conditions (Section I,E,1). In addition, formation of plasmodia during mutagenesis or isolation procedures could cause difficulties. Thus it may be preferable to use a delayed strain. For example, both CLd and an *mt-h npfA1* strain have been used for the isolation of amoebal temperature-sensitive growth mutants (Burland, 1978). Strains carrying *npfA1* have the advantage that selfing and crossing can be controlled by temperature (Section I,D,1) and also that the *npf* mutation is unlinked to *mt* and does not readily revert (Anderson and Dee, 1977). There are no obvious advantages in using heterothallic strains for mutant isolation.

6. Genetics of *Physarum polycephalum*

CLd was originally used for isolation of plasmodial mutants because of difficulties with crossing CL amoebae (Cooke and Dee, 1975). Now that the effects of *matB* on crossing are understood, there need be no such problems with analyzing mutants isolated in CL, and a strain which forms plasmodia in small colonies is clearly desirable when large numbers of clonally formed plasmodia are required for screening. However, CLd also had the advantage that amoebae of mutant clones were always available for analysis after plasmodia had formed. When plasmodia are formed in CL clones, there are few amoebae left, and if the plasmodial mutant fails to sporulate, genetic analysis may be impossible. Thus a delayed strain may still be preferable. The choice will depend on the type of mutant required and the screening method employed.

B. Mutant Analysis

1. CROSSING

Mutants isolated in an *mt-h* strain can be analyzed by crossing with a heterothallic strain or with another *mt-h* strain carrying a complementing *npf* mutation. The latter method has the advantage that haploid plasmodia can be derived from all progeny clones instead of only 50%. If mutants have been isolated in CLd or in heterothallic amoebae, crossing with an *mt-h npfA1* or *mt-h aptA1* strain allows rapid-selfing progeny to be isolated; 25% are expected to be of this type since *npfA1* and *aptA1* are unlinked to *mt*. This is useful for the analysis of the effects on plasmodia of mutations identified in amoebae (Dee, 1973; Anderson, 1977; Burland *et al.*, 1981).

For all crosses involving apogamic strains, it will be necessary to select strains carrying appropriate alleles of *matB* (see Sections I,C,3 and I,E,2) and *fusA* (Section II,A), and the *whi-1* mutation will also be valuable (Section II,E). The *fus* alleles are essential to prevent fusion of selfed and crossed plasmodia. Since *whi-1* is recessive, crosses between apogamic and heterothallic strains can be designed so that selfed plasmodia are white and crossed plasmodia yellow. Crosses between apogamic and heterothallic strains should be carried out at a temperature which inhibits selfing (above 28.5°C for *npfA1* and above 29.5°C for other *mt-h* strains). However, crosses between *mt-h* strains will be successful only at a temperature which is permissive for selfing (see Section I,E,3,*d*).

In addition to the diploid, heterozygous plasmodia formed in a cross, illegitimate plasmodia may occasionally be formed by heterothallic strains (Section I,D,3), and illegitimate heterokaryons (Section II,A) may also occur (Adler, 1975). Plasmodial fusion tests should identify the former, but tests of nuclear DNA content or progeny analysis may be necessary to reveal the latter (Adler, 1975; Anderson, 1976).

Fusion tests to identify crossed plasmodia are carried out using standard tester strains. The scoring of such tests is usually simple and unambiguous (Poulter and Dee, 1968). However, microscopic regions of fusion are occasionally observed between strains of different fusion genotype (for example, *fusA1/fusA2* and *fusA2*). Such tests can invariably be scored eventually as clear nonfusions (Anderson and Dee, 1977); rescoring of all tests after an interval is therefore advisable.

2. ANALYSIS OF PROGENY FROM CROSSES

Amoebae obtained from spores should be recloned to avoid the isolation of mixed cultures. It should also be borne in mind that diploid or aneuploid amoebae may occasionally be present, giving rise to clones in which plasmodia are formed (Section I,E,2). These plasmodia may be heterozygous for various markers, unlike plasmodia formed by apogamic amoebae.

For analysis of plasmodial markers, haploid plasmodia are obtained from *mt-h* clones; 1:1 segregation of alleles is then expected for genes unlinked to *mt*.

3. COMPLEMENTATION TESTS

Complementation of genes expressed in plasmodia can be tested in heterokaryons or in heterozygous diploid plasmodia. The isolation of diploid amoebae from mixtures of strains carrying different *matB* alleles and the same *mt* (Youngman *et al.*, 1979b) may soon allow complementation tests of mutations expressed in amoebae.

C. Storage and Maintenance of Strains

Strains used for *P. polycephalum* genetics are usually stored as amoebal clones and should not be subcultured frequently, as this may lead to complications due to senescence (Adler and Holt, 1974b) or back-mutation. Amoebae cultured on bacterial lawns on agar survive for long periods when refrigerated. Amoebae suspended in a glycerol–water suspension remain viable frozen at −70°C (C. E. Holt, personal communication). Plasmodia are also prone to senescence (Poulter, 1969; McCullough *et al.*, 1973), and plasmodial cultures should be frequently reinitiated from sclerotia or from amoebae if changes in their genetic constitution are to be avoided.

If strains capable of selfing are used, care is necessary to avoid the selection of non-selfing mutants during subculture (Section I,E,1); this can be done by culturing amoebae in conditions in which plasmodia are not normally formed and storing amoebal cultures before plasmodium formation is initiated.

D. Genetic Nomenclature

Geneticists working with *P. polycephalum* have now adopted a uniform system of nomenclature for mutant alleles and strains. The nomenclature rules outlined by D. J. Cooke, R. T. M. Poulter, and A. E. Wheals (unpublished), based on those proposed for bacterial genetics by Demerec *et al.* (1966), have been circulated informally and, on the whole, seem to have been applied without difficulty. Where analysis of mutants has been carried out independently in separate laboratories, different names have sometimes been given to the same locus; this is inevitable, since publication clearly cannot be delayed until all relevant complementation and mapping tests have been done. However, it may be necessary to find some procedure for agreeing on the name to be used once the identity of a locus has been established. For example, since apparently *npfC* = *aptC* = *difA*, the authors of these names should perhaps agree on a single name to avoid further confusion. Personally, as the author of *rac*, I suggest that *matB* be used for this locus in future, since much fuller analyses have been carried out by its originators. However, the proposed substitution of *matA* for *mt* (Youngman *et al.*, 1979a) is another matter; as suggested in the nomenclature rules, changes in established symbols should be avoided since they are liable to cause unnecessary confusion.

ACKNOWLEDGMENTS

I am very grateful to all the authors who have supplied me with copies of their papers prior to publication. I am also indebted to several colleagues for reading the manuscript of this chapter and am particularly grateful to Roger Anderson and Anne Chainey for their helpful comments and to Sheila Mackley for excellent secretarial assistance.

REFERENCES

Adler, P. N. (1975). Control of the differentiated state in *Physarum polycephalum*. Ph.D. Thesis, Massachusetts Inst. of Tech., Cambridge, Mass.

Adler, P. N., and Holt, C. E. (1974a). Genetic analysis in the Colonia strain of *Physarum polycephalum:* Heterothallic strains that mate with and are partially isogenic to the Colonia strain. *Genetics* **78,** 1051–1062.

Adler, P. N., and Holt, C. E. (1974b). Change in properties of *Physarum polycephalum* amoebae during extended culture. *J. Bacteriol.* **120,** 532–533.

Adler, P. N., and Holt, C. E. (1975). Mating type and the differentiated state in *Physarum polycephalum*. *Dev. Biol.* **43,** 240–253.

Adler, P. N., and Holt, C. E. (1977). Mutations increasing asexual plasmodium formation in *Physarum polycephalum*. *Genetics* **87,** 401–420.

Adler, P. N., Davidow, L. S., and Holt, C. E. (1975). Life cycle variants of *Physarum polycephalum* that lack the amoeba stage. *Science* **190,** 65–67.

Aldrich, H. C. (1967). The ultrastructure of meiosis in three species of *Physarum*. *Mycologia* **59,** 127–148.

Anderson, R. W. (1976). Analysis of the amoebal-plasmodial transition in *Physarum polycephalum*. Ph.D. Thesis, University of Leicester, England.

Anderson, R. W. (1977). A plasmodial colour mutation in the Myxomycete *Physarum polycephalum*. *Genet. Res.* **30,** 301–306.

Anderson, R. W. (1979). Complementation of amoebal-plasmodial transition mutants in *Physarum polycephalum*. *Genetics* **91**, 409–419.

Anderson, R. W. and Holt, C. E. (1981). Revertants of selfing (*gad*) mutants in *Physarum polycephalum*. *Dev. Genet.* (in press).

Anderson, R. W., and Dee, J. (1977). Isolation and analysis of amoebal-plasmodial transition mutants in the Myxomycete *Physarum polycephalum*. *Genet. Res.* **29**, 21–34.

Anderson, R. W., Cooke, D. J., and Dee, J. (1976). Apogamic development of plasmodia in the Myxomycete *Physarum polycephalum:* A cinematographic analysis. *Protoplasma* **89**, 29–40.

Burland, T. G. (1978). Temperature-sensitive mutants of *Physarum polycephalum:* A search for cell cycle mutants. Ph.D. Thesis, University of Leicester, England.

Burland, T. G., and Dee, J. (1979). Temperature-sensitive mutants of *Physarum polycephalum*— expression of mutations in amoebae and plasmodia. *Genet. Res.* **34**, 33–40.

Burland, T. G., and Dee, J. (1980). Isolation of cell cycle mutants of *Physarum polycephalum*. *Mol. Gen. Genet.* **179**, 43–48.

Burland, T. G., Chainey, A. M., Dee, J., and Foxon, J. L. (1981). Analysis of development and growth in a mutant of *Physarum polycephalum* with defective cytokinesis. *Dev. Biol.* **85**, 26–38.

Carlile, M. J. (1972). The lethal interaction following plasmodial fusion between two strains of the Myxomycete *Physarum polycephalum*. *J. Gen. Microbiol.* **71**, 581–590.

Carlile, M. J. (1973). Cell fusion and somatic incompatibility in Myxomycetes. *Ber. Dtsch. Bot. Ges.* **86**, 123–139.

Carlile, M. J. (1976). The genetic basis of the incompatibility reaction following plasmodial fusion between different strains of the Myxomycete *Physarum polycephalum*. *J. Gen. Microbiol.* **93**, 371–376.

Carlile, M. J., and Dee, J. (1967). Plasmodial fusion and lethal interaction between strains in a Myxomycete. *Nature (London)* **215**, 832–834.

Collins, O. R. (1975). Mating types in five isolates of *Physarum polycephalum*. *Mycologia* **67**, 98–107.

Collins, O. R., and Haskins, E. F. (1972). Genetics of somatic fusion in *Physarum polycephalum:* The PpII strain. *Genetics* **71**, 63–71.

Collins, O. R., and Tang, H. (1977). New mating types in *Physarum polycephalum*. *Mycologia* **69**, 421–423.

Cooke, D. J. (1974). Studies on the Colonia isolate of *Physarum polycephalum*. Ph.D. Thesis, University of Leicester, England.

Cooke, D. J., and Dee, J. (1974). Plasmodium formation without change in nuclear DNA content in *Physarum polycephalum*. *Genet. Res.* **23**, 307–317.

Cooke, D. J., and Dee, J, (1975). Methods for the isolation and analysis of plasmodial mutants in *Physarum polycephalum*. *Genet. Res.* **24**, 175–187.

Davidow, L. S., and Holt, C. E. (1977). Mutants with decreased differentiation to plasmodia in *Physarum polycephalum*. *Mol. Gen. Genet.* **155**, 291–300.

Dee, J. (1960). A mating-type system in an acellular slime mould. *Nature (London)* **185**, 780–781.

Dee, J. (1962). Recombination in a Myxomycete, *Physarum polycephalum* Schw. *Genet. Res.* **3**, 11–23.

Dee, J. (1966a). Multiple alleles and other factors affecting plasmodium formation in the true slime mould *Physarum polycephalum* Schw. *J. Protozool.* **13**, 610–616.

Dee, J. (1966b). Genetic analysis of actidione-resistant mutants in the Myxomycete *Physarum polycephalum* Schw. *Genet. Res.* **8**, 101–110.

Dee, J. (1973). Aims and techniques of genetic analysis in *Physarum polycephalum*. *Ber. Dtsch. Bot. Ges.* **86**, 93–121.

Dee, J. (1975). Slime moulds in biological research. *Sci. Prog. Oxford*, **62**, 523–542.

6. Genetics of *Physarum polycephalum*

Dee, J. (1978). A gene unlinked to mating-type affecting crossing between strains of *Physarum polycephalum*. *Genet. Res.* **31**, 85-92.

Dee, J., and Poulter, R. T. M. (1970). A gene conferring actidione resistance and abnormal morphology on *Physarum polycephalum* plasmodia. *Genet. Res.* **15**, 35-41.

Dee, J., Wheals, A. E., and Holt, C. E. (1973). Inheritance of plasmodial valine requirement in *Physarum polycephalum*. *Genet. Res.* **21**, 87-101.

Del Castillo, L., Oustrin, M. L., and Wright, M. (1978). Characterization of thermosensitive mutants of *Physarum polycephalum*. *Mol. Gen. Genet.* **164**, 145-154.

Demerec, M., Adelberg, E. A., Clark, A. J., and Hartman, P. E. (1966). A proposal for a uniform nomenclature in bacterial genetics. *Genetics* **54**, 61-76.

Evans, T. E. and Evans, H. H. (1980). Cycloheximide resistance of *Physarum polycephalum*. *J. Bacteriol.* **143**, 897-905.

Fulton, C. (1970). Amoebo-flagellates as research partners: The laboratory biology of *Naegleria* and *Tetramitus*. *Methods Cell Physiol.* **4**, 341-476.

Funderud, S., and Haugli, F. (1977). DNA replication in *Physarum polycephalum*: Characterization of DNA replication products made *in vivo* in the presence of cycloheximide in strains sensitive and resistant to cycloheximide. *Nucleic Acids Res.* **4**, 405-413.

Gingold, E. C., Grant, W. D., Wheals, A. E., and Wren, M. (1976). Temperature-sensitive mutants of the slime mould *Physarum polycephalum* II. Mutants of the plasmodial phase. *Mol. Gen. Genet.* **149**, 115-119.

Goodman, E. M. (1972). Axenic culture of myxamoebae of the Myxomycete *Physarum polycephalum*. *J. Bacteriol.* **111**, 242-247.

Gorman, J. A., and Dove, W. F. (1974). A method of indirect mutant selection in *Physarum polycephalum* using the antibiotic netropsin. *Mol. Gen. Genet.* **133**, 345-351.

Gorman, J. A., and Wilkins, A. S. (1980). Developmental phases in the life cycle of *Physarum* and related Myxomycetes. *In* "Growth and Differentiation in *Physarum polycephalum*". (W. F. Dove and H. P. Rusch, eds.), pp. 157-202. Princeton Univ. Press, Princeton, New Jersey.

Gorman, J. A., Dove, W. F., and Shaibe, E. (1977). Anisomycin sensitive mutants of *Physarum polycephalum* isolated by cyst selection. *Mol. Gen. Genet.* **151**, 253-259.

Gorman, J. A., Dove, W. F., and Shaibe, E. (1979). Mutations affecting the initiation of plasmodial development in *Physarum polycephalum*. *Dev. Genet.* **1**, 47-60.

Hall, L., and Braun, R. (1977). The organisation of genes for transfer RNA and ribosomal RNA in amoebae and plasmodia of *Physarum polycephalum*. *Eur. J. Biochem.* **76**, 165-174.

Hall, L., Turnock, G., and Cox, B. J. (1975). Ribosomal RNA genes in the amoebal and plasmodial forms of the slime mould *Physarum polycephalum*. *Eur. J. Biochem.* **51**, 459-465.

Haugli, F. B. (1971). Mutagenesis, selection and genetic analysis in *Physarum polycephalum*. Ph.D. Thesis, University of Wisconsin, Madison, Wisconsin.

Haugli, F. B., and Dove, W. F. (1972). Mutagenesis and mutant selection in *Physarum polycephalum*. *Mol. Gen. Genet.* **118**, 109-124.

Haugli, F. B., Dove, W. F., and Jimenez, A. (1972). Genetics and biochemistry of cycloheximide resistance in *Physarum polycephalum*. *Mol. Gen. Genet.* **118**, 97-107.

Haugli, F. B., Cooke, D., and Sudbery, P. (1980). The genetic approach in the analysis of the biology of *Physarum polycephalum*. *In* "Growth and Differentiation in *Physarum polycephalum*" (W. F. Dove and H. P. Rusch, eds.), pp. 129-156. Princeton University Press, Princeton, New Jersey.

Holt, C. E., Heunert, H., Hüttermann, A., and Galle, H.-K. (1979). Effect of *matA* on nuclear fusion. Abstract, *Physarum* meeting, Laval University, Quebec.

Honey, N. K. (1979). The mating type locus and differentiation in *Physarum polycephalum*. Ph.D. Thesis, University of Otago, Dunedin, New Zealand.

Honey, N. K., Poulter, R. T. M., and Teale, D. M. (1979). Genetic regulation of differentiation in *Physarum polycephalum. Genet. Res.* **34**, 131-142.
Honey, N. K., Poulter, R. T. M., and Winter, P. J. (1981a). Selfing mutants from heterothallic strains of *Physarum polycephalum.Genet. Res.* **37**, 113-121.
Honey, N. K., Poulter, R. T. M., and Aston, R. J. (1981b). Nonselfing mutants from selfing (het^-) strains of *Physarum polycephalum* (submitted for publication).
Jacobson, D. N. (1979). The role of regulation of cell speed in the behaviour of *Physarum polycephalum* amoebae. *Exp. Cell Res.* **122**, 219-231.
Jacobson, D. N. (1980). Locomotion of *Physarum polycephalum* amoebae is guided by a short range interaction with *E. coli. Exp. Cell Res.* **125**, 441-452.
Jacobson, D. N., and Dove, W. F. (1975). The amoebal cell of *Physarum polycephalum:* Colony formation and growth. *Dev. Biol.* **47**, 97-105.
Kerr, N. S. (1967). Plasmodium formation by a minute mutant of the true slime mould, *Didymium nigripes. Exp. Cell Res.* **45**, 646-655.
Kirouac-Brunet, J., Masson, S., and Pallotta, D. (1981). Multiple allelism at the *matB* locus in *Physarum polycephalum. Can. J. Genet. Cytol.* **23**, 9-16.
Laane, M. M., and Haugli, F. B. (1976). Nuclear behaviour during meiosis in the Myxomycete *Physarum polycephalum. Norw. J. Bot.* **23**, 7-21.
Laane, M. M., Haugli, F. B., and Mellem, T. R. (1976). Nuclear behaviour during sporulation and germination in the Colonia strain of *Physarum polycephalum. Norw. J. Bot.* **23**, 177-189.
Laffler, T. G., and Dove, W. F. (1977). Viability of *Physarum polycephalum* spores and ploidy of plasmodial nuclei. *J. Bacteriol.* **131**, 473-476.
Laffler, T. G., Wilkins, A., Selvig, S., Warren, N., Kleinschmidt, A., and Dove, W. F. (1979). Temperature-sensitive mutants of *Physarum polycephalum:* Viability, growth and nuclear replication. *J. Bacteriol.* **138**, 499-504.
Lane, E. B., and Carlile, M. J. (1979). Post-fusion somatic incompatibility in plasmodia of *Physarum polycephalum. J. Cell Sci.* **35**, 339-354.
Lunn, A., Cooke, D. J., and Haugli, F. B. (1977). Genetics and biochemistry of 5-bromodeoxyuridine resistance in *Physarum polycephalum. Genet. Res.* **30**, 1-12.
McCullough, C. H. R., and Dee, J. (1976). Defined and semi-defined media for the growth of amoebae of *Physarum polycephalum. J. Gen. Microbiol.* **95**, 151-158.
McCullough, C. H. R., Cooke, D. J., Foxon, J. L., Sudbery, P. E., and Grant, W. D. (1973). Nuclear DNA content and senescence in *Physarum polycephalum. Nature (London), New Biology* **245**, 263-265.
McCullough, C. H. R., Dee, J., and Foxon, J. L. (1978). Genetic factors determining the growth of *Physarum polycephalum* amoebae in axenic medium. *J. Gen. Microbiol.* **106**, 297-306.
Martin, G. W., and Alexopoulos, C. J. (1969). "The Myxomycetes." Univ. of Iowa Press, Iowa City, Iowa.
Mir, L., Del Castillo, L., and Wright, M. (1979). Isolation of *Physarum* amoebal mutants defective in flagellation and associated morphogenetic processes. *FEMS Microbiol. Lett.* **5**, 43-46.
Mohberg, J. (1977). Nuclear DNA content and chromosome numbers throughout the life cycle of the Colonia strain of the Myxomycete, *Physarum polycephalum. J. Cell Sci.* **24**, 95-108.
Mohberg, J., and Rusch, H. P. (1971). Isolation and DNA content of nuclei of *Physarum polycephalum. Exp. Cell Res.* **66**, 305-316.
Mohberg, J., Babcock, K. L., Haugli, F. B., and Rusch, H. P. (1973). Nuclear DNA content and chromosome numbers in the Myxomycete *Physarum polycephalum. Dev. Biol.* **34**, 228-245.
Mohberg, J., Dworzak, E., Sachsenmaier, W., and Haugli, F. B. (1980). Thymidine kinase-deficient mutants of *Physarum polycephalum;* relationships between enzyme activity levels and ploidy. *Cell Biol. Int. Rep.* **4**, 137-148.

6. Genetics of *Physarum polycephalum*

Pallotta, D. J., Youngman, P. J., Shinnick, T. M., and Holt, C. E. (1979). Kinetics of mating in *Physarum polycephalum*. *Mycologia* **71,** 68–84.

Poulter, R. T. M. (1969). Senescence in the Myxomycete *Physarum polycephalum*. Ph.D. Thesis, University of Leicester, England.

Poulter, R. T. M., and Dee, J. (1968). Segregation of factors controlling fusion between plasmodia of the true slime mould *Physarum polycephalum*. *Genet. Res.* **12,** 71–79.

Poulter, R. T. M., and Honey, N. K. (1977). Genetic analysis of a cross between two homothallic strains of *Physarum polycephalum*. *Genet. Res.* **29,** 55–63.

Schrauwen, J. A. M. (1981). Post-fusion incompatibility in *Physarum polycephalum*: The involvement of DNA. *Arch. Microbiol.* **129,** 257–260.

Shinnick, T. M., and Holt, C. E. (1977). A mutation (*gad*) linked to *mt* and affecting asexual plasmodium formation in *Physarum polycephalum*. *J. Bacteriol.* **131,** 247–250.

Shinnick, T. M., Pallotta, D. J., Jones-Brown, Y. R., Youngman, P. J., and Holt, C. E. (1978). A gene, *imz*, affecting the pH sensitivity of zygote formation in *Physarum polycephalum*. *Curr. Microbiol.* **1,** 163–166.

Sudbery, P., Haugli, K., and Haugli, F. (1978). Enrichment and screening of heat-sensitive mutants of *Physarum polycephalum*. *Gen. Res.* **31,** 1–12.

Truitt, C. L., and Holt, C. E. (1979). Analysis of an *alc* mutation. *In* "Current Research on Physarum" (W. Sachsenmaier, ed.), Vol. 120, pp. 26–30. University of Innsbruck, Austria.

Von Stosch, H. A., Von Zyl-Pischinger, M., and Dersch, G. (1964). Nuclear phase alternance in the Myxomycete *Physarum polycephalum*. *In* Abstracts, Tenth International Botanical Congress, Edinburgh, pp. 481–482.

Wakasugi, M., and Ohta, J. (1973). Studies on the amoebo-flagellate transformation in *Physarum polycephalum*. *Bot. Mag. (Tokyo)* **86,** 299–308.

Wheals, A. E. (1970). A homothallic strain of the Myxomycete *Physarum polycephalum*. *Genetics* **66,** 623–633.

Wheals, A. E. (1973). Developmental mutants in a homothallic strain of *Physarum polycephalum*. *Genet. Res.* **21,** 79–86.

Wheals, A. E., Grant, W. D., and Jockusch, B. M. (1976). Temperature-sensitive mutants of the slime mould *Physarum polycephalum*.I. Mutants of the amoebal phase. *Mol. Gen. Genet.* **149,** 111–114.

Williams, K. L., Kessin, R. H., and Newell, P. C. (1974). Parasexual genetics in *Dictyostelium discoideum:* Mitotic analysis of acriflavin resistance and growth in axenic medium. *J. Gen. Microbiol.* **84,** 59–69.

Youngman, P. J., Adler, P. N., Shinnick, T. M., and Holt, C. E. (1977). An extracellular inducer of asexual plasmodium formation in *Physarum polycephalum*. *Proc. Natl. Acad. Sci. U.S.A.* **74,** 1120–1124.

Youngman, P. J., Pallotta, D. J., Hosler, B., Struhl, G., and Holt, C. E. (1979a). A new mating compatibility locus in *Physarum polycephalum*. *Genetics* **91,** 683–693.

Youngman, P. J., Anderson, R. W., and Holt, C. E. (1979b). The genetic regulation of zygote formation and zygote differentiation during mating in *Physarum polycephalum*. Abstract, *Physarum* Meeting, Laval University, Quebec.

Youngman, P. J., Anderson, R. W., and Holt, C. E. (1981). Two multiallelic mating compatibility loci separately regulate zygote formation and zygote differentiation in the Myxomycete *Physarum polycephalum.Genetics* **97,** 513–530.

CHAPTER 7

Ploidy throughout the Life Cycle in *Physarum polycephalum*

JOYCE MOHBERG

I.	Introduction	253
II.	Methods of Estimating Ploidy	254
	A. DNA Analyses	254
	B. Chromosome Counts	257
III.	Ploidy throughout the Life Cycle	263
	A. Heterothallic Strains	263
	B. Apogamic Strains	265
	C. "CPF" and "ALF" Strains	266
IV.	Factors Affecting Ploidy	266
V.	Uses of Cultures with Different Ploidy	268
	References	269

I. INTRODUCTION

In spite of the progress that has been made in the areas of cytology and genetics of laboratory-grown strains of *Physarum polycephalum* (for reviews, see other chapters in this volume and in Dove and Rusch, 1980), it was not until recently that direct evidence was obtained either for alternance of ploidy during the life cycle or for occurrence of meiosis within developing spores. It appeared quite likely, from the earliest data on mating types (Dee, 1960, 1962), that *P. polycephalum* must be diploid in the plasmodial stage and haploid in the amoebal stage. This was soon shown directly with *P. flavicomum* and *Perichaena vermicularis* when Ross succeeded in counting chromosomes in plasmodia and amoebae of both species (Ross, 1966, 1967). Since the predominant chromosome number in the *P. polycephalum* plasmodium was about 50 (Ross, 1966; Koevenig and Jackson, 1966), amoebae could be assumed to have 25 chromosomes. Likewise, it seemed reasonably certain from electron microscopic studies of sporulation in several species (Aldrich, 1967; Aldrich and Mims,

1970) that meiosis occurred in the spore after cleavage. Similar results have now been obtained from electron microscopic studies, in this case combined with Feulgen DNA analyses (Arescaldino, 1971; Laane and Haugli, 1976; Laane *et al.*, 1976; Lacorre-Arescaldino, 1977) done on developing spores of the strains of *P. polycephalum* which are actually used for biochemical and genetic experiments.

This chapter has been written with several objectives in mind. First, of course, it attempts to review the literature on ploidy changes during the life cycle of *P. polycephalum*. However, sections are also included which are primarily concerned with methodology, their purpose being to alert both new and more experienced workers to the pitfalls and potentials inherent in using *P. polycephalum* as a research tool. Thus, there is a section on factors which can cause ploidy to change in plasmodia and amoebae. Another section discusses the chromosome spreading protocol in its entirety, together with labor-saving tips acquired through several years' experience. A final section describes the kinds of experiments which can be done to exploit differences in ploidy among cultures at the same stage or at different stages in the life cycle.

II. METHODS OF ESTIMATING PLOIDY

A. DNA Analyses

1. MICROSPECTROPHOTOMETRY

Since Therrien's pioneering efforts (Therrien, 1966), a number of workers have used microspectrophotometry of Feulgen-stained nuclei to estimate ploidy in Myxomycetes (see Collins, 1979, for a review, and P. Mulleavy, Chapter 16, Volume II). The procedure of Darlington and LaCour (1962) can be used to give relative DNA values (Cooke and Dee, 1974), or absolute values if avian nuclei are added as an internal standard (Mohberg, 1977; McCullough *et al.*, 1978).

2. MICROFLUOROMETRY

The DNA content of *Physarum* nuclei, particularly spore and amoeba nuclei, is at the lower limit of sensitivity of most microspectrophotometers, and several laboratories have therefore used microfluorometry instead. Laane and Lie (1975) have found that ordinary Feulgen-stained material fluoresces sufficiently to give 10 times the sensitivity obtained by spectrophotometry. Bovey and Ruch (1972) and Laffler and Dove (1977) have used the better-known fluorescent Schiff reagent, BAO [bis(4-aminophenyl)-1, 3, 4-oxadiazol]. Preliminary studies by Prof. G. Czihak, University of Salzburg, and the author showed different fluorescence decay rates in haploid and diploid nuclei, and one should probably check this parameter whenever using the method with a new strain.

Chromomycin A_3 (Crissman and Tobey, 1974) has been used to stain nuclei for analysis by laser microspectrofluorometry (Turner *et al.*, 1978). Here internal standards should also be used, if possible, because we noted considerable differences in staining intensity from one experiment to the next.

3. CHEMICAL ANALYSES

Although they usually require more starting material than the methods mentioned above, chemical analyses of known numbers of nuclei have the advantage of not requiring equipment beyond that found in any ordinary research laboratory. We have used the Burton diphenylamine reaction (Burton, 1956) with 0.5 M perchloric acid extracts (two extractions, each for 15 minutes at 70°C). Details are given elsewhere (Mohberg and Rusch, 1971). However, it should be added that *Physarum* nuclei cannot be counted in a hemacytometer with overflow grooves because they are so small that they begin to flow out of the chamber before they have settled enough to be counted.

The Keck method (Keck, 1956) can also be used with *Physarum* nuclei provided DNA is extracted with trichloroacetic acid instead of perchloric acid to avoid hydrolysis of dAMP (Hubbard *et al.*, 1970), and the indole color reaction tubes are extracted with chloroform after amyl acetate to prevent development of turbidity (W. Sachsenmaier, personal communication).

Gorman *et al.* (1979) and the author (unpublished data) have used the fluorometric procedure of Kissane and Robins (1958) and have obtained consistent results. However, with calf thymus and salmon sperm DNA standards, they were close to 40% higher than with the Burton and Keck methods.

4. MEASUREMENT OF NUCLEAR DIAMETERS

A quick, rough estimate of DNA content of plasmodial nuclei can be made by measuring diameters with a micrometer. The diameter of a haploid nucleus averages about 3 μm and a diploid nucleus 4.3 μm. There is a linear relationship between nuclear surface area and DNA content, with 40 μm^2 being equivalent to about 1 pg of DNA (Mohberg *et al.*, 1973; Adler and Holt, 1975).

5. NUCLEAR DNA CONTENTS OF DIFFERENT STRAINS

Table I gives the nuclear DNA content of a number of plasmodial and amoebal strains. The list includes only strains for which there are absolute values for DNA content, either from Burton or Feulgen analyses. Certain RS amoeba crosses analyzed by Mohberg *et al.* (1973) are omitted because inoculum is no longer available. Burton and Feulgen values are combined in the table without correction for the approximately 15% lower readings obtained with the Feulgen method

TABLE I

Nuclear DNA Content of Different Sublines and Strains of *Physarum polycephalum*

DNA content (pg per nucleus)	Strains
Plasmodia	
1.0–1.3	M_3a^a, M_3b^b, M_3cIV, V, and VII,[c] RSI,[c] $M_3cVII(IIe)^a$
	RSD crosses (RSD4 × RSD8 = TU291; RSD9 × RSD8 = TU145)[a]
	Heterothallic × heterothallic crosses (LU648 × LU688; LU648 × LU1;[d] LU688 × LU853)[c]
	Heterothallic × apogamic crosses (LU624 × CLd; LU647 × LU5001d; CH188 × LU860; CH207 × LU860)[c]
	Heterothallic × *npf* crosses (CL6129 × LU648) and *npf* × *npf* crosses (CL6129 × CL6136; CL6129 × CL6099)[f]
0.5–0.9	Apogamic LU strains (C50, CL, LU640, LU348)[c]
	Tromsø University mutants (TU84, TU63, TU265)[b]
	EMG[g]
3.6	RSD5 × RSD2[a]
Amoebae	
0.5–0.6	RSD4 (= TU4);[c] C50;[d] CLd-AXE and CLd-BACT;[e] *npf* strains (CL6129, 6136, 6099)[f]

[a] Mohberg *et al*. (1973).
[b] Mohberg *et al*. (1980).
[c] Mohberg and Rusch (1971).
[d] Mohberg (1977).
[e] McCullough *et al*. (1978).
[f] Anderson and Dee (1977).
[g] Turner *et al*. (1978).

(Mohberg, 1977). For more accurate figures, the reader can consult the references given in the table footnote.

Most of the McArdle (M) plasmodia, all of which arose from sclerotia with the exception of $M_3cVII(IIe)$, have a G_2-phase DNA content of 1.0–1.3 pg per nucleus. Similar results are seen with crosses of most RSD amoebae and with crosses of Leicester University (LU) or C. E. Holt (CH) heterothallic amoebae with other heterothallics, with apogamics, or with non-plasmodium-forming (*npf*) mutants. Since evidence from genetic studies and from chromosome counts (Section III,A) indicates that these crosses are predominantly diploid, 1.0–1.3 pg appears to be the diploid G_2-phase or 4C DNA content.

Plasmodial subline EMG2, the apogamic LU strains, and several Tromsø University mutants derived from mutated RSD4 amoebae have DNA contents in the range of 0.5–0.9 pg. Feulgen DNA profiles have shown that all of these cultures have a main DNA peak at 0.5–0.6 pg, but some, for example, C50, have

7. Ploidy throughout the Life Cycle in *Physarum polycephalum*

a significant population of diploid and larger nuclei and hence a higher mean nuclear DNA content.

Except for a very few strains, all exponentially growing amoebae have been found to have a DNA content of 0.5–0.6 pg, indicating that even these nonsynchronous populations are predominantly in G_2 phase. It was deduced from analyses of the crosses RSD5 × RSD2 and RSD5 × RSD6 that RSD2 amoebae are pentaploid.

B. Chromosome Counts

1. SPREADS OF PLASMODIAL NUCLEI

a. Method of Spreading. Although Ross (1966) and Koevenig and Jackson (1966) were able to make chromosome spreads from whole culture material, we have never been successful with anything but highly purified prometaphase nuclei, not even with "metaphase-rich" preparations. The procedure which we now use for preparing the nuclei and spreads is essentially as described by Mohberg *et al.* (1973) but includes some refinements which improve the quality of the final spreads as below.

1. Make four to six petri dish plasmodia as directed by Mohberg and Rusch (1969) and incubate until early prophase of mitosis II.

2. Cut three or four plasmodia in half and begin checking smears of each piece for mitotic stage by means of phase contrast microscopy (McCormick and Nardone, 1970).

3. When a piece reaches prometaphase, dip it in ice water and immediately scrape it into 40-ml nuclear isolation medium (0.25 M sucrose; 0.01 M $CaCl_2$; 0.1% (w/v) Triton X-100; 0.01 M Tris, pH 7.3), avoiding the inoculum spot.

4. Stir for 2 minutes in a 250-ml monel, goblet-shaped blender cup on a Waring Blendor; run through a variable transformer at a speed just at the point of frothing. Do not overblend. The homogenate should contain bits of unbroken plasmodium, and these should settle without centrifuging. Leave in ice until all homogenizations have been done.

5. Gravity-filter each homogenate through one thickness of milk filter, clamped in a two-piece, 55-mm Buchner funnel, and divide each filtrate equally between two 50-ml, conical, polycarbonate or polypropylene centrifuge tubes.

6. With a large-caliber syringe needle (No. 15), underlay filtrates with 10 ml of a 1 M sucrose solution containing the same levels of $CaCl_2$, Triton, and Tris as the homogenizing medium. Stir the interface between the two sucrose layers with the syringe needle (to keep nuclei from aggregating at the interface and being lost with the 50 g pellet in the next step).

7. Centrifuge in swing-out rotor (International No. 269) at 50 g for 15 min-

utes. Withdraw 2 ml of solution from the bottom of each tube with a wide-tipped pipette (to remove clumps not caught by the filter) and centrifuge at 1500 g for 15 to 20 minutes to sediment nuclei.

8. Decant supernatants and combine the two pellets from each nuclear preparation in 40 ml of homogenizing medium with stirring in the 250-ml blender cup at low speed, 45 V, for 1 minute.

9. Centrifuge at 1500 g for 15 minutes. Decant supernatants, leaving nuclear pellets in about 0.2 ml of fluid. (Pellets should be no larger than the head of a pin and thus almost invisible.)

10. Suspend nuclei by vortexing for 1 minute and put two drops on each of four to six microscope slides.

11. Immediately add three drops of 66% (v/v) lactic acid to each slide and mix by rocking the slide. Heat over a microburner to the smoking point. Cool to room temperature. (This step spreads chromosomes.)

12. Add two drops 2% (w/v) orcein in 75% (v/v) acetic acid. Mix and heat slightly. (For Giemsa staining, see Mohberg, 1977.) Cover with No. 1 22 × 50-mm coverslip and press between several layers of paper towel to express excess stain. (Application of extreme pressure at this point does not increase spreading beyond what was accomplished by the hot lactic acid.)

13. Examine one slide (upper and lower edges usually are best) of each nuclear preparation with the phase contrast microscope at a magnification of at least 1000 times. When a good slide, one yielding at least a dozen countable spreads in an hour's scanning, appears, save it and others from the same preparation of nuclei.

14. Photograph within a few days, using phase contrast optics, green filter, and Kodak Technical Pan, Kodak Recordak, or Agfa 25 Professional film.

15. Print the photos at about 2500 times and count by marking chromosomes on the print with a lead pencil mounted in a bacterial plaque counter. Resolve chromosomes which are touching or overlapping by examining the negative with a dissecting microscope.

If a good slide is not found, do not waste further time in searching, but start over with new plasmodia and nuclear preparations.

Despite its length, the procedure is relatively easy and the photography part at least will provide hours of innocent fun once a good nuclear preparation is obtained. This, in turn, requires hitting the "window in time" at which chromosomes spread and stain best and getting clean nuclei. The former is best achieved by making a number of nuclear preparations and the latter by following the above protocol exactly, particularly in regard to conditions of homogenization. (For additional tips on isolating clean nuclei, see the Appendix, Volume II).

The procedure for spreading chromosomes at the precleavage mitosis is the same as for growing plasmodia except that fruiting cultures are always somewhat slimy and hard to homogenize. They are therefore washed with 0.01 M

ethylenediaminetetraacetate (pH 7.5) in 0.25 M sucrose (80 ml per half-plasmodium, 2 minutes of slow stirring and 10 minutes of centrifugation at 1000 g) and "mashed" in homogenization medium, this time containing 0.4% Triton X-100, by four or five passes in a Potter–Elvehjem homogenizer before they are stirred in the blender. From step (4) on, they are handled like growing plasmodia, except that chromosomes stain much more slowly and photography should therefore not be started for at least 2 days.

b. Structure of Mitotic Figures and Chromosomes. In early prophase, chromosomes seem to be joined to one another around the nucleolus so that after the nucleolus disappears, they show a string-bag arrangement (Fig. 1A). The open center persists through metaphase, with suitably oriented spreads showing chromosomes in a wreath (Figs. 1B,D). Chromosomes are predominantly telomeric (Fig. 1C) and seem to be paired (Fig. 1B), even in strains which we now consider to be haploid. Banding (Fig. 1C) is evident until mid-metaphase, particularly in old orcein-stained slides (not shown), where it is almost as pronounced as in spreads (Fig. 1E) stained with Giemsa's after denaturation by heating at 60°C (Hecht *et al.*, 1974). It appears from Figs. 1B and C that there are differences in the lengths and shapes of chromosomes, but these differences disappear later in metaphase (Fig. 2A) when chromosomes become more condensed.

c. Chromosome Numbers. Ross (1966) found that an M_3a-derived culture had a uniform chromosome number of 50 ± 2, whereas one from a biological supply house had a chromosome profile as follows: 14, 2%; 20, 5%; 40, 5%; 50, 58%; 70, 5%; and 90, 25%. Mohberg *et al.* (1973) found a uniform count of close to 50 chromosomes in old M_3 cultures, but up to 75% of the nuclei of younger cultures had 75 to 80 chromosomes. When a similar study was made of heterothallic Leicester strains (Mohberg, 1977), counts were somewhat higher. This was traced to the improved counting procedure, which makes possible resolution of V-, T- and X-shaped structures, previously counted as single chromosomes, into two or three chromosomes. We have accordingly reprinted and counted some of our old negatives of RSD9 × 8 and M_3cVII(IIe). A mean count of about 90 chromosomes has been obtained for both strains. Data for RSD9 × 8 are given in a histogram in Fig. 3 and are considered later.

2. SPREADS OF GERMINATING SPORES AND AMOEBAE

a. Method of Spreading. Ross (1966) spread $HgCl_2$-ethanol-acetic acid-fixed amoebae by drying them rapidly in acetic acid. We used hot lactic acid for spreading in our earlier work (Mohberg *et al.*, 1973) but later found that chromosomes were apparently destroyed by this treatment. More reliable results could

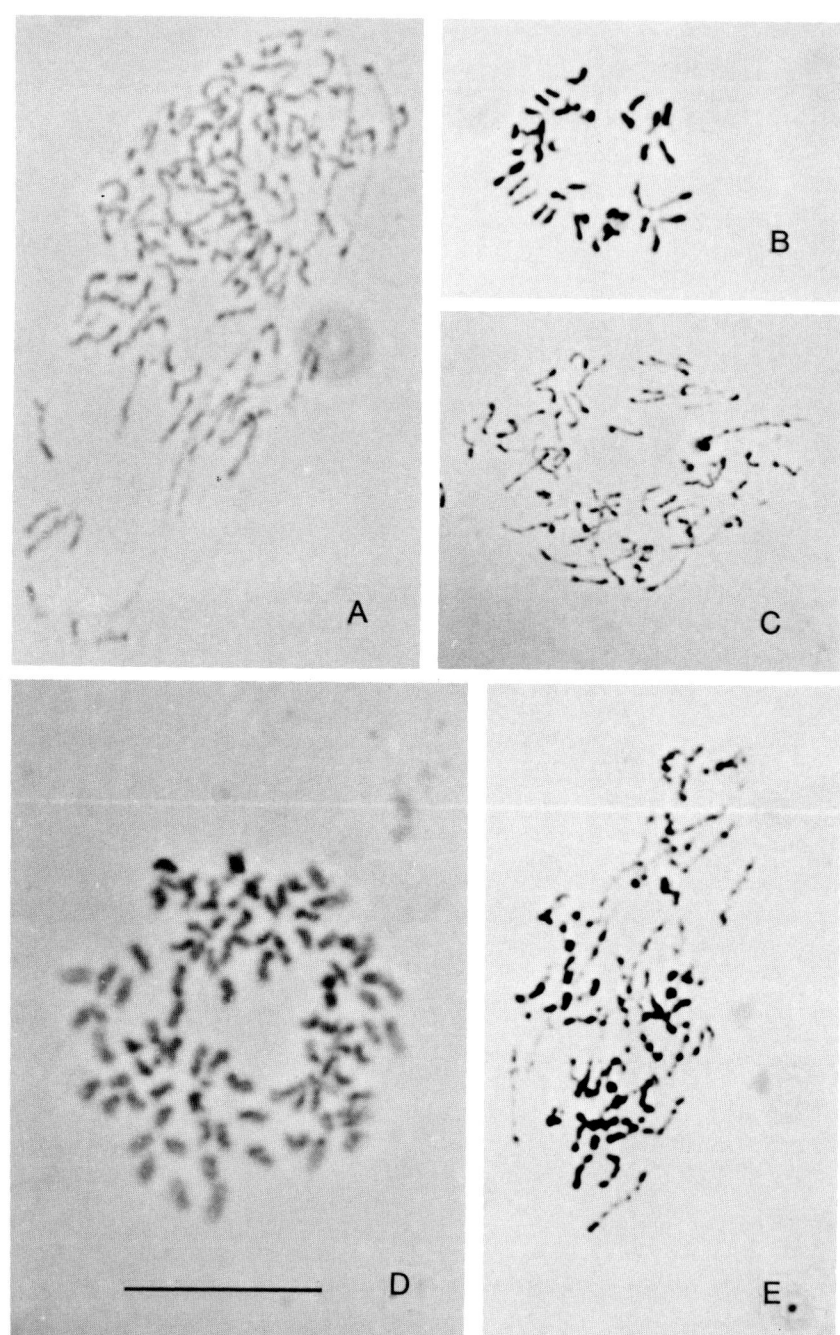

Fig. 1. Spreads of plasmodial chromosomes. (A) LU647 × LU5001d, late prophase; (B) a × i, showing pairing of chromosomes; (C) a × i, prometaphase, showing banding; (D) B173 × B174, precleavage mitosis. (A–D) Acetoorcein stain. (E) CL, Giemsa banding by the method of Hecht *et al.* (1974). All spreads at same magnification. Scale in D, 10 μm. [(B) is reproduced by permission of *Developmental Biology*.]

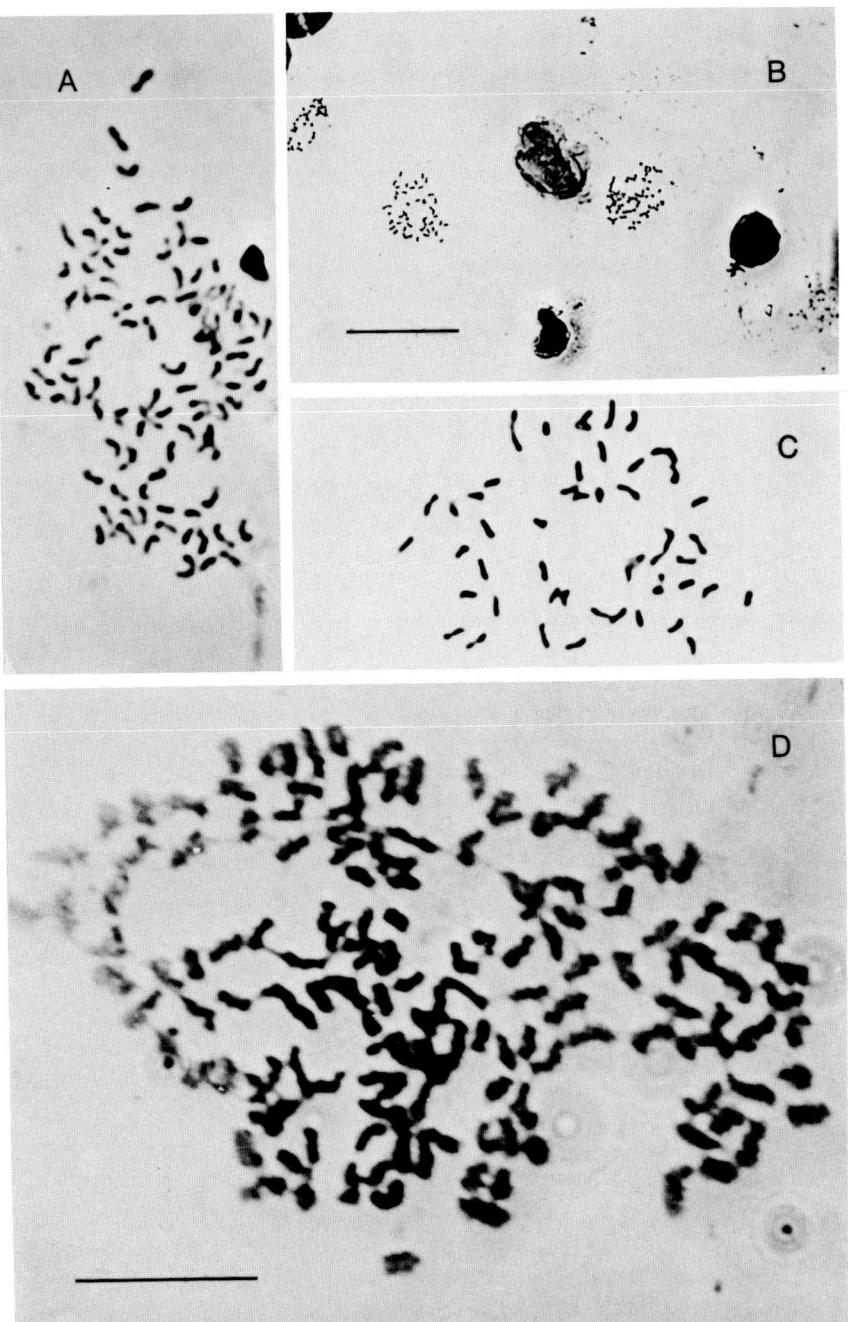

Fig. 2. Chromosomes in RSD9 × RSD8 plasmodium and germinating spores. (A) Plasmodial spread with about 90 chromosomes. (B) Three spreads in a germinating spore squash. Scale, 20 μm. (C) Higher magnification of middle spread of (B). (D) Polyploid nucleus; at least 140 chromosomes. (A, C, D), same magnification. Scale in D, 10 μm. [(C) is reproduced by permission of *J. of Cell Science*.]

be obtained if cells were fixed with formalin before staining (Mohberg, 1977). Axenically grown cells gave still better spreads because interference from partially digested bacteria was eliminated (Mohberg et al., 1980). The complete protocol for spreading and staining chromosomes of germinating spores and amoebae is given by Mohberg (1977).

b. *Structure of Mitotic Spindles and Chromosomes.* Aside from slight banding in an occasional spread, little structure is seen in chromosomes of either growing amoebae (Fig. 4A) or germinating spores (Figs. 2C,D). Presumably this is due to the formalin fixation step, which allows swelling of chromosomes, as shown by comparison of spreads of amoebae and plasmodia in Fig. 4. Whether this also accounts for the difference in size of spore chromosomes in Figs. 2C and 2D is not known.

c. *Chromosome Numbers.* Amoebae of strains RSD4 (or TU4) and CLd have mean chromosome counts of 43.5 ± 4 and 42.7 ± 6, respectively (Mohberg et al., 1980). Germinating spores have a population of nuclei with a similar chromosome count, but diploid and polyploid nuclei are frequently seen, and in spores of apogamic and heterothallic × apogamic strains they can comprise as much as 75% of the total (Mohberg, 1977).

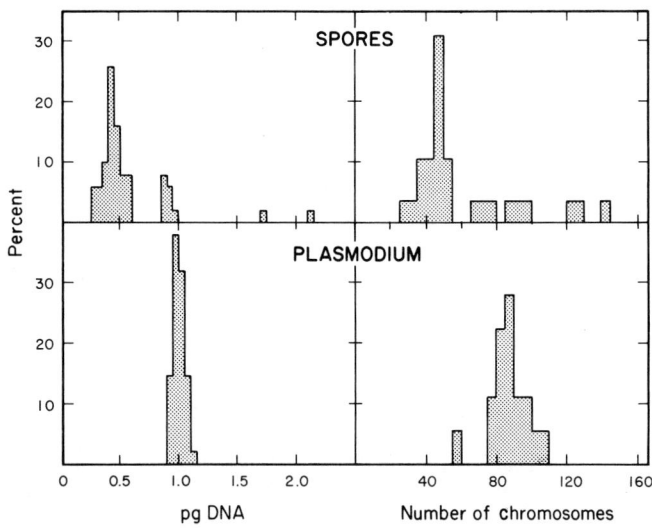

Fig. 3. Feulgen DNA analyses and chromosome counts in germinating spores and plasmodium of RSD9 × RSD8. DNA readings were made of 50 nuclei in each sample. DNA content was calculated by means of an internal standard (chicken erythrocyte nuclei). Data were grouped in 0.05-pg intervals and plotted against frequency (in percent). Chromosome histograms were made from counts of about 20 spreads each of spore and plasmodial nuclei.

7. Ploidy throughout the Life Cycle in Physarum polycephalum

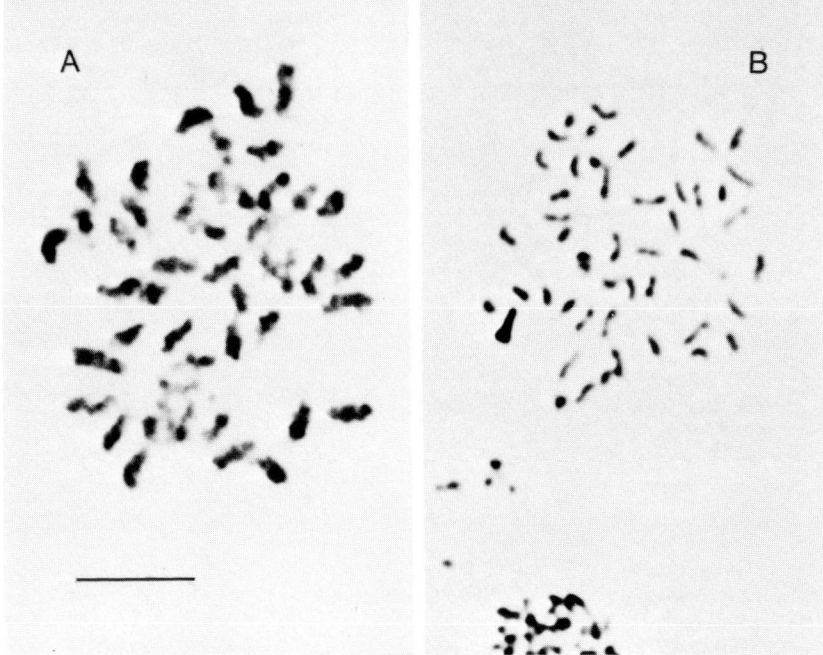

Fig. 4. Chromosomes in RSD4 amoebae and TU84 plasmodium. Spread (A), amoeba; spread (B), plasmodium. Both spreads contain about 50 chromosomes. Scale in (A), 5 μm. [TU84 is derived from a mutant of RSD4 (Haugli and Dove, 1972) and appears to be haploid (Mohberg et al., 1980)]. [(A) is reproduced by permission of Cell Biology International Reports.]

3. SPREADS OF MEIOTIC SPORES

To the author's knowledge, chromosome spreads at the meiotic divisions have not yet been done. However, Laane and Haugli (1976) have obtained good meiotic figures in spores by first soaking them in 5 N HCl for 5 minutes at 20°C to soften the spore coat, then squashing and staining with Feulgen reagent.

III. PLOIDY THROUGHOUT THE LIFE CYCLE

A. Heterothallic Strains

Table II attempts to correlate the changes in morphology and nuclear DNA content which occur during spore formation and maturation. Strains RSD9 × 8 and $M_3cVII(IIe)$ are used as examples because they are the only ones for which there are both electron microscopic and spectrophotometric data.

Growing G_2-phase plasmodia have a DNA content of 1.0–1.3 pg (Table I). In

TABLE II

Changes in Morphology and Nuclear DNA Content during Sporulation in *Physarum polycephalum*

Morphological stage/state[a]	Hours after end of illumination[a,b]	Nuclear DNA content[b,c] (pg per nucleus)		C value
		9 × 8	M_3cVII(IIe)	
Plasmodium				
Growing; G_2 phase	—	1.0	1.4 (1.2)	4
Starving; interphase	—		(1.3)	4[d]
Sporulating; sporangial budding	7		(1.0)	4[d]
Sporulating; late interphase	12			4[d]
Sporulating; precleavage; reconstruction	13			2[d]
Sporulating; at cleavage	13.5			
Spores				
Melanizing	14	1.1	—	
Interphase; premeiotic	15.5			3[d]
Interphase; premeiotic	17.5	—	0.9	4[d]
prophase MI (with SC's)	25 to 40[e,f]	0.9	0.9	3[b]
MI metaphase	43[e]			4[e]
MI anaphase	43[e]			2[e]
MII metaphase	43[e]			2[e]
MII anaphase	43[e]			1[e]
Interphase; postmeiotic	100	0.8	—	3
Interphase	112	0.7	0.7	2
Interphase	12–15 das.	0.6	0.5 (0.7)	2
Germinating	10–15 das.	—	0.5[g]	2

[a] Sauer *et al.* (1969).
[b] Mohberg *et al.* (1973).
[c] Mohberg and Rusch (1971); data in parentheses are for M_3cVII.
[d] Arescaldino (1971).
[e] Laane and Haugli (1976).
[f] Lacorre-Arescaldino (1977).
[g] Mohberg (1977).

RSD9 × 8, and probably also in M_3cVII(IIe), this corresponds to 90 chromosomes (Figs. 2 and 3). Starving plasmodia also have close to the 4C DNA content, indicating that whereas starvation greatly increases the intermitotic time (McCorquodale and Guttes, 1977), it has little effect on the rate of DNA synthesis. In the sporulating plasmodium there is a nuclear division at 11–16 hours (Lacorre-Arescaldino, 1977) after the end of illumination (4 hours' duration), and this is followed by an S phase, so that the 4C state is regained within 3 hours. The next nuclear activity begins at about 25 hours after the end of illumination, at which time synaptonemal complexes (SC's) appear. By 35 hours, SC's are seen in most nuclei. Metaphase of meiosis I (MI) is seen at about 40 hours. It is followed immediately by meiosis II (MII), so that at 45 hours figures of both divisions are seen. Both divisions are intranuclear (Laane and Haugli, 1976).

During the two meiotic divisions, DNA content drops from 4C at MI metaphase to 1C at MII anaphase. There seems to be an S phase at some time between MII and hatching, for whole spores and nuclei of hatching spores contain 0.6 pg of DNA, the haploid G_2 amount. Close to the time of spore germination, there is another synchronous nuclear division (Howard, 1931; Fig. 2B). At this stage, RSD9 × 8 spores contain about 0.5 pg of DNA and 40 chromosomes in the bulk of the nuclei, although larger nuclei are also present. It is not known whether an S phase follows this division. Feulgen analysis of nuclei during the germination period failed to detect a shift in nuclear DNA content (J. Mohberg, unpublished data.) However, there is an increase in thymidine kinase specific activity, and the electrophoretic profile of the enzyme is similar to that seen during plasmodial S phase (Gröbner and Mohberg, 1980).

No data are available on nuclear DNA content or chromosome numbers in amoebae grown from spores of RSD9 × 8 or $M_3cVII(IIe)$, but presumably they are similar to those for RSD4, that is, 0.5–0.6 pg of DNA and a mean chromosome count of 43 (Section II,B,2,c).

Some of Laane and Haugli's observations on sporulation (see Chapter 14, Volume II) agree with certain early results which have been regarded as artifacts in recent years. Carroll and Dykstra (1966) found SC's in precleavage sporangia of *Didymium iridis*,* and Ross (1967) reported that *Perichaena vermicularis* could be induced to undergo meiosis in the precleavage sporangium by appropriate manipulation of the illumination cycle. Laane and Haugli (1976) also found an occasional meiotic figure in sporangia before cleavage, and they have suggested that even in normal sporangia, events in the nucleus can occur independently of morphological changes in the sporangium. They have sketched a number of sequences, also included in Goodman (1980), in which meiosis occurs progressively earlier, relative to cleavage. Most of the schemes give a single viable nucleus per spore, but they differ in the number of degenerating meiotic nuclei which are also present.

B. Apogamic Strains

Cooke and Dee reported in 1974 that in Colonia strains which form plasmodia within amoebal plaques, plasmodia have the same nuclear DNA content as amoebae, both apparently being haploid. This has been confirmed by Feulgen

---

*This result, however, was retracted in Aldrich, H. C., and Carroll, G. C. (1971). *Mycologia* **68**, 308–316.

analyses (Adler and Holt, 1975; Anderson and Dee, 1977; and McCullough *et al.*, 1978). Chromosome counts in plasmodium and germinating spores of C50 showed that the plasmodium had a single peak at about 40 chromosomes, whereas spores had peaks at 40 and 60 chromosomes (Mohberg, 1977).

It is not known how the organism can go through sporulation without changing its nuclear DNA content. Laane *et al.* (1976) have found several differences between spore development in CL and TU145: (1) about half of CL spores were multinucleate after cleavage, making it virtually impossible to get meaningful Feulgen data; (2) very few SC's were found, and these had a poorly developed core structure, as though the constituent chromosomes were only partially homologous; (3) only about 25% of total nuclei seemed to go through meiotic metaphase or anaphase; (4) SC's were seen in 10–25% of spores at germination. On the basis of these results, Laane and co-workers have proposed that apogamic strains undergo a "pseudomeiosis" with the second division suppressed, so that the spore ends up with the same number of chromosomes as the parent plasmodium. Alternatively, and possibly also concomitantly, a normal meiosis might occur in the small population (2% or less) of diploid nuclei which seem present in all apogamic plasmodia. The observations of Lacorre-Arescaldino (1977) are not in complete agreement with those of the Oslo group. For example, she found that SC's appeared later in CL and C50 spores than in those of M_3c sublines but both the number and morphology seemed normal. Lacorre-Arescaldino also noted a tendency of Colonia spores to be multinucleate, and because of this, she was not able to determine whether there were one or two meiotic divisions.

C. "CPF" and "ALC" Strains

"CPF" or "clonal plasmodium-forming" strains (Adler and Holt, 1975) include the apogamic strains, mentioned in the preceding section, and a second group in which spores hatch to give amoebae which are heterozygous for mating type. Both groups have the same DNA content throughout the life cycle, but the latter is diploid in both plasmodium and amoeba. A third variant is the "ALC" or "amoebaless life cycle" mutation (Adler *et al.*, 1975; Adler and Holt, 1977), in which plasmodia hatch directly from spores. Although there is genetic evidence that meiosis occurs in spores of these strains, a direct study with the electron microscope has not been done.

IV. FACTORS AFFECTING PLOIDY

Prolonged culture of plasmodia on agar has been shown by McCullough *et al.* (1973) to cause an increase in nuclear DNA content. Strain B173 × B174 responds particularly rapidly and can undergo a 50-fold increase in nuclear DNA

7. Ploidy throughout the Life Cycle in *Physarum polycephalum* 267

content in approximately as many days. Continued culture of microplasmodia in shaken flasks has the opposite effect, and most old McArdle sublines have a lower chromosome number than newer cultures (Mohberg *et al.*, 1973). M_3c-VII(IIe) seems more prone to change than other M_3 sublines and can drop to haploid during 400 days of growth (Turner *et al.*, 1978). Preliminary results (H. M. Turner and J. Mohberg, unpublished) indicate that there is actually a loss of chromosomes from diploid nuclei, not simply survival of haploid nuclei at the expense of larger nuclei.

Heat shock in prometaphase (15–20 minutes at 37°C) can block nuclear division while allowing DNA synthesis to proceed normally (Brewer and Rusch, 1968), and Guttes (1974) adapted the procedure for use in making cultures with "8- and 16-ploid" nuclei. Although Guttes never, to the author's knowledge, got completely satisfactory chromosome spreads of these huge nuclei, he was able to establish that the heat treatment caused polyploidy, not polyteny (E. Guttes, personal communication).

Since the effect of heat is not uniform throughout the plasmodium, in order to obtain a culture with more uniform ploidy and therefore less tendency to drop to a lower ploidy level, the plasmodium should be cut into pieces after heating and the pieces cultured separately. After the pieces have resumed normal growth, they are checked with the microscope, and those having the most uniform population of large nuclei are saved for further work (personal communication from E. Guttes). Strain TU84, one of the thymidine kinase-deficient mutants (Haugli and Dove, 1972), sclerotizes after heat shock and must be allowed to grow out again before it can be given another shock (Mohberg *et al.*, 1980).

Griseofulvin seems to have an effect similar to that of heat shock and causes diploidization by destroying microtubules so that karyokinesis cannot occur (Hebert *et al.*, 1980).

Whereas most of the treatments mentioned thus far cause an increase in nuclear DNA content, growth in large tray culture has the opposite effect on the CL plasmodium (Turnock, 1979). When measurements of isotope dilution repeatedly showed that neither RNA nor DNA was doubling between MII and MIII in the large plasmodia, nuclear DNA content was followed by Feulgen analysis. Nuclei were found to behave normally until MII prophase, but at MIII prophase over half of them still had not doubled their DNA. The cause of this behavior is not known. The inoculum was not at fault, since petri dish cultures carried in parallel doubled their DNA after both MII and MIII. Likewise, the culture apparatus (Hall and Turnock, 1976) would not seem to be at fault, since it has almost four times the medium capacity of the rocker tray (Mohberg and Rusch, 1969) and M_3cVIII flourishes in it. But regardless of the cause, there is no reason to believe that this is an isolated occurrence, a bizarre characteristic of CL. Rather, it probably occurs to some extent in full-grown (MIII to MIV when started with the usual amount of inoculum) petri dish cultures of all strains. Since this problem is not detected by pulse-labeling techniques, it would seem advisable for anyone studying DNA or RNA synthesis to check the performance of his particular culture or strain by doing bulk chemical analyses or by following the dilution of isotopic labeling (Hall and Turnock, 1976; Turnock, 1979).

Amoebae, like plasmodia, increase in nuclear DNA content during prolonged culture on agar. They must be cloned continually if they are to be used for genetic experiments (Adler and Holt, 1974). P. Mulleavy has found that pure clones with high levels of ploidy can be isolated from established cultures of amoebae of both *Didymium iridis* (Mulleavy, 1979) and *P. polycephalum* (Mulleavy and Evans, 1980).

Diploid amoebae can be obtained by mixing strains of amoebae having the same *matA* genotype and different *matB* genotypes (Youngman *et al.*, 1981). A triploid plasmodium can then be made by mating the diploid amoeba with a haploid amoeba of a different *matA* genotype. Crossing two diploid amoebae, however, results in a diploid heterokaryon (Youngman *et al.*, 1981) and not a tetraploid, as occurs when diploids of *D. iridis* are mated (Mulleavy and Collins, 1979).

Physarum amoebae are not affected by griseofulvin but can be polyploidized by exposure to methyl benzimidazole carbamate or its derivative, nocodazole (Mir and Wright, 1978). Isopropyl N-(3-chlorophenyl)carbamate can presumably be used similarly, since it diploidizes *D. iridis* amoebae (Mulleavy and Collins, 1981). As yet, there have not been observations of benzimidazole derivatives causing metaphase arrest or haploidization, as they can in amoebae of the cellular slime mold *Dictyostelium discoideum* (Welker and Williams, 1980).

V. USES OF CULTURES WITH DIFFERENT PLOIDY

Polyploid strains have thus far been used by only a few laboratory groups. However, they appear to have considerable potential, judging from the types of experiments for which they have been used.

One of the earliest investigations was to compare whole culture protein:DNA ratios in a series of cultures showing a sixfold range in ploidy (Mohberg *et al.*, 1973). Protein:DNA ratios were approximately equal in all strains, implying that if a culture had nuclei with a higher DNA content, it had proportionately fewer of them. Similar results were obtained in a more recent study with polyploids made by heat shock (Mohberg *et al.*, 1980).

Hall *et al.* (1975) used a series of plasmodial cultures with different ploidy— CL, M_3cVIII, and RSD5 × 2—to show that ribosomal DNA makes up a constant fraction of the total genome, regardless of ploidy of the culture. Sudbery (1975) used the same series to determine the relationship between nuclear DNA content and the mitotic delay produced by ultraviolet irradiation. He found that a dose of

2000 ergs/mm^2 at 3 hours after MII delayed MIII for 4 hours in CL, 3.3 hours in M$_3$cVIII, and 2.5 hours in RSD5 × 2.

Several workers have taken advantage of the fact that the G$_2$-phase nucleus of *Physarum* contains a single nucleolus and that the size of the nucleolus increases with ploidy (Mohberg *et al.*, 1973). Grant (1973) used Colonia 501 and Mohberg (1974) and Grainger and Ogle (1978) used RSD5 × 2 for the isolation of nucleoli because nucleoli of both strains were as large as M$_3$c nuclei and hence much easier to separate from cytoplasmic organelles. Similarly, Grant and Poulter (1973) used a polyploid derivative of B173 × B174 as a source of mitochondria because it was easier to free mitochondria of the larger nucleoli.

Polyploids can also increase the range of usefulness of heterokaryons. In their earlier experiments, Guttes and Guttes (1969) made heterokaryons of starved and growing plasmodia because nuclei of the starved culture could be recognized by their small size. Later E. Guttes (1974) polyploidized M$_3$cVII(IIe) by heat shock and then made heterokaryons of the polyploids and the parent culture. This system has the advantage of avoiding problems with incompatibility, since all derivatives should have the same plasmodial fusion type as the parent M$_3$c-VII(IIe).

Jeter *et al.* (1974) studied the exchange of nuclear proteins in a heterokaryon of RSD5 × 2 and RSD9 × 8. They monitored labeling by separating the two sizes of nuclei on a sucrose density gradient (McCormick, 1974) and extracting and electrophoresing the proteins.

D. Cooke and the author (unpublished results) have tried to do genetic studies with heterokaryons of the lysine-requiring haploid, LU348, and the diploid, LU647 × LU5001d (Cooke and Dee, 1975), using Feulgen DNA analysis to monitor ploidy. We found that diploid nuclei always disappeared when nutrients became limiting, as during incubation on lysine-free medium or starvation to induce sporulation or spherulation. If, however, this same heterokaryon or heterokaryons of RSD5 × 2 and RSD9 × 8 or of M$_3$cVII(IIe) and EMG2 were given a change of fresh medium every 24 hours, they were stable for at least a week, indicating that their use in short-term experiments should be valid.

ACKNOWLEDGMENTS

The author wishes to thank Drs. C. E. Holt, P. W. Mulleavy, and P. E. Sudbery for contributing unpublished data and preprints.

REFERENCES

Adler, P. N., and Holt, C. E. (1974). Changes in properties of *Physarum polycephalum* amoebae during extended culture. *J. Bacteriol.* **120**, 532–533.

Adler, P. N., and Holt, C. E. (1975). Mating type and the differentiated state in *Physarum polycephalum*. *Dev. Biol.* **43**, 240–253.

Adler, P. N., and Holt, C. E. (1977). Mutations increasing asexual plasmodium formation in *Physarum polycephalum*. *Genetics* **87**, 401–420.

Adler, P. N., Davidow, L. S., and Holt, C. E. (1975). Life cycle variants of *Physarum polycephalum* that lack the amoeba stage. *Science* **190**, 65–67.

Aldrich, H. C. (1967). The ultrastructure of meiosis in three species of *Physarum*. *Mycologia* **59**, 127–148.

Aldrich, H. C., and Mims, C. W. (1970). Synaptonemal complexes and meiosis in myxomycetes. *Am. J. Bot.* **57**, 935–941.

Anderson, R. W., and Dee, J. (1977). Isolation and analysis of amoebal-plasmodial transition mutants in the myxomycete *Physarum polycephalum*. *Genet. Res.* **29**, 21–34.

Arescaldino, I. (1971). Evolution de la teneur en ADN des noyaux de *Physarum polycephalum* (Myxomycetes) au cours de la sporulation. *C. R. Acad. Sci. Ser. D* **273**, 398–401.

Bovey, F., and Ruch, F. (1972). Cytofluorometric determination of DNA and protein in the nuclei and nucleoli of *Physarum polycephalum* during the intermitotic period. *Histochemie* **32**, 153–162.

Brewer, E. N., and Rusch, H. P. (1968). Effect of elevated temperature shocks on mitosis and on the initiation of DNA replication in *Physarum polycephalum*. *Exp. Cell Res.* **49**, 79–85.

Burton, K. (1956). A study of the conditions and mechanism of the diphenylamine reaction for the colorimetric estimation of deoxyribonucleic acid. *Biochem. J.* **62**, 315–323.

Carroll, G. C., and Dykstra, R. (1966). Synaptinemal complexes in *Didymium iridis*. Mycologia **58**, 166–169.

Collins, O. R. (1979). Myxomycete biosystematics: Some recent developments and future research opportunities. *Bot. Rev.* **45**, 145–201.

Cooke, D. J., and Dee, J. (1974). Plasmodia formation without change in nuclear DNA content in *Physarum polycephalum*. *Genet. Res.* **23**, 307–317.

Cooke, D. J., and Dee, J. (1975). Methods for the isolation and analysis of plasmodial mutants in *Physarum polycephalum*. *Genet. Res.* **24**, 175–187.

Crissman, H. A., and Tobey, R. A. (1974). Cell cycle analysis in 20 minutes. *Science* **184**, 1297–1298.

Darlington, C. D., and LaCour, L. F. (1962). "The Handling of Chromosomes." 4th ed., pp. 159–162. Allen and Unwin, London.

Dee, J. (1960). A mating-type system in an acellular slime mould. *Nature (London)* **185**, 780–781.

Dee, J. (1962). Recombination in a myxomycete, *Physarum polycephalum* Schw. *Genet. Res.* **3**, 11–23.

Dove, W. F., and Rusch, H. P. (1980). "Growth and Differentiation in *Physarum polycephalum*." Princeton Univ. Press, Princeton, New Jersey.

Goodman, E. M. (1980). *Physarum polycephalum:* A review of a model system using a structure-function approach. *Int. Rev. Cytol.* **63**, 1–58.

Gorman, J. A., Dove, W. F., and Shaibe, E. (1979). Mutations affecting the initiation of plasmodial development in *Physarum polycephalum*. *Dev. Genet.* **1**, 47–60.

Grainger, R. M., and Ogle, R. C. (1978). Chromatin structure of the ribosomal RNA genes in *Physarum polycephalum*. Chromosoma **65**, 115–126.

Grant, W. D. (1973). RNA synthesis during the cell cycle in *Physarum polycephalum*. In "The Cell Cycle in Development and Differentiation" (M. J. Ball and F. S. Billett, eds.), pp. 77–109. Cambridge Univ. Press, London and New York.

Grant, W. D., and Poulter, R. T. M. (1973). Rifampicin-sensitive RNA and protein synthesis by isolated mitochondria of *Physarum polycephalum*. *J. Mol. Biol.* **73**, 439–454.

7. Ploidy throughout the Life Cycle in *Physarum polycephalum*

Gröbner, P., and Mohberg, J. (1980). Thymidine kinase enzyme variants in *Physarum polycephalum:* Change of pattern during the life cycle. *Exp. Cell Res.* **126,** 137-142.

Guttes, E. (1974). Continuous nucleolar DNA synthesis in late-interphase nuclei of *Physarum polycephalum* after transplantation into postmitotic plasmodia. *J. Cell Sci.* **15,** 131-143.

Guttes, E., and Guttes, S. (1969). Initiation of mitosis in interphase plasmodia of *Physarum polycephalum* by coalescence with premitotic plasmodia. *Experientia* **25,** 1168-1170.

Hall, L., and Turnock, G. (1976). Synthesis of ribosomal RNA during the mitotic cycle in the slime mould *Physarum polycephalum. Eur. J. Biochem.* **62,** 471-477.

Hall, L., Turnock, G., and Cox, B. J. (1975). Ribosomal RNA genes in the amoebal and plasmodial forms of the slime mould *Physarum polycephalum. Eur. J. Biochem.* **51,** 459-465.

Haugli, F. B., and Dove, W. F. (1972). Mutagenesis and mutant selection in *Physarum polycephalum. Mol. Gen. Genet.* **118,** 109-124.

Hebert, C. D., Steffens, W. L., and Wille, J. J., Jr. (1980). The role of spindle microtubule assembly in the control of mitotic timing in *Physarum. Exp. Cell Res.* **126,** 1-13.

Hecht, F., Wyandt, H. E., and Magenis, R. E. H. (1974). The human cell nucleus; quinacrine and other differential stains in the study of chromatin and chromosomes. *In* "The Cell Nucleus" (H. Busch, ed.), Vol. II, p. 48. Academic Press, New York.

Howard, F. L. (1931). The life history of *Physarum polycephalum. Am. J. Bot.* **18,** 116-132.

Hubbard, R. W., Matthew, W. T., and Dubowik, D. A. (1970). Factors influencing the determination of DNA with indole. *Anal. Biochem.* **38,** 190-201.

Jeter, J. R., Jr., Cameron, I. L., Hart, N. E., and Rusch, H. P. (1974). Nucleo-cytoplasmic interactions and the control of nuclear events in the cell cycle of *Physarum polycephalum. J. Cell Biol.* **63,** 156A.

Keck, K. (1956). An ultramicro technique for the determination of deoxypentose nucleic acid. *Arch. Biochem. Biophys.* **63,** 446-451.

Kissane, J. M., and Robins, E. (1958). The fluorometric measurement of deoxyribonucleic acid in animal tissues with special reference to the central nervous system. *J. Biol. Chem.* **233,** 184-188.

Koevenig, J. L., and Jackson, R. C. (1966). Plasmodial mitoses and polyploidy in the myxomycete *Physarum polycephalum. Mycologia* **58,** 662-667.

Laane, M. M., and Haugli, F. B. (1976). Nuclear behaviour during meiosis in the myxomycete *Physarum polycephalum. Norw. J. Bot.* **23,** 7-21.

Laane, M. M., and Lie, T. (1975). Examination of fungal nuclei with the Feulgen fluorescence method. *Mikroskopie* **31,** 85-90.

Laane, M. M., Haugli, F. B., and Mellem, T. R. (1976). Nuclear behaviour during sporulation and germination in the Colonia strain of *Physarum polycephalum. Norw. J. Bot.* **23,** 177-189.

Lacorre-Arescaldino, I. (1977). Etude ultrastructurale de la meiose dans les souches C_{50}, CL et M_3C de *Physarum polycephalum* (Myxomycetes). *C. R. Acad. Sci. Ser. D* **284,** 1215-1218.

Laffler, T. G., and Dove, W. F. (1977). Viability of *Physarum polycephalum* spores and ploidy of plasmodial nuclei. *J. Bacteriol.* **131,** 473-476.

McCormick, J. J. (1974). Physical separation of nuclei from two independent plasmodia of *Physarum polycephalum* after fusion. *J. Cell Biol.* **62,** 227-231.

McCormick, J. J., and Nardone, R. M. (1970). The effect of nitrogen mustard on the nuclear cycle and DNA synthesis in *Physarum polycephalum. Exp. Cell Res.* **60,** 247-256.

McCorquodale, M. M., and Guttes, E. (1977). Advanced initiation of the first synchronous mitosis following coalescence of starved, UV-irradiated microplasmodia of *Physarum polycephalum. Exp. Cell Res.* **104,** 279-285.

McCullough, C. H. R., Cooke, D. J., Foxon, J. L., Sudbery, P. E., and Grant, W. D. (1973). Nuclear DNA content and senescence in *Physarum polycephalum.Nature (London)*, New Biol. **245,** 263-265.

McCullough, C. H. R., Dee, J., and Foxon, J. L. (1978). Genetic factors determining the growth of *Physarum polycephalum* amoebae in axenic medium. *J. Gen. Microbiol.* **106,** 297-306.

Mir, L., and Wright, M. (1978). Action of antimicrotubular drugs on *Physarum polycephalum*. *Microbios Lett.* **5,** 39-44.

Mohberg, J. (1974). The nucleus of the plasmodial slime molds. *In* "The Cell Nucleus (H. Busch, ed.), Vol. I, pp. 187-218. Academic Press, New York.

Mohberg, J. (1977). Nuclear DNA content and chromosome numbers throughout the life cycle of the Colonia strain of the myxomycete, *Physarum polycephalum*. *J. Cell Sci.* **24,** 95-108.

Mohberg, J., and Rusch, H. P. (1969). Growth of large plasmodia of the myxomycete *Physarum polycephalum*. *J. Bacteriol.* **97,** 1411-1418.

Mohberg, J., and Rusch, H. P. (1971). Isolation and DNA content of nuclei of *Physarum polycephalum*. *Exp. Cell Res.* **66,** 305-316.

Mohberg, J., Babcock, K. L., Haugli, F. B., and Rusch, H. P. (1973). Nuclear DNA content and chromosome numbers in the myxomycete *Physarum polycephalum*. *Dev. Biol.* **34,** 228-245.

Mohberg, J., Dworzak, E., Sachsenmaier, W., and Haugli, F. B. (1980). Thymidine kinase-deficient mutants of *Physarum polycephalum:* Relationships between enzyme activity levels and ploidy. *Cell Biol. Int. Rep.* **4,** 137-148.

Mulleavy, P. (1979). Genetic and cytological studies in heterothallic and non-heterothallic isolates of the myxomycete *Didymium iridis*. Ph.D. Thesis, Univ. of California, Berkeley, California.

Mulleavy, P., and Collins, O. R. (1979). Development of apogamic amoebae from heterothallic lines of a myxomycete, *Didymium iridis*. *Am. J. Bot.* **66,** 1067-1073.

Mulleavy, P., and Collins, O. R. (1981). Isolation of CIPIC-induced and spontaneously-produced diploid myxamoebae in a myxomycete, *Didymium iridis*. A study of mating type heterozygotes. Mycologia **73,** 62-77.

Mulleavy, P., and Evans, T. E. (1980). Comparative sensitivities of an isogenic ploidal series of *Physarum polycephalum* to UV light and ionizing radiation. *J. Cell Biol.* **87,** 325A.

Ross, I. K. (1966). Chromosome numbers in pure and gross cultures of myxomycetes. *Am. J. Bot.* **53,** 712-718.

Ross, I. K. (1967). Growth and development of the myxomycete *Perichaena vermicularis*. II. Chromosome numbers and nuclear cycles. *Am. J. Bot.* **54,** 1231-1236.

Sauer, H. W., Babcock, K. L., and Rusch, H. P. (1969). Sporulation in *Physarum polycephalum:* A model system for studies on differentiation. *Exp. Cell Res.* **57,** 319-327.

Sudbery, P. (1974). Studies on the control of mitosis in the plasmodium of *Physarum polycephalum*. Ph.D. Thesis, Leicester University, England.

Therrien, C. D. (1966). Microspectrophotometric measurement of nuclear deoxyribonucleic acid in two Myxomycetes. *Can. J. Bot.* **44,** 1667-1675.

Turner, H. M., Mohberg, J., and Goodman, E. M. (1978). Characterization of the cell cycle in *Physarum polycephalum* myxamoebae. *J. Cell Biol.* **79,** 12A.

Turnock, G. (1979). Patterns of nucleic acid synthesis in *Physarum polycephalum*. *Prog. Nucleic Acid Res. Mol. Biol.* **23,** 53-104.

Welker, D. L., and Williams, K. L. (1980). Mitotic arrest and chromosome doubling using thiabendazole, cambendazole, nocodazole and benlate in the slime mould *Dictyostelium discoideum*. *J. Gen. Microbiol.* **116,** 397-407.

Youngman, P. J., Anderson, R. W., and Holt, C. E. (1981). Two multiallelic mating compatibility loci separately regulate zygote formation and zygote differentiation in the myxomycete *Physarum polycephalum*. Genetics **97,** 513-530.

CHAPTER 8

Genealogy and Characteristics of Some Cultivated Isolates of *Physarum polycephalum*

JOYCE MOHBERG and KARLEE L. BABCOCK

I.	Introduction	273
II.	Origin of Isolates	274
	A. M (McArdle) Series	274
	B. RS (TU) Series	276
	C. DJ Series	276
	D. Other Wisconsin Isolates	276
	E. Indiana and Colonia (LU and CH) Strains	277
	F. Strains in Use Overseas	277
III.	Characteristics of Isolates	278
	A. Spherule Formation and Germination	278
	B. Spore Formation and Germination	278
	C. Growth Requirements and Drug Sensitivities	279
	D. Suppliers of Inoculum	279
	References	281

I. INTRODUCTION

When we first began recording the histories of our various *Physarum polycephalum* cultures, our main concern was to avoid those which had lost the ability to fruit during long cultivation in the laboratory (Daniel and Baldwin, 1964). Since that time, we have learned that a number of other defects—or attributes, depending upon one's perspective—can occur spontaneously in cultures which are seemingly normal, or at least which have not to anyone's knowledge been mutagenized. In this chapter, we present an abridged version of the family tree of McArdle derivatives (see Babcock and Mohberg, 1975, for a more complete listing), showing the origin of some of the more thoroughly studied sublines, and a table of traits of certain McArdle and Leicester strains. In this way, we hope to convince the newcomer to the field that there are indeed differences among strains and to help him choose the culture which best fits his particular needs. For example, M_3b, which grows rapidly but fruits poorly, is quite satisfactory for studies of mitosis, whereas TU291, which grows as well but spherulates poorly, may cause problems because inoculum stocks are hard to maintain. We also hope to give the more experienced worker some leads on traits which he might like to add to his present culture line.

Originally, we intended to restrict ourselves to McArdle and Colonia derivatives, since they are used for most biochemical studies and we have had firsthand experience with them. However, relatively recent data on mating types (Masson et al., 1979) have suggested that there are also other isolates in circulation. We have therefore contacted several other investigators, particularly C. J. Alexopoulos, J. D. Anderson, and N. Kamiya, all of whom were already working with *Physarum* before the Rusch group started. J. D. Anderson, in collaboration with D. M. Miller, had already made a survey of the dispersal of early *Physarum* strains, including some collected by Howard (1931), and he has very generously put these data at our disposal.

II. ORIGIN OF ISOLATES

A. M (McArdle) Series

The origin of the McArdle sublines of *P. polycephalum* is outlined in Fig. 1, together with the approximate dates (month and year) when spherules were made or germinated. All M sublines are derived from one sclerotium, Wis 1, which M. Backus, University of Wisconsin Department of Botany, collected about 1947 (personal communication from M. Backus to S. G. Jong, American Type Culture Collection). J. W. Daniel grew a piece of the Wis 1 sclerotium axenically in semidefined medium in 1956. Shortly thereafter, J. Dee (Dee, 1960) fruited this culture and isolated amoeba strains a (mating type 1) and i (mating type 2), which eventually gave rise to the plasmodial strains which Carlile uses for his studies of lethal interaction (Carlile and Dee, 1967; Carlile, 1973) and Bradbury's group uses for their studies of chromatin (Bradbury et al., 1973).

Daniel and co-workers succeeded in establishing the M culture in a complex, chemically defined medium in 1959 (Daniel et al., 1963) and in a minimal, defined medium (Daniel and Baldwin, 1964) in 1961.

The Daniel group, which did most of the earlier work on preservation of cultures, saved only a few sublines which grew in chemically defined media for the simple reason that they rated a medium according to the length of time it supported growth; when growth failed, they discarded both medium and culture. In the case of cultures in a semidefined medium, however, new cultures were

8. Genealogy and Characteristics of Some Cultivated Isolates

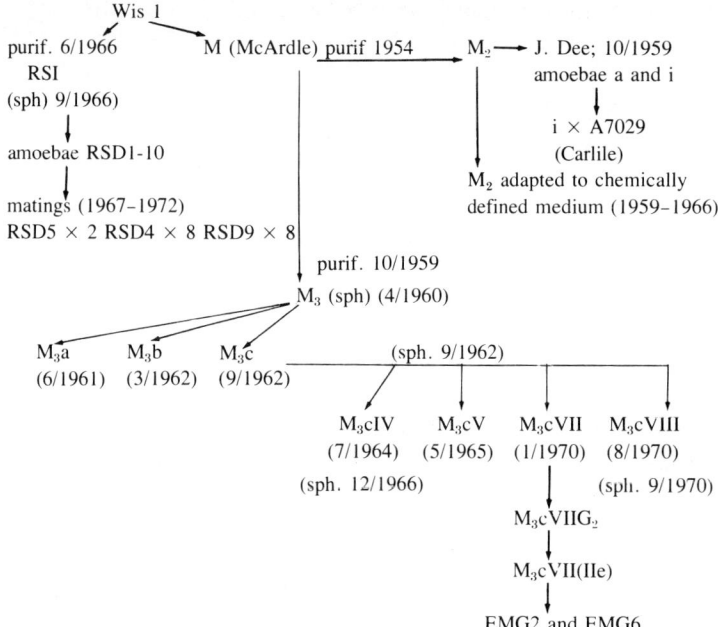

Fig. 1. Origin of major McArdle sublines. Dates (month and year) are given in parentheses.

started at roughly 6-month intervals in order to maintain a high fruiting efficiency. Usually the old culture was spherulated and saved. In order to keep track of this mixture of new and aged cultures, the following system, which is discussed below, was devised.

In 1960, a large supply of spherule strips was made of both M_2 and M_3, which had by that time been growing in semidefined medium for 4 years and for 6 months, respectively. [This was the point at which the McArdle group began using "N + C," semidefined medium with hematin and citrate buffer (Daniel and Baldwin, 1964), and M_3a is the first culture to be grown on N + C only (personal communication from J. W. Daniel).] Each time a strip of spherules was germinated, it was given a letter designation, e.g., M_3a, M_3b, M_3c. By the time M_3c was germinated, the spherule supply was almost exhausted, so M_3c was respherulated to replenish our stocks. Each time an M_3c strip was germinated, it was given a Roman numeral. This continued to M_3cVIII, when again we were almost out of strips, and M_3cVIII was therefore respherulated. M_3cVIII derivatives have not been given any special designation because we have used each one for only a few weeks and then have discarded it.

All of the sublines mentioned above were started from sclerotia or spherules, as far as we know. However, Haugli (1971) isolated several M_3c strains by putting M_3cVII through the life cycle several times by mass inoculation of spores and then selecting for plasmodia homozygous for plasmodial fusion type. M_3c-VII(IIe) has proved the most interesting of these strains, both because it gives high-viability spores and because it appears to drop from diploid to haploid after only a few months in submersed culture (Turner *et al.*, 1978; H. M. Turner and J. Mohberg, unpublished).

B. RS (TU) Series

R. Steeper obtained a new bit of Wis 1 sclerotium from the University of Wisconsin Botany Department in 1966 and began to grow it in axenic culture. This was probably not part of the same sclerotium given to Daniel, because the Botany Department routinely replenishes its supply of *Physarum* inoculum by culturing and sclerotizing on oatmeal (personal communication from Prof. W. Whittingham). Steeper put his cultures through the life cycle twice and then isolated the RSD amoeba clones from second-generation gametes. Some of these clones may have been exposed to ultraviolet irradiation at some time during their isolation/culture, but it is not possible to determine which they might be because pertinent records have been lost (personal communication from R. Steeper). Haugli continued the work with RSD clones (see Lunn *et al.*, 1977, for the TU designations for these clones) and discovered the highly polyploid RSD2 (Haugli, 1971; Mohberg *et al.*, 1973). He also mutagenized RSD4 to produce BUdR- and cycloheximide-resistant mutants (Haugli and Dove, 1972; Haugli *et al.*, 1972). Goodman (1972) adapted RSD4 to axenic growth in protein-containing medium, and McCullough and Dee (1976) used RSD4 for the development of an autoclavable semidefined medium and a chemically defined medium.

C. DJ Series

Jacobson used Steeper's plasmodium as starting material for isolating amoeba clones which would complete the life cycle more reliably than most of the RSD clones did. He then used them for studies of cell motility and colony formation (Jacobson and Dove, 1975). Gorman has mutagenized the DJ clones to produce "selfing" mutants (Gorman *et al.*, 1979).

D. Other Wisconsin Isolates

M. Backus collected a second sclerotium, Wis 2, some 10 years after Wis 1. This isolate never found its way into biochemical studies, which may be fortunate, since it has different mating types from Wis 1 (Dee, 1973).

In addition, in 1959 and 1960, Daniel and his co-workers were studying and distributing an isolate, T, from Turtox and another A, from P. J. Allen, University of Wisconsin Botany Department. The background of these cultures is not known (J. W. Daniel, personal communication), although one might suspect that A came from W. Seifriz at the University of Pennsylvania (Allen and Price, 1950). At any rate, they should not be considered M cultures.

E. Indiana and Colonia (LU and CH) Strains

The Indiana strain came from sclerotia which the General Biological Supply House sent to C. J. Alexopoulos. It was the source of amoebae B173 (mating type 3) and amoebae B174 (Dee, 1966), and has been combined with Colonia to give a number of the genetically marked CH strains, isolated by C. E. Holt's group at the Massachusetts Institute of Technology (Adler and Holt, 1974, 1975).

The Colonia strain was isolated by H. von Stosch from material given him by the Botany Department of the University of Cologne (Dee, this volume, Chapter 6). It is the source of the apogamic Leicester University (LU) strains (Wheals, 1970); Dee, 1973). For further information on LU strains, both apogamic and heterothallic, see Chapter 6.

F. Strains in Use Overseas

Most of the *Physarum* laboratories on the European continent now work with McArdle derivatives brought back from Rusch's laboratory by postdoctoral fellows. For example, W. Sachsenmaier took back M_3b and passed it along to the group at the University of Bonn (personal communication from K. Wohlfarth-Bottermann). However, J. D. Anderson began sending inoculum overseas in the late 1940s, and among the stocks he sent were some derived from material collected by Howard while at the University of Iowa (Howard, 1931). Among the laboratories receiving these isolates was E. Zeuthen's in Copenhagen.

There also seems to be a chance that Howard isolates might still be found in culture collections in Japan, for when N. Kamiya returned to Japan from W. Seifriz's laboratory at the University of Pennsylvania, he took with him a culture which may originally have come from Howard. This culture was used by Kamiya and co-workers for their earlier experiments but has now been replaced by material from Ward Biological (personal communication from Prof. N. Kamiya). T. Kuroiwa (personal communication) at the National Institute for Basic Biology, Okazaki, Japan, uses three strains—one obtained from Kamiya, another from S. Hatano at Nagoya University, and a third which is derived from the "Ochanomizu strain." This last, we assume, is the "Ocha-no-mizu strain" which Hosoda (1980) states came from McArdle by way of J. Ohta. N. Kamiya (personal communication) says that *P. polycephalum* grows in the wild in Japan, but it appears thus far not to have been domesticated.

III. CHARACTERISTICS OF ISOLATES

A. Spherule Formation and Germination

Most of the M sublines and Leicester strains readily form spherules or "microsclerotia" (Daniel and Baldwin, 1964) in shaken culture. Moreover, if the spherules are properly washed to remove slime, they can be dried on filter paper and stored in the cold for years (at least 14 years in the case of M_3b). In contrast, the RSD matings form spherules only if allowed to starve in exhausted nutrient medium with very gentle shaking, apparently because microplasmodia become fragile when they starve and lyse before spherule walls can form. Haugli has been able to obtain preparations which withstood drying, but in our hands the spherules remain viable only if stored in exhausted medium, and then for no longer than 6 months. This makes it quite difficult to maintain inoculum stocks of such key strains as the BUdR-resistant mutants.

The rate of spherule germination, whether fast (12 hours or less) or slow (24 hours or more) is not critical to anyone interested only in starting a new culture. However, studies of germination itself (Mitchell and Rusch, 1973; Forde and Sachsenmaier, 1979; Gröbner and Mohberg, 1980) are not feasible with such strains as $M_3cVII(IIe)$ and CL because replicates may germinate any time from 24 to 48 hours after plating.

B. Spore Formation and Germination

Although many of the isolates listed in Table I will not sporulate on niacin-salts medium (Daniel and Baldwin, 1964), most will produce viable spores if first starved on exhausted half-strength medium for 2 to 3 weeks and then illuminated. An exception is M_3cIV, which forms sporangia on all media but has lost the capacity for spore cleavage during long growth in shaken culture (Mohberg and Rusch, 1971). Another is RSD8 × RSD4 (or TU291), which, like most of the RSD matings, never fruited readily, even when new.

Anyone interested in spore germination per se has several alternatives: (1) Use $M_2cVII(IIe)$ or RSD9 × 8, keeping in mind that IIe may drop to haploid and that RSD9 × 8 does not sphaerulate well. (2) Put an isolate such as M_3b through the life cycle two or three times to improve the spore germination (Gröbner and Mohberg, 1980). (3) Choose one of the heterothallic × heterothallic or heterothallic × apogamic crosses from the Leicester collection, for example, LU647 × LU5001d (Mohberg, 1977). Someone wishing to isolate polyploid amoebae might do better to use the haploid strains, in spite of the low spore germination, because there is a higher percentage of polyploids among the emerging cells than with diploid strains (Mohberg, 1977).

Hosoda (1980) has found that a McArdle derivative which had completely lost the ability to fruit on niacin-salts medium during 8 years of shaken culture had a sporulating efficiency of 90% or more when transferred to oat flake culture. No one seems to have investigated whether growth on oats can also restore the capacity for spore cleavage to aged isolates like M_3cIV.

C. Growth Requirements and Drug Sensitivities

In addition to the differences in tolerance to shaking mentioned in regard to spherulation, certain strains are sensitive to other culture conditions. In our hands, most RSD matings grew better on filter paper than on Millipore membrane, presumably because they were sensitive to the Triton X-100 in the Millipore. B173 × B174 can be grown on agar for only about 2 months before it "senesces," that is, becomes highly polyploid and dies, whereas both CL and M derivatives can be carried in this way for a much longer time before showing ill effects (Poulter, 1969; McCullough et al., 1973).

CL is peculiar in that its nucleic acids do not replicate properly when it is grown in large tray cultures (Hall and Turnock, 1976) but do so when CL is grown in petri dishes or when tray plasmodia are cut into bits (Turnock, 1979).

Several instances of strain differences in sensitivity to agents affecting nucleic acid synthesis have been observed. (1) M_3b is more sensitive to actinomycin D than M_3c is (Sachsenmaier and Becker, 1965), even though they are both derived from the same sclerotium (see Section II,A). (2) DNA synthesis in TU291 is completely blocked by FUdR but is only slowed in M_3b (Mohberg et al., 1980). (3) Ultraviolet irradiation causes twice the mitotic delay in CL as in M_3cVIII (Sudbery, 1974).

A complete investigation of plasmodial fusion types has not been done with the cultures in Table I, but V. Richmond (Dee, 1973) has found that M_3cIV, V, VIII, and RSI all fuse with one another but not with strain a × i. Other compatible combinations are $M_3cVII(IIe)$ and EMG2, and RSD9 × 8 and RSD5 × 2. The last pair makes a particularly convenient heterokaryon because RSD5 × 2 is pale yellow, as well as polyploid, and can thus be recognized by both color and nuclear size. White mutants have been obtained from Colonia. Among them are a polyploid, 501 (Grant, 1973), and the haploids—LU897 and LU 898 (Anderson, 1977)— which can be mated to give a diploid (Gorman et al., 1979).

D. Suppliers of Inoculum

Much of the original McArdle spherule collection is now under the supervision of Dr. T. E. Evans at Case Western University in Cleveland. Several sublines and strains are also available from the American Type Culture Collection

TABLE I

Characteristics of Sublines of *Physarum polycephalum*[a]

Subline or strain	Spherulation	Spherule germination[b]	Sporulation	Spore germination (%)	Special features
Wis 1	Good	—	N + C/2	—	—
M$_3$b	Good	Fast	N + C/2 only	0.1	Fast growth; actinomycin D-sensitive[c]
M$_3$cIV	Good	—	All media	0.1	Sporulation without cleavage
M$_3$cV(AD)	Good	—	All media	1–5	Fast growth; vigorous
M$_3$cVII and M$_3$cVIII[d]	Good	Fast	All media	5–25	All-purpose
M$_3$cVII(IIe)	Good	Slow	N + C/2; 22°C	50–60	Homozygous for plasmodial fusion type[e]
EMG2	Good	Slow	N + C/2	2	Haploid[f]
a × i	Good	Slow	N + C/2	10	Highly heterogeneous ploidy
RSI	Good	—	All media	5–10	All-purpose
RSD9 × RSD8 (TU145)	Poor	Low viability	All media	50–60	Most ready fruiter
RSD4 × RSD8 (TU291)	Poor	Low viability	Infrequent	—	FUdR-sensitive[g]
CL$_{15}$	Good	Slow	N + C/2	0.1–0.2	Haploid
C50	Good	Slow	N + C/2	1–3	Haploid
RSD5 × RSD2	Fair	Very slow	Infrequent	—	Highly polyploid
B173 × B174	Poor	—	All media	—	Fast senescence[h]

[a] Cultures were spherulated in exhausted nutrient medium in shaken flasks. Spherules were germinated in petri dishes on filter paper above one-third strength nutrient medium. Sporulation was done in three ways: (1) Microplasmodia were incubated on niacin-salts medium immediately after fusion (Daniel and Baldwin, 1964). (2) Plasmodia were grown in petri dishes on nutrient medium for 24 hours and then transferred to niacin-salts medium (Haugli et al., 1972). (3) Plasmodia were started on half-strength nutrient medium (N + C/2) and incubated in the dark for at least 10 days before illumination. After fruiting, plates were wrapped in Parafilm to prevent evaporation and were left at room temperature for 10 days before germination. Spores were then washed and plated on bacterial lawns. Plaques were counted after 4 days.

[b] Fast, within 12 hours of plating on N + C/3; slow, 24 hours or more after plating.

[c] Sachsenmaier and Becker, 1965.

[d] M$_3$cVII = M$_3$cVIII, since VII was in culture for only a few weeks before replacement by VIII.

[e] Haugli, 1971.

[f] Turner et al., 1978.

[g] Mohberg et al., 1980.

[h] Poulter, 1969; McCullough et al., 1973.

(ATCC), Rockville, Md. Among these are the plasmodial strains—a × i, RSI, Wis 2, M_3cIV, V, and VIII—and amoebal strains RSD4, RSD8, PpII(+) and PpII(−) (Collins and Haskins, 1972). Dr. S. C. Jong, head of the Mycology Department at ATCC, states that the department is interested in adding to its collection and would be happy to receive additional isolates (personal communication).

It must be remembered that cultures available from both of the above sources are those which spherulate well. Others, such as the RSD crosses, must be obtained as growing plasmodia, usually from Haugli in Tromsö.

ACKNOWLEDGMENTS

We wish to express our gratitude to all those who have responded to our requests for information concerning the cultures they have used—Drs. C. J. Alexopoulos, J. W. Daniel, J. Dee, E. Hosoda, N. Kamiya, T. Kuroiwa, J. R. Steeper, and K. E. Wohlfarth-Bottermann. We particularly wish to thank Dr. J. D. Anderson for putting his as yet unpublished history of *Physarum* isolates at our disposal.

REFERENCES

Adler, P. N., and Holt, C. E. (1974). Genetic analysis in the Colonia strain of *Physarum polycephalum*: Heterothallic strains that mate with and are partially isogenic to the Colonia strain. *Genetics* **78**, 1051-1062.

Adler, P. N., and Holt, C. E. (1975). Mating type and the differentiated state in *Physarum polycephalum*. *Dev. Biol.* **43**, 240-253.

Allen, P. J., and Price, W. H. (1950). The relation between respiration and protoplasmic flow in the slime mold, *Physarum polycephalum*. *Am. J. Bot.* **37**, 393-402.

Anderson, R. W. (1977). A plasmodial colour mutation in the myxomycete *Physarum polycephalum*. *Genet. Res.* **30**, 301-306.

Babcock, K. L., and Mohberg, J. (1975). The pedigree of Wis 1 (McArdle) *Physarum polycephalum*. *Physarum Newsletter* **7**, 3-5.

Bradbury, E. M., Matthews, H. R., McNaughton, J., and Molgaard, H. V. (1973). Sub-nuclear components of *Physarum polycephalum*. *Biochim. Biophys. Acta* **335**, 19-29.

Carlile, M. J. (1973). Cell fusion and somatic incompatibility in myxomycetes. *Ber. Dtsch. Bot. Ges.* **86**, 123-139.

Carlile, M. J., and Dee, J. (1967). Plasmodial fusion and lethal interaction between strains in a Myxomycete. *Nature (London)* **215**, 832-834.

Collins, O. R., and Haskins, E. F. (1972). Genetics of somatic fusion in *Physarum polycephalum*: The PpII strain. *Genetics* **71**, 63-71.

Daniel, J. W., and Baldwin, H. H. (1964). Methods of culture for plasmodial myxomycetes. *Methods Cell Physiol.* **1**, 9-41.

Daniel, J. W., Babcock, K., Sievert, A. H., and Rusch, H. P. (1963). Organic requirements and synthetic media for growth of the myxomycete *Physarum polycephalum*. *J. Bacteriol.* **86**, 324-331.

Dee, J. (1960). A mating-type system in an acellular slime mould. *Nature (London)* **185**, 780-781.

Dee, J. (1966). Genetic analysis of actidione-resistant mutants in the Myxomycete *Physarum polycephalum* Schw. *Genet. Res.* **8,** 101-110.
Dee, J. (1973). Aims and techniques of genetic analysis in *Physarum polycephalum*. *Ber. Dtsch. Bot. Ges.* **86,** 93-121.
Forde, B. G., and Sachsenmaier, W. (1979). Oxygen uptake and mitochondrial enzyme activities in the mitotic cycle of *Physarum polycephalum*. *J. Gen. Microbiol.* **115,** 135-143.
Goodman E. M. (1972). Axenic culture of myxamoebae of the myxomycete *Physarum polycephalum*. *J. Bacteriol.* **111,** 242-247.
Gorman, J. A., Dove, W. F., and Shaibe, E. (1979). Mutations affecting the initiation of plasmodial development in *Physarum polycephalum*. *Dev. Genet.* **1,** 47-60.
Grant, W. D. (1973). RNA synthesis during the cell cycle in *Physarum polycephalum*. *In* "The Cell Cycle in Development and Differentiation" (M. J. Ball and F. S. Billett, eds.), pp. 77-109. Brit. Soc. for Cell Biol. Symp., Cambridge Univ. Press, London and New York.
Gröbner, P., and Mohberg, J. (1980). Thymidine kinase enzyme variants in *Physarum polycephalum:* Change of pattern during the life cycle. *Exp. Cell Res.* **126,** 137-142.
Hall, L., and Turnock, G. (1976). Synthesis of ribosomal RNA during the mitotic cycle in the slime mould *Physarum polycephalum*. *Eur. J. Biochem.* **62,** 471-477.
Haugli, F. B. (1971). Mutagenesis, selection and genetic analysis in *Physarum polycephalum*. Ph.D. Thesis, Univ. Wisconsin, Madision, Wisconsin.
Haugli, F. B., and Dove, W. F. (1972). Mutagenesis and mutant selection in *Physarum polycephalum*. *Mol. Gen. Genet.* **118,** 109-124.
Haugli, F. B., Dove, W. F., and Jimenez, A. (1972). Genetics and biochemistry of cycloheximide resistance in *Physarum polycephalum*. *Mol. Gen. Genet.* **118,** 97-107.
Hosoda, E. (1980). Culture methods and sporulation in *Physarum polycephalum*. *Mycologia* **72,** 500-504.
Howard, F. L. (1931). The life history of *Physarum polycephalum*. *Am. J. Bot.* **18,** 116-132.
Jacobson, D. N., and Dove, W. F. (1975). The amoebal cell of *Physarum polycephalum:* Colony formation and growth. *Dev. Biol.* **47,** 97-105.
Lunn, A., Cooke, D., and Haugli, F. (1977). Genetics and biochemistry of 5-bromouridine resistance in *Physarum polycephalum*. *Genet. Res.* **30,** 1-12.
Masson, S., Kirouac-Brunet, J., and Pallotta, D. (1979). Multiple allelism at the *mat* B locus in *Physarum polycephalum*. 7th North American Myxomycete Conference. Laval University, Quebec, Canada.
McCullough, C. H. R., and Dee, J. (1976). Defined and semi-defined media for the growth of amoebae of *Physarum polycephalum*. *J. Gen. Microbiol.* **95,** 151-158.
McCullough, C. H. R., Cooke, D. J., Foxon, J. L., Sudbery, P. E., and Grant, W. D. (1973). Nuclear DNA content and senescence in *Physarum polycephalum*. *Nature (London), New Biol.* **245,** 263-265.
Mitchell, J. L. A., and Rusch, H. P. (1973). Regulation of polyamine synthesis in *Physarum polycephalum* during growth and differentiation. *Biochim. Biophys. Acta* **297,** 503-516.
Mohberg, J. (1977). Nuclear DNA content and chromosome numbers throughout the life cycle of the Colonia strain of the myxomycete, *Physarum polycephalum*. *J. Cell Sci.* **24,** 95-108.
Mohberg, J., and Rusch, H. P. (1971). Isolation and DNA content of nuclei of *Physarum polycephalum*. *Exp. Cell Res.* **66,** 305-316.
Mohberg, J., Babcock, K. L., Haugli, F. B., and Rusch, H. P. (1973). Nuclear DNA content and chromosome numbers in the myxomycete *Physarum polycephalum*. *Dev. Biol.* **34,** 228-245.
Mohberg, J., Dworzak, E., and Sachsenmaier, W. (1980). Thymidine kinase-deficient mutants of *Physarum polycephalum:* Biochemical characterization. *Exp. Cell Res.* **126,** 351-357.

8. Genealogy and Characteristics of Some Cultivated Isolates

Poulter, R. T. M. (1969). Senescence in the Myxomycete, *Physarum polycephalum*. Ph.D. Thesis, Univ. Leicester, England.

Sachsenmaier, W., and Becker, J. E. (1965). Wirkung von Actinomycin D auf die RNS-Synthese und die synchrone Mitose Tätigkeit in *Physarum polycephalum*. *Monatsh. Chem.* **96,** 754-765.

Sudbery, P. E. (1974). Studies on the control of mitosis in the plasmodium of *Physarum polycephalum*. Ph. D. Thesis, Leicester University, England.

Turner, H. M., Mohberg, J., and Goodman, E. M. (1978). Characterization of the cell cycle of *Physarum polycephalum* myxamoebae. *J. Cell Biol.* **79,** 12A.

Turnock, G. (1979). Patterns of nucleic acid synthesis in *Physarum polycephalum*. *Prog. Nucleic Acid Res. Mol. Biol.* **23,** 53-104.

Wheals, A. E. (1970). A homothallic strain of the Myxomycete *Physarum polycephalum*. *Genetics* **66,** 623-633.

PART IV

The Plasmodial Cell Cycle

CHAPTER 9

Nuclear Organization during the Cell Cycle in the Myxomycete *Physarum polycephalum*

JEAN-G. LAFONTAINE and MONIQUE CADRIN

I.	Introduction	287
II.	Structure of the Chromosomes during the Mitotic Cycle	288
III.	The Nucleolar Cycle	301
	A. Interphase	301
	B. Mitotic Stages	304
	C. Nucleologenesis	307
IV.	The Mitotic Apparatus	309
	References	312

I. INTRODUCTION

Although various ultrastructural studies of the cell cycle in *Physarum polycephalum* have already been carried out, a number of important aspects of nuclear organization in this organism deserve further attention. For instance, relatively little is known about the structural evolution of the nucleus during interphase. Also, although the participation of persisting bodies in the reconstruction of the nucleolus has been considered in several reports, the role of newly synthesized material, such as RNA and proteins, in nucleologenesis remains to be clarified.

In this chapter, available information on the morphological evolution of the nucleus during the mitotic cycle will be reviewed together with results from recent ultracytochemical and radioautographic studies. Since there are few published data on this subject, the gradual transformation of the nucleus from early telophase to late interphase will be emphasized. Much attention will also be devoted to the various steps of nucleolar formation, a subject which has generated much interest in the past. A number of other interesting aspects of the mitotic cycle in *P. polycephalum* have been reviewed by Goodman (1980).

II. STRUCTURE OF THE CHROMOSOMES DURING THE MITOTIC CYCLE

It has become evident, from many ultrastructural investigations of a wide variety of organisms, that nuclear organization is greatly affected by the preservation techniques used (Ris, 1962; Moyne *et al.*, 1975; Puvion and Bernhard, 1975). In the case of *P. polycephalum* and other related myxomycetes, no comparative study has yet appeared on the merit of different fixatives for maintaining adequate nuclear preservation. Since it has been our experience that the compacting effect of glutaraldehyde on nuclei leads to masking of certain of their ultrastructural features, we will occasionally refer to material fixed with osmium tetroxide only, even though we realize that this agent does not provide the best overall preservation either.

In osmium tetroxide-fixed specimens, the early interphase nucleus is highly irregular in outline. Apart from the presence of numerous small globular bodies known to be nucleolar remnants and frequently characterized by the presence of opaque particles (Lord *et al.*, 1977), this nucleus is more or less completely filled with a fine fibrillar material assumed to consist of dense chromatin. Following glutaraldehyde–osmium tetroxide preservation (Fig. 1), the nucleus is slightly more compacted in organization and exhibits a number of dense, uniformly fibrillar masses which, judging from their size and shape, correspond to the nucleolar remnants first described by Guttes *et al.* (1968). On first examination, the remaining portion of the nuclear cavity seems to be uniformly occupied by a pervading fibrillogranular substance, but closer scrutiny reveals the presence throughout this material of several small fibrillar lumps which are best interpreted as being chromosomal in nature.

Several staining techniques have proved useful in obtaining more specific information on the distribution of chromatin in early interphase nuclei while providing further data on the cytochemical makeup of the surrounding fibrillogranular material and on the nucleolar remnants. Treatment of glutaraldehyde-fixed specimens with EDTA, for instance, bleaches chromatin and reveals the distribution of RNP-containing substances (Bernhard, 1969). Contrary to the impression gained from the examination of conventionally processed samples, the early interphase chromatin does not entirely permeate the nuclear cavity but forms instead a very coarse network, the interstices of which are occupied by a fibrillogranular substance (Fig. 2). A number of slightly more transparent circular zones may be recognized throughout this latter material, suggesting the presence of chromatin-containing structures, which we will refer to as "minichromo-

9. Nuclear Organization during the Cell Cycle

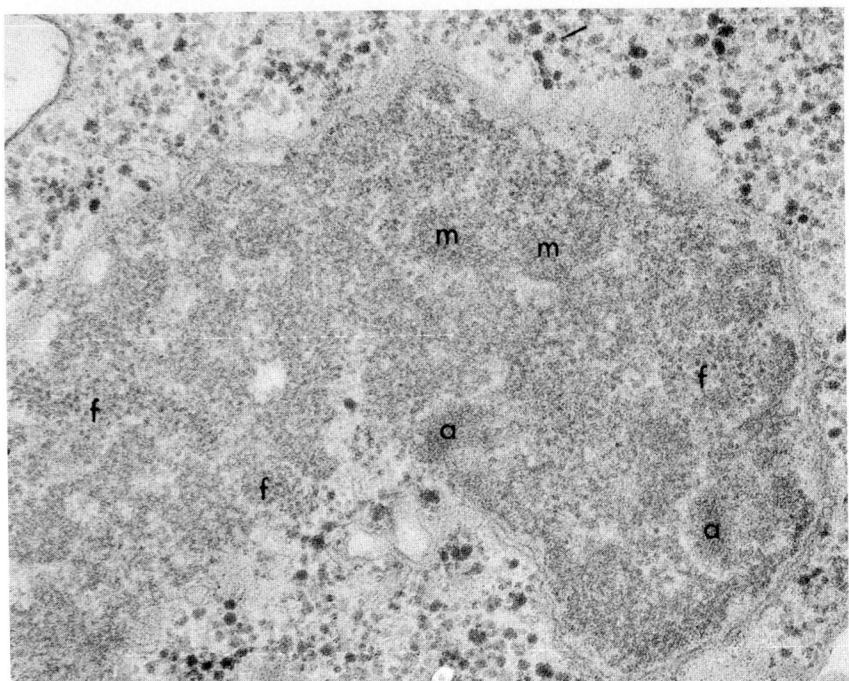

Fig. 1. Micrograph of a very early interphase nucleus. Whereas corresponding specimens preserved with osmium tetroxide show only an almost homogeneous fibrillar texture, double-fixed nuclei are seen to consist of three different types of materials: (1) numerous irregular, denser, fibrillar masses that, without doubt, represent the globular nucleolar remnants or minichromosomes (m); (2) a pervading fibrillogranular material (f) occupying almost all of the remaining nuclear volume; and (3) more amorphous masses (a), certain of which are more easily recognized along the nuclear envelope. These last formations are better seen in EDTA-treated specimens (Fig. 2). Only a few small, more transparent nucleoplasmic zones are present at this stage. Material fixed with glutaraldehyde–osmium tetroxide and stained with uranyl acetate–lead citrate. ×42,000.

somes'' (Seebeck *et al.*, 1979). When sections are subjected to a longer exposure to EDTA, these globular formations are bleached further (Lafontaine *et al.*, 1981). Strangely enough, they then exhibit opaque particles of the type observed in osmium tetroxide-fixed preparations. A quite similar distribution of material is also revealed by first staining samples en bloc with phosphotungstic acid (Sheridan and Barrnett, 1969) and counterstaining the sections with uranyl acetate–lead citrate. The coarse chromatin network now appears denser, whereas most of the pervading material is slightly lighter and contains clusters of opaque particles characteristically associated with prenucleolar material. A perhaps still more precise delineation of the distribution of the three main types of substances present in early interphase nuclei is obtained by fixing samples in formaldehyde

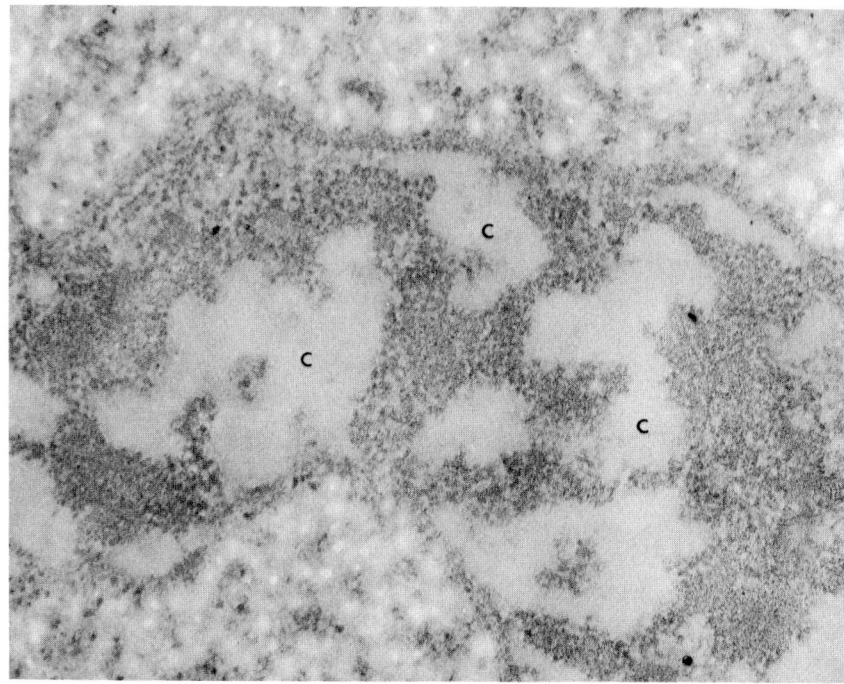

Fig. 2. Micrograph of an early interphase nucleus fixed in glutaraldehyde and treated with EDTA. After such processing, chromatin (c) is seen in the form of a coarse network. One notes in particular that many chromatin lumps of different sizes are closely associated with the nuclear envelope. The remaining regions of the nuclear cavity are filled with a denser material, certain portions of which show a granular texture. Other parts of the interchromatin zones are only partially bleached by EDTA, but judging by their shape and size, they represent minichromosomes. The fact that these regions definitely contain chromatin is demonstrated by subjecting preparations to a longer EDTA treatment. ×32,000. From Lafontaine et al. (1981).

and staining with bismuth (Locke and Huie, 1977). Under these experimental conditions, most meshes of the coarse chromatin network contain a substance, presumably the prenucleolar material, which is studded with opaque granules, whereas other meshes are more transparent and appear to consist of nucleoplasm (M.Cadrin, unpublished observations).

The results furnished by these different cytochemical techniques leave little doubt, therefore, that at the very beginning of interphase, the chromosomes are still in a condensed state and occupy only a portion of the nuclear cavity. As for the interchromatin material, it consists predominantly of the globular nucleolar remnants and fibrillogranular material which have migrated with the chromosomes to the poles of the mitotic nucleus. As will be discussed in Section III,C, a

9. Nuclear Organization during the Cell Cycle

newly synthesized RNA-containing component is apparently also rapidly accumulated in the early interphase nucleus.

Investigation of conventionally processed specimens shows that during the first few minutes of interphase, some redistribution of the material in the nucleus occurs. However, the exact nature of these changes is hard to interpret because of the difficulty of visualizing chromatin. Using the stains referred to above, as well as DAB, a reagent which gives a greater density to DNA than RNA (Anteunis *et al.*, 1973), it is observed that chromatin is now organized into a much looser reticulum extending throughout the nuclear cavity, which also contains an intervening fibrillogranular material (Fig. 3). Digestion of such nuclei with deoxyribonuclease before staining with DAB shows that this pervading substance consists partly of RNA. Judging from available high-resolution radioautographic data (Kuroiwa, 1973), it would appear that incorporation of thymidine in diffuse

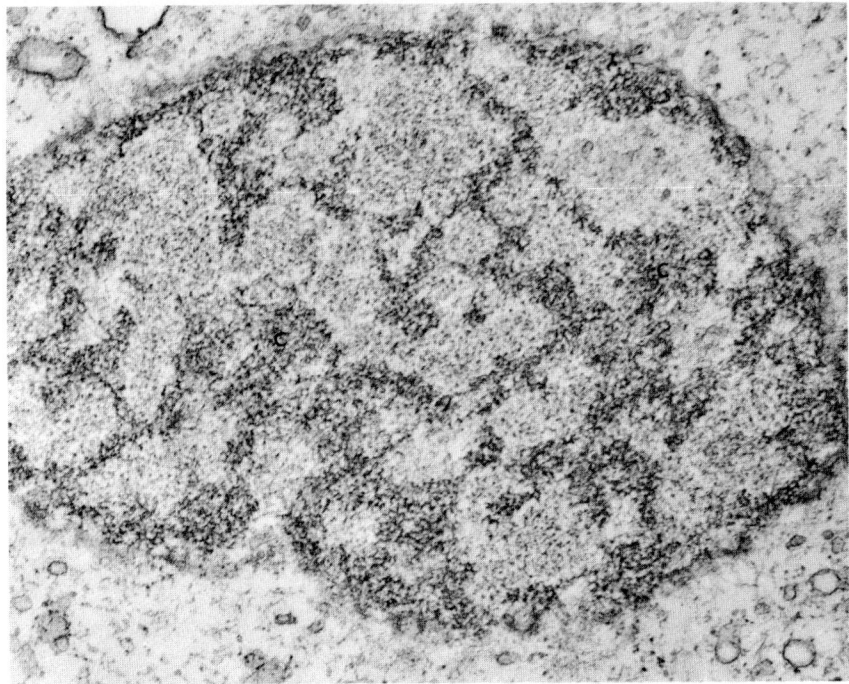

Fig. 3. Nucleus fixed 10 minutes after metaphase in formaldehyde and stained with DAB. In such preparations, chromatin (c) stands out most conspicuously and is then seen to form a much more delicate network than that observed in slightly younger nuclei (Fig. 2). The meshes of this reticulum are occupied by a lighter material which exhibits a fibrillogranular ultrastructure. ×51,000. From Lafontaine *et al.* (1981).

chromatin is initiated as soon as small nucleoplasmic areas appear throughout the early interphase nucleus.

From 10–15 minutes after metaphase, the interphase nucleus enlarges significantly, becomes rounder in contour and, as is well known from earlier electron microscopic studies (Guttes *et al.*, 1968; Goodman and Ritter, 1969; Lord *et al.*, 1977), prenucleolar bodies gradually form. In corresponding DAB-stained specimens it is observed that, apart from these latter masses, the nucleus is characterized by the presence of numerous small, irregular patches of fibrillogranular material, most of which are closely associated with chromatin. The apparent chromatin network observed in slightly younger nuclei has undergone an extensive transformation. Because part of the mass of the chromosomes has unraveled into a diffuse state, most of the nuclear reticulum has, indeed, disappeared, and instead numerous chromocenters varying greatly in size and shape persist. In conventional preparations, it is only about 20 minutes after metaphase (Lafontaine *et al.*, 1981) that chromatin becomes sufficiently conspicuous to be easily differentiated from the surrounding highly irregular prenucleolar masses (Fig. 4). Whether this higher density results from the increased compactness of the chromocenters or from a greater stainability of chromatin at this time is not clear.

Subsequent growth of the interphase nucleus is mostly achieved through expansion of the nucleoplasmic areas, which eventually represent a sizeable portion of the nuclear volume (Fig. 5). The chromocenters remain rather small (ca. 0.1 μm), and only a few of them are still associated with the nucleolar mass, the majority being either dispersed throughout the nucleoplasm or closely appressed to the nuclear envelope. From approximately 50 minutes after metaphase to the end of interphase the chromocenters appear more heterogeneous in size, and fewer of them are associated with the nuclear envelope than previously (Fig. 6).

The onset of prophase in *P. polycephalum* has been found, by phase contrast

Fig. 4. As interphase progresses, the nucleoplasmic zones enlarge noticeably and eventually represent much of the nuclear volume. After double fixation, these areas exhibit a closely knitted, fine fibrillar meshwork throughout which numerous granules of different sizes are present. Contrary to the situation prevailing in slightly younger nuclei equally stained with uranyl acetate–lead citrate, chromatin (c) appears denser than the nucleolar material. It is now organized into lumps, or chromocenters, most of which are associated with the irregular nucleolar masses (n) as well as with the nuclear envelope. ×41,000. From Lafontaine *et al.* (1981).

Fig. 5. Approximately 35 minutes after metaphase, the various nucleolar masses have fused into a unique body which usually takes on a horseshoe configuration. At that time, the nucleolus consists of numerous closely aggregated fibrillar globules and of a pervasive granular material. Only a few chromocenters (c) are still associated with the nucleolar body; they now vary somewhat in size, most of the smaller ones being attached to the nuclear envelope. The greatly enlarged nucleoplasmic zones are richly provided with both fine fibrillar material and clusters of granules. Overall, the texture of the nucleoplasm is noticeably coarser than in previous stages, due to the presence of larger granular aggregates and of very small chromatin lumps. ×41,000. From Lafontaine *et al.* (1981).

9. Nuclear Organization during the Cell Cycle

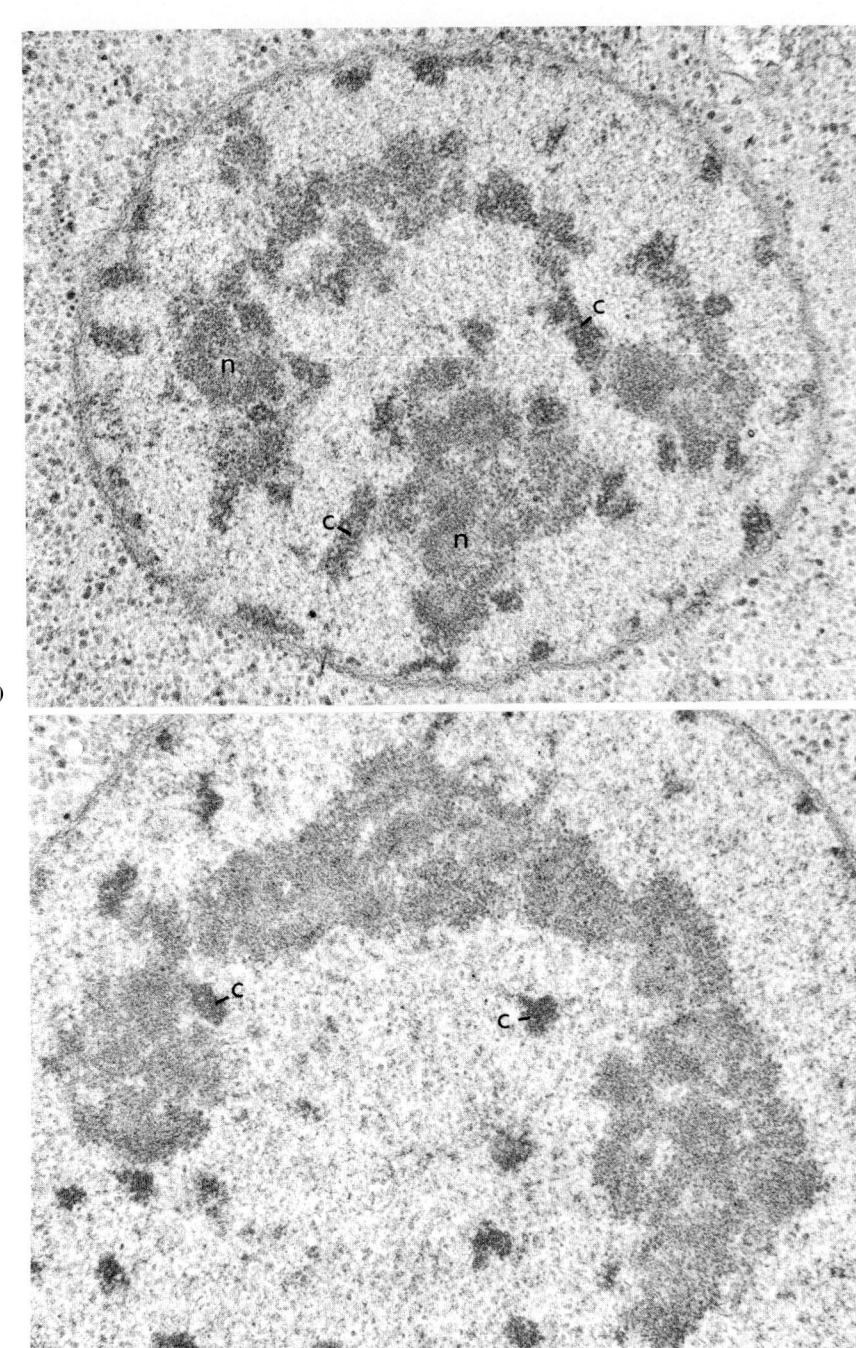

(4)

(5)

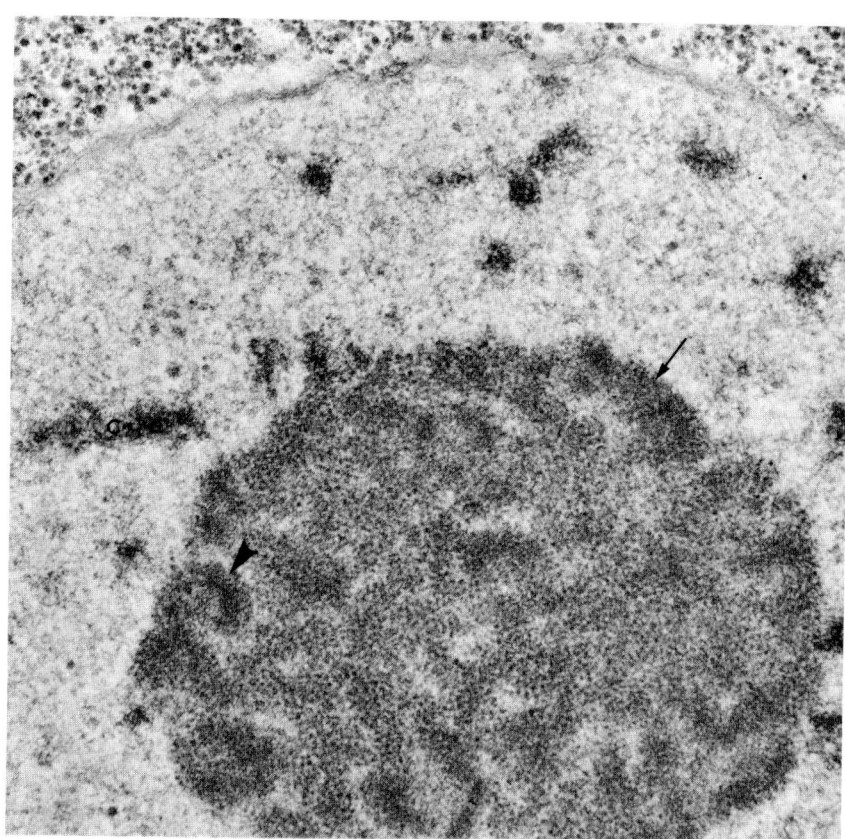

Fig. 6. Portion of a late interphase nucleus from a sample fixed in osmium tetroxide. The chromocenters are very heterogeneous in size, and very few of them are still associated with the nuclear envelope. The nucleoplasm consists of loose, fine, fibrillar material, part of which is presumably diffuse chromatin, as well as of dispersed granules of various sizes. Due undoubtedly to the lesser compacting action of osmium tetroxide, the internal organization of the nucleolus is somewhat clearer than in glutaraldehyde-fixed specimens. The numerous minichromosomes appear to be of different sizes and shapes. Some of them are elongated (arrow) and of uniform density. Others consist of an outer denser region surrounding a more transparent central fibrillar zone (arrowhead). Possibly due to the angle of sectioning, certain minichromosomal profiles are crescent-shaped. A distinctly granular material pervades the remaining portion of the nucleolar mass. ×50,000.

Fig. 7. Part of an early prophase nucleus. The nucleolus has started moving toward the periphery of the nucleus and shows a slightly looser organization than during interphase. As a result, many of the minichromosomes (m) stand out quite conspicuously and most of them are clearly seen to be globular in shape, as in interphase. The granular nucleolar portions are more limited in size than in Fig. 6. At this stage, profiles of the condensing chromosomes are already very irregular in outline and more elongated than during mid-interphase. Material fixed in glutaraldehyde–osmium tetroxide. ×40,000.

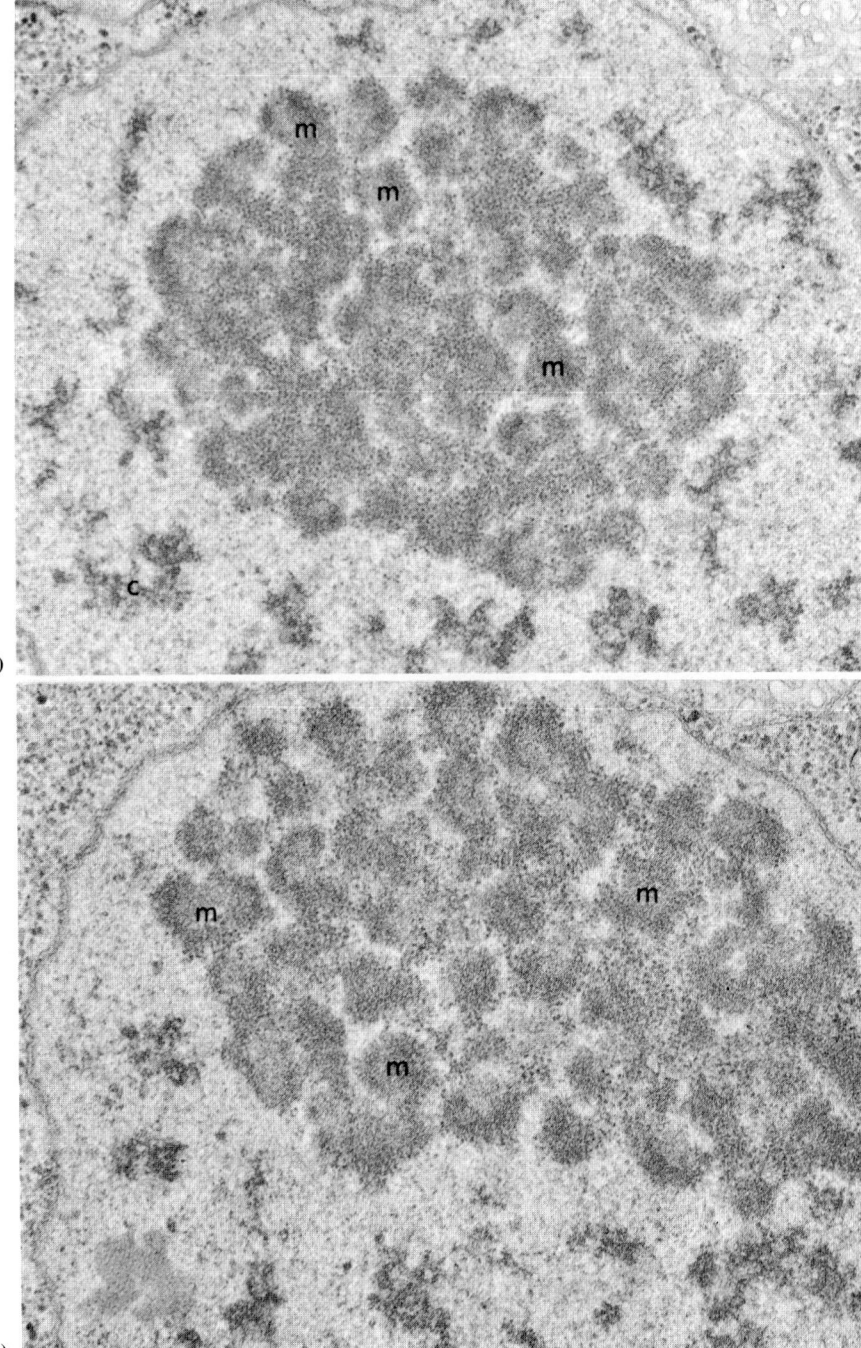

Fig. 8. Mid- to late prophase nucleus. The nucleolus has lost most of its granular component and now consists predominantly of loosely grouped minichromosomes (m). The majority of these latter formations are globular in shape and are clearly seen to be made of a slightly denser peripheral region and a more transparent core portion. The longitudinal and cross-sectional chromosome profiles are very irregular and strongly suggest the presence of twisted structural elements. A somewhat amorphous, irregular body is observed in the lower left corner of the micrograph. Whether this formation is related to the tubule-organizing center is not clear. ×40,000. From Cadrin et al. (1981).

examination of spread preparations, to coincide with the beginning of nucleolar displacement to the nuclear periphery (Guttes *et al.*, 1961; Guttes *et al.*, 1968; Goodman and Ritter, 1969; Wolf *et al.*, 1979). At about that time, or possibly a few minutes earlier, according to our observations, slightly elongated chromatin formations are first recognized using electron microscopy (Fig. 7). In view of the large number of these structures, it is very likely that the numerous chromocenters characterizing late interphase nuclei act as foci for condensation of chromatin into the early prophase chromosomes. Whether these reorganization loci correspond to the centromeric portions of the chromosomes, as in certain other species with small chromosomes (Lord and Lafontaine, 1976), is still not clear. As they first emerge, the early interphase chromosomes take the form of narrow, twisting filaments varying greatly in diameter. It is likely that the wider segments observed correspond to still unraveled chromocentric regions. A few minutes before the eccentrically located nucleolus reaches the nuclear envelope, many profiles of chromosome sections suggest the presence within them of a higher hierarchy of coiling, a feature which has been well documented in many other organisms (Ris and Korenberg, 1979). This coiled organization is especially clear in sufficiently thick (0.15–0.25-μm) preparations, as evidenced by the regular disposition of pointed projections along the mid-prophase chromosomes. By the time the nucleolus has flattened against the nuclear envelope, chromosomes are noticeably more compacted, and longer segments of these chromosomes are represented in ultrathin sections (Fig. 8). Cross-sectional profiles consist of a dense, extremely ragged, peripheral portion surrounding a slightly lighter central region. So far, our rather detailed observations on prophase chromosomes have not furnished any evidence of paired subunits, or chromatids. This problem is now being investigated further by means of ultrathin serial sections.

During prometaphase, while the persisting nucleolar material disperses throughout the nuclear cavity, the chromosomes undergo a striking change in gross organization and transform into extremely compact structures. This further condensation of the mitotic chromosomes has been correlated with concurrent

Fig. 9. In this prometaphase nucleus, the chromosomes are grouped on one side of the nuclear cavity and appear to have aggregated into a more or less continuous mass exhibiting most irregular contours. Although much of the dispersing nucleolar material (n) is still located close to the nuclear envelope, many minichromosomes as well as lumps of fibrillogranular material are observed among the chromosomes (c). The nucleoplasm at this stage shows a much larger population of granules presumably of nucleolar origin. Material fixed in glutaraldehyde–osmium tetroxide. ×50,000.

Fig. 10. Micrograph depicting the close association of the metaphase chromosomes (c) in a more or less continuous mass at the equator of the nucleus. On both sides of the chromosome plate, one notes the presence of persisting nucleolar material (n) having a fibrillogranular texture. Both longitudinal and cross sections of microtubules are observed. A somewhat amorphous structure (arrow), which is believed to represent a portion of the formation illustrated in Fig. 19, is closely associated with a bundle of microtubules. ×50,000.

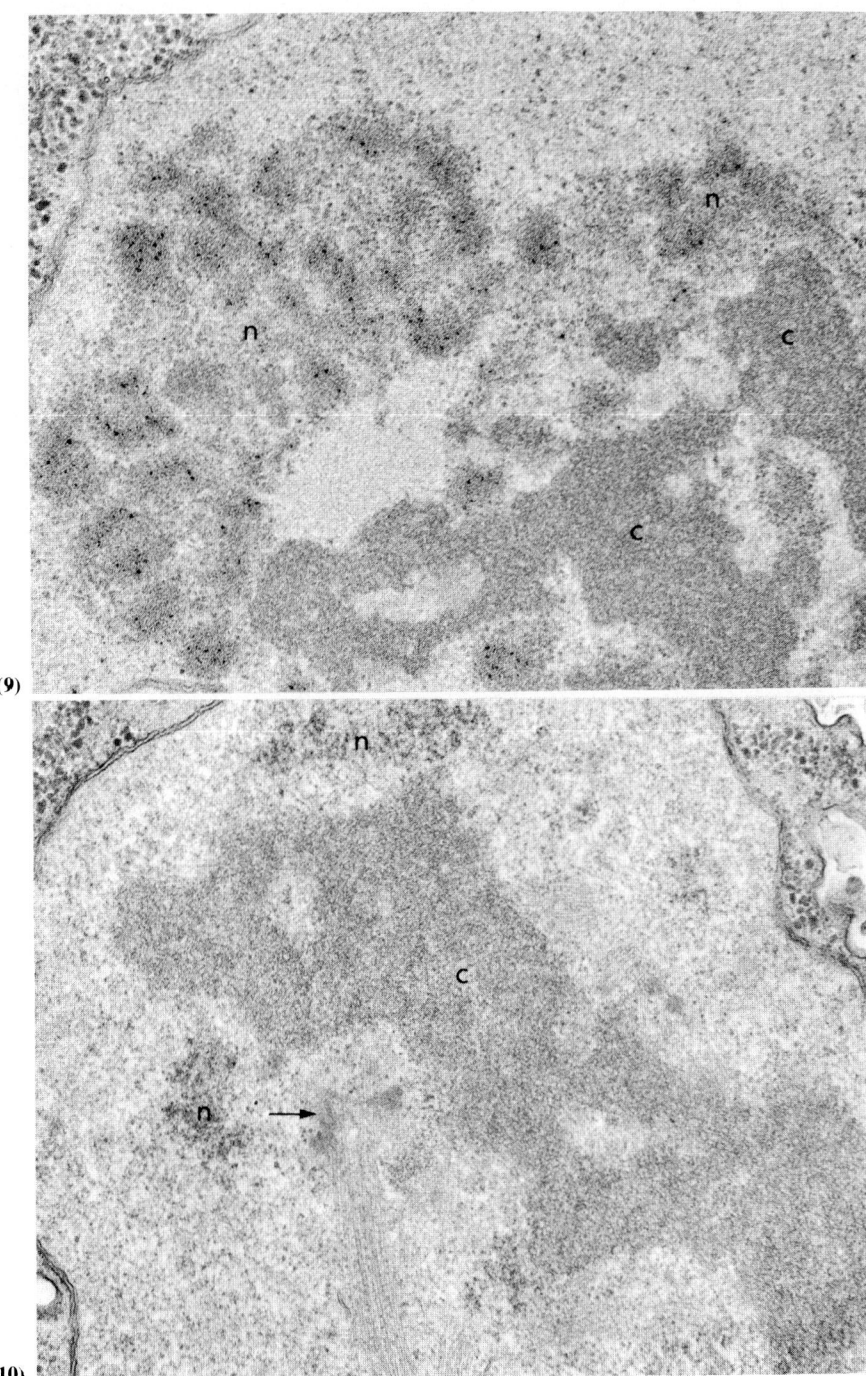

biochemical modifications of chromatin (Bradbury *et al.*, 1973; Chahal *et al.*, 1980). The condensed prometaphase chromosomes generally form complex aggregates, but segments of these are sometimes detected and are found to be roughly 0.25 μm in diameter. In certain cases, however, the chromosomes are so closely appressed as to form continuous masses exhibiting extremely irregular contours (Fig. 9). When they have become well aligned on the equator of the nucleus, the chromosomes are usually tightly grouped, and their constituent chromatids cannot be recognized (Fig. 10).

This lack of distinctiveness of mitotic chromosomes under electron microscopy contrasts sharply to their rodlike appearance in spread preparations (Mohberg *et al.*, 1973; Mohberg, 1974 and this volume, Chapter 7). This may be assumed to be partly due to the greater proximity of the chromosomes during mitotic stages and, as will be discussed in the following section, to be the result of the presence of an enveloping RNP containing matrix substance (Fig. 11).

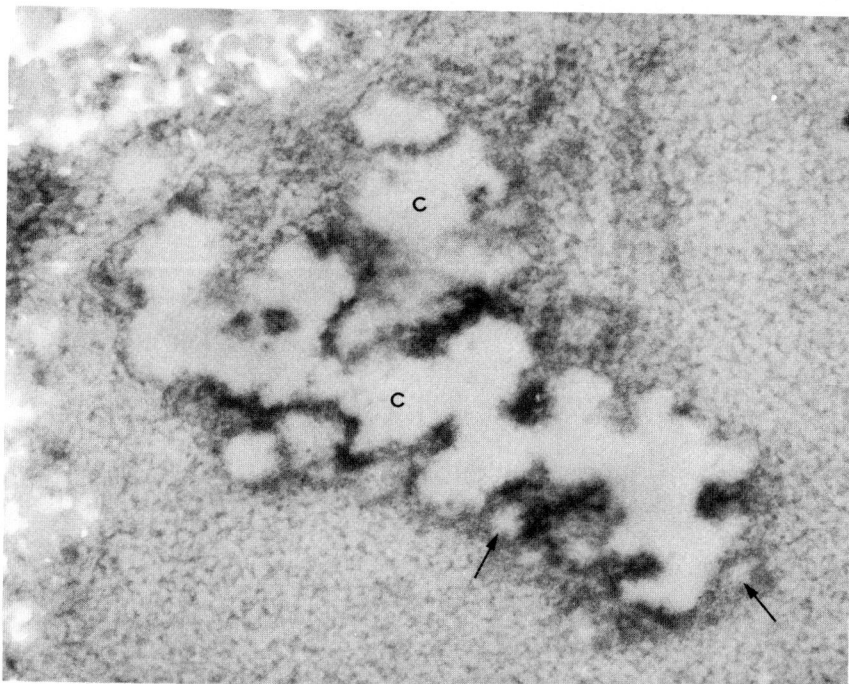

Fig. 11. Metaphase nucleus in a specimen stained with PTA. The chromosomes (c) appear to have aggregated into a rather continuous mass, the contours of which are delineated by an englobing, densely stained substance. The numerous small, light, circular areas mostly localized at the periphery of the chromosomes undoubtedly represent minichromosomes (arrows). The nucleoplasm shows a coarse fibrillogranular texture. Material fixed in formaldehyde. ×56,000. From Cadrin *et al.* (1981).

9. Nuclear Organization during the Cell Cycle

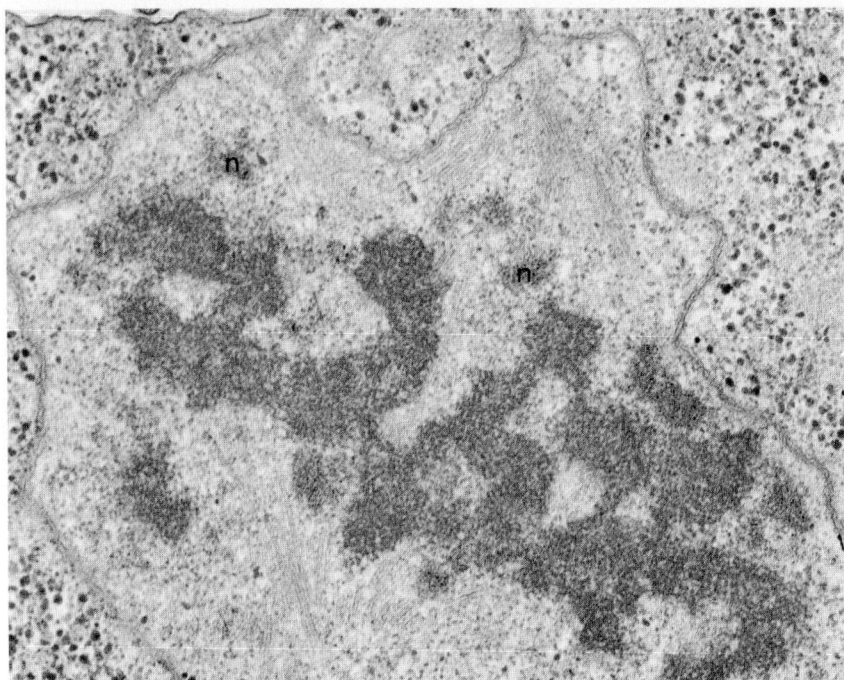

Fig. 12. In this early anaphase nucleus the chromosomes are just separating, and some of their segments may be recognized. Loose granular material and several nucleolar remnants (n) are present at the periphery of the chromosomes. In the upper pole region, bundles of microtubules are observed to converge in a pocketlike region of the nucleus. ×40,000.

As they start separating, the anaphase chromosomes are generally grouped into two rather regular plates (Fig. 12). Very soon thereafter, however, the chromosomes redistribute along the spindle and quite often form two more or less continuous masses moving toward the poles of the nucleus (Lord *et al.*, 1977). A better delineation of the contours of chromosomes is achieved, however, when anaphase nuclei are processed with the bismuth technique because of the more selective staining of the different structural elements present throughout the spindle area (Fig. 13). By late anaphase, the chromosomes and associated material have reached opposite poles of the nuclear cavity. As revealed by both electron microscopic (Guttes *et al.*, 1968; Ryser, 1970; Lord *et al.*, 1977) and phase contrast observations (Wolf *et al.*, 1979), the elongated telophase nucleus becomes a dumbbell in shape (Fig. 14). Within a very short period, the irregular nuclear envelope breaks down in its equatorial region, thus giving rise to the two daughter interphase nuclei.

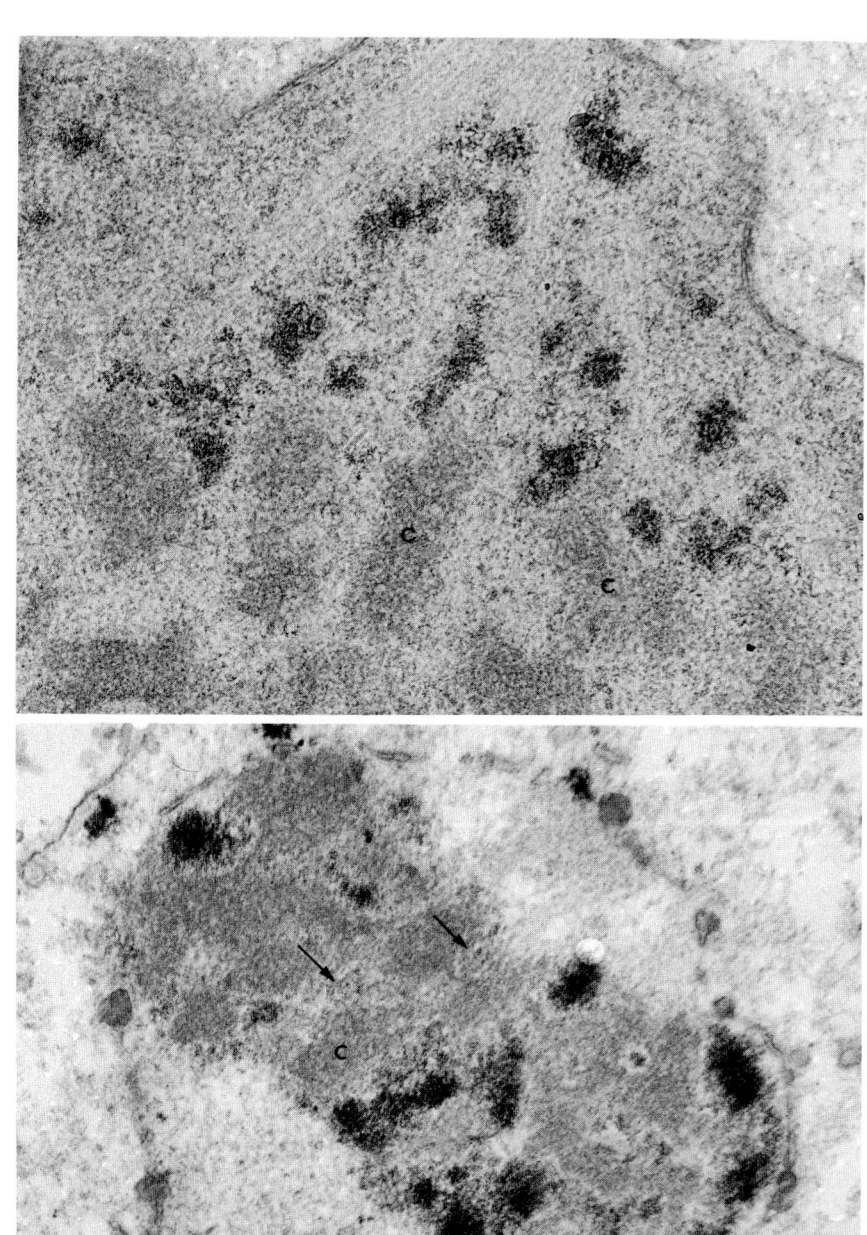

III. THE NUCLEOLAR CYCLE

A. Interphase

Given the unique features of the nucleolar cycle in *P. polycephalum*, we will first describe the ultrastructure of the mid-interphase nucleolus and then, for clarity of presentation, proceed with its disorganization at late prophase and its reconstruction during the first half of interphase.

In their original electron microscopic investigation of the mitotic cycle in this organism, Guttes *et al.* (1968) described the interphase nucleolus in double-fixed specimens as consisting essentially of a meshwork of fibers interspersed with conspicuous clusters of opaque particles. The authors also noted that these opaque granules are often grouped so as to suggest that they may be components of threadlike nucleolar subunits having a diameter of approximately 0.1 μm. Undoubtedly due to the lesser clumping provided by osmium tetroxide fixation alone, it was later observed that, apart from the fine fibrillar zones often studded with opaque particles, the interphase nucleolus also contains a pervasive granular substance. This substance corresponds to one of the key components present in this organelle in all organisms studied so far (Lord *et al.*, 1977). The opaque particles, it should be pointed out, are generally restricted to the fibrillar zones of animal and plant (Lafontaine and Lord, 1973) cell nucleoli and are also both denser and larger than the preribosomes forming the granular zones. When the opaque particles are absent (Fig. 6), the composite organization of the nucleolus can be still better analyzed. It is then evident that a large portion of the nucleolar mass exhibits a distinctly granular texture and that numerous smaller scattered areas are constituted of closely packed, fine, fibrillar material. Careful examination of these zones strongly suggests, moreover, that they consist of filamentous structures approximately 0.1 μm in diameter, some of which also appear to be folded into various types of globulelike formations. Judging from

---

Fig. 13. Portion of an anaphase nucleus slightly more advanced than that illustrated in Fig. 12. The two chromosome groups (c) are now well separated and, in this bismuth-stained specimen, they appear as rod-shaped structures approximately 0.25 μm in width. A number of density-stained nucleolar remnants (n) as well as patches of granular material are scattered throughout the polar region of the nucleus. Several bundles of microtubules may be recognized. ×51,000. From Cadrin *et al.* (1981).

Fig. 14. Micrograph of a telophase nucleus from a specimen stained with bismuth. Although the chromosomes (c) are closely grouped within the polar region of the nucleus, a thin layer of intervening lighter material may be detected between them (arrows). Several much denser globular bodies, the minichromosomes (m), are also observed. The nuclear envelope is in the process of pinching at the equator, thus giving a dumbbell appearance to the nucleus. The interzone consists predominantly of light fibrillar material throughout which a few granules are scattered. ×40,000. From Cadrin *et al.* (1981).

certain of the configurations observed, some of these structures appear to be spherical units approximately 0.5 μm in width and to consist of a denser outer shell surrounding a more transparent core. Others appear cup-shaped.

Following bismuth staining, which is believed to reveal nucleoproteins (Brown and Locke, 1978), the nucleolus exhibits a rather heterogeneous overall appearance (Fig. 15). This is because of the presence of numerous more intensely stained areas having a geometry quite similar to that of the coarse, threadlike and globular units seen in conventional preparations (Fig. 6). The remaining nucleolar zones show a density comparable to that of the chromocenters. Different results are obtained with the PTA staining technique depending on whether the material was fixed in formaldehyde or glutaraldehyde. In the case of glutaraldehyde-fixed specimens, the chromocenters and the nucleolus as a whole stain rather densely. Following formaldehyde fixation, however, the chromocenters and fibrillar zones of the nucleolus remain transparent. As a result, the nucleolus exhibits a quite heterogeneous and rather unexpected organization (Fig. 16). Its fibrillar regions appear elongated, crescent-shaped, or circular, and because of the presence of a central denser portion, some of these zones take on a doughnutlike appearance. The remaining nucleolar regions appear extremely complex in structure. Following counterstaining with uranyl acetate–lead citrate, the fibrillar nucleolar areas as well as the chromocenters become as dense as in conventional preparations. Therefore, it is evident that the observed transparency of chromatin is not due to selective extraction of material during fixation with formaldehyde.

Confirmation that these PTA-negative nucleolar areas contain chromatin is obtained by exposing glutaraldehyde-fixed specimens to EDTA. Under these experimental conditions, the bleached nucleolar regions are similarly distributed throughout the nucleolar mass and closely resemble in size and shape the PTA-negative zones. Considering that these globular fibrillar zones have also been found by high-resolution radioautography to incorporate thymidine (Ryser *et al.*, 1973), there can be little doubt that they consist predominantly of chromatin and

---

Fig. 15. Portion of an interphase nucleus. Following bismuth staining, the nucleolus is seen to consist of numerous dense minichromosomes (m) and of a pervading lighter material which undoubtedly corresponds to the granular nucleolar substance observed in conventional preparations. The chromocenters (c) are also quite clearly delineated by this stain, as are certain particulate elements throughout the nucleoplasm. ×59,000. From Cadrin *et al.* (1981).

Fig. 16. Micrograph of a nucleus stained with PTA. The chromocenters (c) remain rather transparent following staining and their contours are delineated by fibrillar and granular nucleoplasmic components. The nucleolar minichromosomes (m) are either globular or crescent-shaped and are clearly seen to be made of an outer layer of dense material surrounding a light portion. Certain minichromosomes also appear to contain a dense central core. The zones between the minichromosomes are complex in appearance and seem to consist mostly of alveolated, filamentous structures. ×54,000. From Cadrin *et al.* (1981).

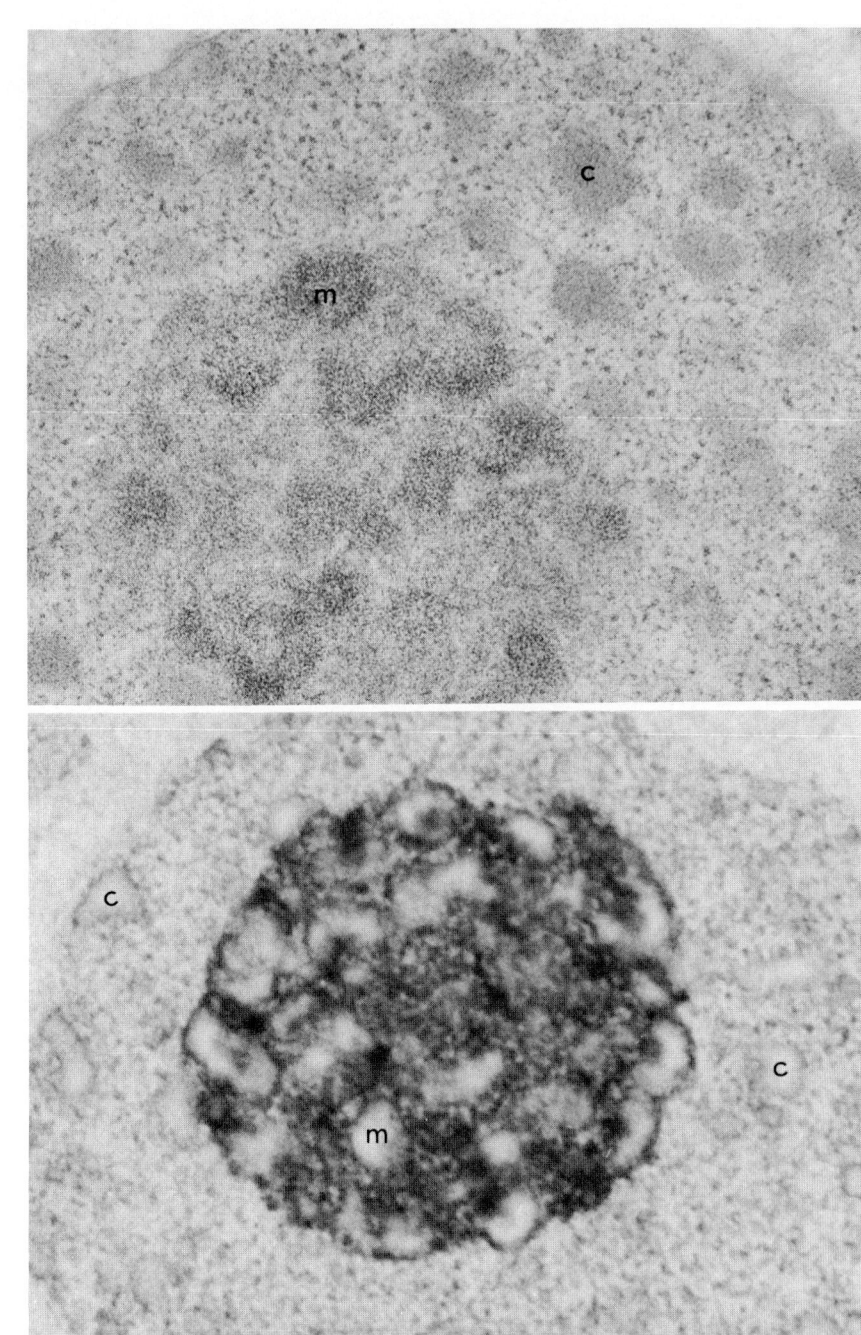

that they correspond, therefore, to the minichromosomes characterized biochemically by Seebeck *et al.* (1979).

Other cytochemical tests were also exploited to gain further insight into nucleolar organization. Deoxyribonuclease has unfortunately proved rather unpredictable, as also reported by other authors in the case of both animal (Bouteille *et al.*, 1974) and plant cells (Luck and Lafontaine, 1980). Digestion with ribonuclease, however, proved much more effective. After a 4-hour hydrolysis of formaldehyde-fixed specimens, the nucleolus exhibits a coarse, dense reticulum, the interstices of which are occupied by loose fibrillogranular material. It is evident from these results that this enzyme induces an unraveling of the minichromosomes, which take on an elongated configuration, as well as of the RNP particles present throughout the granular zones, which transform into fine, twisted fibrils. Following pepsin hydrolysis, the granular nucleolar regions and the chromocenters are more rapidly extracted than the minichromosomes. In the case of proteinase K digestion, the granular nucleolar portions are first extracted and the chromocenters show the greatest resistance to the action of this enzyme. These variations in enzymatic susceptibility of the minichromosomes most likely indicate that their protein content is markedly different from that of the chromocenters. Biochemical data indicating that nucleolar chromatin is, indeed, different from the remainder of the chromatin have been reported for *P. polycephalum* (Allfrey *et al.*, 1978), as well as for various other organisms (Hubbell *et al.*, 1979; Lischwe *et al.*, 1979).

During later interphase, the volume of the nucleolus increases rapidly (Bovey and Ruch, 1972), resulting undoubtedly from the increased synthesis of both nucleolar DNA (Guttes and Guttes, 1969; Newlon *et al.*, 1973) and ribosomal RNA (Hall and Turnock, 1976).

B. Mitotic Stages

At the beginning of prophase, the nucleolus is known from phase contrast observations (Guttes *et al.*, 1961; Wolf *et al.* 1979) to migrate toward the periphery of the nucleus. Sometime before it disperses at late prophase, this organelle undergoes progressive loosening of its mass, so that its minichromosomal components become more and more conspicuous (Figs. 7, 8). By late prophase, the granular component of the nucleolus becomes less abundant and the crescentlike nucleolar mass takes on the appearance of a coarse reticulum. Close examination reveals that, apart from the numerous minichromosomes, this loose network also consists of convoluted, coarse, filamentous elements approximately 0.1 μm in diameter and exhibiting a distinctly fibrillogranular texture.

In order to characterize in greater detail the various stages of nucleolar disorganization and to identify better the different materials from this organelle which

9. Nuclear Organization during the Cell Cycle

persist during the mitotic stages, appropriate plasmodial samples were processed according to the bismuth, PTA, and EDTA techniques. Bismuth staining was found to be particularly suitable for revealing the distribution of the two main structural components of the nucleolus at the time this organelle shows a loose organization (Fig. 17). It is observed in such preparations that, even though the minichromosomes have started moving apart, they are still embedded in a continuous mass of light material which is much more conspicuous than in conventional preparations. Part of this intervening substance is present throughout the nucleolar mass up to late prophase. In PTA-stained specimens (Fig. 18), this material takes on a more complex appearance and resembles that observed at interphase (Fig. 16). As for the now well-separated minichromosomes, their contours are more clearly delimited than at earlier stages, and in formalin-fixed specimens, they are better seen to consist of a densely stained, thin outer portion surrounding a light central zone.

During the prometaphase period, the organization and distribution of the nucleolar material vary significantly from nucleus to nucleus in a given plasmodial sample. In certain nuclei the persisting minichromosomes are grouped into bead-like aggregates, associated with a diffuse coating of granular material, which are located on one side of the nucleus (Fig. 9). In neighboring nuclei, most of these nucleolar remnants have separated into distinct units and have scattered throughout the outer portion of the nuclear cavity.

By metaphase, numerous nucleolar remnants are observed throughout the peripheral portion of the nucleus and in still greater concentration, perhaps, in the region between the nuclear poles and the chromosome plate (Fig. 10). Both high-resolution radioautographic (Lord *et al.*, 1977) and ultracytochemical data (Ploton and Gontcharoff, 1979) indicate that these persisting bodies contain chromatin. When metaphase nuclei are stained with bismuth, one also notes, apart from the intensely reacting minichromosomes, the presence of numerous small, irregular aggregates of fibrillogranular material which give the nucleoplasm a very coarse texture. Although a larger concentration of this material, presumably of nucleolar origin, is observed in proximity to the peripheral minichromosomes, some of it is also present around the centrally located chromosomes as well as between them. That the bismuth staining material present between the chromosomes is not only rich in basic proteins but also contains significant amounts of RNA is demonstrated by its intense reaction with DAB following deoxyribonuclease digestion. The presence of a permeating basic protein-containing substance throughout the metaphase plate is also convincingly demonstrated using the PTA technique (Fig. 11).

At the beginning of anaphase, a larger concentration of minichromosomes is still observed on the sides of the chromosome plates facing the nuclear poles (Fig. 12). As anaphase progresses, the chromosomes redistribute throughout a larger portion of the spindle area, and many nucleolar fragments are then found

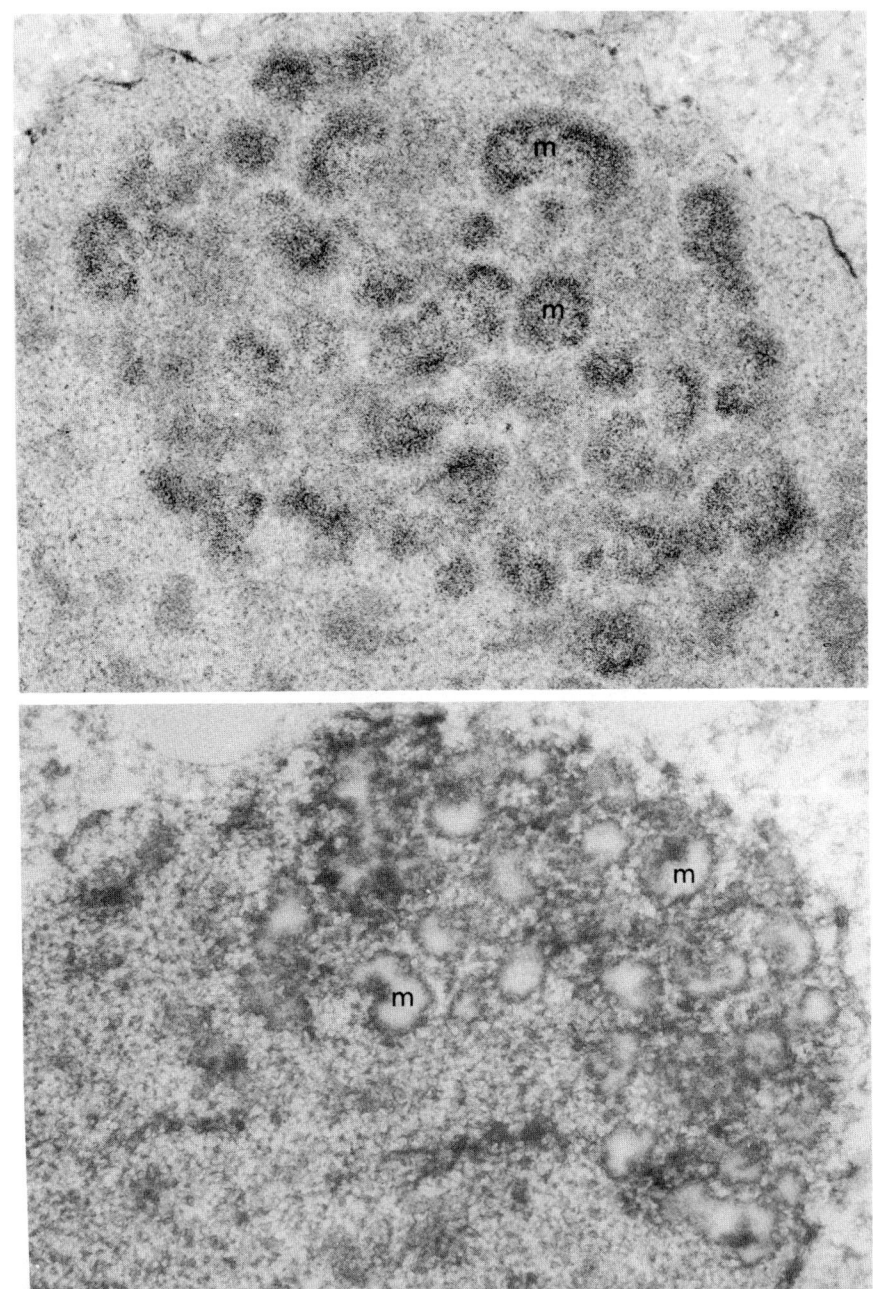

trapped within the more or less continuous mass of chromatin thus formed. Appropriate planes of sectioning show that numerous nucleolar remnants are still also scattered at the periphery of the two chromatin masses as well as throughout the polar regions of the spindle (Fig. 13). By the end of anaphase, however, the majority of the minichromosomes and most of the diffuse RNP-containing material are instead localized through the distal portion of the two polar masses of chromatin (Cadrin et al., 1981). A similar distribution of material is observed during telophase (Fig. 14).

C. Nucleologenesis

Very early in interphase, as indicated in Section I, material within the nucleus is redistributed and the fibrillar nucleolar remnants, or minichromosomes, are then observed to disperse throughout the nuclear cavity. Examination of such nuclei also reveals the presence of a pervading fibrillogranular substance (Fig. 1) which, judging from its reaction to both bismuth and EDTA (Lafontaine et al., 1981), consists predominantly of RNP. Early phase contrast observations (Guttes et al., 1961), as well as subsequent ultrastructural investigations (Guttes et al., 1968; Goodman and Ritter, 1969; Lord et al., 1977), have revealed that, during the first 10 to 15 minutes of interphase, the nucleolar remnants rapidly coalesce into a few irregular masses often referred to as the "prenucleolar bodies." These studies did not indicate, however, whether the preexisting nucleolar remnants are the only material participating in the early growth of these prenucleolar masses. In view of the presence of noticeable amounts of pervading fibrillogranular material in early interphase nuclei, its possible role in nucleologenesis must be considered. This issue is especially relevant since the enlarging prenucleolar masses acquire a granular coating (Fig. 4) just about the time the surrounding fibrillogranular material starts disappearing. To account for the progressive accumulation of granular material at the periphery of the prenucleolar masses, it may be assumed that this material does not totally, or not at all, originate from the preexisting particulate substance that pervades the early interphase nucleus (Figs. 1, 3). It is, indeed, known from biochemical investigations that ribosomal RNA synthesis begins immediately following mitosis (Nygaard, et al., 1960; Sachsenmaier and Rusch, 1964; Schiebel, 1973).

Fig. 17. Late prophase nucleus in a specimen stained with bismuth. The nucleolus shows a looser organization than in interphase and, apart from the numerous densely stained minichromosomes (m), is seen still to consist of relatively large zones of intervening material. The nucleoplasm exhibits a noticeably more granular texture than in early prophase. ×50,000. From Cadrin et al. (1981).

Fig. 18. Portion of a slightly more advanced prophase nucleus than that depicted in Fig. 17. Following staining with PTA, the late prophase nucleolus is seen to contain mostly minichromosomes (m) and intervening material now present in reduced quantity. The nucleoplasm shows a much coarser texture than during earlier stages. ×40,000. From Cadrin et al. (1981).

High-resolution radioautographic observations have also revealed that at early interphase, uridine incorporation is taking place mostly within the minichromosomes (Lafontaine et al., 1981). It is, therefore, most likely that the peripheral granular material characterizing the prenucleolar masses some 15 minutes after metaphase results, at least partly, from the elaboration of preribosomes within the agglomerating minichromosomes and not necessarily from coalescence of the preexisting fibrillogranular substance.

By the time large nucleoplasmic zones have formed, the pervading fibrillogranular substance has completely disappeared, and the prenucleolar bodies have fused into one or two larger masses which are typically horseshoe in shape (Fig. 5). These nucleolar masses are rather heterogeneous in structure and, apart from a thin, irregular granular layer, they consist mostly of closely packed minichromosomes. As the initially scattered minichromosomes fuse into prenucleolar bodies, which in turn give rise to a single nucleolus, there is a marked increase in the number of radioautographic grains over these organelles following exposure to tritiated uridine (Rué and Gontcharoff, 1976; Lafontaine et al., 1981). These results agree with biochemical data indicating that the activity of ribosomal RNA synthesis increases as interphase proceeds (Hall and Turnock, 1976).

It is evident, from the above electron microscopic observations, that the nucleolar remnants play a key role in nucleolar formation. That this should be so is easy to conceive since these small bodies correspond to distinct fragments of the dispersing late prophase nucleolus, which may be shown by high-resolution radioautography as well as by various cytochemical techniques to consist of the same major components as chromatin. It is most reasonable to assume, therefore, that these remnants are biochemically identical to the minichromosomes isolated from mature interphase nucleoli and therefore contain the ribosomal genes (Seebeck et al., 1979). Contrary to the situation which prevails in higher organisms, in which only a small portion of the nucleolar mass is preserved during the mitotic stages, presumably in the form of specific chromosomal segments known as "nucleolar secondary constrictions" (discussed in Lafontaine and Lord 1974), the *P. polycephalum* minichromosomes represent well-defined nucleolar portions, the morphological integrity of which appears to be maintained during these stages. In view of the incorporation of these numerous functionally active minichromosomes in the daughter nuclei, nucleologenesis is initiated at the beginning of interphase and is, undoubtedly, not limited to a simple aggregation of these globular bodies but must also involve the participation of newly formed material, such as RNA and proteins. Considered as a whole, the process of nucleolar formation in *P. polycephalum* is thus intermediate between that which is observed in most eukaryotes (reviewed in Lafontaine, 1974; Giménez-Martín et al., 1977) and that which has been reported in certain organisms in which the nucleolus persists as such during the mitotic stages (Leedale, 1966, 1968; Chaly et al., 1977).

IV. THE MITOTIC APPARATUS

A distinctive feature of the cell cycle in *P. polycephalum* is the formation of microtubules as early as the beginning of prophase. In a first ultrastructural investigation, Guttes *et al.* (1968) noted that a spindle pole appears in early prophase as a small, plaquelike bundle of fibrous material close to the periphery of the disintegrating nucleolus. It was later reported (Sakai and Shigenaga, 1972) that one or two dense, rather amorphous areas associated with fibrous and granular material are present still earlier in prophase in the center of the nucleus. When two amorphous regions are observed, they are adjacent to each other. The number of elongating microtubules associated with amorphous centers was found by these latter authors to increase during prophase, whereas the centers themselves moved away from each other and also became progressively smaller. From these observations, as well as from more recent ones (Laane and Haugli, 1974; Wille and Steffens, 1979), it appears that one such center is first formed in early prophase and divides subsequently into two units, the microtubule-organizing regions.

Other investigators (Blessing, 1972; Tanaka, 1973) have uncovered a similar amorphous structure that is associated with bundles of radiating microtubules and projects into a cup-shaped depression of the early prophase nucleolus. This characteristic disposition of the microtubule-organizing center on one side of the nucleolus persists during subsequent growth of the spindle. At late prophase, however, once the nucleolus has started dispersing, microtubules form a double cone pointing to opposite sides of the nuclear cavity. According to Wille and Steffens (1979), the duplicated mitotic centers have almost completely moved to the poles of the spindle, which then appears to be formed exclusively of a few interpolar microtubules grouped into small bundles. The above observations have led to the interesting suggestion that the unusual spatial relationship of the spindle with the nucleolus is related to movement of this organelle during prophase (Blessing, 1972). If this interpretation is correct, the pressure of the elongating spindle on one side of the nucleolus would account for the displacement of this organelle during prophase and for its eventual change in shape as it becomes closely appressed against the nuclear envelope. Recent investigations using griseofulvin, a substance known to interfere with microtubule formation, have revealed that, under appropriate conditions, this drug also prevents the appearance of the tubule-organizing region. As a result, no spindle is formed and a mitotic delay occurs. Interestingly, nucleolar migration is also affected (Hebert *et al.*, 1980).

By prometaphase, favorable preparations reveal the presence of an elongated, slightly curved structure, 0.05–0.07 μm in thickness, consisting of fine fibrillar material which stains less intensely than either the chromosomes or the nucleolar remnants, Quite similar formations are sometimes observed at the periphery of

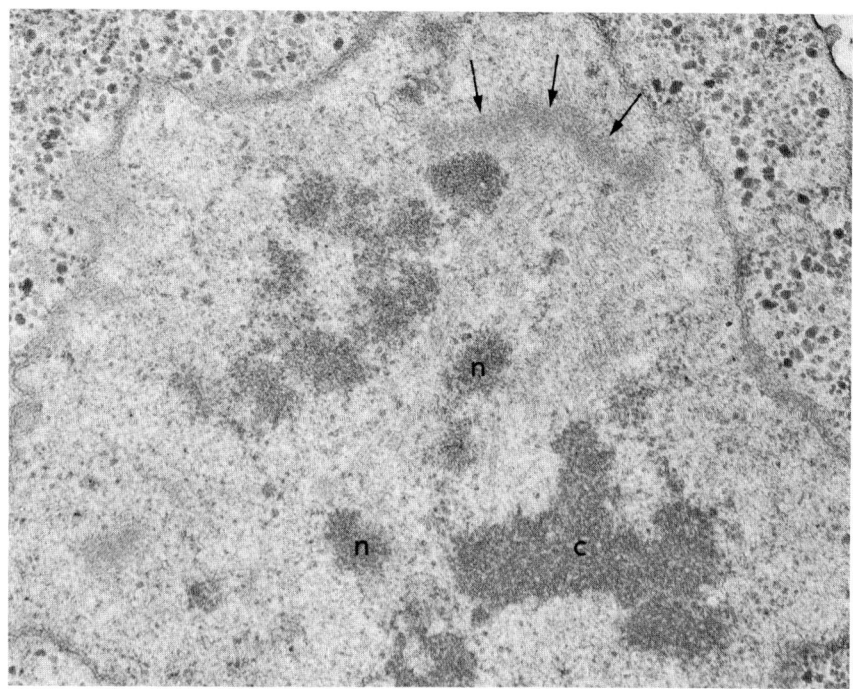

Fig. 19. Micrograph of a portion of a prometaphase nucleus. Besides profiles of chromosomes (c) and several nucleolar remnants (n), one notes the presence of an elongated, slightly curved structure in the polar region of the nucleus (arrows). A number of microtubules are seen converging toward this structure, and cross sections of several others may be detected in its immediate neighborhood. ×50,000.

the still closely grouped nucleolar remnants but at some distance from the nuclear envelope (Fig. 19). When these remnants have sufficiently dispersed, numerous tightly assembled microtubules may be seen to radiate from the concave side of this filamentous structure and to extend toward the central region of the nuclear cavity. The above observations are compatible with the view that the microtubule-organizing center originates from the nucleolus and become visible in the nucleoplasm shortly after dispersion of this organelle at prometaphase (Laane and Haugli, 1974).

Once the chromosomes have aligned on the metaphase plate, apart from the interpolar microtubules, similar elements converging on the kinetochores may also be observed. The kinetochores in *P. polycephalum* have been reported to be disc-shaped (Ryser, 1970) or platelike (Lord *et al.*, 1977; Wille and Steffens, 1979), and only a few microtubules seem to arise from the outer unit of these bilayered structures. In appropriate sections, pairs of kinetochores are seen to be regularly placed along the metaphase plate, the distance between each paired unit being approximately 0.2 μm. Preliminary observations in this laboratory of metaphase nuclei digested with deoxyribonuclease and stained with DAB reveal that kinetochores consist partly of RNA. Based on their intense reaction with DAB under these experimental conditions, anaphase and telophase kinetochores also appear to contain a relatively high concentration of RNA. Similar cytochemical findings were reported for a variety of animal and plant cells (Braselton, 1975; Esponda, 1978; Rieder, 1979) processed according to Bernhard's (1969) EDTA technique.

As in late prophase, not all metaphase microtubules are oriented along the spindle axis. In the polar regions, at least, both longitudinal and cross sections of these structural elements are observed. Judging from certain of our micrographs of metaphase nuclei, the reason is that microtubules are sometimes organized in a whorllike fashion in the polar portions of the spindle.

At anaphase, the nucleus elongates and becomes very irregular in contour as the two groups of chromosomes gradually move apart. Microtubule bundles are typically found to converge on a broad area in opposite pocketlike projections of the envelope. As also appears to be true of plasmodial nuclei of related species (Aldrich, 1969), no structural differentiation of the envelope at these focal areas has been observed, contrary to the situation prevailing in a variety of other organisms with an intranuclear spindle (Robinow and Marak, 1966; Luykx, 1970; Moens and Rapport, 1971; Aist and Williams, 1972; McCully and Robinow, 1976). Although interzonal microtubules were not observed in earlier studies (Guttes *et al.*, 1968; Goodman and Ritter, 1969; Lord *et al.* 1977), this is apparently because they are more unstable, even in the presence of glutaraldehyde, than those attached to the kinetochores (Wille and Steffens, 1979). As the chromosome sets reach the poles of the nucleus, the apex of the spindle appears to push against the nuclear envelope, which is commonly seen to become discontinuous in this region. Concomitantly, there is a rapid breakdown of the microtubules in the interzonal region of the late anaphase nucleus.

The equatorial portion of the anaphase nucleus takes on a very different appearance depending on the mode of preservation and staining used. In osmium tetroxide-fixed material, it is characterized by loosely distributed particles of the type seen in the protoplasm (Lord *et al.* 1977). Although no cytochemical analysis of these particles has ever been carried out, most of them presumably consist of glycogen (Goodman and Rusch, 1969; Aldrich and Blackwell, 1976), whereas the remaining smaller ones are ribosomes. According to the observations of Guttes *et al.* (1968) on double-fixed specimens, the equatorial nuclear region is similar to the cytoplasm in texture and also exhibits a central body composed of fibers and of amorphous material resembling that seen in proximity to the poles of the developing late prophase spindle. Wille and Steffens (1979) report that this nuclear region contains short, closely aggregated bundles of microtubules and that, at telophase, these depolymerizing elements surround a conspicuous mid-body. A further indication that the interzone of anaphase and telophase nuclei possesses still unresolved ultrastructural characteristics is furnished by the examination of specimens stained with PTA. In such preparations the central portion of the nucleus exhibits a coarse fibrillogranular texture and, in approximate sections, is seen to contain a dense, irregular body having a honeycomb organization.

ACKNOWLEDGMENTS

Preparation of this chapter was aided by research grants from the Quebec Ministry of Education and the Natural Sciences and Engineering Research Council (Canada). The authors wish to thank Dr. A. Lord, who participated in the early phase of our work on *Physarum*. We are also particularly grateful to Dr. B. T. Luck for his valuable advice on some of the cytochemical techniques used in our studies. The competent technical assistance of Mrs. D. Michaud and Mr. S. Gugg is gratefully acknowledged.

REFERENCES

Aist, J. R., and Williams, P. H. (1972). Ultrastructure and time course of mitosis in the fungus *Fusarium oxysporum. J. Cell Biol.* **55,** 368–389.
Aldrich, H. C. (1969). The ultrastructure of mitosis in myxamoebae and plasmodia of *Physarum flavicomum. Am. J. Bot.* **56,** 290–299.
Aldrich, H. C., and Blackwell, M. (1976). Resistant structures in the Myxomycetes. *In* "The Fungal Spore, Form and Function" (D. J. Weber and W. M. Hess, eds.), pp. 413–461. Wiley, New York.
Allfrey, V. G., Johnson, E. M., Sun, I. Y. C., Littau, V. C., Matthews, H. R., and Bradbury, E. M. (1978). Structural organization and control of the ribosomal genes in *Physarum. Cold Spring Harbor Symp. Quant. Biol.* **42,** 505–514.
Anteunis, A., Pouchelet, M., Robineaux, R., and Vial, M. (1973). Ultrastructure des acides nucléiques et en particulier de l'ADN nucléolaire des L 929, après coloration par la 3-3'-diaminobenzidine (DAB) oxydée. *C. R. Hebd. Seances Acad. Sci., Ser. D* **277,** 1169–1171.
Bernhard, W. (1969). A new staining procedure for electron microscopical cytology. *J. Ultrastruct. Res.* **27,** 250–265.
Blessing, J. (1972). Ultrastructural changes of the nucleolus from the end of G_2-phase until prometaphase in *Physarum polycephalum. Cytobiologie* **6,** 342–350.
Bouteille, M., Laval, M., and Dupuy-Coin, A. M. (1974). Localization of nuclear functions as revealed by ultrastructural autoradiography and cytochemistry. *In* "The Cell Nucleus" (H. Busch, ed.), Vol. I, pp. 3–71. Academic Press, New York.
Bovey, F., and Ruch, F. (1972). Cytofluorometric determination of DNA and protein in the nuclei and nucleoli of *Physarum polycephalum* during the intermitotic period. *Histochemie* **32,** 153–162.
Bradbury, E. M., Inglis, R. J., Matthews, H. R., and Sarner, N. (1973). Phosphorylation of very lysine rich histone in *Physarum polycephalum:* Correlation with chromosome condensation. *Eur. J. Biochem.* **33,** 131–139.
Braselton, J. P. (1975). Ribonucleoprotein staining of *Allium cepa* kinetochores. *Cytobiologie* **12,** 148–151.

Brown, G. L., and Locke, M. (1978). Nucleoprotein localization by bismuth staining. *Tissue Cell* **10**, 365-388.

Cadrin, M., Lord, A., and Lafontaine, J. G. (1981). The nucleolar cycle in the myxomycete *Physarum polycephalum*. I. Ultracytochemical characteristics of the mature nucleolus and of the persisting nucleolar material during the mitotic stages. *Can. J. Bot.* **59**, 1134-1147.

Chahal, S. S., Matthews, H. R., and Bradbury, E. M. (1980). Acetylation of histone H4 and its role in chromatin structure and function. Nature *(London)* **287**, 76-79.

Chaly, N., Lord, A., and Lafontaine, J. G. (1977). A light- and electron-microscope study of nuclear structure throughout the cell cycle in the euglenoid *Astasia longa*, (Jahn). *J. Cell Sci.* **27**, 23-45.

Esponda, P. (1978). Cytochemistry of kinetochores under electron microscopy. *Exp. Cell Res.* **114**, 247-252.

Giménez-Martín, G., de la Torre, C., Lopez-Saez, J. F., and Esponda, P. (1977). Plant nucleolus: Structure and physiology *Cytobiologie* **14**, 421-462.

Goodman, E. M., and Rusch, H. P. (1969). Glycogen in *Physarum polycephalum*. *Experientia* **25**, 580.

Goodman, E. M. (1980). *Physarum polycephalum:* A review of a model system using a structure-function approach. *Int. Rev. Cytol.* **63**, 1-58.

Goodman, E. M., and Ritter, H. (1969). Plasmodial mitosis in *Physarum polycephalum*. *Arch. Protistenkd.* **111**, 161-169.

Guttes, E., and Guttes, S. (1969). Replication of nucleolus-associated DNA during 'G$_2$ phase' in *Physarum polycephalum*. *J. Cell Biol.* **43**, 229-236.

Guttes, E., Guttes, S., and Rusch, H. P. (1961). Morphological observations on growth and differentiation of *Physarum polycephalum* grown in pure cultures. *Dev. Biol.* **3**, 588-614.

Guttes, S., Guttes, E, and Ellis, R. A. (1968). Electron microscope study of mitosis in *Physarum polycephalum*. *J. Ultrastruct. Res.* **22**, 509-529.

Hall, L., and Turnock, G. (1976). Synthesis of ribosomal RNA during the mitotic cycle in the slime mould *Physarum polycephalum*. *Eur. J. Biochem.* **62**, 471-477.

Hebert, C. D., Steffens, W. L., and Wille, J. J. (1980). The role of spindle microtubule assembly in the control of mitotic timing in *Physarum:* Induction of a novel type of tubular structure by griseofulvin treatment. *Exp. Cell Res.* **126**, 1-13.

Hubbell, H. R., Rothblum, L. I., and Hsu, T. C. (1979). Identification of a silver binding protein associated with the cytological silver staining of actively transcribing nucleolar regions. *Cell Biol. Int. Rep.* **3**, 615-622.

Kuroiwa, T. (1973). Fine structure of interphase nuclei. VI. Initiation and replication sites of DNA synthesis in the nuclei of *Physarum polycephalum* as revealed by electron microscopy autoradiography. *Chromosoma* **44**, 291-299.

Laane, M. M., and Haugli, F. B. (1974). Division centres in mitotic nuclei of *Physarum polycephalum* plasmodia. *Norw. J. Bot.* **21**, 309-318.

Lafontaine, J. G. (1974). Ultrastructural organization of plant cell nuclei. *In* "The Nucleus" (H. Busch, ed.), Vol. I, pp. 149-185. Academic Press, New York.

Lafontaine, J. G., and Lord, A. (1973). An ultrastructural and radioautographic investigation of the nucleolonemal component of plant interphase nucleoli. *J. Cell Sci.* **12**, 369-383.

Lafontaine, J. G., and Lord, A. (1974). A correlated light- and electron-microscope investigation of the structural evolution of the nucleolus during the cell cycle in plant meristematic cells (*Allium porrum*). *J. Cell Sci.* **16**, 63-93.

Lafontaine, J. G., Cadrin, M., and Lord, A. (1981). The nucleolar cycle in the myxomycete *Physarum polycephalum*. II. Mode of formation of the nucleolus. *Can. J. Bot.* **59**, 1148-1158.

Leedale, G. F. (1966). The Euglenophyceae. *In* "The Chromosomes of the Algae" (M. B. E. Godward, ed.) pp. 78-95. Arnold, London.

Leedale, G. F. (1968) The nucleus of *Euglena*. *In* "The Biology of *Euglena*" (D. E. Buetow, ed.), Vol. I, pp. 185-242. Academic Press, New York.

Lischwe, M. A., Smetana, K., Olson, M. O. J., and Busch, H. (1979). Proteins C 23 and B 23 are the major nucleolar silver staining proteins. *Life Sci.* **25**, 701-708.

Locke, M., and Huie, P. (1977). Bismuth staining for light and electron microscopy. *Tissue Cell* **9**, 347-371.

Lord, A., and Lafontaine, J. G. (1976). An ultrastructural and radioautographic study of the chromocentric interphase nucleus in plant meristematic cells (*Raphanus sativus*). *J. Cell Sci.* **21**, 193-207.

Lord, A., Nicole, L., and Lafontaine, J. G. (1977). Ultrastructural and radioautographic investigation of the nucleolar cycle in *Physarum polycephalum:* Characterization of DNA -containing subunits. *J. Cell Sci.* **23**, 25-42.

Luck, B. T., and Lafontaine, J. G. (1980). An ultracytochemical study of nucleolar organization in meristematic plant cells (*Allium porrum*). *J. Cell Sci.* **43**, 37-58.

Luykx, P. (1970). Cellular mechanisms of chromosome distribution. *Int. Rev. Cytol., Suppl.* **2**, 1-171.

McCully, E. K., and Robinow, C. F. (1976). Mitosis in the fission yeast *Schizosaccharomyces pombe:* A comparative study with light and electron microscopy. *J. Cell Sci.* **9**, 475-507.

Moens, P. B., and Rapport, E. (1971). Spindles, spindle plaques, and meiosis in the yeast *Saccharomyces cerevisiae* (Hansen). *J. Cell Biol.* **50**, 344-361.

Mohberg, J. (1974). The nucleus of the plasmodial slime molds. *In* "The Cell Nucleus" (H. Busch, ed.), Vol. I, pp. 187-218. Academic Press, New York.

Mohberg, J., Babcock, K. L., Haugli, F. B., and Rusch, H. P. (1973). Nuclear DNA content and chromosome numbers in the myxomycete *Physarum polycephalum. Dev. Biol.* **34**, 228-245.

Moyne, G., Bertaux, O., and Puvion, E. (1975). The nucleus of *Euglena*. I. An ultracytochemical study of the nucleic acids and nucleoproteins of synchronized *Euglena gracilis* Z. *J. Ultrastruct. Res.* **52**, 362-376.

Newlon, C. S., Sonenshein, G. E., and Holt, C. E. (1973). Time of synthesis of genes for ribosomal ribonucleic acid in *Physarum. Biochemistry* **12**, 2338-2345.

Nygaard, O. F., Guttes, S., and Rusch, H. P. (1960). Nucleic acid metabolism in a slime mold with synchronous mitosis. *Biochim. Biophys. Acta* **38**, 298-306.

Ploton, D., and Gontcharoff, M. (1979). Etude ultrastructurale du cycle nucléolaire de *Physarum polycephalum* après mise en évidence préférentielle des RNP par la réaction régressive à l'EDTA. *Exp. Cell Res.* **118**, 418-423.

Puvion, E., and Bernhard, W. (1975). Ribonucleoprotein components in liver cell nuclei as visualized by cryoultramicrotomy. *J. Cell Biol.* **67**, 200-214.

Rieder, C. L. (1979). Localization of ribonucleoprotein in the trilaminar kinetochore of Pt K_1. *J. Ultrastruct. Res.* **66**, 109-119.

Ris, H. (1962). Interpretation of ultrastructure in the cell nucleus. *Symp. Int. Soc. Cell Biol.* **1**, 69-88.

Ris, H., and Korenberg, J. (1979). Chromosome structure and levels of chromosome organization. *In* "Cell Biology: A Comprehensive Treatise" (D. M. Prescott and L. Goldstein, eds.), Vol. 2, pp. 267-361. Academic Press, New York.

Robinow, C. F., and Marak, J. (1966). A fiber apparatus in the nucleus of the yeast cell. *J. Cell Biol.* **29**, 129-151.

Rué, G., and Gontcharoff, M. (1976). Etude radioautographique du métabolisme de l'ARN ribosomal au cours de la mitose du myxomycète *Physarum polycephalum. C. R. Hebd. Seances Acad. Sci., Ser. D* **283**, 829-831.

Ryser, U. (1970). Die Ultrastruktur der Mitosekerne in den Plasmodien von *Physarum polycephalum. Z. Zellforsch. Mikrosk. Ana.* **110**, 108-130.

9. Nuclear Organization during the Cell Cycle

Ryser, U., Fakan, S., and Braun, R. (1973). Localization of ribosomal RNA genes by high resolution autoradiography. *Exp. Cell Res.* **78,** 89–97.

Sachsenmaier, W., and Rusch, H. P. (1964). The effect of 5-fluoro-2' deoxyuridine on synchronous mitosis in *Physarum polycephalum. Exp. Cell Res.* **36,** 124–133.

Sakai, A., and Shigenaga, M. (1972). Electron microscopy of dividing cells. IV. Behaviour of spindle microtubules during nuclear division in the plasmodium of the myxomycete, *Physarum polycephalum.* Chromosoma **37,** 101–116.

Schiebel, W. (1973). The cell cycle of *Physarum polycephalum. Ber. Dtsch. Bot. Ges.* **86,** 11–38.

Seebeck, T., Stalder, J., and Braun, R. (1979). Isolation of a minichromosome containing the ribosomal genes from *Physarum polycephalum. Biochemistry* **18,** 484–490.

Sheridan, F., and Barrnett, R. (1969). Cytochemical studies on chromosome ultrastructure. *J. Ultrastruct. Res.* **27,** 216–229.

Tanaka, K. (1973). Intranuclear microtubule organizing center in early prophase nuclei of the plasmodium of the slime mold, *Physarum polycephalum. J. Cell Biol.* **57,** 220–224.

Wille, J. J., and Steffens, W. L. (1979). Fine structure of plasmodial nuclei in the slime mold *Physarum polycephalum.* I. Comparison of diploid and haploid vegetative mitosis. *Protoplasma* **101,** 165–180.

Wolf, R., Wick, R., and Sauer, H. (1979). Mitosis in *Physarum polycephalum:* Analysis of time-lapse films and DNA replication of normal and heat-shocked macroplasmodia. *Eur. J. Cell Biol.* **19,** 49–59.

CHAPTER 10

Chromosome Organization and Chromosomal Proteins in *Physarum polycephalum*

HARRY R. MATTHEWS and E. MORTON BRADBURY

I.		Introduction	318
II.		Nucleosome Structure	318
	A.	Introduction	318
	B.	*Physarum* Nucleosomes	319
	C.	Nucleosome DNA Length	320
	D.	Fluorescence Labeling	321
	E.	Peak A	322
	F.	Preferential Digestion of DNA Coding for rRNA (rDNA)	322
III.		Nucleolar Chromatin	325
	A.	Sequence Organization of rDNA	325
	B.	Isolation of Nucleolar Chromatin	327
	C.	Fractionation of Nucleolar Chromatin	329
	D.	Nucleolar Histones	329
IV.		Histones	330
	A.	Isolation Procedures	330
	B.	Total Histone	331
	C.	Histone H1	333
	D.	Histone H2A	339
	E.	Histones H2B and H3	340
	F.	Histone H4	340
V.		Postsynthetic Modifications of Histones	341
	A.	Phosphorylation of Histone H1	341
	B.	Acetylation of Histone H4	346
	C.	Chromosome Structural Transitions	351
VI.		Nuclear Histone Kinase	353
	A.	Cell Cycle Dependence of Activity	353
	B.	Activation of Histone Kinase	354
	C.	Three Nuclear Histone Kinases	354
	D.	Sites of Phosphorylation *in Vitro*	355
	E.	Cell Cycle Role of Histone Kinase	357

	VII.	Non-Histone Chromosomal Proteins	359
	A.	HMG (High Mobility Group) Proteins	359
	B.	Enzymes	359
	C.	Contractile Proteins	360
		References	361

I. INTRODUCTION

In the field of chromosome structure and function, *Physarum polycephalum* has clearly lived up to its early promise of being an important model organism that would lead the way for research on eukaryotes. This is due partly to its naturally synchronous mitotic cycle, as in the work on postsynthetic modifications of histones, and partly to unforeseen factors such as the rDNA "minichromosome" and the taking up of biologically active proteins from the culture medium. Work on the biochemistry of *Physarum* chromosomes has naturally concentrated on the areas where *Physarum* has specific advantages over mammalian cells, and this emphasis is reflected in the coverage of the subject below. We have also concentrated on the aspects not covered elsewhere in this volume. DNA replication and repair are discussed by Evans (this volume, Chapter 11), and RNA metabolism is described by Braun and Seebeck (this volume, Chapter 12). The account below also illustrates how work in one area (histone phosphorylation) can unexpectedly overlap into another (cell cycle controls). Such correlations can be made in *Physarum* partly because of the wealth of experimental data available for the organism.

II. NUCLEOSOME STRUCTURE

A. Introduction

The subunit structure of chromatin in eukaryotes in general is now well established (McGhee and Felsenfeld, 1980). This subunit, the nucleosome, contains two molecules of the histones H2A, H2B, H3, and H4, one molecule of histone H1, and a variable amount of DNA (154–245 base pairs, or bp) depending on the organism, though for most somatic tissues the DNA content is in the range of 195 ± 5 bp of DNA. Nuclease digestion of the DNA in the nucleosome results in the displacement of histone H1 when the digestion proceeds from 168 to 146 bp of DNA, leaving a very regular chromatin core particle. The core particle contains the histone octamer 2[H2A, H2B, H3, H4] and 146 bp of DNA (Lutter, 1979). Neutron scatter studies of this core particle have shown unambiguously that the histone octamer forms a core around which the DNA is coiled with a pitch of approximately 3.0 nm to give a flat disc 11.0 nm in diameter by 5.5–6.0 nm in thickness (Suau *et al.*, 1977). It is now thought that the chromatosome, the constant structural component of the nucleosome, is larger than the core particle and contains two turns of DNA with about 80 bp DNA per turn and a pitch of 2.7-3.0 nm, the histone octamer, and histone H1, which "seals" the two turns of DNA, as shown in Fig. 1 (Thoma *et al.*, 1979; Allan *et al.*, 1980). The DNA joining these structural units varies in length from very short up to about 80 bp, with the typical value for a higher eukaryote being about 30 bp. Other noncovalent forces govern the interactions of core particles with one another and with non-histone proteins. These interactions are thought to involve the well-defined basic N-terminal regions of all the histones and the very basic C-terminal half of histone H1. These basic domains are not required for the core particle structure and contain all the sites of major postsynthetic modifications: acetylations of the core histones and phosphorylation of histone H1. These concepts have been discussed in more detail elsewhere (Bradbury and Matthews, 1981).

B. *Physarum* Nucleosomes

Physarum chromosomal subunits, nucleosomes, can be visualized in the electron microscope (Johnson *et al.*, 1976) and isolated by partial micrococcal nuclease digestion. Micrococcal nuclease preferentially cleaves the linker DNA to give short strings of nucleosomes from monomers upward. In *Physarum* these tend to aggregate, but they can be separated by zone sedimentation in sucrose gradients containing 0.35 M NaCl (Johnson *et al.*, 1978a). The NaCl removes ionically linked proteins such as H1 histone (Section IV,C) and HMG proteins

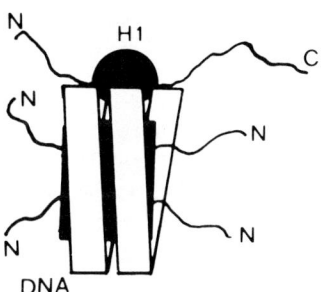

Fig. 1. Diagram of a chromatosome. The broad band represents two turns of DNA (168 bp), with most of it wrapped around the core of histones (H2A.H2B.H3.H4)$_2$ and the ends of the DNA interacting with the globular region of histone H1. The N- and C-terminal regions of H1 are shown free, as are the N-terminal regions of H2A and H2B. The N-terminal regions of H3 and H4 are released by 0.6 M NaCl, and it is possible that these regions may also be released by acetylation. The "free" regions probably interact with "linker" DNA or other chromatosomes or non-histone proteins. Up to about 30 bp of linker DNA join chromatosomes together in a continuous "string." The diameter of the DNA coil shown is 11.0 nm; the thickness of the histone core is 5.5 nm.

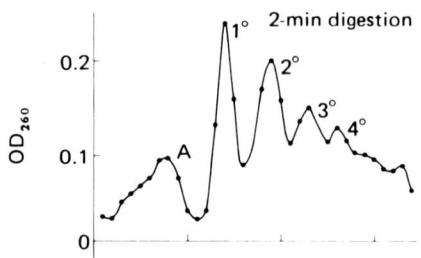

Fig. 2. Sucrose gradient centrifugation of nucleosomes. *Physarum* nuclei were digested with micrococcal nuclease and were sheared in a Potter homogenizer to release chromosome fragments. The soluble material sedimented through a sucrose gradient. Sedimentation is from left to right. "A" is peak A; 1°, 2°, 3°, and 4° are mono-, di-, tri-, and tetranucleosomes, respectively.

(Section VII,A). Figure 2 shows mono- to tetranucleosomes and a slowly sedimenting peak, peak A, that will be discussed in Section II,E. The nucleosomes contain DNA and protein in a ratio of about 1:2, and all the core histones are present. After prolonged digestion, core particles are obtained with a DNA length of approximately 145 bp per particle (Jalouzot *et al.*, 1980), as found for higher eukaryotes. The peaks from the sucrose gradient can be isolated and shown to be substantially pure components by rerunning on a fresh sucrose gradient or by isolating DNA and analyzing its length by gel electrophoresis (Johnson *et al.*, 1978a,b). Proteins can be isolated from each peak and analyzed by SDS gel electrophoresis. Histone H1 is absent, and the mono- and oligonucleosome peaks contain core histones with little other protein. Peak A contains, or cosediments with, a number of non-histone proteins as well as core histones (Johnson *et al.*, 1978a).

C. Nucleosome DNA Length

The length of DNA in a nucleosome varies. In *Physarum* it has been measured by digesting nuclei with micrococcal nuclease for varying periods of time, isolating the DNA, and running it out by gel electrophoresis with sequenced DNA fragments as length markers. It was found that the effect of random nuclease digestion at the ends of strings of nucleosomes could be eliminated by proper analysis of the data, and the values of repeat length were then found to depend on the time of digestion (Table I). The variation probably represents *in vivo* heterogeneity of linker DNA lengths, although migration of core particles during digestion has not been ruled out. Similar results were subsequently obtained for other organisms (Lohr *et al.*, 1977). Compton *et al.* (1976) reported a single value of 176 bp for *Physarum* but did not give experimental details or data.

No detailed cell cycle studies of nucleosome repeat length have been reported, although Vogt and Braun (1976a) showed that no change in repeat length occurred specifically at metaphase (Hozier and Kaus, 1976).

Jalouzot et al. (1980) have studied the kinetics of digestion of Physarum chromatin by nucleases. They found that micrococcal nuclease did not distinguish between chromatins from different stages of the cell cycle, in agreement with earlier reports (Vogt and Braun, 1976a; Hozier and Kaus, 1976). However, micrococcal nuclease did show an increased rate of digestion of newly synthesized DNA, with the DNA being "hypersensitive" for about 2 minutes after synthesis, "sensitive" from about 2 minutes to somewhat longer than 10 minutes, and then "normal" until synthesis began again.

D. Fluorescence Labeling

Physarum nucleosomes have been labeled by incubating microplasmodia in normal semidefined growth medium (Daniel and Baldwin, 1964) supplemented with 40 μg/ml pyrene-labeled Physarum total histone (Prior et al., 1980). The pyrene label is located on the single cysteine in Physarum H3. Individual pyrene-H3 molecules fluoresce blue (blue monomer fluorescence), but if the pyrene groups on two H3 molecules can interact strongly, they fluoresce green (green eximer fluorescence). As microplasmodia grow in the presence of pyrene-H3, their cytoplasm becomes fluorescent (blue monomer fluorescence). When growth is continued in the presence of unlabeled histone, the blue fluorescence is chased into the nuclei, where it gradually becomes green eximer fluorescence. Nucleosomes were prepared from green fluorescent nuclei, and the green eximer fluorescence was found in the nucleosomes. Histone H3 from these nucleosomes contained the fluorescent label, now blue again after dissociation.

The experiments of Prior et al. (1980) provided new evidence that H3, at least, is distributed conservatively during chromatin replication. They also showed unequivocally that Physarum could take up and use a biologically active

TABLE I

Physarum Nucleosome DNA Length[a]

Time of digestion with micrococcal nuclease (min)	Nucleosome DNA length (bp)
1	190
2	189
10	180
60	172

[a] From Johnson et al. (1976).

protein. About 4% of the added label was recovered in nucleosomes. We had previously speculated that *Physarum* could take up added histone kinase, and these new data add support to that speculation (see Section VI,E).

E. Peak A

Micrococcal nuclease digestion of *Physarum* nuclei (Mohberg and Rusch, 1971) produces the usual sedimentation pattern (Section II,A) plus a peak between the DNA fragments at the top of the gradient and the monomer peak (Fig. 2; Staron *et al.*, 1977; Johnson *et al.*, 1978a,b; Jalouzot *et al.*, 1980). This material was named "peak A" by Staron *et al.* (1977).

Johnson *et al.* (1978a) showed that peak A contained DNA with the same length, about 140 bp, as core particles. Peak A also contained histones and non-histone proteins with a protein:DNA ratio of 1:0. It appeared to be partially depleted of histones H3 and H4. The sedimentation coefficient of about 5 S compared with 11 S for mononucleosomes suggested that peak A particles are more extended than mononucleosomes. This was confirmed by electron microscopy (Johnson *et al.*, 1978b). Peak A particles could be produced from monomers, and it was concluded that unfolding of *Physarum* nucleosomes passes through a stable intermediate form that can be isolated as peak A.

The distribution of sequences coding for rRNA was determined by hybridization of *Physarum* rRNA to peak A particles, mononucleosomes, and other nucleosome fractions. Peak A contained a higher proportion of these transcribed sequences than the nucleosome fractions (Johnson *et al.*, 1978a). Nucleosomes were isolated from microsclerotia, but peak A particles were not detected (Johnson *et al.*, 1978b). These data show that active (in microplasmodia) sequences coding for rRNA give rise to peak A particles more readily than the bulk of chromatin and that inactive sequences (in microsclerotia) do not produce peak A particles. It was concluded either that (1) peak A particles represent the structure of actively transcribing chromatin or that (2) actively transcribing chromatin has a structure, different from that of inactive chromatin, that readily forms peak A particles on micrococcal nuclease digestion. The extended form of peak A particles is consistent with the electron microscopic observations that transcribing rRNA genes are extended (Foe *et al.*, 1976; Woodcock *et al.*, 1976; Scheer *et al.*, 1977). The detailed structure of peak A particles and its relationship to active and inactive chromatin structures have not been reported.

F. Preferential Digestion of DNA Coding for rRNA (rDNA)

The digestion of active chromatin has been followed, in *Physarum*, using hybridization to 19 and 26 S rRNA or to restriction fragments of rDNA. (rDNA

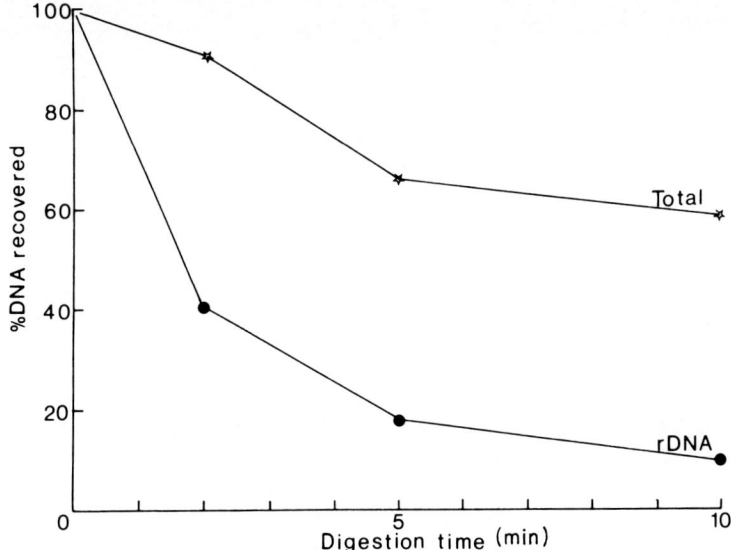

Fig. 3. Preferential digestion of rDNA chromatin. *Physarum* nuclei were digested with micrococcal nuclease, and DNA was extracted. Total DNA concentration was determined by the diphenylamine method, and rDNA was determined by hybridization to 19 S + 26 S rRNA. Data from Johnson *et al.* (1978b).

is discussed in more detail in Section III,A and in Chapter 12, this volume.) Micrococcal nuclease digests rDNA chromatin in microplasmodial nuclei at a substantially greater rate than the bulk of chromatin is digested (Johnson *et al.*, 1978a,b; but see also Stalder *et al.*, 1979; Fig. 3). In addition, the pattern of digestion is different. Johnson *et al.* (1978b) showed that rRNA-complementary sequences are found preferentially in peak A and mononucleosomes and are underrepresented in dinucleosomes and higher oligomers (Fig. 4). Stalder *et al.* (1979) used gel electrophoresis to study the DNA released by micrococcal nuclease followed by phenol extraction and found that the rRNA-complementary sequences occurred not only in the bands of bulk DNA but also in the regions between the major DNA bands. Clearly, at least some of the chromatin containing rDNA can be isolated in the form of nucleosomes, but the nucleosomes may not be in continuous strings and the structure may be a dynamic one that changes during the course of digestion. The structure of rDNA chromatin certainly differs from that of the bulk of chromatin (Matthews, 1977a).

Johnson *et al.* (1979) localized the rDNA-specific structure more precisely by using restriction nuclease fragments of rDNA as hybridization probes. They found that restriction fragments derived from the nontranscribed regions of rDNA hybridized poorly to peak A particles but hybridized well to oligonucleo-

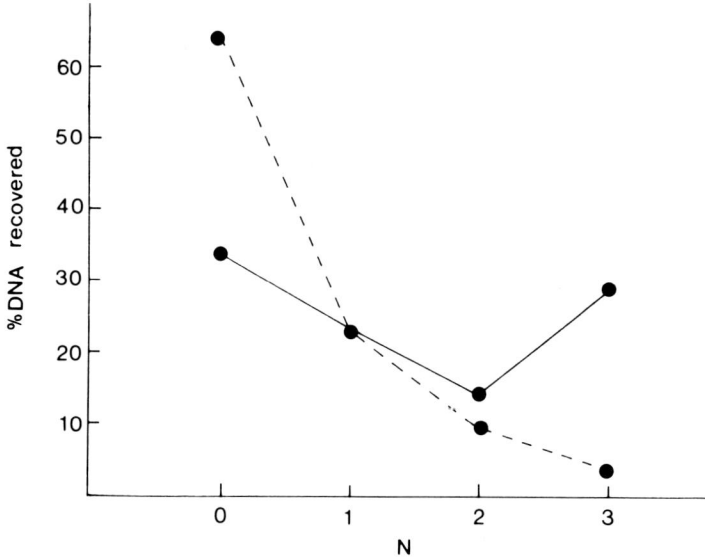

Fig. 4. DNA and rDNA in nucleosome fractions. *Physarum* nuclei were digested with micrococcal nuclease and the chromatin fragments were separated by sedimentation, as in Fig. 2. DNA was extracted from each peak and the amounts of DNA and rDNA measured, as in Fig. 3. Solid line, DNA; dashed line, rDNA, N (0, 1, 2, 3) refers to peaks A, 1, 2 (3 + 4 + higher oligomers). Data from Johnson *et al.* (1978b).

somes, whereas restriction fragments derived from the transcribed regions of rDNA hybridized well to peak A particles and poorly to oligonucleosomes. This confirms the previous observations (Johnson *et al.*, 1978a,b) and shows that the structure of actively transcribing sequences is precisely localized on those sequences and does not spread appreciably into adjacent nontranscribed sequences.

Deoxyribonuclease I (DNase I) digests chromatin to give a characteristic pattern of single-strand DNA fragments in lengths of integer multiples of about 10.4 bases (Lutter, 1979). Actively transcribing chromatin is more highly sensitive to DNase I than is the bulk of chromatin, as is the case for micrococcal nuclease. The effect is much greater for DNase I, however (Weintraub and Groudine, 1976). Stalder *et al.* (1978) reported that *Physarum* rRNA-complementary sequences are more highly sensitive to DNase I than is the bulk of DNA in chromatin, but Swofford (1979) found very little difference in sensitivity, as if the *Physarum* nucleolus were protecting the rDNA (see Chapter 12). Jalouzot *et al.* (1980) showed that *Physarum* has a class of DNA sequences that is highly sensitive to DNase I. It is implied that these are the active sequences, although they have not been characterized as such.

III. NUCLEOLAR CHROMATIN

A. Sequence Organization of rDNA

The vast majority, if not all, of the genes for the large ribosomal RNAs (5.8 S, 19 S, 26 S) are located in the nucleolus and, in *Physarum*, are on separate DNA molecules (rDNA) (Molgaard *et al.*, 1976; Hall and Braun, 1977; Campbell *et al.*, 1979). There are approximately 190 rDNA molecules per haploid genome in plasmodia, spherules, amoebae, and spores (Ryser and Braun, 1974; Hall and Braun, 1977; Affolter and Braun, 1978). No evidence has been obtained for a copy of the rDNA sequence integrated into the normal chromosome DNA, but the presence of an integrated copy has certainly not been ruled out. An integrated copy does exist in *Tetrahymena*, which also has separate rDNA molecules (Yao and Gall, 1977). Each *Physarum* rDNA molecule is a giant palindrome of about 60 kb containing two each of the genes for 5.8 S, 19 S, and 26 S rRNA (Vogt and Braun, 1976b; Molgaard *et al.*, 1976; Campbell *et al.*, 1979). Each 26 S gene contains two inserted sequences that are absent in the native 26 S rRNA (Campbell *et al.*, 1979; Gubler *et al.*, 1979). Transcription begins about 4 kb from the ends of the 19 S genes, proceeds toward the ends of the rDNA, and probably terminates at or close to the ends of the 26 S genes, as shown in Fig. 5 (Grainger and Ogle, 1978; Sun *et al.*, 1979; Gubler *et al.*, 1980). About half of the molecule comprises a large central spacer that appears not to be transcribed.

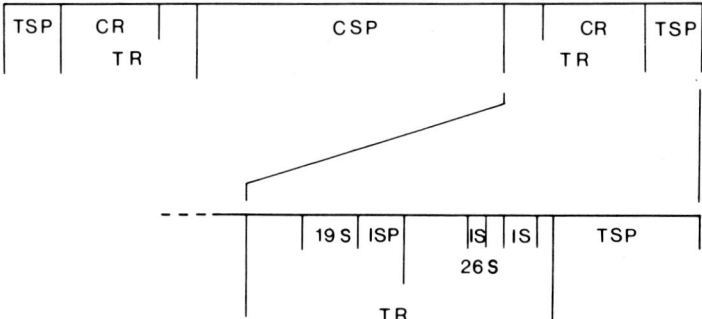

Fig. 5. Sequence organization of rDNA. The upper horizontal line represents one rDNA molecule of about 60 kb. The regions indicated are: TSP, terminal spacer (5kb); CR, "coding" region (9 kb); TR, transcribed region (11 kb); CSP, central spacer (28 kb). The lower horizontal line represents the right-hand end of one rDNA molecule. The regions indicated are: TSP, terminal spacer (5 kb); TR, transcribed region (11 kb); 26 S, region coding for 26 S rRNA plus inserted sequences (5.6 kb); ISP, internal spacer (1.7 kb); 19 S, region coding for 19 S r RNA (2.1 kb); IS, inserted sequences (0.7 kb and 1.2 kb). All the DNA lengths are approximate, but they have been determined by a number of methods, which give consistent results.

The nontranscribed ends of the rDNA molecule have an intriguing structure. Restriction nuclease fragments from the ends show length heterogeneity (Molgaard et al., 1976; Campbell et al., 1979), and Johnson (1980) has shown the presence of sequence heterogeneity as well. Of particular interest is Johnson's finding (1980) that the terminal sequences contain short, single-stranded regions or gaps perhaps just one nucleotide long. These gaps are associated with a 50-bp sequence. The gaps cannot be ligated but can be labeled by nick translation, which labels a terminal repeat sequence beginning CCCTA. The 50-bp sequence is inverted and repeated several times at each end of each rDNA molecule, although the number of repeats varies and there is a variable amount of other sequences interspersed with the repeats. The repeated sequences are different from those in the center of the rDNA molecule (Vogt and Braun, 1976b; Hardman et al., 1979a,b), but they are found in other regions of *Physarum* DNA (Johnson, 1980). Johnson's conclusions are summarized in Fig. 6.

Physarum rDNA also contains proteins that cannot be removed ionically. These proteins can be labeled with I^{125} and are found specifically associated with the nontranscribed spacer regions at the ends of the molecules (E. M. Johnson, unpublished). The rDNA can be isolated from nucleoli and purified as a high-density satellite on CsCl gradients (Bradbury et al., 1973a). Resolution can be

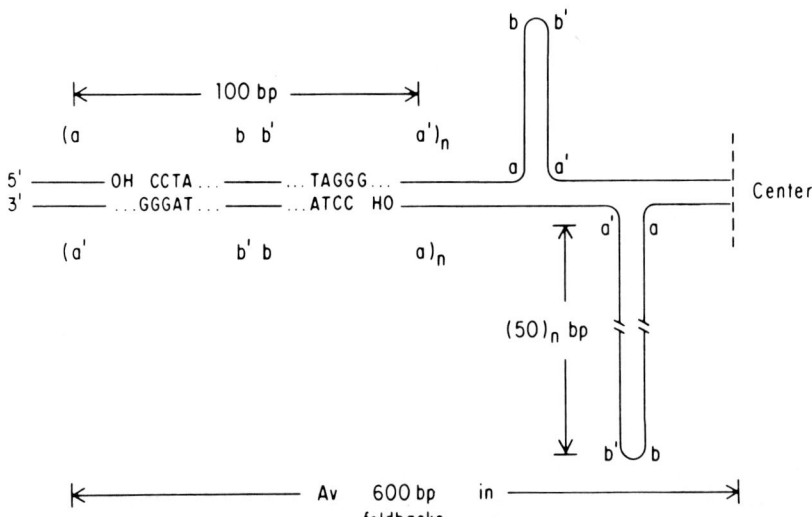

Fig. 6. Structure of the ends of rDNA molecules. The diagram shows the left-hand end of the molecule and summarizes the appearance in the electron microscope of spread rDNA molecules. There is no evidence that the hairpin loops occur *in vivo*. Additional information was obtained from partial sequence studies. a ...b represents a specific sequence of about 50 bp, and b' ...a' is its complementary sequence. Notice the single-strand gaps in some of the a... b sequences. Tightly bound protein(s) are found in this region. Johnson (1980).

enhanced with netropsin (Matthews *et al.*, 1978) or bisbenzimid H332258 (Seebeck *et al.*, 1979, which increases the density difference between rDNA and main band on the gradients. In contrast to a number of other systems, the average nucleotide composition determined from density or thermal denaturation measurements is the same for the transcribed and nontranscribed regions (Steer *et al.*, 1978).

C. Cooney, in our laboratory and in collaboration with E. M. Johnson, has begun a study of 5-methylcytidine (5-MeC) in *Physarum* rDNA. Evans and Evans (1970) reported that approximately 5% of the cytidines in the *Physarum* nuclear satellite (rDNA) were in the form of 5-MeC, about the same proportion as in main band DNA. We attempted to confirm this result by pulse-labeling plasmodia with [methyl-^{3}H]methionine, isolating nuclear DNA and analyzing it by CsCl density gradient centrifugation. Figure 7 shows the gradient with measurements of optical density and radioactivity. Preliminary measurements had shown that under these labeling conditions no label could be detected in unmodified nucleosides. The position of rDNA was confirmed by analyzing fractions from the preparative gradient by analytical ultracentrifugation. The results suggest that GC-rich DNA and rDNA in particular are methylated more slowly than AT-rich DNA. Preliminary results obtained by direct measurement by HPLC of 5-MeC content agree with this conclusion, but the current data are complicated by impurities and incomplete digestion. The apparent discrepancy between our data, which indicate that less than 5% of the cytidines in rDNA are methylated, and those of Evans and Evans (1970) may be due to the different labeling protocols used. Further work is needed to establish the existence and timing of methylation of cytidine in *Physarum* rDNA.

The presence of 5-MeC can also be inferred by comparing restriction nuclease digests obtained with *Hpa*II or with *Msp*I. *Hpa*II recognizes and cleaves the sequence CCGG; *Msp*I recognizes and cleaves both CCGG and C MeC GG. Thus, MeC will make some sites resistant to *Hpa*II but not to *Msp*I. Gel electrophoresis patterns of digests of rDNA with these enzymes are almost identical (C. Cooney, unpublished; Reilly *et al.*, 1980), although we observe a small amount of apparently undigested rDNA in the *Hpa*II digest. This is being investigated.

Methylation of cytidine is inversely correlated with transcription in higher eukaryotes. It remains to be seen if this correlation occurs in *Physarum* rDNA which has both transcribed regions in an "active" chromatin conformation and nontranscribed regions in an "inactive" conformation (Johnson *et al.*, 1979).

B. Isolation of Nucleolar Chromatin

Crude preparations of nucleoli contain both rDNA and nucleoplasmic DNA, but as the nucleoli are progressively purified the proportion of nucleoplasmic

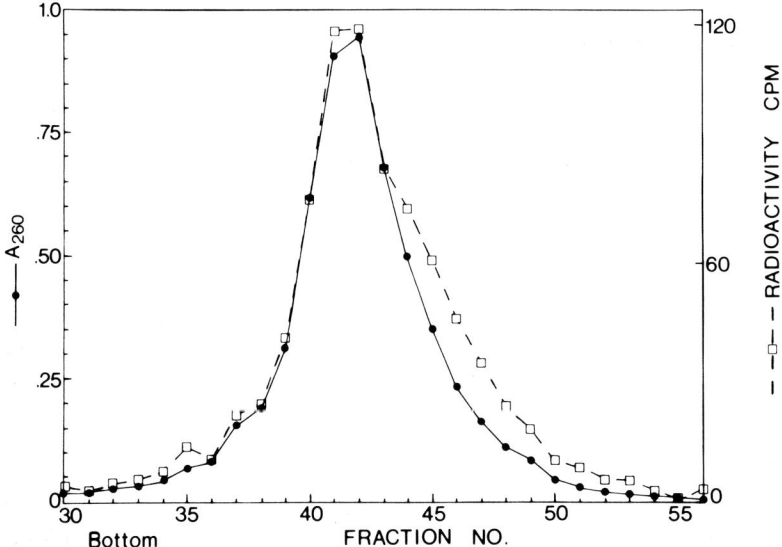

Fig. 7. Distribution of 5-methylcytosine in main band and rDNA. *Physarum* plasmodia were grown in the presence of [methyl-^{3}H]methionine under conditions in which DNA was labeled only in 5-methylcytosine. Nuclear DNA was isolated and separated by preparative CsCl centrifugation. The radioactivity profile shows the presence of 5-methylcytosine in DNA, which is shown by the optical density profile.

DNA falls, consistent with the notion that nucleoli contain only rDNA. Hence, *Physarum* nucleoli offer a potential source of active chromatin coding for specific RNAs. *Physarum* nucleoli can be prepared by vigorous homogenization of whole plasmodia in a homogenizing medium containing 1 mM CaCl$_2$ (Mohberg and Rusch, 1970), or isolated *Physarum* nuclei may be sheared with a Polytron homogenizer (Bradbury *et al.*, 1973c), a sonicator, or a French press (Campbell *et al.*, 1979). Neither method is very reproducible with microplasmodia, and if macroplasmodia are used, the amounts available are limited (Affolter *et al.*, 1979). Nevertheless, substantial success has been achieved with these methods. Nucleoli can be purified either by zone sedimentation in a zonal rotor (Bradbury *et al.*, 1973c) or by isopycnic centrifugation in Percoll (Seebeck *et al.*, 1979). For some experiments, these nucleoli can be used directly. It is essential, however, to characterize them carefully since a small contamination with nuclei or chromatin fragments has a drastic effect on the purity of the ''chromatin.'' For example, one nucleus plus 99 nucleoli (that is, 99% pure microscopically) gives about 1 pg nucleoplasmic DNA plus a maximum of $100 \times 2 \times 10^{-14}g = 2$ pg rDNA, so the rDNA chromatin is only 67% pure at best. The best method of characterization is to run a CsCl gradient of a sample of nucleolar DNA which will show rDNA (density = 1.713 gm/ml) separated from nucleoplasmic DNA (density = 1.703 gm/ml).

Physarum nucleolar chromatin can be solubilized with EDTA either for electron microscopy (Grainger and Ogle, 1978) or for biochemical experiments, although typical problems with obtaining reproducible high yields are found. Seebeck *et al.* (1979) introduced the Percoll method for purifying nucleoli and treated Percoll-purified nucleoli with 10 m*M* EDTA. This lysed at least some of the nucleoli, and the soluble material was characterized by zone sedimentation on sucrose gradients. Chromatin containing rDNA was detected by hybridization to labeled rRNA or by an assay for RNA polymerase activity. Both methods revealed a sharp peak sedimenting with a sedimentation coefficient of about 100 S [pure rDNA has a sedimentation coefficient of 38.5 S (Molgaard *et al.*, 1976)]. The polymerase activity was totally resistant to α-amanitin, and the RNA product hybridized exclusively to rDNA. Seebeck *et al.* (1979) concluded from these and other data that they had isolated a minichromosome containing active rDNA chromatin separated from the rest of the nucleolus and from nucleoplasmic chromatin.

Transcription in this rDNA chromatin complex is stimulated by a specific phosphoprotein that has been isolated by Kuehn *et al.* (1979).

C. Fractionation of Nucleolar Chromatin

The nuclease digestion studies described in Section II,F (Johnson *et al.*, 1979) showed that the structure of the transcribed regions of rDNA chromatin is different from that of the nontranscribed spacer region. These regions can be separated by restriction nuclease digestion of rDNA, and it was shown (H. R. Matthews and E. M. Johnson, unpublished) that the restriction nuclease *Hind*III will digest the rDNA in intact nucleoli. In those experiments, EDTA and other nondenaturing methods failed to solubilize the digested chromatin, but more recent reports (C. Prior and E. M. Johnson, unpublished) describe the solubilization of digested chromatin using colcemid and β-galactosidase. The solubilized chromatin was fractionated by sucrose gradient centrifugation, which gave separated peaks of chromatin each containing only one restriction fragment of the rDNA. If this method has sufficient reproducibility for routine use, it should provide the basis for important studies of the differences in chromatin structure between adjacent active and inactive sequences.

D. Nucleolar Histones

No definitive study of nucleolar histones has been carried out. Mohberg and Rusch (1970) found a normal histone complement in isolated nucleoli. S. S.

Chahal (unpublished) also found all the major histones in isolated nucleoli harvested in the G_2 phase. These G_2 phase nucleoli had unusually high acetylation of H4 histone (Section V,B) and a reduced amount of histone H1, although it did appear to be phosphorylated as expected at this time in the cycle. However, in neither of these studies was the purity of the nucleoli with respect to contamination with nonnucleolar chromatin rigorously established (see Section III,B), nor is it clear whether the overall levels of H3 and H4 are reduced, as they are in peak A nucleosomes.

Nucleolar chromatin isolated by EDTA solubilization of nucleoli bands in metrizamide like a DNA-protein complex (Seebeck et al., 1979), and the presence of rRNA- and rDNA-complementary sequences in nucleosomes after brief micrococcal nuclease digestion of nuclei implies that nucleolar chromatin contains histones. The structural differences between active and inactive chromatin inferred from nuclease digestion studies may reflect differences in histone modifications and/or amounts or differences in non-histone proteins, but these have yet to be measured. Studies on nucleolar proteins are in progress.

IV. HISTONES

A. Isolation Procedures

Mohberg and Rusch (1969) documented the fact that, in *Physarum*, conventional histone isolation procedures (acid or NaCl extraction of nuclei) gave poor yields and high degradation, probably due to the high polysaccharide content of *Physarum* nuclear preparations. The yield of the acid extraction procedure can be increased by adding urea to the extraction medium (N. Tanphaichitr, unpublished). Mohberg and Rusch (1969) introduced the use of extraction with 1 M $CaCl_2$ for histone isolation, and this method has been used for most of the published work on *Physarum* histones. Prior et al. (1980) added citric acid (30 mM, pH 4.6) to the 1 M $CaCl_2$ extraction medium to improve the yield slightly. Digestion of nuclei with DNase I does not release histones in substantial amounts (J. Waterborg, unpublished). Mohberg and Rusch (1969) found that 28% guanidinium chloride (GuCl) could be used to extract histones, although the yield was reduced, but R. Braun (unpublished) has successfully used 40% GuCl for extraction of nuclei to obtain histone H1.

The $CaCl_2$ supernatant contains many proteins besides histones. Mohberg and Rusch (1969) precipitated all the proteins with 25% trichloracetic acid (TCA) and then extracted the TCA pellet with 0.02 N H_2SO_4. Care is required to make this extraction quantitative, but in our experience this method works well. The 0.02 N H_2SO_4 supernatant can be used directly for electrophoresis or the histones can be precipitated with acetone·HCl, washed, and dried (Johns, 1976). More re-

TABLE II

Physarum Histones

Mammalian equivalent	Band no.[a]	Molecular weight			Other distinguishing properties[e]
		a^b	b^c	c^d	
H1	1	24,500	—[f]	22,700	DNA aggregation, soluble in 5% PCA
H2A	3[g]	14,000	17,500	15,700	Binds Triton X-100
H2B	4	14,500	15,600	15,000	Separates from H3 on two-dimensional gels
H3	4	14,500	15,100	14,300	Contains cysteine
H4	6[h]	10,500	11,400	10,500	Comigrates with mammalian H4

[a] From Mohberg and Rusch (1969).
[b] From Jockusch and Walker (1974).
[c] Calculated from data in Johnson et al. (1978a) by Corbett et al. (1980).
[d] From Tyrsin et al. (1977a).
[e] Amino acid analyses are given in Tables III and IV.
[f] Not determined by Johnson et al. (1978a), but Fischer and Laemmli (1980) found a value of 27,500.
[g] Band 2 of Mohberg and Rusch (1969) contains a number of bands, probably including HMG proteins (Matthews et al., 1979).
[h] Band 5 of Mohberg and Rusch (1969) is a minor component of molecular weight 13,000, which Tyrsin et al. (1977b) identified as an H2B-like protein. A similar conclusion was obtained using specific stains (S. Miller and H. Matthews, unpublished).

cently, we have abandoned the TCA precipitation procedure and acidified the 1 M $CaCl_2$ extract directly after spinning out the nuclei. The extract is then centrifuged to remove nucleic acid and acidic proteins and "desalted" on Sephadex G-25 in 0.25 N HCl. The void volume peak (total histone) is precipitated with acetone (Matthews et al., 1979). Fischer and Laemmli (1980) also avoided the TCA precipitation by using a column of DNA–cellulose to isolate histone H1 from a desalted 0.2 M $CaCl_2$ extract of *Physarum* nuclei (Mohberg and Rusch, 1970).

B. Total Histone

Mohberg and Rusch (1969) analyzed total *Physarum* histone by polyacrylamide gel electrophoresis in acetic acid and urea. They identified four major bands and several minor bands (Table II). One of the major bands could be fractionated with ethanol·HCl into two components, so that a total of five major components was described. Three of these components were isolated by preparative gel electrophoresis and their amino acid compositions determined (Tables III

TABLE III

Amino Acid Composition of Histone H1 from *Physarum polycephalum*[a]

Amino acid	Mohberg and Rusch (1969)	Tyrsin et al. (1977a)	Corbett (1979)	Chambers (1980)	Fischer and Laemmli (1980)	
Asp	5.4	5.2	4.4	5.7	3.6	5.9
Thr	5.4	3.5	7.6	7.0	7.0	6.5
Ser	6.2	14.4	10.2	10.8	11.7	10.9
Glu	9.1	10.9	7.8	11.7	7.2	8.6
Pro	10.0	4.0	10.8	9.1	12.6	11.4
Gly	6.7	17.6	4.8	6.7	6.6	6.6
Ala	17.4	6.2	17.5	16.7	20.8	20.0
Cys	0.0	0.0	0.0	0.0	0.0	0.0
Val	3.3	2.6	2.6	2.4	1.7	2.7
Met	0.0	0.0	0.6	0.4	0.0	0.0
Ile	2.5	2.0	1.9	1.8	1.9	2.0
Leu	3.3	2.7	3.3	4.0	2.8	3.2
Tyr	0.0	1.3	1.2	0.0	0.5	0.8
Phe	1.2	2.2	1.4	2.2	0.6	0.8
His	2.5	2.5	3.0	3.5	0.8	0.3
Lys	18.2	14.3	18.5	18.1	20.7	17.9
Arg	1.7	1.9	4.4	n.d.[b]	1.7	1.6
Unidentified acidic	7.1	8.7	—	—	—	—

[a] Data in moles/100 moles.
[b] n.d., Not determined.

and IV). Subsequently, similar gel electrophoresis profiles were obtained by Bradbury et al. (1973c), and Jockusch and Walker (1974) published SDS–gel electrophoresis data from which they obtained values for the molecular weights of the major components (Table II). More recently, Chahal et al. (1980) carried out a comprehensive gel electrophoresis analysis including the use of gels containing Triton X-100. Figure 8 shows a two-dimensional gel electrophoresis pattern of *Physarum* total histone in which Triton X-100 was used in the first dimension and cetyltrimethylammonium bromide in the second (R. Mueller, unpublished; Bonner et al., 1980). Notice the major off-diagonal spot which is typical of histone H2A, the separation of the spots assigned to H2B and H3, and the excellent separation of acetylated species of H4.

Total histone has been fractionated by preparative gel electrophoresis or by chromatography on Bio-Gel P10 or P60 columns eluted with 10 to 20 mM HCl containing up to 0.1 M NaCl (Corbett et al., 1977). Histone H1 can be satisfactorily resolved by elution with 20 mM HCl from a fairly short column (60 cm), but resolution of H4 requires 75 mM NaCl, 20 mM HCl, and a longer column

(100 cm) (Chahal *et al.*, 1980). Histone H2A can be resolved on a 1.5-m column, but H3 and H2B elute together under the conditions used so far (Figs. 9 and 10).

C. Histone H1

1. PURIFICATION AND CHARACTERIZATION

The first component of total histone to elute from Bio-Gel columns is analogous to histone H1 in higher eukaryotes. It can be further purified by chromatography on Bio-Rex 70, where it elutes at about 9% GuHCl (Corbett *et al.*, 1977), or by chromatography on DNA–cellulose, where *Physarum* H1 is released by 0.65 M NaCl (Fischer and Laemmli, 1980). Like other H1 molecules, *Physarum* H1 can

TABLE IV

Amino Acid Composition of *Physarum* histones H2A and H4

H2A (moles per 100 moles)		Amino acid	H4[b] (moles per 100 moles)
a[a]	b[b]		
5.9	6.8	Asp	6.7
2.0	4.7	Thr	7.9
2.6	7.3	Ser	3.8
8.9	11.9	Glu	7.8
3.3	5.3	Pro	2.5
9.5	10.5	Gly	13.7
7.2	11.2	Ala	3.2
—	0.0	Cys	0.3[c]
4.9	3.8	Val	5.0
0.0	0.1	Met	1.0
4.9	3.7	Ile	4.2
7.2	7.4	Leu	6.3
0.0	2.0	Tyr	3.6
0.9	1.3	Phe	2.3
4.9	2.4	His	2.7
21.9	14.4	Lys	9.4
9.5	7.2	Arg	13.7
—	6.2	[d]	—

[a] From Mohberg and Rusch (1969).
[b] Miller and Matthews (unpublished).
[c] This should probably be 0.0.
[d] Unidentified acidic amino acid reported by Mohberg and Rusch and probably representing Ser, Thr, and Tyr modified during hydrolysis.

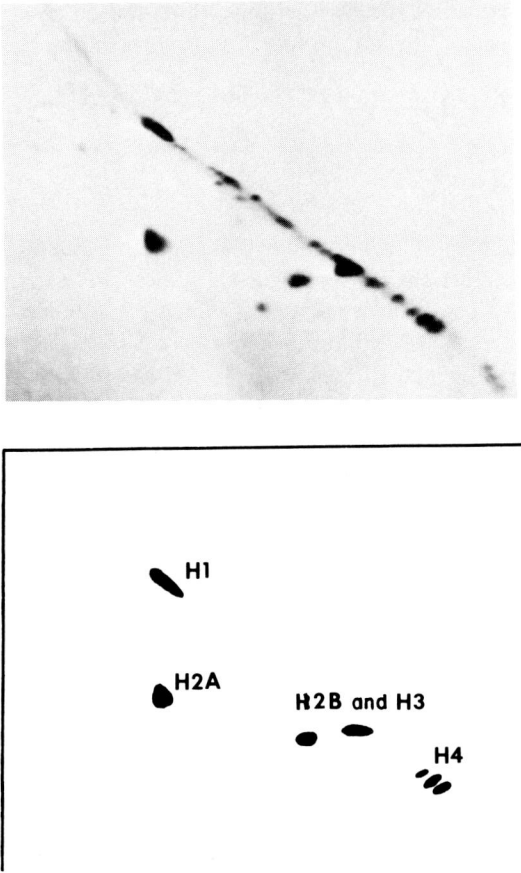

Fig. 8. Two-dimensional gel electrophoresis of CaCl$_2$-soluble nuclear proteins. The first dimension, horizontal, contained 8 M urea and 8 mM Triton X-100 in 1 M acetic acid, and the second dimension contained 6 M urea and 0.15% cetyltrimethylammonium bromide also in 1 M acetic acid (Bonner et al., 1980). H1, H2A, and H4 are clearly identified, but the distinction between H3 and H2B is not yet clear. Several of the other spots correspond to spots observed by Bonner et al. (1980) in electrophoresis of HeLa nuclear acid-extracted proteins, for example, A24 (a ubiquitin–H2A complex described by Goldknopf and Busch (1977) and "z," which was noted but not identified by Bonner et al. (1980).

be released from nuclei by lower ionic strength than the core histones, and *Physarum* H1 is soluble in 5% perchloric acid (Mohberg and Rusch, 1969, 1970). On most gel electrophoresis systems the band for H1 is broader than would be expected for a single species, even after treatment with alkaline phosphatase. Similarly, chromatography on Bio-Rex 70 shows evidence of substantial microheterogeneity which is reduced but not abolished after removal of phosphate. This apparent microheterogeneity has not been characterized further.

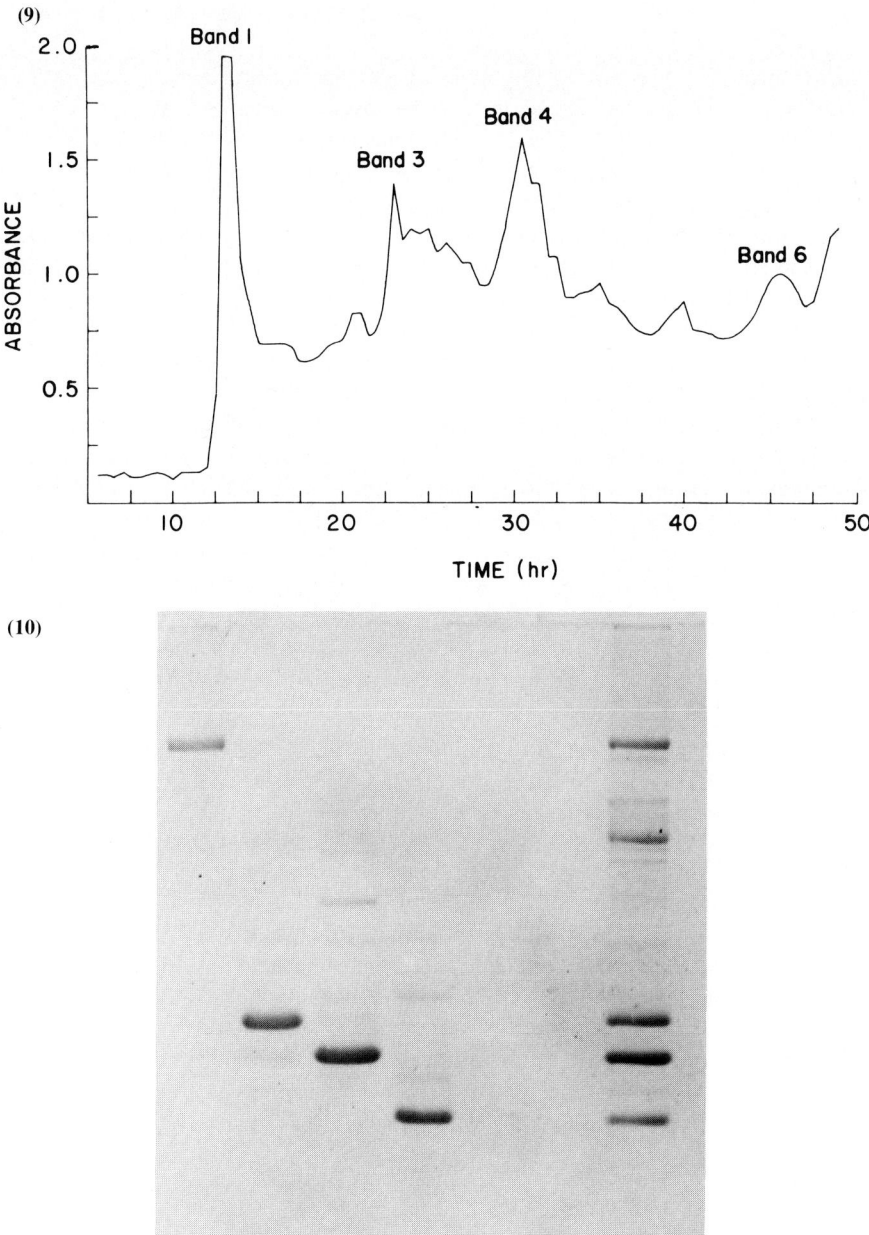

Fig. 9. Elution of *Physarum* histone from a BioGel P60 column. *Physarum* histones were prepared as described by Mohberg and Rusch (1969) and eluted from a column of BioGel P60 with 10 mM HC1 and 20 mM NaCl (Corbett *et al.*, 1977). Better resolution of band 6 (H4) can be obtained with BioGel P10 eluted with 10 mM HCl and 75 mM NaCl. Chahal *et al.* (1980).

Fig. 10. Gel electrophoresis of fractions from the column profile shown in Fig. 9. From left to right, the tracks show band 1 (H1), band 3 (H2A), band 4 (H2B + H3), and band 6 (H4). The track on the far right shows total histone. Corbett *et al.* (1977).

Physarum H1 runs considerably more slowly than mammalian H1's on most gel systems. This is probably due to both the lower lysine content and the larger molecular weight of *Physarum* H1. The effect of the lysine content can be eliminated by SDS-gel electrophoresis at pH 10 (Panyim and Chalkley, 1971), and under these conditions *Physarum* H1 runs nearly as fast as calf H1 (R. J. Inglis, unpublished), implying a molecular weight only slightly greater than that of mammalian H1. Jockusch and Walker (1974), Tyrsin *et al.* (1977a), and Fischer and Laemmli (1980) measured the molecular weight of *Physarum* H1 using mammalian histones as molecular weight markers and obtained values of 24,500, 22,700, and 27,500, respectively. None of these values is documented in any detail, although more data are presented by Jockusch and Walker (1974) than by the others. The unweighted mean of the three values is 24,900 and the approximate value of 25,000 is the best overall estimate currently available, although this may be subject to systematic error, particularly in view of the difficulties encountered with very basic proteins (Panyim and Chalkley, 1971).

Physarum H1 has been isolated by preparative gel electrophoresis (Mohberg and Rusch, 1969; Fischer and Laemmli, 1980); by gel filtration (Tyrsin *et al.*, 1977a; Matthews and Bradbury, quoted in Corbett, 1979); and by gel filtration followed by ion exchange chromatography on Bio-Rex 70 (Chambers, 1980). The reported amino acid analyses are given in Table III. The values taken from Corbett (1979) are the means of determinations on three independent preparations, each shown to be electrophoretically pure. Fischer and Laemmli (1980) reported data for two H1 bands, one of which was thought to be the phosphorylated form, the other being nonphosphorylated (Section V,A,2). The "unidentified" acidic amino acid reported by Mohberg and Rusch (1969) probably arises largely from hydrolysis of serine and threonine, since serine + threonine + unidentified acidic amino acid from Mohberg and Rusch = 18.7 moles/100 moles compared with 17.8 moles/100 moles for serine + threonine from our data (Corbett, 1979). The values for glycine and alanine appear to be transposed in the data from Tyrsin *et al.* (1977a). Apart from these "corrections," the amino acid analyses show substantial similarities to high lysine, alanine, proline, and serine and to very low methionine tyrosine and phenylalanine. The lysine content is lower than that of calf thymus H1, but the overall composition is sufficiently similar to support the identification of *Physarum* H1 as the equivalent of mammalian H1.

2. PARTIAL CHYMOTRYPTIC DIGESTION

Mammalian histone H1 is attacked by chymotrypsin initially at the phenylalanine residue near the center of the molecule [... Gly-Ser^{105}-Phe ... in RTL-3; Cole (1977)]. This gives two fragments: an N-terminal fragment, which is slowly degraded to smaller fragments, and a C-terminal fragment, which is more stable and also carries most of the microheterogeneity observed in the intact

10. Chromosome Organization and Chromosomal Proteins in *Physarum polycephalum*

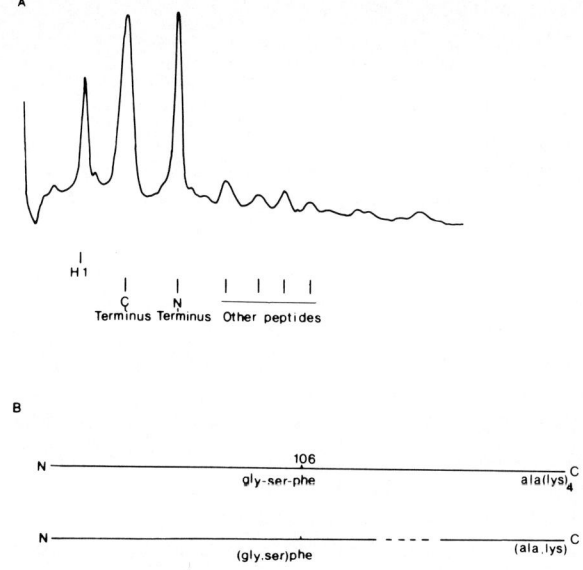

Fig. 11. (A) Partial chymotryptic digest of *Physarum* H1. *Physarum* H1 was partially digested with chymotrypsin, and the fragments were separated by gel electrophoresis. The C-terminal and N-terminal fragments were identified by C-terminal analyses. The N-terminal fragment comigrated with the analogous fragment from calf thymus H1. (Chambers, 1980). (B) Partial amino acid sequence information for *Physarum* H1. The upper horizontal line shows rabbit thymus H1 subfraction RTL3 (Cole, 1977); the lower horizontal line shows *Physarum* H1.

molecule. The central globular region of H1 is highly conserved, and if *Physarum* H1 is truly analogous to mammalian H1's, then it should also have an exposed phenylalanine residue somewhere near the center of the molecule.

This hypothesis was tested by partial chymotryptic digestion of *Physarum* H1 under conditions in which good specificity for the central phenylalanine residue in calf thymus H1 is expressed. Samples were taken from the digest at various times and analyzed by gel electrophoresis. Figure 11 shows gel scans illustrating the course of the digest. *Physarum* H1 has a major site which is sensitive to chymotrypsin. One of the fragments produced comigrates with the N-terminal fragment from a partial digest of calf H1 and is slowly degraded to smaller fragments. The other major fragment shows signs of microheterogeneity and is comparatively stable. Larger amounts of the fragments were obtained from a preparative digest, and the major components were separated by gel filtration. Each fragment was partially digested with carboxypeptidase to determine the C-terminal amino acids. Undigested *Physarum* histone released lysine and alanine under these conditions. The less stable chymotryptic fragment released phenylalanine, glycine, and serine; and the more stable chymotryptic fragment released lysine and alanine, like the undigested H1. Consequently, *Physarum* H1 has an exposed phenylalanine residue, and the fragment containing residues from the N-terminus to the exposed phenylalanine comigrates with the analogous fragment from calf thymus H1. The remaining fragment contains most of the apparent microheterogeneity and migrates more slowly than the analogous fragment from calf thymus H1.

The data are consistent with *Physarum* H1 having a conserved N-terminal half and an extended C-terminal region. A similar conclusion was drawn from the nuclear magnetic resonance spectrum of *Physarum* H1 (P. D. Cary, unpublished).

3. INTERACTION WITH DNA

Histone H1 is thought to interact directly with DNA in chromatin (Fig. 1) and to be involved in chromosome packing, particularly chromosome condensation in mitosis. The ability of H1 to crosslink chromatin or DNA fibers can be seen from the effect of ionic strength. Chromatin isolated in 0.15 M NaCl forms a precipitate which redissolves at low ionic strength. If the ionic strength is now raised, the chromatin precipitates again and then redissolves as the ionic strength reaches 0.5 to 0.6 M NaCl. These macroscopic changes are correlated with changes in the nuclear magnetic resonance (NMR) spectrum of the chromatin. In broad terms, at low ionic strength a fairly sharp NMR spectrum, due to H1, is observed. The spectrum broadens as the chromatin aggregates and then sharpens up again as H1 is released and the chromatin goes back into solution (Bradbury *et al.*, 1973b). Neither the precipitation nor the NMR spectra are observed with chromatin previously depleted of H1. Similar precipitation behavior and NMR spectra are observed with H1 and DNA alone.

In dilute solutions of H1 and DNA (10–100 μg/ml of DNA) the "precipitation" can be measured from the turbidity of the solution, which then provides an index for the interaction of H1 with DNA (Matthews, 1977b). Fig. 12 shows turbidity as a function of NaCl concentration for chicken erythrocyte chromatin, for chicken erythrocyte chromatin depleted of H1 and H5, for DNA and chicken erythrocyte histone H1, and for DNA and *Physarum* histone H1. All the curves show low turbidity at low NaCl concentration and at high NaCl concentration with maximum turbidity at about 0.2 M NaCl. The turbidity in the absence of H1 and H5 is very small and is due to incomplete depletion. There are detailed differences between the other curves due to the different concentrations, the presence of other histones in chromatin, and the differences between chicken erythrocyte H1 and *Physarum* H1. In some respects, *Physarum* H1 behaves like chicken erythrocyte H5. It will be shown below that the turbidity also depends on the phosphorylation state of the H1 (Matthews and Bradbury, 1978). For the present, it should be noted that *Physarum* H1 has DNA cross-linking properties similar to those of chicken erythrocyte very lysine-rich histones and to those of calf thymus H1 (Corbett *et al.*, 1980).

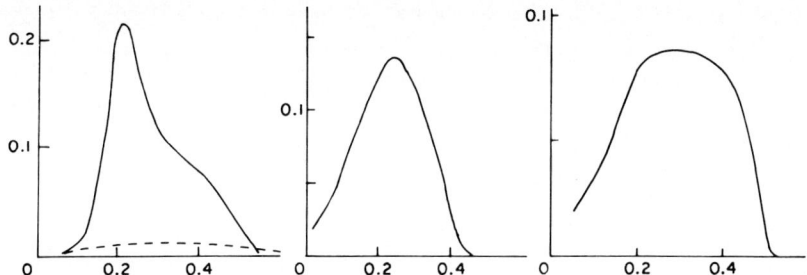

Fig. 12. Turbidity studies of H1·DNA interaction (Corbett *et al.*, 1980) Turbidity was measured as a function of salt concentration using a continuous flow method. Left: Solid line: chicken erythrocyte chromatin; dashed line, chicken erythrocyte chromatin depleted of H1 and H5 (100 μg DNA/ml). Center: Calf thymus DNA + chicken erythrocyte H1 (100 μg DNA/ml). Right: Calf thymus DNA + *Physarum* microplasmodial H1 (80 μg DNA/ml). Matthews and Bradbury (1978).

Physarum H1 binds to DNA cellulose and elutes at about 0.65 *M* NaCl, similar to other H1 histones (Fasy *et al.*, 1979; Fischer and Laemmli, 1980).

D. Histone H2A

Histone H2A can be isolated by chromatography of *Physarum* total histone on BioGel P60 or P10 (Fig. 9) (Corbett *et al.*, 1977). The main reason for identifying this fraction as an H2A histone is its electrophoretic mobility in the presence of Triton X-100 (Chahal *et al.*, 1980). On most gel systems, calf thymus H2A runs slightly faster than H2B and H3, but in the presence of Triton X-100 its hydrophobic character causes it to bind Triton X-100 and run very slowly, much more slowly than calf H1 (Borun *et al.*, 1977). On most gel systems *Physarum* H2A runs slightly slower than *Physarum* H2B and H3, which approximately co-migrate with calf H3, but like calf H2A, *Physarum* H2A binds Triton X-100 and runs even more slowly than *Physarum* H1 in the presence of Triton X-100. Fig. 8 shows *Physarum* H2A as a spot far from the diagonal in the gel system, with Triton X-100 in the first dimension and CTAB in the second dimension (Bonner *et al.*, 1980).

The amino acid composition of *Physarum* band 3, now identified as an H2A, was reported by Mohberg and Rusch (1969), and we have made measurements on three separate preparations. Table IV shows the results. As in the case of H1, the unidentified acidic amino acid found by Mohberg and Rusch (1969) probably arises from degradation of serine, threonine, and tyrosine and accounts for the low values they found for these amino acids. Like calf H2A, *Physarum* H2A has about 10% glycine and no methionine. However, its lysine content appears rather high, which may account for its low mobility in acid–urea gels (Panyim and Chalkley, 1969) and in SDS gels (Laemmli, 1970). In some other respects, the composition seems to have similarities both to H2A and to H2B, so the assignment of "H2A" to this histone rests heavily on its behavior on Triton X-100 gel electrophoresis.

A final point about the Triton gels. Most mammalian H2A's are resolved into two subspecies, but there is little or no evidence of this microheterogeneity in *Physarum* (R. D. Mueller, unpublished).

E. Histones H2B and H3

These histones elute together from Bio-Gel P60 columns, but Mohberg and Rusch (1969) fractionated them with ethanol·HCl and they separate slightly in some gel electrophoresis systems (e.g., Fig. 8). The amino acid analyses of the separated fractions were not reported by Mohberg and Rusch (1969) and we have not separated these fractions on a preparative scale to date, although experiments with affinity chromatography are in progress.

H3 + H2B from *Physarum* comigrates approximately with calf H3 in several gel electrophoresis systems, but the presence of H3 in this fraction is inferred mainly from the presence of cysteine. *Physarum* total histone reacts with labeled *p*-chloromercuribenzoate, and all the label appears in the H3 band (W. Schiebel, personal communication). Prior *et al.* (1980) labeled *Physarum* total histone with iodoacetoxypyrene, which is a fluorescent label specific for cysteine. All the fluorescence bound to *Physarum* total histone was found in the H3 band. Quantitation of this reaction showed that *Physarum* H3 contains one cysteine per molecule.

The second component of band 4 (H3 + H2B) is identified as an H2B by a process of elimination. The amino acid composition of band 4 is consistent with that of H3 + H2B from calf thymus. As in the case of H2A, the lysine content is somewhat higher in *Physarum* H3 + H2B than in calf H3 + H2B, suggesting, since H3 is probably conserved, that both H2A and H2B are more lysine-rich in *Physarum*. Sea urchin H2A and H2B have extended basic N-terminal regions (Von Holt *et al.*, 1979).

Tyrsin *et al.* (1977b) also came to the conclusion that *Physarum* band 4 contained H2B as one component.

F. Histone H4

Histone H4 can be isolated by chromatography of *Physarum* total histone on Bio-Gel P10 in 20 mM HCl and 75 mM NaCl (Chahal *et al.*, 1980). It comigrates with calf H4 in all the gel electrophoresis systems used to date and was identified by Mohberg and Rusch (1969) as an H4 histone. Table IV shows its amino acid composition (mean of three independent preparations). The similarity between this and the amino acid composition of calf thymus H4 (Johns, 1976) is striking.

In gel electrophoresis systems that are sensitive to the charge on a protein, H4 splits into two to five bands representing zero to four acetyl groups on the N-terminal lysines. *Physarum* H4 shows this behavior. The proportion of H4 in the more highly acetylated species of H4 from cultured cells can be dramatically increased by the addition of 7 mM sodium butyrate to the cell culture medium (Riggs *et al.*, 1977). This concentration of butyrate interferes with the normal nuclei preparation in *Physarum,* so its effect on H4 acetylation has not been determined. However, even 2 mM butyrate has a substantial effect on increasing the proportion of H4 in the more highly acetylated species (Chahal *et al.*, 1980). This further confirms the identification of H4 in *Physarum*.

V. POSTSYNTHETIC MODIFICATIONS OF HISTONES

A. Phosphorylation of Histone H1

1. CELL CYCLE DEPENDENCE OF H1 PHOSPHATE (^{32}P)

Physarum incorporates ^{32}P into histones if plasmodia are grown in the presence of labeled inorganic phosphate. The potassium phosphate normally present in the growth medium can be omitted. Most of the incorporation, over a long period, is into H1 histone. Looking back, we can now see that there is also substantial incorporation into H2A, at a fairly constant level in the cell cycle, and a mitosis-specific incorporation into H3 or H2B. There is also substantial ^{32}P incorporation into a minor component running slightly faster than H1 and, except in mitosis, into non-histone material near the top of the gel (Bradbury *et al.*, 1973c).

The incorporation of ^{32}P into H1 was measured through the cell cycle using [^{3}H]lysine and ^{32}P to label plasmodia continuously from fusion to harvesting. Total histone was isolated and fractionated by acrylamide gel electrophoresis and the ^{32}P/^{3}H ratio in the H1 band measured. Figure 13 shows the results (Bradbury *et al.*, 1973c). These data were used to propose a correlation between H1 phosphate content and the "initial mechanism" of chromosome condensation in mitosis. Later studies of histone kinase supported the association of H1 phosphorylation with the initial phase of chromosome condensation (Section VI), and it is now widely accepted that H1 phosphorylation is associated with chromosome condensation in mitosis. The exact time course of phosphorylation and the stage of mitosis at which it is important remain controversial. The data of Fig. 13 show an increase in H1 phosphate during the mid- to late G_2 phase. The main differences, to be outlined in the next two sections, concern the drop in H1 phosphate. Gurley and his colleagues (Gurley *et al.*, 1978a), working with CHO cells, find that the drop in H1 phosphate occurs after metaphase. They suggested that the earlier *Physarum* data might be affected by phosphatase activity during histone isolation (Bradbury *et al.*, 1973c; Gurley *et al.*, 1978b). R. Inglis, in our laboratory,

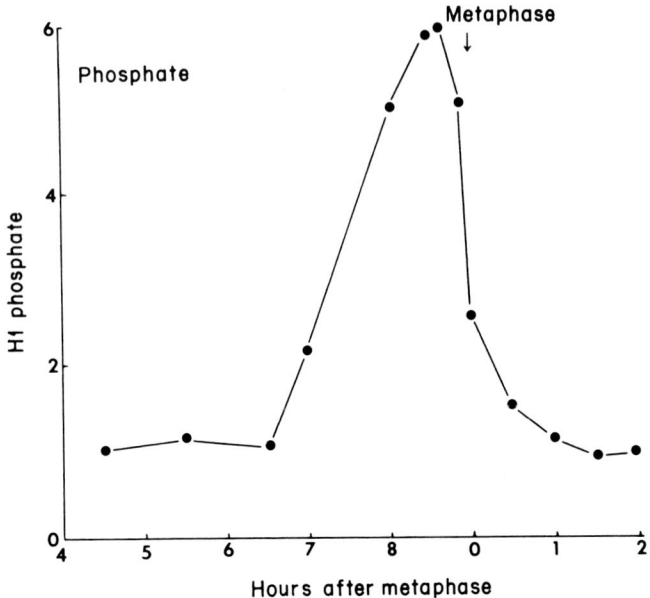

Fig. 13. Phosphate content of *Physarum* H1 histone in the cell cycle. The phosphate content of *Physarum* H1 was determined from long-term labeling with [^{3}H]lysine and [^{32}p]phosphate and normalized to 1.0 in the late S phase. (Mammalian cells have approximately 1 phosphate per H1 molecule in the S phase.) Bradbury *et al.* (1973c).

showed this to be incorrect by isolating histone from metaphase plasmodia using the phosphatase inhibitor 50 mM sodium bisulfite. The inhibitor had no effect on the results, although we showed that the phosphatase present in metaphase nuclei is inhibited 50% by 50 mM sodium bisulfite. It is possible that *Physarum* avoids the problems of phosphatase at metaphase by having an intranuclear mitosis. Fischer and Laemmli (1980), using SDS gels, obtained data that they interpret as showing the complete absence of H1 dephosphorylation in *Physarum*, and they suggest that Bradbury *et al.* (1973c) may have been selectively losing phosphorylated H1 during the S phase. This suggestion is inconsistent with the gel patterns, particularly the relative amounts of H1, reported by Bradbury *et al.* (1973c). More work is required to resolve the differences between these two experimental approaches in *Physarum*.

2. CELL CYCLE DEPENDENCE OF H1 PHOSPHATE (SDS GEL MOBILITY)

Mohberg and Rusch (1970) and Bradbury *et al.* (1973c) observed differences in the electrophoretic mobility of H1 isolated from different stages of the mitotic cycle, using gels containing acetic acid and urea. Fischer and Laemmli (1980) studied *Physarum* H1 using gel electrophoresis in SDS. In acid–urea gels the changes were assumed to be due to changes in net charge, probably due to phosphorylation. In SDS gels the changes must be assumed to be due to changes in the SDS-binding capacity. Treatment of slowly migrating H1 with alkaline phosphatase increased its mobility on SDS gels but not to the maximum mobility observed for newly synthesized H1. The basis of the alkaline phosphatase-resistant decrease in mobility is unknown, as is the mechanism whereby phosphorylation of H1 has a major effect on the SDS-binding capacity. It is interesting to speculate that this effect may reflect the *in vivo* function of H1 phosphorylation.

Fischer and Laemmli (1980) found that H1 labeled with [^{14}C]lysine in the S phase and extracted immediately migrated rapidly, although the total H1 showed both a slow and a fast band. At later times, the mobility of H1 labeled in the previous S phase was lower, reaching a minimum early in mitosis. In the next S phase, this H1 remained in the low mobility position. On the basis of these and other experiments, Fischer and Laemmli (1980) concluded that newly synthesized H1 was phosphorylated during the G_2 phase and then remained phosphorylated, although some turnover of phosphate apparently occurred. This interpretation is inconsistent with the conclusions of Bradbury *et al.* (1973c) that H1 phosphate falls dramatically in late mitosis. It also differs from the situation in CHO cells, where H1 phosphate also falls dramatically after metaphase (Gurley *et al.*, 1978a). The reason for these inconsistencies is not yet clear, but it should be resolved by parallel studies on acid–urea and SDS gels of H1's with differing phosphate contents and by extractions, in parallel, by different methods. This is particularly necessary because the more recent paper (Fischer and Laemmli, 1980) contains no detailed, quantitative cell cycle studies.

3. COMPARISON WITH MAMMALIAN CELLS

The most detailed cell cycle studies in mammalian cells have been carried out by Gurley *et al.* (1978), although Lake (Lake and Salzman, 1972; Lake *et al.*, 1972; Lake, 1973) and Chalkley's group (e.g., Balhorn *et al.*, 1971, 1972a,b; Sherod *et al.*, 1975; Jackson *et al.*, 1975, 1976; Tanphaichitr *et al.*, 1976) made important observations earlier (see also Langan and Hohman, 1974; Ajiro *et al.*, 1975; Langan, 1978a,b; Dolby *et al.*, 1979). Gurley *et al.* (1978a) synchronized CHO cells by extended treatment with hydroxyurea and isoleucine deprivation and then used Colcemid to block the cells in metaphase. They harvested cells at various times after release from hydroxyurea and analyzed them by electron microscopy, which allowed them to measure the proportion of cells in a number of stages of the G_2 phase and mitosis. H1 was isolated from the cells and analyzed by gel electrophoresis. The phosphate content of the H1 can be deduced approximately from these gel electrophoresis profiles and plotted as a function of time after hydroxyurea block (Fig. 14). These data are not directly comparable

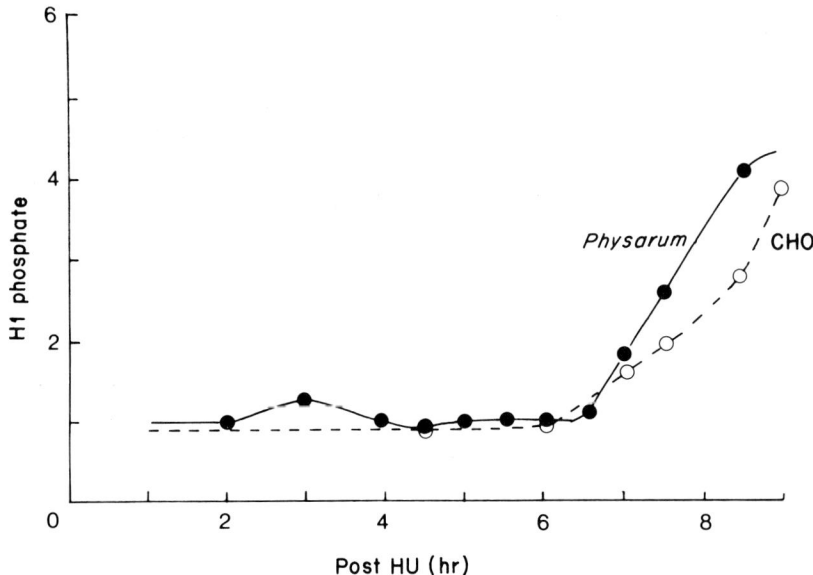

Fig. 14. Comparison of histone H1 phosphate content in *Physarum* and a CHO cell line (Bradbury *et al.*, 1973c; Gurley *et al.*, 1978a). CHO cells were synchronized by isoleucine deprivation and prolonged treatment with hydroxyurea (HU). Histone H1 was isolated and analyzed by gel electrophoresis at various times after release from hydroxyurea. The cells were not very synchronous by *Physarum* standards, and the cell cycle was further confused by adding Colcemid so that the cells accumulated in metaphase (Gurley *et al.*, 1978a). However, at each time point the fraction of cells at each of several stages recognized by electron microscopy was determined. This allowed us to construct an imaginary *Physarum* culture with the same proportion of nuclei at each stage and to calculate its average H1 phosphate content. The results shown in the figure indicate a remarkable similarity between *Physarum* H1 and CHO cell H1. The small difference in early prophase may be real, or it may reflect experimental uncertainties such as the effect of Colcemid.

with the *Physarum* data because *Physarum* is much more synchronous and is not blocked at metaphase. However, the *Physarum* data (Bradbury *et al.*, 1973c; Fig. 13) can be used to calculate the phosphate content that would be observed in a mixture of cell cycle stages equivalent to that obtained by Gurley *et al.* (1978a) at each time point. Fig. 14 shows the result of this calculation and allows a direct comparison of the CHO cell data and the *Physarum* data. Clearly, the two cell types have a very similar time course of H1 phosphorylation. It is likely, however, that the maximum phosphate content in CHO cells occurs at the point of Colcemid arrest, which, if this is equated with metaphase, is slightly later in mitosis than the maximum phosphate content in *Physarum*. This apparently minor difference may be an artifact due to the use of Colcemid or it may reflect the fact that *Physarum* has no G_1 phase (cf. Rao and Sunkara, 1980) in this stage of its life cycle and so must be ready for the S phase immediately after metaphase.

In CHO cells, the H1 phosphate content falls very rapidly after release of the cells from Colcemid arrest.

Tanphaichitr et al. (1976) used $ZnCl_2$ to inhibit histone phosphatase in growing cells and showed that chromosome decondensation was not inhibited. This result supports the original suggestion that H1 phosphorylation is involved in the initiation of chromosome condensation. Much more significantly, Matsumoto et al. (1980) have isolated a mutant mammalian cell line that is blocked in the late G_2 phase and fails to phosphorylate H1. Finally, it should be mentioned that the reports on the absence of H1 from the condensed chromosomes of *Tetrahymena* micronuclei are incorrect, and H1 (like H3) has now been found in this system (Gorovsky and Keevert, 1975; Gorovsky et al., 1974; M. Gorovsky, unpublished communication).

4. INTERACTION WITH DNA

The foregoing data suggest that H1 phosphorylation in the G_2 phase is involved with initiating chromosome condensation, and this is supported by studies of histone kinase described below. However, it has been pointed out that the

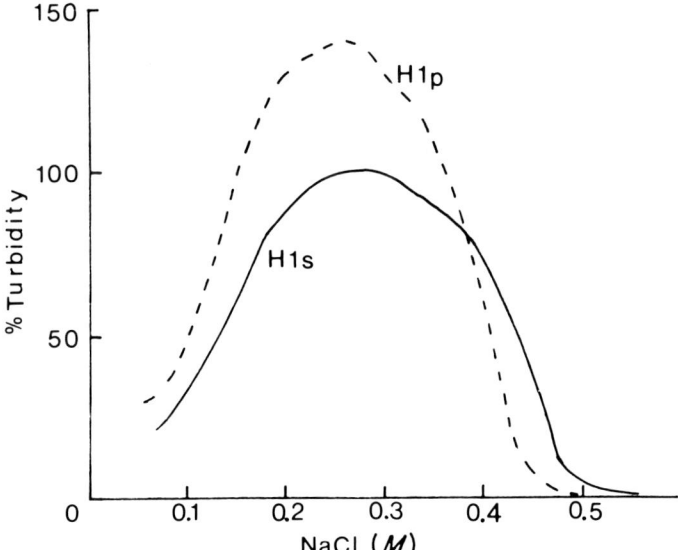

Fig. 15. Interaction of S phase- and prophase-histone H1 with DNA. Histone H1 was isolated from *Physarum* plasmodia in the S phase or prophase. The turbidity of H1·DNA solutions was studied (cf. Fig. 12), and the figure shows that prophase H1 (H1p) aggregated or condensed DNA much better than S phase H1 (H1s), at NaCl concentrations below 0.35 M (Corbett et al., 1980). Experiments with enzymically phosphorylated H1 and alkaline-phosphatase-treated H1 suggest that the difference between S phase H1 and prophase H1 can be accounted for by the high phosphate content of prophase H1. Matthews and Bradbury (1978).

simple electrostatic effect of phosphorylation of H1 is to reduce its strength of binding to DNA. This effect can be seen by physical studies of the interaction of phosphorylated H1 and DNA (Adler *et al.*, 1971, 1972; Rattle *et al.*, 1977; Fasy *et al.*, 1979; Dolby *et al.*, 1979; Fischer and Laemmli, 1980). Some authors have gone on to postulate that the reduced strength of binding would cause chromosome decondensation, contrary to the indications given by *in vivo* studies (Louie and Dixon, 1973; Jerzmanowski and Staron, 1980). There is no detailed basis for such arguments (Matthews, 1980a), but neither is there a detailed experimental picture of the effect of phosphorylation of H1 on chromatin structure, although rapid progress can be expected here.

An indication of the effect of phosphorylation on the H1-induced aggregation of DNA has been obtained from turbidity studies (Matthews and Bradbury, 1978). Studies with H1 isolated from *Physarum* plasmodia in either the S phase (low phosphate) or early prophase (high phosphate) have been reported (Corbett *et al.*, 1980). Figure 15 shows that both H1's show the typical pattern of aggregation of H1 + DNA at 0.1–0.4 M NaCl. However, the phosphorylated H1 shows much greater ability to aggregate DNA at moderate salt concentrations. This shows that the interaction of H1 with DNA is more complex than a simple electrostatic interaction and that phosphorylation at the sites associated with early prophase enables H1 to aggregate DNA more effectively, thus supporting the idea that early prophase H1 phosphorylation is involved with the initiation of chromosome condensation. At high salt concentrations, highly phosphorylated H1 is released from DNA, and so the turbidity falls before the turbidity of low-phosphorylated H1 + DNA.

B. Acetylation of Histone H4

1. INCORPORATION OF LABELED ACETATE

Histone H4 is modified after synthesis by acetylation of the N-terminal lysines in all eukaryotes studied. Allfrey (1977; Allfrey *et al.*, 1964) has suggested that this is associated with transcription, and Dixon (Sung and Dixon, 1970) has suggested that this is associated with DNA replication. *Physarum* plasmodia incorporate ^{14}C from [1-^{14}C]acetate into H4. After short pulses, less than 10 minutes, the label is found only in the acetylated species, although after longer pulses it is found in all species of H4. Amino acid analysis showed that incorporation after longer pulses included incorporation into several unmodified amino acids (Corbett *et al.*, 1977). The different acetylated forms of H4 can be separated by gel electrophoresis. Figure 16 shows incorporation of [^3H]lysine and [^{14}C]acetate into the different forms of *Physarum* H4 as separated by gel electrophoresis. There are a number of minor bands preventing accurate measurements on gels containing total histone, so subsequent experiments have been carried out with H4 purified by chromatography. The gel electrophoresis conditions can then be optimized for H4 fractionation and good separation obtained

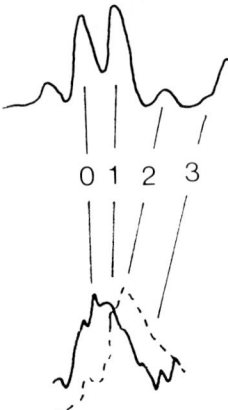

Fig. 16. Incorporation of [^{14}C]acetate into *Physarum* H4 histone (Corbett *et al.*, 1977). Total *Physarum* histone was fractionated by gel electrophoresis. The part of the gel containing H4 was stained and scanned, and a duplicate gel was sliced and its radioactivity determined. The upper trace shows the scan of the stained gel; electrophoresis is from right to left. The lower traces show radioactivity incorporated: solid line for [^{3}H]lysine, dotted line for [^{14}C]acetate. After some small corrections (V. J. Robinson and H. R. Matthews, unpublished), the relative net incorporation of [^{14}C]acetate into bands 0 to 3 was 0, 1.0, 2.2, 2.8 (^{14}C/^{3}H). Other bands on the gel overlap the H4 region, so subsequent work was carried out with prepurified H4. Chahal *et al.* (1980).

(Chahal *et al.*, 1980). Preliminary results from pulse labeling with [^{3}H]acetate show a high turnover of acetate in the diacetyl form, ac$_2$H4, during the S phase and a high turnover, relative to the number of molecules, of acetate in the tetraacetyl form, ac$_4$H4 in the G$_2$ phase (S. S. Chahal, H. R. Matthews, and E. M. Bradbury, unpublished).

2. CELL CYCLE CHANGES IN H4 ACETATE

Figure 17 shows an example of the separation of *Physarum* H4 into the five forms ac$_n$H4 (n = 0, 1, 2, 3, 4) by gel electrophoresis and illustrates the computer procedure that determines the proportion of H4 in each form, fully corrected for overlapping of the peaks. Histone H4 was isolated from *Physarum* plasmodia at various stages of the cell cycle and analyzed. Fig. 18 shows the changes in overall level of acetate through the cell cycle and also the total changes in more highly acetylated forms, ac$_n$H4 (n = 2, 3, 4), through the cell cycle. Fig. 19 shows the changes in two individual forms of H4, ac$_3$H4 and ac$_4$H4. The changes in each form are given by Chahal *et al.* (1980).

3. INVERSE CORRELATION WITH H1 PHOSPHATE AND CHROMOSOME CONDENSATION

The most striking feature of Figs. 18 and 19 is the sharp drop in acetate content in mitosis. This has also been observed in mammalian cells (D'Anna *et al.*, 1977). In *Physarum*, the proportion of ac$_4$H4 falls almost to zero (0.36%

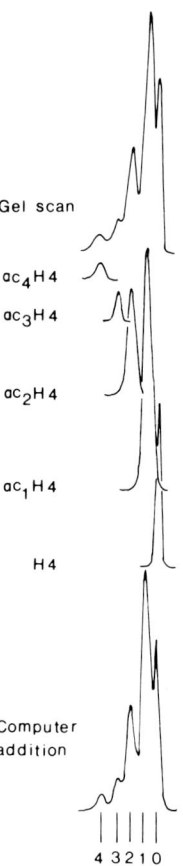

Fig. 17. Computer analysis of histone H4 acetylation. *Physarum* H4 histone from the S phase was isolated by chromatography and then fractionated by gel electrophoresis. The top profile is a densitometer scan of the stained gel. Electrophoresis is from left to right. The middle five profiles show the individual Gaussian components determined by computer-aided analysis, and the bottom profile shows the sum of these components, which compares with the experimental scan at the top.

compared with 3.6% in the S phase) in prophase, just when the H1 phosphate content reaches its maximum value. Consequently, acetylation of H4 appears to be inversely correlated with H1 phosphorylation. There is as yet no other evidence linking the two histone modifications or linking deacetylation with chromosome condensation. Such evidence may be forthcoming from studies of histone deacetylase activity. J. H. Waterborg (unpublished), in our laboratory, has prepared a synthetic substrate for histone H4 deacetylase by isolating the N-terminal 23-residue peptide from calf thymus H4 and chemically acetylating it with [^{3}H]acetic anhydride. This was suggested to us by V. G. Allfrey. The

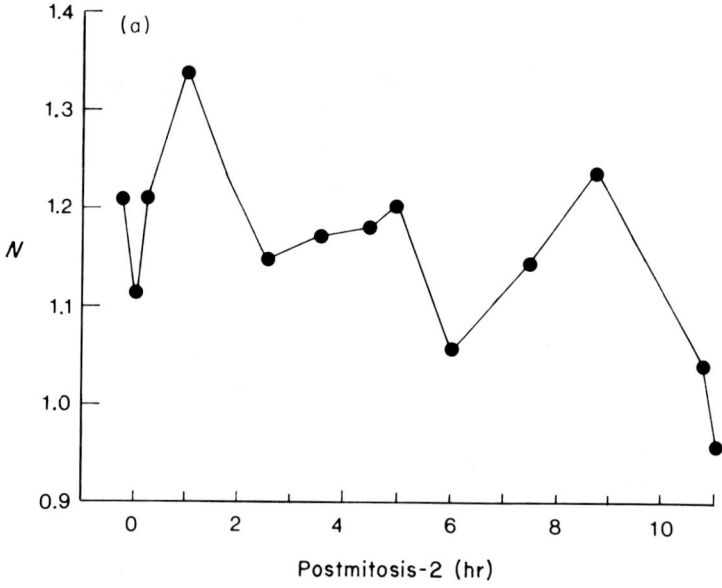

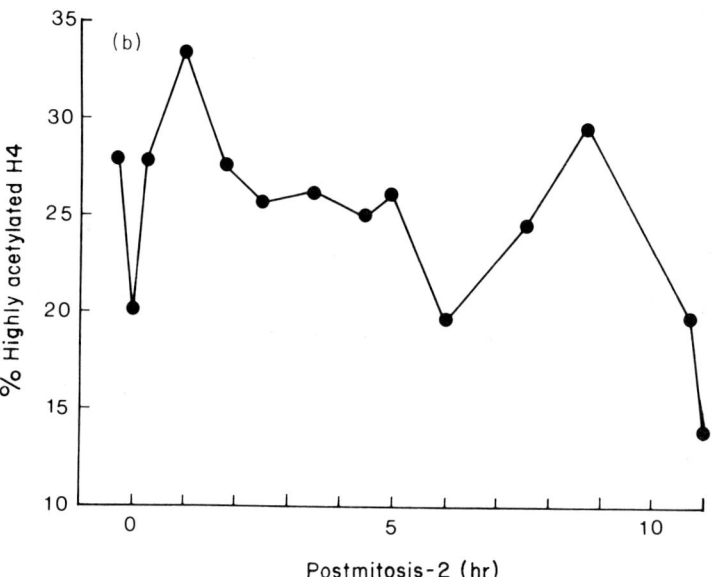

Fig. 18. (a) Acetate content of *Physarum* histone H4 in the cell cycle. The average number of acetates per H4 molecule was calculated from the data of Chahal *et al.* (1980) as $N = \sum_{n=1}^{4} n \cdot P(n)/\sum_{n=0}^{4} P(n)$, where $P(n)$ is the percent of total H4 occurring as the species with n acetates. Mitosis occurred at 0 hours and again at 11 hours. Notice the dip in mitosis and the S phase and G_2 phase peaks. (b) Highly acetylated H4 in the cell cycle. The percent of H4 in the four highly acetylated forms was calculated as $\sum_{n=2}^{4} P(n)$. Notice the dip in mitosis and the S and G_2 phase peaks.

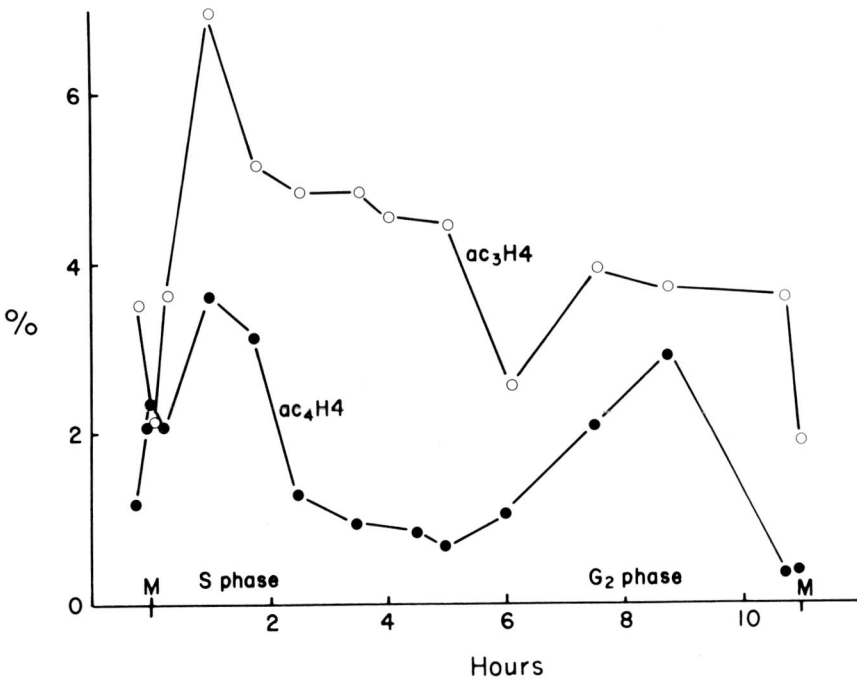

Fig. 19. Percent of total H4 histone in ac_3H4 or ac_4H4 forms throughout the cell cycle.

N-terminal peptide contains lysines only at positions 5, 8, 12, and 16 (and methyllysine at position 20), which are the positions acetylated *in vivo*, so chemical acetylation of this peptide is limited to the *in vivo* sites. Moreover, this N-terminal region is not thought to be part of the globular protein core of the nucleosome (Cary *et al.*, 1978), so the isolated peptide may be a good substitute for chromatin as far as assays for deacetylase are concerned. *Physarum* nuclei possess enzyme(s) that deacetylate this substrate (J. H. Waterborg, unpublished). It will be interesting to see how the cell cycle dependence of this enzyme and acetyltransferase (Cousens and Gallwitz, 1979; Sures and Gallwitz, 1980) compares with that of H1 histone kinase.

4. CORRELATION OF AC_4H4 WITH TRANSCRIPTION

Transcription in the *Physarum* plasmodial cell cycle is biphasic, with a minimum in mitosis and a second minimum in the early G_2 phase (Mittermayer *et al.*, 1964). There is a variety of evidence that the types of RNA synthesized during the S phase and the G_2 phase are different, and most of the evidence supports the hypothesis that transcription during the S phase produces predominantly heterogeneous nuclear RNA (hnRNA), which gives rise to messenger

RNA (mRNA), whereas transcription during the G_2 phase produces predominantly ribosomal RNA (discussed by Braun and Seebeck in Chapter 12 of this volume and by Matthews, 1981). The acetate content of H4 also shows a biphasic pattern through the cell cycle (Fig. 18), but the maxima and minima are not very marked. However, the amount of ac_4H4 alone does show a very marked biphasic pattern (Fig. 19) that is positively correlated with transcription (Chahal et al., 1980). The data suggest that the previously observed correlations of H4 acetylation with transcription (Allfrey, 1977) reflected specific changes in ac_4H4.

Since G_2 phase transcription in *Physarum* is probably concentrated on the genes for ribosomal RNA (rDNA) (e.g., Grant, 1972), the level of ac_4H4 on these genes should be normal or low in the S phase and strikingly high in the G_2 phase. This can be tested since ribosomal chromatin is localized in nucleoli and, in one experiment so far, has been found to be correct (S. S. Chahal, H. R. Matthews, and E. M. Bradbury, unpublished).

5. CORRELATION OF AC_2H4 TURNOVER WITH DNA REPLICATION

If ac_4H4 is correlated with transcription, where does this leave Dixon's proposal that H4 acetate is associated with chromosome replication? Dixon's (Sung and Dixon, 1970; Louie et al., 1973) proposal involved turnover of acetate groups on H4, so the effects might be difficult to see when looking at acetate content. However, pulse labeling with [^{3}H]acetate shows high turnover of ac_2H4, specifically in the S phase (S. S. Chahal, H. R. Matthews, and E. M. Bradbury, unpublished), and the acetate content data are consistent with this finding. These results correlate with data from duck erythroid cells (Ruiz-Carrillo et al., 1975) and from hepatoma cell cultures (Jackson et al., 1976), which suggest that H4 enters the nucleus in the diacetyl form and is then deacetylated after binding to DNA during chromosome replication.

C. Chromosome Structural Transitions

During the mitotic cycle in *Physarum,* as in other eukaryotes, there are at least four different metastable chromosome structures: metaphase chromosomes; replicating chromatin; nonreplicating chromatin active in transcription (perhaps approximately equated with euchromatin); and nonreplicating chromatin inactive in transcription (perhaps approximately equated with heterochromatin) (Bradbury et al., 1981). Replicating chromatin passes through a stage in which it has incomplete nucleosomes and an incomplete complement of nucleosomes. Metaphase chromosomes and nonreplicating chromatin both have a nucleosome structure, as

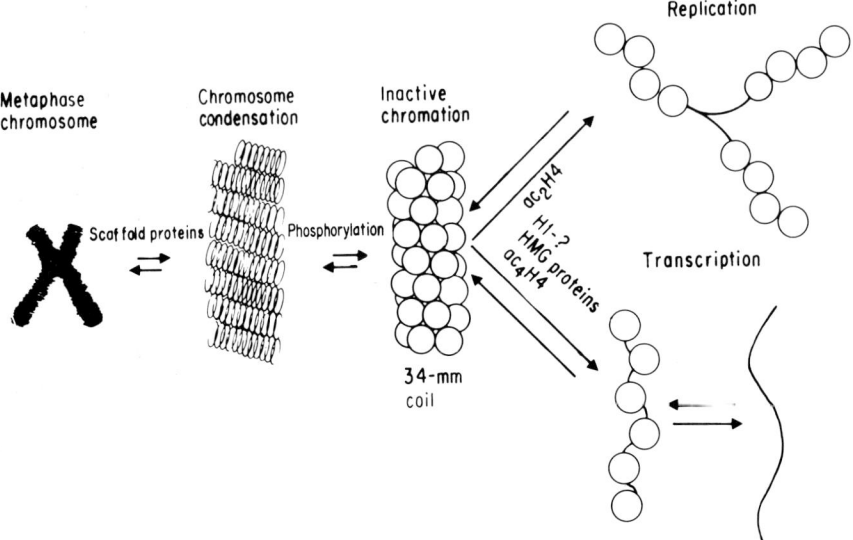

Fig. 20. Chromosome structural transitions. The various structural states are shown in diagrammatic form only.

far as micrococcal nuclease digestion goes (Johnson *et al.*, 1976; Vogt and Braun, 1976a), but actively transcribing chromatin has a more open nucleosome structure as judged by its nuclease sensitivity (Stalder *et al.*, 1978, 1979; Johnson *et al.*, 1978a,b) and a tendency to be isolated as peak A in gradients containing nucleosomes (Johnson *et al.*, 1978a,b). Transitions between these structures probably involved non-histone proteins such as polymerases, "scaffold" proteins (Adolph *et al.*, 1977; Paulson and Laemmli, 1977), and HMG proteins (a class of relatively abundant non-histone proteins, Goodwin *et al.*, 1978), but the foregoing data on histone modification suggest that it may be a major factor in initiating the structural transition to metaphase chromosomes and may be closely involved in all the other transitions. Chahal *et al.* (1980) have proposed a scheme summarizing the involvement of histone modification in chromosome structural transitions. A slightly modified scheme is shown in Fig. 20, which brings out the probable involvement of ac_4H4 in transcription and ac_2H4 in replication. A major feature of the scheme is the central role of the "30-nm" coil of nucleosomes, in which the histones are essentially unmodified. This 30-nm coil can be further coiled or condensed by phosphorylation of H1. Conversely, it can be opened out for DNA processing by acetylation of H4. Further work is necessary to characterize the different structural states in more detail and to identify the other factors involved in the structural transitions.

VI. NUCLEAR HISTONE KINASE

A. Cell Cycle Dependence of Activity

Of the chromosomal enzymes, the polymerases have been discussed elsewhere in this book, and some others that have been studied in *Physarum*, such as thymidine kinase and polyadenosine diphosphoribose polymerase [poly(ADPR)-polymerase], are discussed in the next section. This section is devoted to nuclear histone kinase because of its relation to histones, because of the amount of information available, and because of the evidence that its activity is important for the initiation of mitosis.

The cell cycle dependence of *Physarum* nuclear histone kinase was determined by assaying 0.4 M NaCl extracts of nuclei, and very large changes were observed. Fig. 21 shows that the kinase activity was low (but not zero) during the S phase and increased 15-fold during the early and mid-G_2 phase. In the late G_2 phase it fell very rapidly, to reach its low level again just before the next S phase. The data suggested that this increase in kinase activity was the immediate cause of the rise in H1 histone phosphate described above (Fig. 13) (Bradbury *et al.*, 1974a). There was also a correlation between the histone kinase activity and the results of plasmodial fusion experiments and heat shock experiments in *Physarum* (Section VI,E).

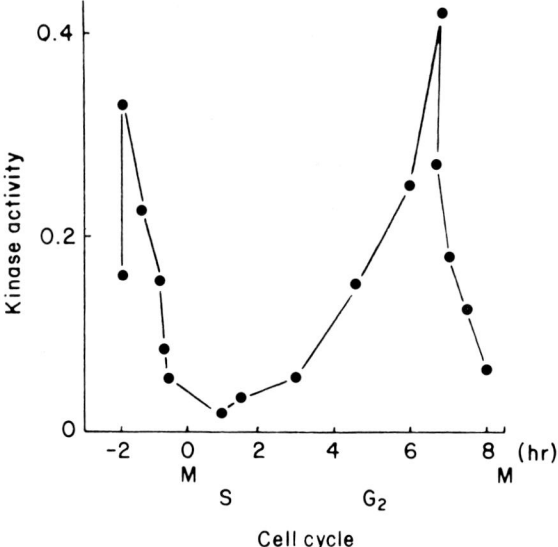

Fig. 21. Cell cycle changes in nuclear histone kinase.

B. Activation of Histone Kinase

The original data on enzyme activity did not address the question: "Is the change in activity due to enzyme synthesis or to another mechanism, such as activation or transport?" This question was approached using the technique of deuterium labeling and fractionation on Metrizamide density gradients. The enzyme was located in the gradient from its activity, and the resolution was sufficient to separate normal enzyme from enzyme grown in the presence of deuterated amino acids, which are more dense. Plasmodia grown from fusion to mitosis-2 in normal medium, then transferred to deuterium-containing medium and harvested in the G_2 phase, showed two major peaks of histone kinase activity on a Metrizamide gradient. One coincided with that of normal enzyme; the other was taken to be deuterium-labeled enzyme. The two peaks were of comparable size, showing that the main cause of the 15-fold increase in activity was not enzyme synthesis but rather an activation process (Mitchelson *et al.*, 1978).

It has been suggested that the equilibration of amino acid pools might be too slow for the deuterium-labeling experiment, but this is probably not the case since the peak of labeled enzyme was as sharp as the peak of normal enzyme. Slowly equilibrating pools would give a range of densities for labeled enzyme, which was not observed. There were also two smaller peaks on the Metrizamide gradients, probably due to different levels of hydration of the kinase, but in terms of labeling they both behaved like the major peak.

The search for an activation mechanism has so far been unsuccessful. *Physarum* has cyclic AMP-dependent kinase activity in the cytoplasm (A. M. Campbell, unpublished), but the histone kinase activity extracted from nuclei is not affected by cyclic AMP (Matthews, 1980b). The total activity also appeared not to be affected by the heat-stable inhibitor of cyclic AMP-dependent protein kinases (Ashby and Walsh, 1972; Cohen *et al.*, 1977). However, as described below, the kinase activity can be resolved into three components. One of these components remains unaffected by cyclic AMP and the inhibitor; the other two components are unaffected by cyclic AMP but do respond to the inhibitor, and evidence is accumulating that one of these components is analogous to the catalytic subunit of mammalian protein kinase (Chambers, 1980). Chambers (1980) also looked for an effect of alkaline phosphatase on *Physarum* histone kinase using alkaline phosphatase bound to polyacrylamide beads (suggested to us by Dr. S. Shall). He was unable to demonstrate a specific effect due to the phosphatase, suggesting that the kinase is not activated by phosphorylation, but this cannot be regarded as a rigorous conclusion.

C. Three Nuclear Histone Kinases

Hardie *et al.* (1976) showed that *Physarum* nuclear histone kinase could be separated into three components by ion-exchange chromatography but were unsuccessful in purifying any of the components further. Chambers (1980) confirmed the fractionation of DEAE-cellulose and obtained a small degree of purification by prior gel filtration on Sepharose 6B. However, the enzyme has remained unstable except in crude extracts and has consequently resisted efforts to purify it. Three peaks are obtained for histone kinase on Metrizamide gradients, but these peaks do not correspond to the different enzymes (Mitchelson *et al.*, 1978). Rather, each separated enzyme runs on Metrizamide, like the major peak from unfractionated enzyme. Small quantities of a nuclear histone kinase have been purified from Ehrlich ascites cells (Schlepper and Knippers, 1975).

The three *Physarum* kinases were assayed through the separation procedures using H1 histone substrate. In the case of one of the enzymes, the most strongly bound to DEAE-cellulose, H1 histone is a better substrate than any of the other histones and much better than protamine. In contrast, the other enzyme that binds to DEAE-cellulose, called "kinase A," phosphorylates histone H2B twice as well as histone H1 and protamine three-times as well as H1. H2A is about half as good as H1, whereas H3 and H4 alone are very poor substrates. There is evidence of H2A phosphorylation in *Physarum*, but the substrate specificity measured with isolated histones may not be a good guide to their availability as substrates in nucleosomes (Hardie *et al.*, 1976). The differential behavior of the enzymes with protamine as substrate can be used to assay for both enzymes in a mixture of the two enzymes or to distinguish readily the enzymes (Mitchelson *et al.*, 1978).

The cell cycle dependence of the two enzymes that bound to DEAE-cellulose was measured (Hardie *et al.*, 1976), and both showed a large peak of activity in the G_2 phase, similar to that observed with the total activity. However, the precise timing was different, with kinase A peaking one hour later than the strongly bound kinase. Hardie *et al.* (1976) speculated that the enzymes might phosphorylate different sites sequentially on H1. The substrate specificity and cell cycle dependence of the enzyme that did not bind to DEAE-cellulose have not been measured.

D. Sites of Phosphorylation *in Vitro*

1. LOCATION ON CALF HISTONE H1

Chambers (1980), partly in our laboratory and partly in collaboration with T. A. Langan, phosphorylated calf H1 with *Physarum* kinase A. The phosphorylated H1 was subjected to peptide analysis and a major phosphorylation site, serine-37, was found. This is the same site specificity shown by the catalytic subunit of cyclic AMP-dependent protein kinase (Langan, 1969, 1978a), and since *Physarum* kinase A is also inhibited by the inhibitor of cyclic AMP-

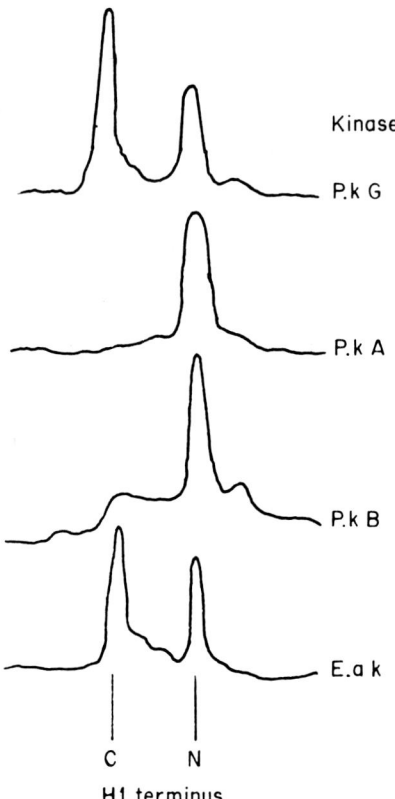

Fig. 22. Phosphorylation of calf thymus H1 histone by various kinases. Calf thymus H1 was phosphorylated *in vitro* by separated *Physarum* kinases or Ehrlich ascites growth-associated kinase (a gift from Dr. T. A. Langan) and [γ-^{32}P]ATP. The labeled H1 from each incubation was partly digested with chymotrypsin, and the fragments were separated by gel electrophoresis. The gels were autoradiographed and the autoradiographs scanned. The top scan shows H1 phosphorylated by *Physarum* kinase G (not bound to DEAE–cellulose); the next used *Physarum* kinase A (analogous to the catalytic subunit of cyclic AMP-dependent protein kinase); the next used *Physarum* kinase B (strongly bound to DEAE-cellulose); and the bottom scan used Ehrlich ascites growth-associated kinase. Electrophoresis was from right to left, and the two major peaks are the N- and C-terminal halves of H1. Chambers (1980).

dependent protein kinase, *Physarum* kinase A is tentatively regarded as analogous to the catalytic subunit of cyclic AMP-dependent protein kinase (Chambers, 1980).

In a later study, the three *Physarum* kinases and Ehrlich ascites kinase (Langan, 1978b) were used to phosphorylate calf thymus H1 histone. The phosphorylated histone was partially digested with chymotrypsin, as described above (Section IV,C,2), and the fragments were separated by gel electrophoresis. The gels were autoradiographed, and scans of the autoradiographs are shown in Fig. 22. *Physarum* kinase A phosphorylated only the N-terminal fragment, residues 1–106, consistent with serine-37 being the major phosphorylation site. *Physarum* kinase B (which binds strongly to DEAE–cellulose and is specific for H1) phosphorylated mainly the N-terminal peptide but showed some phosphorylation in the C-terminal region. *Physarum* kinase G (which does not bind to DEAE–cellulose) and Ehrlich ascites kinase G phosphorylated H1 at sites in both peptides in a very similar manner (Chambers, 1980).

2. LOCATION ON *PHYSARUM* HISTONE H1

Physarum histone H1 was phosphorylated in the same way with the separated *Physarum* enzymes. Partial chymotryptic digestion was carried out, and the N- and C-terminal peptides were separated. The site specificity showed some differences from the case of calf thymus H1, with kinase A showing substantial phosphorylation in the C-terminal peptide as well as in the N-terminal peptide and kinase B showing more phosphorylation in the C-terminal peptide than in the N-terminal peptide. These results imply that the C-terminal region of *Physarum* contains a phosphorylation site or sites that is absent in calf thymus H1. *Physarum* kinase G showed phosphorylation of *Physarum* H1 very similar to that of calf thymus H1 (Chambers, 1980).

E. Cell Cycle Role of Histone Kinase

The cell cycle behavior of histone kinase activity correlates with plasmodial fusion experiments and heat shock experiments (Rusch *et al.*, 1966; Brewer and Rusch, 1968; Chin *et al.*, 1972; Bradbury *et al.*, 1974a). The data suggested that histone kinase activity was a controlling or rate-limiting factor in the timing of mitosis, and attempts were made to affect the timing of mitosis directly by adding exogenous histone kinase. Prior *et al.* (1980) have subsequently shown that *Physarum* will take up and use exogenous protein in an intact form, in the case of histone H3. The kinase preparations available were from Ehrlich ascites cells, but Fig. 23 shows that even this heterologous kinase was very effective in changing the time of mitosis, bringing it forward by up to 60 minutes. Various control substances, including inactivated kinase preparations, were without effect (Bradbury *et al.*, 1974b; Inglis *et al.*, 1976; Trakht *et al.*, 1980). Antibody to the histone kinase preparation specifically delayed mitosis when added to plasmodia (Zanker *et al.*, 1977). There is, however, a major difficulty with these experiments, namely, that the histone kinase was only partly purified (Inglis *et al.*, 1976), and so, the possibility that the effect was due to another component of the preparation cannot be ruled out. Nevertheless, an important indication that

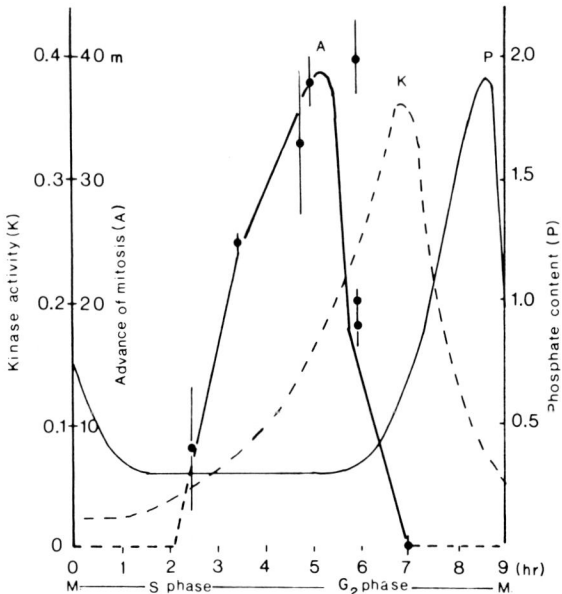

Fig. 23. Control of mitosis by histone kinase. The thin solid line (P) shows the change in histone H1 phosphate content throughout the *Physarum* cell cycle. The broken line (K) shows the change in histone kinase throughout the cycle. The heavy solid line (A) shows the advance of mitosis produced by adding histone kinase at various times in the cycle. Bradbury *et al.* (1973c, 1974a,b).

histone kinase is the active ingredient is provided by the cell cycle dependence of the advancement of mitosis. If histone kinase is added during the normal rise of histone kinase activity, it is very effective in advancing mitosis; if addition takes place after the normal rise in activity, then no effect is observed; and if addition takes place before the normal rise, then no effect is observed. The data are consistent with histone kinase playing a "triggering" role in the cell cycle, in which it is unable to act before the end of DNA synthesis (perhaps because of high phosphatase activity) and in which a number of events, not normally rate-limiting, are triggered by histone kinase activity and result in mitosis (Matthews *et al.*, 1976; Matthews, 1980a). This scheme is consistent with Sachsenmaier's proposal of a trigger substance occupying nuclear sites if these sites are equated with the phosphorylation sites on H1 histone (Sachsenmaier, 1976).

It is possible, for example with cycloheximide, to block mitosis much later in the cell cycle than the maximum of histone kinase activity (e.g., Tyson *et al.*, 1979). This does not mean that histone kinase is not normally the rate-limiting step. Rather, it confirms that there are other steps before mitosis that are triggered by histone kinase (Bradbury *et al.*, 1974a).

VII. NON-HISTONE CHROMOSOMAL PROTEINS

A. HMG (High Mobility Group) Proteins

Many eukaryotes possess a small group of proteins bound to chromatin by weak ionic linkages. They can be isolated by washing nuclei or chromatin with 0.35 M NaCl and fractionating the extract by precipitation with trichloracetic acid. Several HMG proteins have been extensively characterized (Goodwin *et al.*, 1978), and they are of particular interest because they are released from chromatin by mild digestion with DNase-1 (Levy-W. *et al.*, 1977) in parallel with the digestion of active genes. It has been reported that HMG proteins are necessary for the expression of DNase-1 sensitivity by active genes (Weisbrod and Weintraub, 1979; Weisbrod *et al.*, 1980). *Physarum* contains a group of HMG-like proteins (Matthews *et al.*, 1979), but they have not yet been extensively characterized.

B. Enzymes

Nucleic acid polymerases are discussed elsewhere in this volume (T. Evans, Chapter 11, and Braun and Seebeck, Chapter 12). Braun and Seebeck also discuss proteins that stimulate or inhibit transcription (Sauer *et al.*, 1969; Ernst and Sauer, 1977; Atmar *et al.*, 1978; Hildebrandt *et al.*, 1979; Kuen *et al.*, 1979). Also in relation to transcription, Christensen *et al.* (1977) have described a nuclear RNP protein in *Physarum*. Magun (1979) has studied cytoplasmic DNA-binding phosphoproteins.

Poly(ADPR)polymerase activity has been found associated with *Physarum* chromatin, and its activity was inversely correlated with DNA replication (Brightwell *et al.*, 1975). Wielckins *et al.* (1979) have distinguished two kinds of ADPR linkage: a hydroxylamine-resistant linkage which is found to be synthesized in the S phase and a hydroxylamine-sensitive linkage which increases sharply at the S/G$_2$ boundary. Very little ADPR appears to be bound to histone H1 *in vivo*. The function of protein modification by ADPR is still not clear.

An interesting and potentially useful enzyme for cell cycle studies is thymidine kinase (see also Tyson, Chapter 3, this volume). Thymidine kinase is normally a peak enzyme with low activity through most of the mitotic cycle in *Physarum* plasmodia and high activity just before and during metaphase (Sachsenmaier and Ives, 1965; Sachsenmaier *et al.*, 1967). The peak of activity is due to enzyme synthesis, as shown by density labeling experiments (Oleinick, 1972). At a low temperature, 22°C, thymidine kinase behaves as a "step" enzyme, increasing in activity in mitosis (Wright and Tollon, 1979a). Wright and Tollon (1979a) interpret their data in terms of continuous synthesis and degradation of thymidine kinase, with an increased rate of synthesis in mitosis. At low temperatures, the rate of degradation seems to be sharply reduced, giving the appearance of a step enzyme. Various agents that block or delay mitosis have a similar effect on thymidine kinase (Sachsenmaier *et al.*, 1967, 1970; Oleinick, 1972; Wright and Tollon, 1979b). However, it is possible to uncouple the peak of thymidine kinase from mitosis using heat shock, so that the peak of thymidine kinase occurs in the absence of mitosis (Wright and Tollon, 1979b). This suggests that thymidine kinase synthesis and mitosis have a common initiation event but that mitosis can be specifically inhibited after that event.

Thymidine kinase can be separated into a number of enzyme variants (Grobner and Sachsenmaier, 1976). Some of the components can be interconverted by phosphorylation/dephosphorylation (Grobner, 1979a,b). Mohberg *et al.* (1980) have described thymidine kinase-deficient mutants of *Physarum*. Woertz and Sachsenmaier (1979) have described two forms of deoxyadenosine kinase that vary during the cell cycle in *Physarum*.

C. Contractile Proteins

Physarum nuclei contain nuclear proteins analogous to actin and myosin (Jockusch *et al.*, 1970, 1971, 1973, 1974). *Physarum* actin has an amino acid sequence very similar to that of mammalian actin (Vandekerckhove and Weber, 1978), although the actomyosin complexes differ (D'Haese and Hinssen, 1978). *Physarum* actin copurifies with RNA polymerase II (Smith *et al.*, 1979). It has also been identified in gel electrophoresis studies of *Physarum* nuclear nonhistone proteins (LeStourgeon and Rusch, 1973; LeStourgeon *et al.*, 1971, 1973a,b, 1975; Nations *et al.*, 1974). In these studies the effects of changes in growth conditions, particularly starvation, were determined and changes in a number of protein bands were observed. In particular, *Physarum* nuclear actin increased in concentration in proliferative cell states, and a Mg^{2+}-dependent actomyosin-binding protein was present only in proliferative states and always appeared before mitosis occurred on refeeding. Myosin was found in nuclei and was enriched in chromosomes (LeStourgeon *et al.*, 1975).

ACKNOWLEDGMENTS

We are grateful to Dr. V. G. Allfrey, Dr. E. M. Johnson, Dr. T. A. Langan, Dr. R. Braun, and Dr. J. Tyson for discussions and access to unpublished work and to our colleagues for critical reading of the manuscript. We are grateful to E. M. Johnson for Fig. 6, to C. Cooney for Fig. 7, and to R. Mueller for Fig. 8. Our work was supported by the Science Research Council and the Cancer Research Campaign, both of the United Kingdom, by NATO, and recently by NIH grant GM 26901.

REFERENCES

Adler, A. J., Shaffhaussen, B., Langan, T. A., and Fasman, G. D. (1971). Altered conformational effects of phosphorylated lysine-rich histone (f-l) in f-l deoxyribonucleic acid complexes: Circular dichroism and immunological studies. *Biochemistry* **10**, 909–913.

Adler, A. J., Langan, T. A., and Fasman, G. D. (1972). Complexes of deoxyribonucleic acid with lysine-rich (F1) histone phosphorylated at two separate sites: Circular dichroism studies. *Arch. Biochem. Biophys.* **153**, 769.

Adolph, K. W., Cheng, S. M., and Laemmli, U. K. (1977). Role of nonhistone proteins in metaphase chromosome structure. *Cell* **12**, 805–816.

Affolter, H.-U., and Braun, R. (1978). Ribosomal DNA in spores of *Physarum polycephalum*. *Biochim. Biophys. Acta* **519**, 118–124.

Affolter, H. U., Behrens, K., Seebeck, T., and Braun, R. (1979). Large scale isolation of ribosomal DNA from giant surface cultures of *Physarum polycephalum*. *FEBS. Lett.* **107**, 340–342.

Ajiro, K., Borun, T., and Cohen, L. (1975). Phosphorylation sites of histone 1 (F1) in relation to the cell cycle. *Fed. Proc., Fed. Am. Soc. Exp. Biol.* **34**, 581.

Allan, J., Hartman, P. G., Crane-Robinson, C., and Aviles, F. X. (1980). The structure of histone H1 and its location in chromatin. *Nature (London)* **288**, 675–679.

Allfrey, V. G. (1977). Post-synthetic modifications of histone structure. *In* "Chromatin and Chromosome Structure" (H. J. Li and R. A. Eckhardt, eds.), p. 167. Academic Press, New York.

Allfrey, V. G., Faulkener, R. M., and Mirsky, A. E. (1964). Acetylation and methylation of histones and their possible role in the regulation of RNA synthesis. *Proc. Nat. Acad. Sci. U.S.A.* **51**, 786–794.

Ashby, C. D., and Walsh, D. A. (1972). Characterization of the interaction of a protein inhibitor with adenosine 3′, 5′-mono-phosphate-dependent protein kinases. I. Interaction with the catalytic subunit of the protein kinase. *J. Biol. Chem.* **247**, 6637–6242.

Atmar, V. J., Daniels, G. R., and Kuehn, G. D. (1978). Polyamine stimulation of phosphorylation of nonhistone acidic protein in nuclei and nucleoli from *Physarum polycephalum*. *Eur. J. Biochem.* **90**, 29–37.

Balhorn, R., Riecke, O., and Chalkley, R. (1971). Rapid electrophoretic analysis for histone phosphorylation. A reinvestigation for phosphorylation of lysine-rich histone during rat liver regeneration. *Biochemistry* **10**, 3952–3959.

Balhorn, R., Balhorn, M., Morris, H. P., and Chalkley, R. (1972a). Comparative high-resolution electrophoresis of tumor histones: Variation in phosphorylation as a function of cell replication rate. *Cancer Res.* **32**, 1775–1784.

Balhorn, R., Bardwell, J., Sellers, L., Granner, D., and Chalkley, R. (1972b). Histone phosphorylation and DNA synthesis are linked in synchronous cultures of HTC cells. *Biochem. Biophys. Res. Commun.* **46**, 1326–1333.

Bonner, W. M., West, M. H. P., and Stedman, J. D. (1980). Two-dimensional gel analysis of histones in acid extracts of nuclei, cells, and tissues. *Eur. J. Biochem.* **109**, 17–23.

Borun, T. W., Ajiro, K., Zweidler, A., Dolby, T. W., and Stephens, R. E. (1977). Studies of human histone messenger RNA. II. The resolution of fractions containing individual human histone messenger RNA species. *J. Biol. Chem.* **252**, 173–180.

Bradbury, E. M., and Matthews, H. R. (1981). Structure and function of chromatin. *In* "The Fungal Nucleus" (K. Gull and S. Oliver, eds., Cambridge Univ. Press, New York (in press).

Bradbury, E. M., Maclean, N., and Matthews, H. R. (1981). "Chromosome Structure and Function." Blackwells, Oxford.

Bradbury, E. M., Matthews, H. R., McNaughton, J., and Molgaard, H. V. (1973a). Sub-nuclear components of *Physarum polycephalum*. *Biochim. Biophys. Acta* **335**, 19–29.

Bradbury, E. M., Carpenter, B. G., and Rattle, H. W. E. (1973b). Magnetic resonance studies of deoxyribonucleoprotein. *Nature (London)* **241**, 123-126.

Bradbury, E. M., Inglis, R. J., Matthews, H. R., and Sarner, N. (1973c). Phosphorylation of very-lysine-rich histone in *Physarum polycephalum:* Correlation with chromosome condensation. *Eur. J. Biochem.* **33**, 131-139.

Bradbury, E. M., Inglis, R. J., and Matthews, H. R. (1974a). Control of cell division by very lysine-rich histone phosphorylation: Correlation with chromosome condensation. *Nature (London)* **247**, 257-261.

Bradbury, E. M., Inglis, R. J., Matthews, H. R., and Langan, T. A. (1974b). Molecular basis of control of mitotic cell division in eukaryotes. *Nature (London)* **249**, 553-556.

Brewer, E. N., and Rusch, H. P. (1968). Effect of elevated temperature shocks on mitosis and on the initiation of DNA replication in *Physarum polycephalum*. *Exp. Cell Res.* **49**, 79-86.

Brightwell, M. D., Leech, C. E., O'Farrell, M. K., Whish, W. J. D., and Shall, S. (1975). Poly (adenosine diphosphate ribose) polymerase in *Physarum polycephalum*. *Biochem. J.* **147**, 119-129.

Campbell, G. R., Littau, V. C., Melera, P. W., Allfrey, V. G., and Johnson, E. M. (1979). Unique sequence arrangement of ribosomal genes in the palindromic rDNA molecule of *Physarum polycephalum*. *Nucleic Acids Res.* **6**, 1433-1447.

Cary, P. D., Moss, T., and Bradbury, E. M. (1978). High resolution proton magnetic resonance studies of chromatin core particles. *Eur. J. Biochem.* **89**, 475-482.

Chahal, S. S., Matthews, H. R., and Bradbury, E. M. (1980). Acetylation of histone H4 and its role in chromatin structure and functions. *Nature (London)* **287**, 76-79.

Chambers, T. (1980). Histone kinase. Ph.D. Thesis, C.N.A.A., Portsmouth, England.

Chin, B., Friedrich, P. D., and Bernstein, I. A. (1972) Stimulation of mitosis following fusion of plasmodia in the myxomycete *Physarum polycephalum*. *J. Gen. Microbiol.* **71**, 93-101.

Christensen, M. E., Beyer, A. L., Walker, B., and LeStourgeon, W. M. (1977). Identification of NG-NG-dimethyl-arginine in a nuclear protein from the lower eukaryote *Physarum polycephalum* homologous to the major proteins of mammalian 40S ribonucleoprotein particles. *Biochem. Biophys. Res. Commun.* **74**, 621-629.

Cohen, P., Nimmo, G. A., and Antoniw, J. F. (1977). Specificity of a protein phosphatase inhibitor from rabbit skeletal muscle. *Biochem. J.* **162**, 435-444.

Cole, R. D. (1977). *In* "The Molecular Biology of the Mammalian Genetic Apparatus" (P. Tso, ed.), pp. 93-104. Elsevier, Amsterdam.

Compton, J. L., Bellard, M., and Chambon, P. (1976). Biochemical evidence of variability in the DNA repeat length in the chromatin of higher eukaryotes. *Proc. Nat. Acad. Sci. U.S.A.* **73**, 4382-4386.

Corbett, S. (1979). Histone modification in *Physarum polycephalum*. Ph.D. Thesis, C.N.A.A., Portsmouth, England.

Corbett, S., Miller, S., Robinson, V. J., Matthews, H. R., and Bradbury, E. M. (1977). *Physarum polycephalum* histones. *Biochem. Soc. Trans.* **5**, 943-946.

Corbett, S., Bradbury, E. M., and Matthews, H. R. (1980). Histone H1 from prophase aggregates DNA better than histone H1 from S phase. *Exp. Cell Res.* **128**, 127-132.

Cousens, D. G., and Gallwitz, D. (1979). Different accessibilities in chromatin to histone acetylase. *J. Biol. Chem.* **254**, 1716-1723.

D'Anna, J. A., Tobey, R. A., Barham, S. S., and Gurley, L. R. (1977). A reduction in the degree of H4 acetylation during mitosis in Chinese hamster cells. *Biochem. Biophys. Res. Commun.* **77**, 187-202.

D'Haese, J., and Hinssen, H. (1978). Contraction properties of isolated slime mould actomyosin. I. Comparison of thread models made of natural, recombined, and hybridised actomyosins from slime mould and muscle. *Protoplasma* **95**, 273-296.

Daniel, J. W., and Baldwin H. H. (1964). Methods for culture of plasmodial myxomycetes. *Methods Cell Physiol.* **1,** 9-41.

Dolby, T. W., Ajiro, K., Borun, T., Gilmour, R. S., Zweidler, A., Cohen, L., Miller, P., and Nicolini, C. (1979). Physical properties of DNA and chromatin isolated from G1 and S phase HeLa S-3 cells. Effects of histone H1 phosphorylation and stage specific non-histone chromosomal proteins in the molar ellipticity of native and reconstituted nucleoproteins during thermal denaturation. *Biochemistry* **18,** 1333-1343.

Ernst, G. H., and Sauer, H. W. (1977). A nuclear elongation factor of transcription from *Physarum polycephalum in vitro. Eur. J. Biochem.* **74,** 253-261.

Evans, H. H., and Evans, T. E. (1970). Methylation of the deoxyribonucleic acid of *Physarum polycephalum* at various periods during the mitotic cycle. *J. Biol. Chem.* **245,** 6436-6441.

Fasy, T. M., Inoue, A., Johnson, E. M., and Allfrey, V. G. (1979). Phosphorylation of H1 and H5 histones by cyclic AMP dependent protein kinase reduced DNA binding. *Biochim. Biophys. Acta* **564,** 322-334.

Fischer, S. G., and Laemmli, U. K. (1980). Cell cycle changes in *Physarum polycephalum* histone H1 phosphate: Relationship to deoxyribonucleic acid binding and chromosome condensation. *Biochemistry* **19,** 2240-2246.

Foe, V. E., Wilkinson, L. E., and Laird, C. D. (1976). Comparative organisation of active transcription units in *Oncopeltus fasciatus. Cell* **9,** 131 146.

Goldknopf, I. L., and Busch, H. (1977). Isopeptide linkage between nonhistone and histone 2A polypeptides of chromosomal conjugate-protein A24. *Proc. Natl. Acad. Sci. U.S.A.* **74,** 864-868.

Goodwin, G. H., Walker, J. M., and Johns, E. W. (1978). The high mobility group (HMG) chromosomal non-histone proteins. *In* "The Cell Nucleus" (H. Busch, ed.), pp. 181-219. Academic Press, New York.

Gorovsky, M. A., and Keevert, J. B. (1975). Absence of histone F1 in a mitotically dividing, genetically inactive nucleus. *Proc. Nat. Acad. Sci. U.S.A.* **72,** 2672-2776.

Gorovsky, M. A., Keevert, J. B., and Pleger, G. L. (1974). Histone F1 of *Tetrahymena* macronuclei: Unique electrophoretic properties and phosphorylation of F1 in an amitotic nucleus. *J. Cell Biol.* **61,** 134-145.

Grainger, R. M., and Ogle, R. C. (1978). Chromatin structure of the ribosomal RNA genes in *Physarum polycephalum. Chromosoma* **65,** 115-126.

Grant, W. D. (1972). The effect of alpha-amanitin and $(NH_4)_2SO_4$ on RNA synthesis in nuclei and nucleoli isolated from *Physarum polycephalum* at different times during the cell cycle. *Eur. J. Biochem.* **2,** 94-98.

Grobner, P. (1979a). Thymidine kinase (EC 2.7.1.21)enzyme variants in *Physarum polycephalum: In vitro* interconversion of the enzyme variants. *J. Biochem.* **86,** 1595-1606.

Grobner, P. (1979b). Thymidine kinase (EC 2.7.1.21) enzyme variants in *Physarum polycephalum:* Kinetics and properties of the enzyme variants. *J. Biochem.* **86,** 1607-1614.

Grobner, P., and Sachsenmaier, W. (1976). Thymidine kinase enzyme variants in *Physarum polycephalum:* Change of pattern during the synchronous mitotic cycle. *FEBS. Lett.* **72,** 181-184.

Gubler, U., Wyler, T., and Braun, R. (1979). The gene for 26S rRNA in *Physarum* contains two insertions. *FEBS. Lett.* **100,** 347-350.

Gubler, U., Wyler, T., Seebeck, T., and Braun, R. (1980). Processing of ribosomal precursor RNAs in *Physarum polycephalum. Nucleic Acids Res.* **8,** 2647-2664.

Gurley, L. R., D'Anna, J. A., Barham, S. S., Deaven, L. L., and Tobey R. A. (1978a). Histone phosphorylation and chromatin structure during mitosis in Chinese Hamster cells *Eur. J. Biochem.* **84,** 1-16.

Gurley, L. R., Tobey, R. A., Walters, P. A., Hilderbrand, C. E., Hohman, P. G., D'Anna, J. A.,

Barham, S. S., and Deaven, L. L. (1978b). *In* "Cell Cycle Regulation" (J. R. Jeter, I. L. Cameron, G. M. Padilla, and A. M. Zimmerman, eds.), pp. 37-60. Academic Press, New York.

Hall, L., and Braun, R. (1977). The organisation of genes for transfer RNA and ribosomal RNA in amoebae and plasmodia of *Physarum polycephalum*. *Eur. J. Biochem.* **76**, 165-174.

Hardie, D. G., Matthews, H. R., and Bradbury, E. M. (1976). Cell-cycle dependence of two nuclear histone kinase enzyme activities. *Eur. J. Biochem.* **66**, 37-42.

Hardman, N., Jack, P. L., Brown, A. J. P., and McLachlan, A. (1979a). Distribution of inverted repeat sequences in nuclear DNA from *Physarum polycephalum*. *Eur. J. Biochem.* **94**, 179-187.

Hardman, N., Jack, P. L., Brown, A. J. P., and McLachlan, A. (1979b). Characterisation of ribosomal satellite in total nuclear DNA from *Physarum polycephalum*. *Biochim. Biophys. Acta* **562**, 365-376.

Hildebrandt, A., Mengel, R., and Sauer, H. (1979). Characterisation of an endogenous transcription inhibitor from *Physarum polycephalum*. *Z. Naturforsch.* **34C**, 76-86.

Hozier, J. C., and Kaus, R. (1976). Sub-unit structure of chromosomes in mitotic nuclei of *Physarum polycephalum*. *Chromosoma* **57**, 95-102.

Inglis, R. J., Langan, T. A., Matthews, H. R., Hardie, D. G., and Bradbury, E. M. (1976). Advance of mitosis by histone phosphokinase. *Exp. Cell Res.* **97**, 418-425.

Jackson, V., Shires, A., Chalkley, R., and Granner, D. K. (1975). Studies on highly metabolically active acetylation and phosphorylation of histones. *J. Biol. Chem.* **250**, 4856-4863.

Jackson, V., Shires, A., Tanphaichitr, N., and Chalkley, R. (1976). Modifications to histones immediately after synthesis. *J. Mol. Biol.* **104**, 471-483.

Jalouzot, R., Briane, D., Ohlenbusch, H. H., Wilhelm, M. L., and Wilhelm, F. X. (1980). Kinetics of nuclease digestion of *Physarum polycephalum* nuclei at different stages of the cell cycle. *Eur. J. Biochem.* **104**, 423-450.

Jerzmanowski, A., and Staron, K. (1980). Mg as a trigger of condensation-decondensation transition of chromatin during mitosis. *J. Theor. Biol.* **82**, 41-46.

Jockusch, B. M., and Walker, I. O. (1974). The preparation and preliminary characterisation of chromatin from the slime mold *Physarum polycephalum*. *Eur. J. Biochem.* **48**, 417-425.

Jockusch, B. M., Brown, D. F., and Rusch, H. P. (1970). Synthesis of a nuclear protein in G2 phase. *Biochem. Biophys. Res. Commun.* **38**, 279-283.

Jockusch, B. M., Brown, D. F., and Rusch, H. P. (1971). Synthesis and some properties of an actin-like nuclear protein in the slime mold *Physarum polycephalum*. *J. Bacteriol.* **108**, 705-714.

Jockusch, B. M., Ryser, U., and Behnke, O. (1973). Myosin-like protein in *Physarum* nuclei. *Exp. Cell Res.* **76**, 464-466.

Jockusch, B. M., Becker, M., Hindenrach, I., and Jockusch, H. (1974). Slime mould actin: Homology to vertebrate actin and presence in the nucleus. *Exp. Cell Res.* **89**, 241-246.

Johns, E. W. (1976). Fractionation and isolation of histones. *In* "Subnuclear Components" (G. D. Bernie, ed.). Butterworth, London.

Johnson, E. M. (1980). A family of inverted repeat sequences containing specific single-strand gaps at the termini of the *Physarum* ribosomal gene palindrome. *Cell* **22**, 857-886.

Johnson, E. M., Littau, V. C., Allfrey, V. G., Bradbury, E. M., and Matthews, H. R. (1976). The sub-unit structure of chromatin from *Physarum polycephalum*. *Nucleic Acids Res.* **3**, 3313-3329.

Johnson, E. M., Allfrey, V. G., Bradbury, E. M., and Matthews, H. R. (1978a). Altered nucleosome structure containing DNA sequences complementary to 19S and 26S ribosomal RNA in *Physarum polycephalum*. *Proc. Natl. Acad. Sci. U.S.A.* **75**, 1116-1120.

Johnson, E. M., Matthews, H. R., Littau, V. C. Lothstein, L., Bradbury, E. M., and Allfrey, V. G. (1978b). The structure of chromatin containing DNA complementary to 19S and 26S ribosomal

RNA in active and inactive stages of *Physarum polycephalum*. *Arch. Biochem. Biophys.* **191,** 537-550.

Johnson, E. M., Campbell, G. R., and Allfrey, V. G. (1979). Different nucleosome structures on transcribing and non-transcribing ribosomal gene sequences. *Science* **206,** 1192-1194.

Kuehn, G. D., Affolter, U.-U., Atmar, V. J., Seebeck, T., Gubler, U., and Braun, R. (1979). Polyamine-mediated phosphorylation of a nucleolar protein from *Physarum polycephalum* that stimulates rRNA synthesis. *Proc. Nat. Acad. Sci. U.S.A.* **76,** 2541-2545.

Laemmli, U. (1970). Cleavage of structural proteins during assembly of the head of bacteriophage T4. *Nature (London)* **227,** 680-685.

Lake, R. S. (1973). F1-histone phosphorylation in metaphase chromosomes of cultured Chinese hamster cells. *Nature (London), New Biol.* **242,** 145-146.

Lake, R. S., and Salzman, N. P. (1972). Occurence and properties of a chromatin-associated F-1 histone phosphokinase in mitotic Chinese hamster cells. *Biochemistry* **11,** 4817-4826.

Lake, R. S., Goidl, J. A., and Salzman, N. P. (1972). F1 histone modification at metaphase in Chinese hamster cells. *Exp. Cell Res.* **73,** 113-121.

Langan, T. A. (1969). Phosphorylation of liver histone following the administration of glucagon and insulin. *Proc. Nat. Acad. Sci. U.S.A.* **64,** 1276-1283.

Langan, T. A. (1978a). Isolation of histone kinases. *Methods Cell Biol.* **19,** 143-152.

Langan, T. A. (1978b). Methods for the assessment of site-specific histone phosphorylation. *Methods Cell Biol.* **19,** 127-142.

Langan, T. A., and Hohman, P. (1974). Phosphorylation of threonine and serine residues of lysine-rich histone in growing cells. *Fed. Proc., Fed. Am. Soc. Exp. Biol.* **33,** 1597.

LeStourgeon, W. M., and Rusch, H. P. (1973). Localization of nucleolar and chromatin residual acidic protein changes during differentiation in *Physarum polycephalum*. *Arch. Biochem. Biophys.* **155,** 144-158.

LeStourgeon, W. M., Wallace, M., and Rusch, H. P. (1971). Nuclear acidic protein changes during differentiation in *Physarum polycephalum*. *Science* **174,** 1233-1236.

LeStourgeon, W. M., Nations, C., and Rusch, H. P. (1973a). Temporal synthesis and intranuclear accumulation of the nuclear acidic proteins during periods of chromatin reactivation in *Physarum polycephalum*. *Arch. Biochem. Biophys.* **159,** 861-872.

LeStourgeon, W. M., Wray, W., and Rusch, H. P. (1973b). Functional homologies of acidic chromatin proteins in higher and lower eukaryotes. *Exp. Cell Res.* **79,** 487-492.

LeStourgeon, W. M., Forer, A., Yang, Y-Z., Bertram, J. S., and Rusch, H. P. (1975). Contractile proteins: Major components of nuclear and chromosome non-histone proteins. *Biochim. Biophys. Acta* **379,** 529-552.

Levy-W., B., Wong, N. C. W., and Dixon, G. H. (1977). Selective association of the trout-specific H6 protein with chromatin regions susceptible to DNase II: Possible location of HMG-T in the spacer region between core nucleosomes. *Proc. Nat. Acad. Sci. U.S.A.* **74,** 2810-2814.

Lohr, D., Cordieu, J., Tatchell, K., Kovak, R. T., and van Holde, K. E. (1977). A comparative subunit structure of HeLa, yeast and chicken erythrocyte chromatin. *Proc. Nat. Acad. Sci. U.S.A.* **74,** 79-83.

Louie, A. J., and Dixon, G. H. (1973). Kinetics of phosphorylation of testes histones and their possible role in determining chromosomal structure. *Nature (London), New Biol.* **243,** 164-168.

Louie, A. J., Candido, E. P. M., and Dixon, G. H. (1973). Enzymatic modifications and their possible role in regulating the binding of basic proteins to DNA and controlling chromosome structure. *Cold Spring Harbor Symp. Quant Biol.* **38,** 803-819.

Lutter, L. C. (1979). Precise location of DNase I cutting sites in the nucleosome core determined by high resolution gel electrophoresis. *Nucleic Acids Res.* **6,** 41-56.

McGhee, J. D., and Felsenfeld, G. (1980). Nucleosome structure. *Annu. Rev. Biochem.* **49,** 1115-1156.

Magun, B. (1979). Changes in cytoplasmic DNA-binding phosphoproteins during the cell cycle of *Physarum polycephalum*. *Cell Differ.* **8,** 157-172.

Matsumoto, Y., Hasuda, H., Mita, S., Marunouchi, T., and Yamada, M. (1980). Evidence for the involvement of H1 histone phosphorylation in chromosome condensation. *Nature (London)* **284,** 181-183.

Matthews, H. R. (1977a). The structure of transcribing chromatin. *Nature (London)* **267,** 203-204.

Matthews, H. R. (1977b). Phosphorylation of H1 and chromosome condensation. *In* "The Organisation and Expression of the Eukaryotic Genome" (E. M. Bradbury and K. Javaherian, eds.), pp. 67-80. Academic Press, New York.

Matthews, H. R. (1980a). Chromosome condensation in mitosis. *J. Theor. Biol.* **83,** 367-368.

Matthews, H. R. (1980b). Modification of histone H1 by reversible phosphorylation and its relation to chromosome condensation and mitosis. *In* "Protein Phosphorylation in Regulation" (P. Cohen, ed.), pp. 235-254. Elsevier, Amsterdam.

Matthews, H. R. (1981). Chromatin proteins and progress through the cell cycle. *In* "The Cell Cycle" (P. John, ed.) pp. 223-246. Cambridge Univ. Press, London.

Matthews, H. R., and Bradbury, E. M. (1978). The role of histone H1 phosphorylation in the cell cycle: Turbidity studies of H1-DNA interaction. *Exp. Cell Res.* **111,** 343-351.

Matthews, H. R., Hardie, D. G., Inglis, R. J., and Bradbury, E. M. (1976). The molecular basis of control of mitotic cell division. *Life Sci. Res. Rep.* **1,** 395-408.

Matthews, H. R., Johnson, E. M., Steer, W. M., Bradbury, E. M., and Allfrey, V. G. (1978). The use of netropsin with CsCl gradients for the analysis of DNA and its application to restriction nuclease fragments of ribosomal DNA from *Physarum polycephalum*. *Eur. J. Biochem.* **82,** 569-576.

Matthews, H. R., Chahal, S. S., Miller, S., Inglis, R. J., and Bradbury, E. M. (1979). *Physarum* chromatin. *In* "Current Research on *Physarum*" (W. Sachsenmaier, ed.), pp. 51-58. Univ. of Innsbruch, Austria.

Mitchelson, K., Chambers, T., Bradbury, E. M., and Matthews, H. R. (1978). Activation of histone kinase in G2 phase of the cell cycle in *Physarum polycephalum*. *FEBS. Lett.* **92,** 339-342.

Mittermayer, C., Braun, R., and Rusch, H. P. (1964). RNA synthesis in the mitotic cycle of *Physarum polycephalum*. *Biochim. Biophys. Acta* **91,** 399-405.

Mohberg, J., and Rusch, H. P. (1969). Isolation of the nuclear histones from the myxomycete *Physarum polycephalum*. *Arch. Biochem. Biophys.* **134,** 577-589.

Mohberg, J., and Rusch, H. P. (1970). Nuclear histones in *Physarum polycephalum* during growth and differentiation. *Arch. Biochem. Biophys.* **138,** 418-432.

Mohberg, J., and Rusch, H. P. (1971). Isolation and DNA content of nulcei of *Physarum polycephalum*. *Exp. Cell Res.* **66,** 305-316.

Mohberg, J., Dworzak, E., and Sachsenmaier, W. (1980). Thymidine kinase-deficient mutants of *Physarum polycephalum:* Biochemical characterization. *Exp. Cell Res.* **126,** 351-357.

Molgaard, H. V., Matthews, H. R., and Bradbury, E. M. (1976). Organisation of genes for ribosomal RNA in *Physarum polycephalum*. *Eur. J. Biochem.* **68,** 541-549.

Nations, C., LeStourgeon, W. M., Magun, B. E., and Rusch, H. P. (1974). The rapid intranuclear accumulation of preexisting proteins in response to high plasmodial density in *Physarum polycephalum*. *Exp. Cell Res.* **88,** 207-215.

Oleinick, N. L. (1972). The radiation sensitivity of mitosis and the synthesis of thymidine kinase in *Physarum polycephalum:* A comparison to the sensitivity to Actinomycin D and cycloheximide. *Radiat. Res.* **51,** 638-653.

Panyim, S., and Chalkley, R. (1969). High resolution acrylamide gel electrophoresis of histones. *Arch. Biochem. Biophys.* **130,** 337-346.

Panyim, S., and Chalkley, R. (1971). The molecular weights of vertebrate histones exploiting a modified sodium dodecyl sulfate electrophoretic method. *J. Biol. Chem.* **246,** 7557-7560.

Paulson, J. R., and Laemmli, U. K. (1977). The structure of histone-depleted metaphase chromosomes. *Cell* **12,** 817-828.
Prior, C. P., Cantor, C. R., Johnson, E. M., and Allfrey, V. G. (1980). Incorporation of exogenous pyrene-labeled histone into *Physarum* chromatin: A system for studying changes in nucleosomes assembled *in vivo*. *Cell* **20,** 597-608.
Rao, P. N., and Sunkara, P. S. (1980). Correlation between the high rate of protein synthesis during mitosis and the absence of G1 period in V79-8 cells. *Exp. Cell Res.* **125,** 507-510.
Rattle, H. W. E., Langan, T. A., Danby, S. E., and Bradbury, E. M. (1977). Studies on the role and mode of operation of the very-lysine-rich histones in eukaryote chromatin. Effect of A and B site phosphorylation on the conformation and interaction of histone H1. *Eur. J. Biochem.* **81,** 499-505.
Reilly, J. G., Braun, R. and Thomas, C. A. (1980). Methylation in *Physarum* DNA. *FEBS Lett.* **116,** 181-184.
Riggs, M. G., Whittaker, R. G., Neumann, J. R., and Ingram, V. M. (1977). n-Butyrate causes histone modification in HeLa and Friend erythroleukaemia cells. *Nature (London)* **268,** 462-464.
Ruiz-Carrillo, A., Waugh, L. J., and Allfrey, V. G. (1975). Processing of newly synthesized histone molecules: Nascent histone H4 chains are reversibly phosphorylated and acetylated. *Science* **190,** 117-128.
Rusch, H. P., Sachsenmaier, W., Behrens, K., and Gruiter, V. (1966). Synchronization of mitosis by the fusion of the plasmodia of *Physarum polycephalum*. *J. Cell Biol.* **31,** 204-209.
Ryser, U., and Braun, R. (1974). The amount of DNA coding for rRNA during differentiation (spherulation) in *Physarum polycephalum*. *Biochim. Biophys. Acta* **361,** 33-36.
Sachsenmaier, W. (1976). Control of synchronous nuclear mitosis in *Physarum polycephalum*. *In* "The Molecular Basis of Circadian Rythms" (J. W. Hastings and H.-G. Schweiger, eds.), pp. 409-420. Dahlem Konferenzen, Berlin.
Sachsenmaier, W., and Ives, D. H. (1965). Periodische anderungen der thymidinkinase aktivitat im synchronen mitose cyclus von *Physarum polycephalum*. *Biochem. Z.* **343,** 399-406.
Sachsenmaier, W., Von Fournier, D., and Gurtler, K. F. (1967). Periodic thymidine kinase production in synchronous plasmodia of *Physarum polycephalum:* Inhibition by actinomycin and actidion. *Biochem. Biophys. Res. Commun.* **27,** 655-660.
Sachsenmaier, W., Bohnert, E., Clausnizer, B., and Nygaard, O. F. (1970). Cycle dependent variation of X-ray effects on synchronous mitosis and thymidine kinase induction in *Physarum polycephalum*. *FEBS. Lett.* **10,** 185-189.
Sauer, H. W., Goodman, E. M., Babcock, K. L., and Rusch, H. P. (1969). Polyphosphate in the life cycle of *Physarum polycephalum* and its relation to RNA synthesis. *Biochim. Biophys. Acta* **195,** 401-409.
Scheer, U., Trendelenburg, M. F., Krohne, G., and Franke, W. W. (1977). Lengths and patterns of transcriptional units in the amplified nucleoli of oocytes of *Xenopus laevis*. *Chromosoma* **60,** 147-167.
Schlepper, J., and Knippers, R. (1975). Nuclear protein kinases from murine cells. *Eur. J. Biochem.* **60,** 209-220.
Seebeck, T., Stalder, J., and Braun, R. (1979). Isolation of a minichromosome containing the ribosomal genes from *Physarum polycephalum*. *Biochemistry* **18,** 484-490.
Sherod, D., Johnson, G., Balhorn, R., Jackson, V., Chalkley, R., and Granner, D. (1975). The phosphorylation region of lysine-rich histone in dividing cells. *Biochim. Biophys. Acta* **381,** 337-347.
Smith, S. S., Kelly, K. H., and Jokhusch, B. M. (1979). Actin co-purifies with RNA polymerase II. *Biochem. Biophys. Res. Commun.* **86,** 161-166.
Stalder, J., Seebeck, T., and Braun, R. (1978). Degradation of the ribosomal genes by DNase I in *Physarum polycephalum*. *Eur. J. Biochem.* **90,** 391-395.

Stalder, J., Seebeck, T., and Braun, R. (1979). Accessibility of the ribosomal genes to micrococcal nuclease in *Physarum polycephalum. Biochim. Biophys. Acta* **561**, 452–463.

Staron, K., Jerzmanowski, A., Tyniec, B., Urbanska, A., and Toczo, K. (1977). Nucleoprotein chromatin sub-unit from *Physarum polycephalum. Biochim. Biophys. Acta* **475**, 131–138.

Steer, W. M., Molgaard, H. V., Bradbury, E. M., and Matthews, H. R. (1978). Ribosomal genes in *Physarum polycephalum:* Transcribed and non-transcribed sequences have similar base compositions. *Eur. J. Biochem.* **88**, 599–605.

Suau, P., Kneale, G. G. Braddock, G. W., Baldwin, J. P., and Bradbury, E. M. (1977). A low resolution model for the chromatin core particle by neutron scattering. *Nucleic Acids Res.* **4**, 3769–3786.

Sun, I. Y.-C., Johnson, E. M., and Allfrey, V. G. (1979). Initiation of transcription of ribosomal DNA sequences in isolated nuclei of *Physarum polycephalum:* Studies using nucleoside 5′ gamma S triphosphates and labelled precursors. *Biochemistry* **18**, 4556–4563.

Sung, M. T., and Dixon, G. H. (1970). Modification of histones during spermiogenesis in trout: A molecular mechanism for altering histone binding to DNA. *Proc. Nat. Acad. Sci. U.S.A.* **67**, 1616–1623.

Sures, I., and Gallwitz, D. (1980). Histone-specific acetyltransferases from calf thymus. Isolation, properties and substrate specificity of three different enzymes. *Biochemistry* **19**, 943–951.

Swofford, L. K. (1979). Chromatin structure of the ribosomal genes in *Physarum polycephalum*. Ph.D. Thesis, Cornell Univ., Ithaca, New York.

Tanphaichitr, N., Moore, K. C., Granner, D., and Chalkley, R. (1976). Relationship between chromosome condensation and metaphase lysine-rich histone phosphorylation. *J. Cell Biol.* **69**, 43–50.

Thoma, F., Koller, Th., and Klug, A. (1979). Involvement of histone H1 in the organisation of the nucleosome and of the salt-dependent superstructure of chromatin. *J. Cell Biol.* **83**, 403–427.

Trakht, I. N., Grozdova, I. D., Gulyaev, N. N., Severin, E. S., and Gnuchev, N. V. (1980). Effect of some protein kinases, cyclic nucleotides and specific phosphorylation inhibitors on the onset time of mitosis in the myxomycete *Physarum polycephalum. Biokhimiya (Moscow)* **45**, 788–793.

Tyrsin, Yu. A., Krasheninnikov, I. A., and Tyrsina, E. G. (1977a). Histones from *Physarum polycepahlum. Biokhimiya (Moscow)* **42**, 898–905.

Tyrsin, Yu. A., Krasheninnikov, I. A., and Tyrsina, E. G. (1977b). Homology of histone fractions H2B from calf thymus and P4B from *Physarum polycephalum. Biokhimiya (Moscow)* **42**, 1056–1061.

Tyson, J., Garcia-Herdugo, G., and Sachsenmaier, W. (1979). Control of nuclear division in *Physarum polycephalum*. Comparison of cycloheximide pulse treatment, UV irradiation, and heat shocks. *Exp. Cell Res.* **119**, 87–98.

Vandekerckhove, J., and Weber, K. (1978). The amino-acid sequence of *Physarum* actin. *Nature (London)* **276**, 720–721.

Vogt, V., and Braun, R. (1976a). Repeated structure of chromatin in metaphase nuclei of *Physarum. FEBS. Lett.* **64**, 190–192.

Vogt, V. and Braun, R. (1976b). Structure of ribosomal DNA in *Physarum polycephalum. J. Mol. Biol.* **106**, 567–587.

Von Holt, C., Strickland, W. N., Brandt, W. F., and Strickland, M. (1979). More histone structures. *FEBS. Lett.* **100**, 201–218.

Weintraub, H., and Groudine, M. (1976). Chromosomal subunits in active genes have an altered conformation. *Science* **193**, 848–856.

Weisbrod, S., and Weintraub, H. (1979). Isolation of a sub class of nuclear proteins responsible for conferring a DNase I-sensitive structure on globin chromatin. *Proc. Nat. Acad. Sci. U.S.A.* **76**, 630–635.

Weisbrod, S., Groudine, M., and Weintraub, H. (1980). Interaction of HMG 14 and 17 with actively transcribed genes. *Cell* **19,** 289–302.

Wielckens, K., Sachsenmaier, W., and Hilz, H. (1979). Protein-bound mono (adenosine-diphosphate-ribose) levels during the cell cycle of the slime mold *Physarum polycephalum*. *Hoppe Seyler's Z. Physiol. Chem.* **360,** 39–43.

Woertz, G., and Sachsenmaier, W. (1979). Deoxyadenosine kinase in *Physarum polycephalum*. In "Current Research on *Physarum*" (W. Sachsenmaier, ed.), pp. 123–129. Univ. of Innsbruch, Austria.

Woodcock, C. L. F., Frado, L. L.-Y., Hatch, C. L., and Ricciardiello, L. (1976). Fine structure of active ribosomal genes. *Chromosoma* **58,** 33–39.

Wright, M., and Tollon, Y. (1979a). *Physarum* thymidine kinase: A step or a peak enzyme depending on temperature of growth. *Eur. J. Biochem.* **96,** 177–182.

Wright, M., and Tollon, Y. (1979b). Regulation of thymidine kinase (EC 2.7.1.21) synthesis during the cell cycle of *Physarum polycephalum* by the heat-sensitive system which triggers mitosis and S phase. *Exp. Cell Res.* **122,** 273–280.

Yao, M.-C., and Gall, J. G. (1977). A single integrated gene for ribosomal RNA in a eukaryote, *Tetrahymena pyriformis*. *Cell* **12,** 121–132.

Zanker, K., Inglis, R. J., Matthews, H. R., and Bradbury, E. M. (1977). Cross-reacting nuclear non-histone antigens from *Physarum polycephalum* and Ehrlich ascites cells. *Biochem. Soc. Trans.* **5,** 953–957.

CHAPTER 11

Organization and Replication of DNA in *Physarum polycephalum*

THOMAS E. EVANS

I.	The Groundwork	371
II.	Early Advances	372
	A. Interphase and Mitosis	372
	B. Temporal Order of DNA Replication	372
	C. Protein Synthesis Required for DNA Replication	372
	D. Other (Nonchromosomal) DNAs	373
	E. *In Vitro* Studies	374
	F. Details of the Primary Structure	374
III.	Current Research Areas	374
	A. Genome Organization	374
	B. DNA Replication	380
	References	386

I. THE GROUNDWORK

Work reported in the early papers from the Rusch group established *Physarum polycephalum* as a valuable organism for use in experimental biology. Daniel and Rusch (1961) described the axenic culture of *P. polycephalum* on semidefined liquid medium; Guttes *et al.* (1961) developed straightforward cytological methods which they used to follow the progress of nuclei through the division cycle; and Nygaard *et al.* (1960) reported that a discrete S period occurred immediately following mitosis, so that the timing of nuclear DNA synthesis could be accurately predicted by following the cytological changes of the nuclei through division. This latter result (synchronous division immediately followed by the S period) opened a number of avenues for DNA research.

II. EARLY ADVANCES

In the following decade, biochemical and molecular biological studies using *P. polycephalum* made important and often pioneering contributions to the general knowledge of DNA metabolism. In most instances, researchers took direct advantage of the ability to carry out straightforward biochemical analyses of the nuclear division cycle of the plasmodium. Various aspects of the work to be described below have been reviewed recently by Turnock (1979), Goodman (1980), and Holt (1980). Therefore, except to state selected conclusions for the purpose of establishing an information base for this chapter, this material will not be reevaluated.

A. Interphase and Mitosis

Sachsenmaier and Rusch (1964) used the antimetabolite fluorodeoxyuridine to demonstrate that completion of DNA replication is required before (vegetative) nuclear division can occur. Similarly, their work indicated that a postreplication interphase period (G_2) was indispensable for the completion of the normal mitotic cycle. Other groups have subsequently used fluorodeoxyuridine to investigate additional aspects of DNA metabolism, confirming and extending the observations of Sachsenmaier and Rusch (Rao and Gontcharoff, 1960; Nygaard *et al.*, 1973).

B. Temporal Order of DNA Replication

A landmark molecular biological study was carried out by Braun *et al.* (1965). Density-labeling DNA with the thymidine analog bromodeoxyuridine, these workers were able to demonstrate that DNA replicated in one portion of the S period would again replicate in the comparable portion of the subsequent S period. The results of this so-called anniversary experiment were confirmed and extended by Braun and Wili (1969). The general experimental approach was also important in the later studies of Muldoon *et al.* (1971; see Section II,C) and Wille and Kauffman (1975) (see Chapter 3 on periodic phenomena by Tyson, this volume).

C. Protein Synthesis Required for DNA Replication

Cummins and Rusch (1966) studied the consequence of inhibiting protein synthesis (with cycloheximide) on the progression of the S period. Their work supported the idea that continued protein synthesis is required for the completion of nuclear DNA replication. This result was confirmed and extended by Muldoon *et al.* (1971), who showed that a number of discrete steps of DNA replication could be defined by transferring segments of S-phase plasmodia onto cycloheximide-containing medium at closely spaced intervals and following the yield of DNA replication using radioisotope techniques. Furthermore, the temporal identity of each step of DNA replication was found to be constant from one S period to the next, using the density-shift protocol of Braun et al. (1965). The mechanism by which cycloheximide inhibits DNA replication will be considered in Section III,B,1,c.

D. Other (Nonchromosomal) DNAs

In the course of analyzing the buoyant density distributions of DNA preparations by centrifugation in CsCl, two minor DNA components were observed by several groups. These satellite DNAs were subsequently characterized as to their physical and chemical properties, their localization within the plasmodium, and their times of replication during interphase.

Nucleolar DNA was first reported as a heavy satellite by Braun et al. (1965) and as an organellar DNA by Guttes and Guttes (1969). Studies by Braun and Evans (1969) and by Holt and Gurney (1969) showed that this DNA fraction could be isolated from nuclei and was replicated throughout interphase. DNA prepared from purified nucleoli was greatly enriched for the satellite DNA (Zellweger et al., 1972; Bradbury et al., 1974; Grainger and Ogle, 1978). The molecular architecture of this DNA fraction has been worked out by several groups. These papers are considered in this volume, Chapter 12, by Braun and Seebeck.

The existence of mitochondrial DNA in *P. polycephalum* was initially inferred by an autoradiographic study by Guttes and Guttes (1964) and in *Didymium nigripes* by an ultrastructural and autoradiographic study by Schuster (1965). The G_2 synthesis of DNA (Sachsenmaier, 1964) was subsequently found to be principally due to the replication of a light satellite DNA (Evans, 1966). The identity of this satellite as mitochondrial DNA resulted from several types of experiments: autoradiography (Guttes and Guttes, 1964; Schuster, 1965, Guttes et al., 1967; Kessler, 1969); cytology (Schuster, 1965, Guttes et al., 1966; Stockem, 1968; Kuroiwa, 1974; Kuroiwa et al., 1978); isolation from purified mitochondria (Evans and Suskind, 1971; Bohnert, 1977); and biosynthesis *in vitro* using purified mitochondria (Brewer et al., 1967). Other results have shown that the molecule (1) comprises 5–10% of the total plasmodial DNA and has a low guanine + cytosine (G + C) content (Evans, 1966; Guttes et al., 1967; Braun and Evans, 1969; Holt and Gurney, 1969; Evans and Suskind, 1971); (2) is replicated throughout the nuclear division cycle (Evans, 1966; Guttes et al., 1967; Braun and Evans, 1969; Holt and Gurney, 1969); (3) is probably circular *in vivo* with a molecular weight of about 40×10^6 (Sonenshein and Holt, 1968; Bohnert, 1977); and (4) is structurally homogeneous (Evans and Suskind, 1971).

Kuroiwa et al. (1978) have presented evidence that mitochondria have a synchronous doubling cycle linked temporally with the nuclear division cycle; such a result could lead to interesting new research on mitochondrial DNA replication.

E. *In Vitro* Studies

In the mid-1960s, Brewer and co-workers demonstrated the *in vitro* synthesis of nuclear and mitochondrial DNA using purified preparations of these organelles. Of particular interest was the observation that the rate and yield of DNA synthesis using isolated nuclei reflected the time in interphase at which the nuclei were isolated (Brewer and Rusch, 1965). That is, "S phase nuclei" were far more active in DNA synthesis *in vitro* than were "G_2 phase nuclei." The general features of this *in vitro* nuclear DNA synthesis have been confirmed by others, and a good deal of work has been carried out on the circumstances of DNA synthesis in various *in vitro* systems (see Section III,B,2).

F. Details of the Primary Structure

The usual four bases are found in the various DNA fractions of *P. polycephalum* (Evans and Evans, 1970). The actual quantitative relationships among these principal components of both major nuclear DNA and mitochondrial DNA were determined by Evans and Suskind (1971) and found to be in agreement with base compositions predicted by buoyant density in CsCl and by thermal transition midpoints.

A minor methylated base (5-methylcytosine) was detected by Evans and Evans (1970), who described various aspects of the biochemistry involved with the methylation of cytosine residues *in situ*. This work was later extended by Evans *et al.* (1973) and more recently by Jeter *et al.* (1980), as discussed in Section III,A,4.

Nearest neighbor frequency analyses of the major nuclear and mitochondrial DNAs have been carried out. The data for G-ending dinucleotides (in particular, CpG) gave substantial support to the hypothesis that the mitochondrion evolved from a prokaryotic life form (Cummins *et al.*, 1967). The complete study has not been published; a summary of the data is presented in Table I.

III. CURRENT RESEARCH AREAS

A. Genome Organization

General features of the organization of the genome of *P. polycephalum* are being described using three approaches: renaturation analyses, DNase sensitivity

TABLE I

Nearest Neighbor Frequency Patterns in *Physarum* Nuclear and Mitochondrial DNA[a]

Doublet pairs	Native nuclear DNA	Denatured nuclear DNA	Native mitochondrial DNA	Denatured mitochondrial DNA
ApA, UpU	0.113, 0.114	0.130, 0.133	0.146, 0.130	0.147, 0.127
CpA, UpG	0.046, 0.059	0.056, 0.060	0.040, 0.054	0.052, 0.055
GpA, UpC	0.063, 0.049	0.050, 0.050	0.047, 0.043	0.042, 0.043
CpU, ApG	0.045, 0.074	0.067, 0.064	0.061, 0.054	0.051, 0.056
GpU, ApC	0.058, 0.044	0.044, 0.050	0.043, 0.049	0.050, 0.053
GpG, CpC	0.073, 0.048	0.055, 0.052	0.031, 0.033	0.033, 0.034
UpA	0.071	0.067	0.117	0.101
ApU	0.072	0.063	0.107	0.102
CpG	0.021	0.021	0.016	0.016
GpC	0.030	0.043	0.025	0.026
Adenylic acid	0.312	0.300	0.350	0.352
Guanylic acid	0.228	0.200	0.155	0.160
Cytidylic acid	0.169	0.195	0.156	0.156
Uridylic acid	0.291	0.305	0.340	0.330
Guanine + cytosine	0.397	0.395	0.311	0.316

[a] Dinucleotide frequencies are calculated from the analysis of RNA transcripts using *Escherichia coli* RNA polymerase. The data on the sequences ApG, GpG, CpG, and UpG of the reaction primed by native DNA were reported previously (Cummins *et al.*, 1967). The fractional content for each nucleotide is computed from the frequency data for the dinucleotides (Josse *et al.*, 1961). (Unpublished data of J. E. Cummins and T. E. Evans.)

of chromatin, and site-specific cleavage of DNA by restriction endonucleases. For a related review, see Chapter 10, in this volume.

1. DENATURATION–RENATURATION EXPERIMENTS

A quantitative evaluation of the complexity of a DNA preparation can most readily be obtained using the C_0t analysis as developed and described by Britten and Kohne (1968). Using this method, Fouquet *et al.* (1974) studied the renaturation kinetics of CsCl-purified chromosomal DNA. In addition to a rapidly renaturing fraction (not studied in detail by these authors), two principal fractions were observed: a middle repetitive fraction renaturing with a corrected $C_0t_{1/2}$ of 0.5 mole·sec and a unique fraction that renatures with $C_0t_{1/2}$ of about 500 mole·sec. When DNA fractions replicated at different times in the S period were analyzed, it was found that little or no repetitive DNA was replicated in the first quarter of S. As pointed out by the authors, the average buoyant density of DNA replicated early in the S period is measurably greater than that of bulk chromosomal DNA (Evans and Brewer, 1970; Braun and Ruedi-Wili, 1971), even though the rDNA fraction (which is a high G + C satellite) is not synthesized at this time

(Zellweger et al., 1972). Thus it may be that in this organism, single-copy genes are relatively rich in G + C content.

Hardman et al. (1980) carried out a detailed study of renaturation kinetics of nuclear DNA, confirming the earlier work referred to above. In addition, they looked at the reaction products after $C_0 t = 10$ mole·sec renaturation of 0.9×10^6-dalton single-stranded DNA (a $C_0 t$ of 10 mole·sec is sufficient to allow the renaturation of nearly all DNA species except for the unique sequence fraction). The investigators observed duplex structures averaging 590 bp in length separated by single-stranded gene-size regions of about 930 nucleotide residues. Although these values are averages (with large variations around the interduplex distances), they roughly correspond to the proportions of middle-repetitive and unique sequences estimated by $C_0 t$ analysis of nuclear DNA. It seems clear from the electron micrographs that middle-repetitive sequences are inserted between unique-sequence DNA in a more or less random pattern.

Reports describing a rapidly renaturing foldback fraction of nuclear DNA have been presented by Hardman and Jack (1977, 1978) and Hardman et al. (1979); results summarized here are principally from the most recent paper. DNA is denatured with alkali, diluted, and spread for electron microscopy (total $C_0 t$ allowed estimated to be 5×10^{-5} mole·sec). Nearly all molecules observed contained at least one hairpin or foldback region, which presumably represents the presence of an inverted repeat within one DNA strand. These rapidly renaturing segments represent about 6% of the total nuclear DNA. Inserts may be present within the arm of one or both of the repeat sequences that make up the stems of the hairpin, resulting in "bubbled" hairpins. The repeated regions range in size from 2×10^4 to 2×10^6 daltons in simple hairpin structures and up to 2×10^7 in bubbled hairpins. The authors propose that in most cases the lengths of the stems and the loops bounded by the stems fit an arithmetic series of whole-integer multiples of 360 nucleotide "monomer" lengths. The function of these inverted repeated sequences is not known.

Denaturation–renaturation experiments have also been carried out on the two satellite DNA fractions. Britten and Smith (1971) and Evans and Suskind (1971) demonstrated that the ribosomal and mitochondrial DNAs renature easily. In the case of ribosomal DNA, this renaturability is a direct consequence of its sequence organization (see Chapter 12, this volume).

2. NUCLEOSOMES

Initial studies of chromatin in *P. polycephalum* were principally concerned with separating and characterizing the protein components (Jockusch and Walker, 1974). However, following the discovery that gel electrophoresis of micrococcal nuclease digests of chromatin (or isolated nuclei) reveals a series of discrete DNA fragments, a number of studies of the DNA component of chromatin appeared. The first of these was a report by Jerzmanowski et al. (1976). In this study, isolated nuclei were incubated with micrococcal nuclease, and the DNA was extracted and electrophoresed on slab gels. The individual bands were visualized by fluorescence of ethidium bromide-DNA complexes upon excitation with UV light. The subunit structure observed was quite similar to that obtained from calf thymus nuclei, although the molecular weights of the DNA fragments from the mold were somewhat smaller. Other experiments using micrococcal nuclease by Vogt and Braun (1976) and by Hozier and Kaus (1976) gave similar results, and in addition demonstrated that the same DNA patterns were obtained from nuclei isolated in interphase and in mitosis. Also, Stalder and Braun (1978) showed that DNA patterns were similar in nuclei isolated from either plasmodia or amoebae.

Use of a different nuclease led to a different result, however. Jalouzot *et al.* (1980) reported that chromatin in nuclei isolated at different times in the cycle varies substantially in susceptibility to digestion by DNase I. Metaphase chromosomes are the most resistant, demonstrating linear digestion kinetics, whereas interphase nuclei have a more sensitive initial component which accounts for about 15% of the limit digest. What this sensitive component represents awaits further experimentation.

On the basis of additional kinetic digestion data, Jalouzot *et al.* (1980) suggest that newly replicated chromatin is more susceptible to nuclease digestion than bulk chromatin. Along these same lines, Butler *et al.* (1978), studying nucleolar chromatin, proposed that newly replicated fractions were more sensitive to micrococcal nuclease, although the data presented are not convincing. Further experiments are needed in order to test these conclusions.

Johnson *et al.* (1976) presented a detailed analysis of the sizes of the micrococcal nuclease digestion products from microplasmodial nuclei and included an electron micrograph confirming the "beads on a string" configuration inferred by the nuclease data. Their calculations led to the values of 159 bp in the protected region, with a variable spacing region of 13–31 bp (repeat lengths of 172–190). Compton *et al.* (1976) obtained a repeat length of 171 for multimers of five or more.

Staron *et al.* (1977) purified the micrococcal nuclease digestion products of chromatin by sedimentation in sucrose gradients. They were able to resolve peaks that contained monomer through tetramer-size pieces of DNA as well as a more slowly sedimenting peak (peak A). The various nucleosome oligomers contain a constant proportion of RNA, whereas peak A contains substantially greater RNA than the other peaks. Johnson *et al.* (1978a) reported similar digestion products after sucrose gradient analysis but found monomer-length DNA in both peak A and the monomer nucleosome peak. They also found that 19 S and 26 S RNA hybridized preferentially to peak A (discussed further in Section III,A,3,c). In a subsequent paper, Johnson *et al.* (1978b) presented electron micrographs of various fractions taken from preparative sucrose gradients. Their observations indicate that peak A is of a different, more extended structure than the nucleosomes found in other peak fractions of the gradients. Presumably this structure is more susceptible to endogenous nuclease activity which could well have degraded the DNA component during the purification effected by Staron *et al.* (1977). Experiments using probes prepared from various cloned sequences of the *P. polycephalum* genome should reveal the relationship of peak A to other chromatin structures.

The study of chromatin structure in *P. polycephalum* seems likely to continue to yield new and important information, particularly on the dynamics of the formation and stability of nucleosomes throughout the nuclear division cycle, i.e., at times of inactivity as well as at times of DNA replication, RNA transcription, and chromosomal condensation. In a relatively recent study, Prior *et al.* (1980) have taken advantage of several unique features of this organism in such a way that nucleosomes made both *in vitro* and *in vivo* were fluorescently tagged with a pyrene-labeled H3 histone preparation. The conclusion of this elegant study is that once nucleosomes are formed, they are stable and are distributed conservatively upon chromosomal replication.

3. NUCLEOLAR NUCLEOSOMES

Three approaches have been taken in order to investigate the structure of nucleolar chromatin. These will be summarized briefly here. A detailed consideration of ribosomal DNA (rDNA) structure is presented in Chapter 12 by Braun and Seebeck.

a. Nucleolar Isolation. Grainger and Ogle (1978) prepared nuclear and nucleolar fractions from plasmodia, incubated these with micrococcal nuclease, extracted the DNA, and compared the molecular weights (as inferred by the electrophoretic mobilities) of the various fragments. By this criterion, the authors concluded that the nucleosomes from the two organellar preparations were identical. The nucleolar preparations were quite pure in that the DNA prepared therefrom was essentially all ribosomal (as judged by buoyant density in CsCl and by molecular weight analyses).

b. G_2 Labeling. Butler *et al.* (1978) took advantage of the fact that rDNA replicates during the G_2 period. By preparing gel slices and determining their radioactivity by liquid scintillation counting, they were able to compare the total DNA pattern from nuclease digests (ethidium bromide–DNA fluorescence) with a specific pattern as determined by the period of radioactive precursor incorporation. Their results showed that peaks of radioactivity correspond to the positions of dye–DNA fluorescence, whether the nuclei were prepared from plasmodia labeled in S or in G_2.

11. Organization and Replication of DNA in *Physarum polycephalum*

c. RNA–DNA Hybridization. Two groups have used labeled RNA probes to identify complementary DNA sequences in micrococcal nuclease digests. Johnson *et al.* (1978a,b) separated the initial digestion products using sucrose gradient sedimentation and characterized the peak fractions as to their content of rDNA sequences and the molecular weight of the DNA fragments. By this analysis, rDNA sequences were detected predominantly in peak A (discussed earlier) and the monomer and dimer fractions. These results suggested to Johnson *et al.* that nucleosomes containing rDNA are especially sensitive to micrococcal nuclease digestion, although the size of the DNA segments protected by the nucleosomal structure is the same as in bulk chromosomal DNA.

A similar, highly specific hybridization response in the more nuclease-susceptible fractions was observed by Stalder *et al.* (1979). In this instance, however, the investigators purified DNA from the nuclear digests, ran the fragments on slab gels, and eluted the DNA fractions from slices of the entire gel. Each fraction was then assayed for DNA content and for hybridizability to ^{3}H-labeled ribosomal RNA. Their results again indicated that after modest digestion, most of the hybridization occurred in the monomer-length fraction (which would contain DNA from both peak A and the mononucleosome fractions as prepared by Johnson *et al.*). In addition, however, the results indicate a continuum of DNA fragment lengths, with the highest specific activity of hybridization seen in the region between the monomer- and dimer-length fractions.

The results of these two groups are consistent, but the conclusions reached regarding the basic nucleosome structure of rDNA sequences are different, principally because Johnson *et al.* analyzed peak fractions (or pooled fractions) from their sucrose gradients, whereas Stalder *et al.* analyzed each individual fraction from their preparative slab gels. It seems clear from both investigations that the structure, and quite possibly the size, of rDNA nucleosomes is different from that of the nucleosomes of chromosomal DNA.

4. RECOGNITION SEQUENCES

It is evident that site-specific protein–DNA interactions are crucial to the various DNA functions and their regulation. It is not yet evident, however, that in large genomes, unique or nearly unique sites are constructed on the basis of variations in the primary sequence of the usual four deoxyribonucleotides alone. One variable in DNA chemistry that could play a role in increasing site specificity is base methylation. Although there is no direct evidence on this point, some basic work has been carried out and will be described briefly below.

A quantitative analysis of the formation and distribution of 5-methylcytosine (5-MC) in the DNA of *P. polycephalum* has appeared (Evans and Evans, 1970). 5-MC was the only minor methylated base detected (in a ratio of 1:12–24 relative to C), and was formed by the methylation of C residues throughout interphase. In a subsequent paper, Evans *et al.* (1973) presented data supporting the unexpected conclusion that the methylation of any one DNA strand apparently continued for several generations after its original synthesis.

Jeter *et al.* (1980) have confirmed the quantitative values for 5-MC and additionally have discovered 6-methylaminopurine (ratio of 1 : 1000 relative to A). Further, they have noted that *P. polycephalum* DNA is quite resistant to type II restriction endonucleases that have CpG (but not 5MCpG in their recognition sites, suggesting that a significant fraction of the CpG dinucleotides detected previously using *in vitro* polymerase methods are actually 5-MCpG *in vivo*. In fact, about 1 in 15 C residues are in the dinucleotide CpG (Cummins *et al.*, 1967; Table I), a ratio quite similar to that of 5-MC : C. A possible association of the CpG sequence with RNA-DNA primer junctions is noted in Section III, B, 1, *b*.

5. OUTLOOK

Clearly, the use of cloning technology should play a major role in future progress in this area. Although there are severe limitations to possible cytological studies of genome organization (e.g., tiny chromosomes, lack of salivary glands), other uses of cloned DNAs from various repetitive classes can be imagined: to generate probes for identifying restriction fragments, for ultrastructural hybridization studies, etc.

B. DNA Replication

1. DNA REPLICATION *IN VIVO*

a. Replication Intermediates. Single-strand molecular weight analyses of nuclear DNA were first carried out by McGrath and Williams (1967). Applying their technique of direct lysis in a dilute sodium hydroxide layer on top of an alkaline sucrose gradient, these investigators obtained values of 1.5×10^7 and 4×10^7 daltons for S and G_2 phase DNA, respectively. Brewer (1972) extended this work using both neutral and alkaline sucrose gradient analyses. His results with alkaline gradients for pulse-labeled S phase DNA and mature (G_2) DNA were in agreement with the McGrath and Williams values. Brewer also found that the sedimentation of pulse-labeled, double-stranded DNA increased during a chase period and then decreased to the G_2 value (2.3×10^8) after about 30 minutes. An extension of this work was reported by Brewer *et al.* (1974) in which the rate of strand elongation was determined to be 0.3×10^6 daltons/min. Their data indicated that initiation of replicons occurred as late as 2 hours after the beginning of the S period, when S (as defined by label incorporation or mass DNA increase) is nearly over, and that the maturation of these strands extended well into the G_2 period, suggesting that strand growth by ligation may account for some of the molecular weight increases observed in the pulse–chase experiments.

Data presented by Funderud and Haugli (1975) and later confirmed by Funderud et al. (1978a,b,c) represent further progress in the description of in vivo DNA replication intermediates in P. polycephalum. (1) Okasaki-type fragments of about 10^5 daltons were detected by short labeling pulses throughout the S period. (2) These short pieces increased in molecular weight in a discontinuous fashion. (3) Full-size, single-stranded pieces of more than 10^8 daltons were formed, apparently from the fusion within a population of relatively long-lived intermediates (ca. 2×10^7 daltons). Although these authors also used alkaline gradients in their analyses, their methods differed in several ways from those of Brewer. The pulse-labeling times were substantially shorter (30 seconds versus 10 minutes), and the nuclei (isolated using $MgCl_2$ rather than $CaCl_2$ in the homogenizing medium) were lysed in a test tube and then layered onto the alkaline sucrose gradients.

What about the directionality of replication? Funderud et al. (1978a) addressed this question by sequentially pulse-labeling DNA at the beginning of the S period with bromodeoxyuridine and [^{3}H]deoxyadenosine, isolating nuclei, irradiating the nuclei with U C /U V light, and analyzing the single-stranded DNA molecular weight using alkaline sucrose gradient centrifugation. Their results gave clear support to the bidirectional replication model. This work was extended to include replication throughout the S period in a second paper using the same experimental approach (Funderud et al., 1978b). Their results gave independent support to the hypothesis that the 2×10^7-dalton intermediate actually represents the replicon size in P. polycephalum. They labeled early initiation sites by a bromodeoxyuridine pulse at the beginning of S followed by [^{3}H]deoxyadenosine pulse–chases at various times in the S period. After U C /U V photolysis, the tritium-labeled regions were always associated with replicon-sized pieces (that is, 2×10^7 daltons). This result, in the context of the other experiments reported in this exciting paper, strongly suggested to the authors that all sites of initiation are primed at the beginning of S but that full activation of the various replicons occurs in spatially related groups and that various groups become activated at different times throughout S.

Direct evidence for the 2×10^7 size of replicons in P. polycephalum came from the electron microscopic observations of Funderud et al. (1979) and Gillespie and Hardman (1979) (center-to-center distances of replicating loops and regions, respectively). Pointing the way toward new concepts of the DNA replication process, the Gillespie and Hardman (1979) report gives clear evidence of multiple "microbubbles" (representing multiple initiation sites?) within the putative replicons.

It is, of course, not known how the structures noted above, which are seen and/or analyzed as isolated dechromatinized DNA, are related to the replication complexes that must exist *in vivo* in a milieu of binding proteins, nucleosomes, etc. Two groups have looked unsuccessfully for possible associations between replicating DNA and the nuclear membrane (Oppenheim and Wahrman, 1973; Wille and Steffens, 1979).

Beach *et al.* (1980) have reported experiments in which replication origins have been isolated. Their intention is to clone such origins, leading to a determination of their primary structure(s). Among many other possibilities, such molecules might be used for the preparation of probes for "R loop" localization of origins within replicons, as well as for the localization of origins within various biochemical or cytological subnuclear preparations.

b. RNA Primers. Using brief pulses and high concentrations of [^{3}H]thymidine, Waqar and Huberman (1973) were able to demonstrate incorporation into a low-molecular-weight DNA fraction (ca. 4 S) that appeared by buoyant density analysis in Cs_2SO_4 to have an RNA component. By microinjecting α-^{32}P-labeled deoxyribonucleoside triphosphates, Waqar and Huberman (1975) again found incorporation into DNA molecules that behaved in Cs_2SO_4 as if they contained some RNA. In this latter report, transfer of the label to ribonucleotides could be effected by appropriate nuclease digestion of the isolated DNA. The authors proposed that such RNA–DNA joints resulted from RNA primers involved in the synthesis of Okasaki-type pieces of DNA. It is interesting to note that transfer from dGTP to ribonucleotides was far more efficient than from any other deoxyribonucleotide, and further, that the transfer to any other base was more or less equal. However, since the frequency of the dinucleotide CpG is very low in eukaryotes in general and *P. polycephalum* in particular (Cummins *et al.*, 1967; Table I), it appears that there may well be a nonrandom sequence at the juncture of RNA primers and newly replicated DNA strands.

c. Protein Synthesis. As first described by Cummins and Rusch (1966) and by Muldoon *et al.* (1971), addition of cycloheximide to the growth medium results in the inhibition of DNA replication. Such inhibition has generally been considered to be secondary to the inhibition of protein synthesis effected by cycloheximide and, as such, implies the need for concomitant protein synthesis (and perhaps the synthesis of specific proteins) for the continuation and/or control of DNA replication. However, such coupled macromolecular synthesis would be required only for chromosomal DNA replication in that cycloheximide has little or no effect on the synthesis of nucleolar or mitochondrial DNA (Werry and Wanka, 1972; T. Evans and H. Evans, unpublished).

The kinetics and mechanism of the cycloheximide-mediated inhibition of DNA synthesis were studied in detail by Evans *et al.* (1976). They found that cycloheximide blocks the molecular weight increase of replicating DNA molecules, suggesting that the inhibition of precursor incorporation effected by the drug is not the result of a specific inhibition of DNA initiation. Funderud and Haugli (1977a) also found that cycloheximide blocks elongation of progeny strands *in vivo*. In another portion of their study, Evans *et al.* (1976) confirmed the results of Bersier and Braun (1974), who found that cycloheximide causes an expansion of the nucleotide triphosphate pools in *P. polycephalum*. Additionally, by measuring the specific radioactivity of the TTP pools from plasmodia that had been incubated in medium containing [^{3}H]thymidine, Evans *et al.* (1976) concluded that the conversion of [^{3}H]thymidine to TTP was markedly inhibited by cycloheximide. In a later study, Evans and Evans (1980) determined that the effect of cycloheximide on DNA replication is at least as rapid as the effect on protein synthesis and that the level of inhibition obtained is quite similar.

Cycloheximide-resistant strains have been isolated by Funderud and Haugli (1977a) and by Evans and Evans (1980). As expected, the drug does not affect protein synthesis or DNA synthesis in these strains. Also, the cycloheximide-resistant mutant described by Evans and Evans was insensitive to the drug as far as pool expansion and precursor utilization were concerned.

Thus the drug has several effects in this system in addition to inhibiting protein synthesis. So far, these effects have not been separated, and the relationships between them are unclear. As pointed out by Evans *et al.* (1976), the regulation of DNA replication as well as the pool effects might be affected by the inhibition of protein synthesis per se.

2. DNA REPLICATION *IN VITRO*

As is clear from the foregoing discussion, a great deal has been learned about DNA replication by studying the molecular events that occur in the whole organism. Another approach is to work with isolated replication systems *in vitro*, so that individual components can be identified, purified, and used to reconstruct the whole process. This goal would be greatly aided by, if not dependent on, the availability of appropriate conditional mutants. Although this latter circumstance has yet to be obtained, progress is being made (see Chapter 6, this volume).

The work on *in vitro* replication systems has continued since the original article on isolated nuclei by Brewer and Rusch (1965). These efforts utilize three types of *in vitro* systems: (1) isolated nuclei; (2) plasmodial homogenates and subnuclear extracts; and (3) purified proteins.

a. Isolated Nuclei. The report of Brewer and Rusch (1965) indicated that isolated nuclei could support the incorporation of deoxyribonucleotides into high-molecular-weight DNA. Of particular interest was the fact that the amount of activity reflected the time in interphase at which the nuclei were isolated (high activity in S phase nuclei, low activity in G_2 phase nuclei). This correlation was later confirmed by Schiebel and Schneck (1974) and by Funderud and Haugli (1977b). These two groups also demonstrated that the incorporation observed represented a continuation of DNA replication initiated *in vivo;* no initiation events could be demonstrated in this *in vitro* system. However, both Okasaki-size and higher-molecular-weight replication intermediates were reported by Funderud and Haugli (1977b), with the patterns observed in alkaline sucrose gradients being quite similar to those seen after short labeling periods *in vivo.*

Each group of investigators referred to above optimized various conditions for the *in vitro* synthesis of DNA with isolated nuclei (or crude homogenates, in the case of Brewer and co-workers). A dependence on all four deoxyribonucleotide triphosphates and magnesium ion was found, along with a strong stimulation by ATP. Other inclusions are EGTA and a buffer at about pH 7.5. As noted by Schiebel and Schneck (1974) and Brewer and Ting (1975), the temperature optimum is approximately 33°–37°C. Spermine stimulates endogenous DNA synthesis in isolated nuclei (Brewer and Rusch, 1966; Schiebel and Schneck, 1974; Brewer and Ting, 1975); as will be indicated in Section III,B,2,c, such stimulation may reflect the neutralization of an inhibitory substance. Brewer (1975) has shown that a further increase in the rate and yield of this process can be effected by substituting dextran for spermine in the homogenizing medium.

b. Homogenates and Nuclear Extracts. Brewer and co-workers have studied *in vitro* DNA synthesis using plasmodial homogenates (Brewer and Ting, 1975; Brewer, 1975) and gently extracted or disrupted nuclei (Brewer and Busacca, 1979). The incubation conditions for assaying DNA synthesis are similar to those used for purified nuclei except that dextran is included (Brewer, 1975). In all cases, physiological control is maintained; that is, the activity of the preparations reflects the time in interphase at which the homogenates were made.

TABLE II

Comparative Characteristics of DNA Polymerase Activities

	Murakami-Murofushi et al. (1976)	Baer and Schiebel (1978)	Choudhry and Cox (1979)
Component optima[a]			
Mg^{2+}	7 mM	5 mM	10 mM
K^+		150 mM	80 mM
pH	7.6	7.2–8.3 (7.5)	7.0
Temperature	(30)	37–45 (27)	(37)
Template preference[b]	ss > ds	act >> ss > ds	Act only

[a] Values in parentheses are conditions employed in standard assay and do not necessarily imply optima.

[b] dsDNA, ds; ssDNA, ss; DNase-activated DNA, act.

Replication intermediates observed in these systems include a slowly sedimenting component (referred to as "Okasaki pieces") as well as a much larger fraction. Since these two components are found in approximately equal amounts (as judged by label incorporation), and since the size of the large component is variable whereas the size of the small pieces is rather constant, Brewer (1975) proposed that DNA chain elongation in *P. polycephalum* occurs by continuous growth of one strand (presumably the leading strand) and discontinuous growth of the other (via Okasaki pieces and their subsequent ligation).

The replication intermediates observed by Brewer (1975) in homogenates are qualitatively similar to the pattern seen by Funderud and Haugli in isolated nuclei (1977b) as well as in short pulses *in vivo* (Funderud and Haugli, 1975). Funderud and Haugli, however, interpret their results in terms of discontinuous synthesis on both strand. Definitive experiments on this point have yet to be reported.

c. Purified Proteins. The cornerstone for any *in vitro* DNA replication system in which defined components are used is the DNA polymerase itself. Three groups have reported the purification from plasmodia of one major DNA polymerase activity. In addition to requiring the four deoxyribonucleotides and magnesium ion, the enzyme is inhibited by various -SH reagents. Some of the characteristics of this polymerase are presented in Table II. The enzyme resembles the alpha polymerase isolated from mammalian cells (Choudhry and Cox, 1979).

Baer and Schiebel (1978) showed that although addition of spermidine to crude enzyme preparations stimulates activity, this effect is virtually eliminated after the enzyme is purified using an anion-exchange (DEAE) column. Murakami-Murofushi *et al.* (1976) have described a cytoplasmic inhibitor of the polymerase which may be separated from the polymerase by DEAE–Sephadex. This inhibitor binds to the template DNA. It seems likely from these results that the stimulatory effect of polyanions (such as spermidine) on crude DNA polymerase systems is the result of the competitive displacement from the DNA template of basic protein inhibitors of DNA synthesis. Such DNA-binding proteins are not likely to be histones, as implied by other experiments described by Murakami-Murofushi *et al.* (1976).

Zanker and Schiebel (1978) obtained evidence for different molecular weight species of polymerase activities in partially purified enzyme preparations. Baer and Schiebel (1978) and Schiebel and Baer (1979) have extended this work, and have suggested that various molecular weight fractions predominate depending upon the growth conditions at the time of harvest.

The molecular weight of the principal species of polymerase obtained from "late-log" cultures is reported by Schiebel and Baer (1979) to be 116,000, which agrees well with the value assigned by Choudhry and Cox (1979) to the DNA polymerase that they have purified to apparent homogeneity (112,000).

Starting with 3.4-kg wet weight of microplasmodia, Choudhry and Cox (1979) obtained 5.1 mg of pure enzyme. Some of this material has been used as antigen with the successful production of rabbit antibody of high specificity and enzyme-neutralizing activity. In addition, the authors have radioiodinated some of their enzyme and have developed a sensitive (5–50 ng) radioimmunoassay for the analysis of *P. polycephalum* DNA polymerase. The authors are now in a position to determine at what time during interphase the polymerase is synthesized and, in addition, whether the various molecular weight species observed by Schiebel and co-workers contain immunologically related components. By combining the radioimmunoassay with an *in vitro* protein-synthesizing system, Choudhry and Cox face the exciting prospect of identifying the specific mRNA and subsequently isolating the cloned sequences of the DNA polymerase gene itself.

In addition to the work on polymerase isolation, two other proteins have been purified from *P. polycephalum* that may be involved with DNA replication. As indicated above, Murakami-Murofushi *et al.* (1976) have reported the purification of a cytoplasmic inhibitor of DNA polymerase. Brewer (1979) has reported the purification (using washed nuclei as the *in vitro* assay system) of a nuclear protein that is a possible component of the DNA replication complex *in vivo*. Whether this protein will be active in a purified-protein system remains to be tested.

Using DNA cellulose columns, Magun (1976, 1979) has described the phosphorylation levels of DNA-binding proteins at different times during interphase. The results show that some of the DNA-binding proteins undergo changes in phosphorylation state, suggesting a possible role in the regulation of DNA function. It is not known whether any of the purified proteins described in this section are phosphorylated.

3. OUTLOOK

Although the results of recent work concerning the mechanism of DNA replication in *P. polycephalum* have not had the opportunity to stand the test of time and independent experimentation, it seems clear that information concerning this process is accumulating rapidly. The *in vivo* analyses and work in isolated systems will be greatly aided by cloning technology and the isolation of mutant strains, bringing the molecular genetic approach to full fruition.

REFERENCES

Baer, A., and Schiebel, W. (1978). Deoxyribonucleic acid polymerase from *Physarum polycephalum*. Properties of the major cytoplasmic activity in exponentially growing microplasmodia. *Eur. J. Biochem.* **86,** 77–84.

Beach, D., Piper, M., and Shall, S. (1980). Isolation of newly-initiated DNA from the early S phase of the synchronous eukaryote, *Physarum polycephalum*. *Exp. Cell Res.* **129,** 211–221.

11. Organization and Replication of DNA in *Physarum polycephalum*

Bersier, D., and Braun, R. (1974). Effect of cycloheximide on pools of deoxyribonucleoside triphosphates. *Exp. Cell Res.* **84,** 436-440.
Bohnert, H. J. (1977). Size and structure of mitochondrial DNA from *Physarum polycephalum. Exp. Cell Res.* **106,** 426-430.
Bradbury, E. M., Matthews, H. R., McNaughton, J., and Molgaard, H. V. (1974). Sub-nuclear components of *Physarum polycephalum. Biochim. Biophys. Acta* **335,** 19-29.
Braun, R., and Evans, T. E. (1969). Replication of nuclear satellite and mitochondrial DNA in the mitotic cycle of *Physarum. Biochim. Biophys. Acta* **182,** 511-522.
Braun, R., and Ruedi-Wili, H. (1971). Early replicating DNA of *Physarum* is denser than late replicating DNA. *Experientia* **27,** 1412.
Braun, R., and Wili, H. (1969). Time sequence of DNA replication in *Physarum. Biochim. Biophys. Acta* **174,** 246-252.
Braun, R., Mittermayer, C., and Rusch, H. P. (1965). Sequential temporal replication of DNA in *Physarum polycephalum. Proc. Nat. Acad. Sci. U.S.A.* **53,** 924-931.
Brewer, E. N. (1972). DNA replication in *Physarum polycephalum. J. Mol. Biol.* **68,** 401-412.
Brewer, E. N. (1975). DNA replication by a possible continuous-discontinous mechanism in homogenates of *Physarum polycephalum* containing dextran. *Biochim. Biophys. Acta* **402,** 363-371.
Brewer, E. N. (1979). Isolation of a stimulatory factor for nuclear DNA replication. *Biochim. Biophys. Acta* **564,** 154-161.
Brewer, E. N., and Busacca, P. A. (1979). DNA synthesis in a sub-nuclear preparation isolated from *Physarum polycephalum. Biochem. Biophys. Res. Commun.* **91,** 1352-1357.
Brewer, E. N., and Rusch, H. P. (1965). DNA synthesis by isolated nuclei of *Physarum polycephalum. Biochem. Biophys. Res. Commun.* **21,** 235-241.
Brewer, E. N., and Rusch, H. P. (1966). Control of DNA replication: Effect of spermine on DNA polymerase activity in nuclei isolated from *Physarum polycephalum. Biochem. Biophys. Res. Commun.* **25,** 579-584.
Brewer, E. N., and Ting, P. (1975). DNA replication in homogenates of *Physarum polycephalum. J. Cell. Physiol.* **86,** 459-470.
Brewer, E. N., De Vries, A., and Rusch, H. P. (1967). DNA synthesis by isolated mitochondria of *Physarum polycephalum. Biochim. Biophys. Acta* **145,** 686-692.
Brewer, E. N., Evans, T. E., and Evans, H. H. (1974). Studies on the mechanism of DNA replication in *Physarum polycephalum. J. Mol. Biol.* **90,** 335-342.
Britten, R. J., and Kohne, D. E. (1968). Repeated sequences in DNA. *Science* **161,** 529-540.
Britten, R. J., and Smith, J. F. (1971). The nuclear satellite in slime mold. *In* "Carnegie Institution Year Book 69." Dept. of Terrestrial Magnetism, (E. T. Bolton, ed.), pp. 518-521. Carnegie Institution, Washington, D.C.
Butler, M. J., Davies, K. E., and Walker, I. O. (1978). The structure of nucleolar chromatin in *Physarum polycephalum. Nucleic Acids Res.* **5,** 667-678.
Choudhry, M. K., and Cox, R. A. (1979). Purification of DNA polymerase from *Physarum polycephalum* and the production of specific antibodies. *In* "Proc. 4th Eur. *Physarum* Workshop" (W. Sachsenmaier, ed.), pp. 71-76. Univ. of Innsbruck, Austria.
Compton, J. L., Bellard, M., and Chambon, P. (1976). Biochemical evidence of variability in the DNA repeat length in the chromatin of higher eukaryotes. *Proc. Nat. Acad. Sci. U.S.A.* **73,** 4382-4386.
Cummins, J. E., and Rusch, H. P. (1966). Limited DNA synthesis in the absence of protein synthesis in *Physarum polycephalum. J. Cell Biol.* **31,** 577-583.
Cummins, J. E., Rusch, H. P., and Evans, T. E. (1967). Nearest neighbor frequencies and the phylogenetic origin of mitochondrial DNA in *Physarum polycephalum. J. Mol. Biol.* **23,** 281-284.

Daniel, J. W., and Rusch, H. P. (1961). The pure culture of *Physarum polycephalum* on a partially defined soluble medium. *J. Gen. Microbiol.* **25,** 47-59.

Evans, H. H., and Evans, T. E. (1970). Methylation of the deoxyribonucleic acid of *Physarum polycephalum* at various periods during the mitotic cycle. *J. Biol. Chem.* **245,** 6436-6441.

Evans, H. H., Evans, T. E., and Littman, S. (1973). Methylation of parental and progeny DNA strands in *Physarum polycephalum. J. Mol. Biol.* **74,** 563-572.

Evans, H. H., Littman, S. R., Evans, T. E., and Brewer, E. N. (1976). Effects of cycloheximide on thymidine metabolism and on DNA strand elongation in *Physarum polycephalum. J. Mol. Biol.* **104,** 169-184.

Evans, T. E. (1966). Synthesis of a cytoplasmic DNA during the G_2 interphase of *Physarum polycephalum. Biochem. Biophys. Res. Commun.* **22,** 678-683.

Evans, T. E., and Brewer, E. N. (1970). Relative rates of synthesis of DNA fractions of different densities during the mitotic cycle. *Physarum Newsletter* **2,** #2 (Appendix).

Evans, T. E., and Evans, H. H. (1980). Cycloheximide resistance in *Physarum polycephalum. J. Bacteriol.* **143,** 897-905.

Evans, T. E., and Suskind, D. (1971). Characterization of the mitochondrial DNA of the slime mold *Physarum polycephalum. Biochim. Biophys. Acta* **228,** 350-364.

Fouquet, H., Bierweiler, B., and Sauer, H. W. (1974). *Eur. J. Biochem.* **44,** 407-410.

Funderud, S., and Haugli, F. (1975). DNA replication in *Physarum polycephalum:* Characterization of replication products *in vivo. Nucleic Acids Res.* **2,** 1381-1390.

Funderud, S., and Haugli, F. (1977a). DNA replication in *Physarum polycephalum*: Characterization of DNA replication products made *in vivo* in the presence of cycloheximide in strains sensitive and resistant to cycloheximide. *Nucleic Acids Res.* **4,** 405-413.

Funderud, S., and Haugli, F. (1977b). DNA replication in *Physarum polycephalum*: Characterization of replication products made in isolated nuclei. *Biochem. Biophys. Res. Commun.* **74,** 941-948.

Funderud, S., Andreassen, R., and Haugli, F. (1978a). DNA replication in *Physarum polycephalum:* Bidirectional replication of DNA within replicons. *Nucleic Acids Res.* **5,** 713-721.

Funderud, S., Andreassen, R., and Haugli, F. (1978b). DNA replication in *Physarum polycephalum:* UV photolysis of maturing 5-bromo-deoxyuridine substituted DNA. *Nucleic Acids Res.* **5,** 3303-3313.

Funderud, S., Andreassen, R., and Haugli, F. (1978c). Size distribution and maturation of newly replicated DNA through the S and G_2 phases of *Physarum polycephalum.* Cell **15,** 1519-1526.

Funderud, S., Andreassen, R., and Haugli, F. (1979). DNA replication in *Physarum polycephalum:* Electron microscopic and autoradiographic analysis of replicating DNA from defined stages of the S-period. *Nucleic Acids Res.* **6,** 1417-1431.

Gillespie, D. A. F., and Hardman, N. (1979). Microbubbles in replicating nuclear deoxyribonucleic acid from *Physarum polycephalum. Biochem. J.* **183,** 477-480.

Goodman, E. M. (1980). *Physarum polycephalum*: A review of a model system using a structure-function approach. *Int. Rev. Cytol.* **63,** 1-58.

Grainger, R. M., and Ogle, R. C. (1978). Chromatin structure of the ribosomal RNA genes in *Physarum polycephalum. Chromosoma* **65,** 115-126.

Guttes, E., and Guttes, S. (1964). Thymidine incorporation by mitochondria in *Physarum polycephalum. Science* **145,** 1057-1058.

Guttes, E., and Guttes, S. (1969). Replication of nucleolus-associated DNA during "G_2 phase" in *Physarum polycephalum. J. Cell Biol.* **43,** 229-236.

Guttes, E., Guttes, S., and Rusch, H. P. (1961). Morphological observations on growth and differentiation of *Physarum polycephalum* grown in pure culture. *Dev. Biol.* **3,** 588-614.

Guttes, E. W., Hanawalt, P. C., and Guttes, S. (1967). Mitochondrial DNA synthesis and the mitotic cycle in *Physarum polycephalum. Biochim. Biophys. Acta* **142,** 181-194.

Guttes, S., Guttes, E., and Hadek, R. (1966). Occurrence and morphology of a fibrous body in the mitochondria of the slime mold *Physarum polycephalum*. *Experientia* **22**, 452-454.
Hardman, N., and Jack, P. L. (1977). Characterization of foldback sequences in *Physarum polycephalum* nuclear DNA using the electron microscope. *Eur. J. Biochem.* **74**, 275-283.
Hardman, N., and Jack, P. L. (1978). Periodic organisation of foldback sequences in *Physarum polycephalum* nuclear DNA. *Nucleic Acids Res.* **5**, 2405-2423.
Hardman, N., Jack, P. L., Brown, A. J. P., and McLachlan, A. (1979). Distribution of inverted repeat sequences in nuclear DNA from *Physarum polycephalum*. *Eur. J. Biochem.* **94**, 179-187.
Hardman, N., Jack, P. L., Fergie, R. C., and Gerrie, L. M. (1980). Sequence organization in nuclear DNA from *Physarum polycephalum*. *Eur. J. Biochem.* **103**, 247-257.
Holt, C. E. (1980). The nuclear replication cycle in *Physarum polycephalum*. *In:* "Growth and Differentiation in *Physarum polycephalum*" (W. F. Dove and H. P. Rusch, eds.), pp. 9-63. Princeton Univ. Press, Princeton, New Jersey.
Holt, C. E., and Gurney, E. G. (1969). Minor components of the DNA of *Physarum polycephalum*. *J. Cell Biol.* **40**, 484-496.
Hozier, J. C., and Kaus, R. (1976). Subunit structure of chromosomes in mitotic nuclei of *Physarum polycephalum*. *Chromosoma* **57**, 95-102.
Jalouzot, R., Briane, D., Ohlenbusch, H. H., Wilhelm, M. L., and Wilhelm, F. X. (1980). Kinetics of nuclease digestion of *Physarum polycephalum* nuclei at different stages of the cell cycle. *Eur. J. Biochem.* **104**, 423-431.
Jerzmanowski, A., Staron, K., Tyniec, B., Bernhardt-Smigielska, J., and Toczko, K. (1976). Subunit structure of *Physarum polycephalum* chromatin. *FEBS Lett.* **62**, 251-254.
Jeter, J., Jr., Gama-Sosa, M., Gehrke, C., Kuo, K., Wang, R., and Ehrlich, M. (1980). Distribution of M5C in the DNA of *Physarum polycephalum*. *J. Cell Biol.* **87**, 112a.
Jockusch, B. M., and Walker, I. O. (1974). The preparation and preliminary characterization of chromatin from the slime mould *Physarum polycephalum*. *Eur. J. Biochem.* **48**, 417-425.
Johnson, E. M., Littau, V. C., Allfrey, V. G., Bradbury, E. M., and Matthews, H. R. (1976). The subunit structure of chromatin from *Physarum polycephalum*. *Nucleic Acids Res.* **3**, 3313-3329.
Johnson, E. M., Allfrey, V. G., Bradbury, E. M., and Matthews, H. R. (1978a). Altered nucleosome structure containing DNA sequences complementary to 19S and 26S ribosomal RNA in *Physarum polycephalum*. *Proc. Nat. Acad. Sci. U.S.A.* **75**, 1116-1120.
Johnson, E. M., Matthews, H. R., Littau, V. C., Lothstein, L., Bradbury, E. M., and Allfrey, V. G. (1978b). The structure of chromatin containing DNA complementary to 19S and 26S ribosomal RNA in active and inactive stages of *Physarum polycephalum*. *Arch. Biochem. Biophys.* **191**, 537-550.
Josse, J., Kaiser, A. D., and Kornberg, A. (1961). Enzymatic synthesis of deoxyribonucleic acid. VIII. Frequencies of nearest neighbor base sequences in deoxyribonucleic acid. *J. Biol. Chem.* **236**, 864-875.
Kessler, D. (1969). Mitochondrial DNA in *Physarum polycephalum*. *J. Cell Biol.* **43**, 68A.
Kuroiwa, T. (1974). Studies on mitochondrial structure and function in *Physarum polycephalum*. III. Electron microscopy of a large amount of DNA released from a central body in mitochondria by trypsin digestion. *J. Cell Biol.* **63**, 299-306.
Kuroiwa, T., Hizume, M., and Kawano, S. (1978). Studies on mitochondrial structure and function in *Physarum polycephalum*. IV. Mitochondrial division cycle. *Cytologia* **43**, 119-136.
McGrath, R. A., and Williams, R. W. (1967). Interruptions in single strands of the DNA in slime mold and other organisms. *Biophys. J.* **7**, 309-317.
Magun, B. E. (1976). Cytoplasmic DNA-binding phosphoproteins of *Physarum polycephalum*. *Exp. Cell Res.* **103**, 219-231.

Magun, B. E. (1979). Changes in cytoplasmic DNA-binding phosphoproteins during the cell cyle of *Physarum polycephalum*. *Cell Differ.* **8**, 157–172.
Muldoon, J. J., Evans, T. E., Nygaard, O. F., and Evans, H. H. (1971). Control of DNA replication by protein synthesis at defined times during the S period in *Physarum polycephalum*. *Biochim. Biophys. Acta* **247**, 310–321.
Murakami-Murofushi, K., Nagano, H., and Mano, Y. (1976). A cytoplasmic inhibitor of DNA polymerase from the plasmodia of *Physarum polycephalum*. *J. Biochem. (Tokyo)* **80**, 735–741.
Nygaard, O. F., Guttes, S., and Rusch, H. P. (1960). Nucleic acid metabolism in a slime mold with synchronous mitosis. *Biochim. Biophys. Acta* **38**, 298–306.
Nygaard, O. F., Brewer, E. N., Evans, T. E., and Wolpaw, J. R. (1973). Correlation between sensitivity to ionizing radiation and DNA replication in *Physarum polycephalum*. *Adv. Radiat. Res. Biol. Med.* **2**, 989–995.
Oppenheim, A., and Wahrman, J. (1973). DNA-membrane association during the mitotic cycle of *Physarum polycephalum*. *Exp. Cell Res.* **79**, 287–294.
Prior, C. P., Cantor, C. R., Johnson, E. M., and Allfrey, V. G. (1980). Incorporation of exogenous pyrene-labeled histone into *Physarum* chromatin. A system for studying changes in nucleosomes assembled *in vivo*. *Cell* **20** 597–608.
Rao, B., and Gontcharoff, M. (1969). Functionality of newly-synthesized DNA as related to RNA synthesis during mitotic cycle in *Physarum polycephalum*. *Exp. Cell Res.* **56**, 269–274.
Sachsenmaier, W. (1964). Zur DNS- und RNS-synthese im teilungscyclus synchroner plasmodien von *Physarum polycephalum*. *Biochem. Z.* **340**, 541–547.
Sachsenmaier, W., and Rusch, H. P. (1964). The effect of 5-fluoro-2'-deoxyuridine on synchronous mitosis in *Physarum polycephalum*. *Exp. Cell Res.* **36**, 124–133.
Scheibel, W., and Baer, A. (1979). DNA polymerase of *Physarum polycephalum*. *In* "Proc. 4th Eur. *Physarum* Workshop" (W. Sachsenmaier, ed.), pp. 77–81. Univ. of Innsbruck, Austria.
Schiebel, W., and Schneck, U. (1974). DNA replication in isolated nuclei of synchronously growing *Physarum polycephalum*. *Hoppe-Seyler's Z. Physiol. Chem.* **355**, 1515–1525.
Schuster, F. L. (1965). Deoxyribose nucleic acid component in mitochondria of *Didymium nigripes*, a slime mold. *Exp. Cell Res.* **39**, 329–345.
Sonenshein, G. E., and Holt, C. E. (1968). Molecular weight of mitochondrial DNA in *Physarum polycephalum*. *Biochem. Biophys. Res. Commun.* **33**, 361–367.
Stalder, J., and Braun, R. (1978). Chromatin structure of *Physarum polycephalum* plasmodia and amoebae. *FEBS Lett.* **90**, 223–227.
Stalder, J., Seebeck, T., and Braun, R. (1979). Accessibility of the ribosomal genes to micrococcal nuclease in *Physarum polycephalum*. *Biochim. Biophys. Acta* **561**, 452–463.
Staron, K., Jerzmanowski, A., Tyniec, B., Urbanska, A., and Toczko, K. (1977). Nucleoprotein chromatin subunit from *Physarum polycephalum*. *Biochim. Biophys. Acta* **475**, 131–138.
Stockem, W. (1968). Uber den DNS- und RNS-gehalt der mitochondrien von *Physarum polycephalum*. *Histochemie* **15**, 160–183.
Turnock, G. (1979). Patterns of nucleic acid synthesis in *Physarum polycephalum*. *Prog. Nucleic Acid Res. Mol. Biol.* **23**, 53–104.
Vogt, V. M., and Braun, R. (1976). Repeated structure of chromatin in metaphase nuclei of *Physarum*. *FEBS Lett.* **64**, 190–192.
Waqar, M. A., and Huberman, J. A. (1973). Evidence for the attachment of RNA to pulse-labeled DNA in the slime mold *Physarum polycephalum*. *Biochem. Biophys. Res. Commun.* **51**, 174–180.
Waqar, M. A., and Huberman, J. A. (1975). Covalent linkage between RNA and nascent DNA in the slime mold *Physarum polycephalum*. *Biochim. Biophys. Acta* **383**, 410–420.

Werry, P. A. T. J., and Wanka, F. (1972). The effect of cycloheximide on the synthesis of major and satellite DNA components in *Physarum polycephalum. Biochim. Biophys. Acta* **287,** 232–235.

Wille, J. J., Jr., and Kauffman, S. A. (1975). Premature replication of late S period DNA regions in early S nuclei transferred to late S cytoplasm by fusion in *Physarum polycephalum. Biochim. Biophys. Acta* **407,** 158–173.

Wille, J. J., Jr., and Steffens, W. L. (1979). Cycle specific association of nascent chromatin with nuclear envelope components in *Physarum polycephalum. Nucleic Acids Res.* **6,** 3323–3339.

Zanker, K. S., and Schiebel, W. (1978). The heterogeneity of cytoplasmic deoxyribonucleic acid polymerase from *Physarum polycephalum. Biochem. J.* **171,** 445–451.

Zellweger, A., Ryser, U., and Braun, R. (1972). Ribosomal genes of *Physarum:* Their isolation and replication in the mitotic cycle. *J. Mol. Biol.* **64,** 681–691.

CHAPTER 12

RNA Metabolism

RICHARD BRAUN and THOMAS SEEBECK

I.	Introduction	393
II.	Properties of DNA and RNA	394
	A. DNA	394
	B. RNA	396
III.	RNA Synthesis	400
	A. *In Vivo* Transcription	400
	B. RNA Polymerases	400
	C. Regulatory Factors	404
	D. *In Vitro* Transcription	407
	E. Chromatin Structure and Transcription	411
IV.	RNA Degradation	413
	A. *In Vivo* Degradation	413
	B. RNA-Degrading Enzymes	414
V.	Transcription in the Mitotic Cycle	416
	A. *In Vivo* Transcription	416
	B. *In Vitro* Transcription	420
VI.	Transcription in Differentiation	425
	A. *In Vivo* Transcription	425
	B. *In Vitro* Transcription	426
VII.	Concluding Remarks	428
	References	429

I. INTRODUCTION

Physarum and to a lesser extent *Didymium* have become important organisms in which several basic problems of cell biology are being studied. RNA as a mediator between the genome and the organism's phenotypic appearance is pivotal in questions of genome expression and differentiation. It is also important when the organization of the genome is an issue and when the cell cycle is under study. For investigations of RNA metabolism, *Physarum* is attractive in being easy to grow axenically in rather large amounts; many nucleic acid precursors are taken up rapidly; and no rigid cell wall surrounds the cytoplasm. Difficulties can be encountered in the preparation of DNA or RNA because of the high level of nucleases present and the polysaccharide slime, which at high concentrations makes solutions very viscous.

The nucleic acid metabolism of *Physarum* has been dealt with quite extensively in several reviews. The metabolism of precursors and the biochemistry of nucleic acid synthesis are discussed in a lucid review by Turnock (1979). He also gives much thought to the interrelation of the synthesis of DNA, RNA, and proteins. The volume on *Physarum* by Dove and Rusch (1980) contains a thoughtful review on DNA and the cell cycle by Holt, as well as an informative discussion of transcription by Melera. Our own review attempted to emphasize the biological significance of transcription and its regulation (Seebeck and Braun, 1980). Here we would like to discuss both synthesis and degradation of RNA in *Physarum* and to relate these biochemical processes to the mitotic cycle and to differentiation.

II. PROPERTIES OF DNA AND RNA

A. DNA

All data reviewed here pertain to *Physarum polycephalum*, since no observations on the DNA of other acellular slime molds have been published. Holt (1980) has compiled an extensive and critical review, so that we will limit ourselves to those points which are essential for understanding the process of transcription.

The macromolecular composition of rapidly growing microplasmodia is about 1:22.5:120 for DNA:RNA:protein (Plaut and Turnock, 1975). Three types of DNA can easily be distinguished on the basis of buoyant density (Braun and Evans, 1969): nuclear chromosomal DNA (approx. 90%), nuclear satellite DNA or ribosomal DNA (approx. 2%), and mitochondrial DNA (approx. 10%).

1. CHROMOSOMAL DNA

Haploid, nonreplicated nuclei of *P. polycephalum* contain 0.28 pg of DNA or 63 times more than *Escherichia coli* (Mohberg, 1977; Holt, 1980). Taking the most likely haploid chromosome number to be 40 (see Mohberg, this volume, Chapter 7), this gives 4×10^9 daltons or 6000 kilobase pairs (kbp) of DNA per chromosome. In CsCl gradients, nuclear DNA of several strains of *P. polycephalum* bands at a density of 1.700 ± 0.003 gm/ml, corresponding to 41% guanine + cytosine. For one strain of *P. polycephalum*, a slightly higher value has been published (1.705) which requires further confirmation (Holt, 1980). Strains with an aberrant DNA density could be of interest with respect to the

12. RNA Metabolism

fusion killing phenomenon (Carlile and Dee, 1967). Perhaps DNA modification, similar to that known from bacteria, also protects DNA in acellular slime molds. Although several DNases have been tentatively characterized in *P. polycephalum* extracts (Polman *et al.*, 1974), none of the enzymes has so far been shown to cut DNA as specifically as do bacterial restriction enzymes. The fact that foreign DNA can get into eukaryotic nuclei, as shown for yeast (Hinnen *et al.*, 1978), should further encourage the search for site-specific DNases in organisms such as the acellular slime molds.

The complexity of a genome can be estimated by measuring the speed of reannealing of denatured DNA (Britten and Kohne, 1968). Older data had suggested that *Physarum* contained about 45% moderately repeated sequences, 55% unique sequences, and no highly repeated elements (Britten and Smith, 1970; Fouquet *et al.*, 1974a). More recently, Hardman *et al.* (1980) have studied the organization of chromosomal DNA in more detail. They found 6% foldback DNA, 31% repeated sequences, and 63% single-copy sequences. If all unique-sequence DNA were coding for proteins, *P. polycephalum* would code for three times as many proteins as *Dictyostelium discoideum*, since both organisms have about the same proportion of single-copy sequences (Firtel and Bonner, 1972) and *P. polycephalum* has about three times more DNA per nucleus than does *D. discoideum*. The repeats were found to be 590 nucleotides long on average, interspersed between unique sequences 930 bp in length. There are likely to be about 80 families of repeats, each with about 1800 repeats. This organization resembles that found originally in *Xenopus* (Davidson *et al.*, 1973).

The foldback DNA mentioned above has been studied in considerable detail by electron microscopy of rapidly renatured molecules (Hardman and Jack, 1977, 1978; Hardman *et al.*, 1979). On single large DNA molecules, there can be several foldback structures. By length and frequency measurements, it was found that they occur once in about every 7000 base pairs. The foldback structures can be either simple hairpins without loops or hairpins with loops. When plotting lengths of hairpins and loops versus their frequency, diagrams resembling Poisson distributions are obtained. The hairpins vary from about 100 to several hundred base pairs; many of the loops are a few thousand base pairs long. It will be interesting to see whether the hairpins actually fall into a regular arithmetic length series, as suggested by Hardman *et al.* (1979). The function of foldback DNA, found in many eukaryotes, is not clearly established.

2. rDNA

Ribosomal DNA (rDNA) is present exclusively, or at least nearly exclusively, in the nucleolus as free extrachromosomal DNA. In CsCl, rDNA bands at a density of 1.712 ± 0.002 gm/ml, corresponding to 54% guanine + cytosine. For each haploid unreplicated genome there are 190 free rDNA molecules, each coding for two molecules of 5.8 S rRNA, 19 S rRNA, and 26 S rRNA. The coding sequences are arranged head to head with a large central, presumably untranscribed, region (Molgaard et al., 1976; Vogt and Braun, 1976; Hall and Braun, 1977). A detailed restriction enzyme analysis of the noncoding area revealed that it is satellitelike and contains many repeated elements, some of which are themselves inverted repeats (P. Ferris, V. W. Vogt, and R. Braun, unpublished data). The DNA region coding for the 26 S rRNA contains two intervening sequences or introns 0.5 kb and 1.0 kb in length. This was shown by electron microscopy and R loop mapping by both Campbell et al. (1979) and Gubler et al. (1979). The rDNA has a molecular weight of 38×10^6 or 58 kb. For reviews, see Molgaard (1978), Braun and Seebeck (1979), and Holt (1980).

B. RNA

Though there has been considerable interest in the process of transcription in acellular slime molds for several years, individual RNA species have not been analyzed in much detail. There is at least one technical reason for this: large RNA molecules are notoriously hard to isolate intact from these organisms, presumably because of the large amount of ribonucleases present (see Section IV,B,1).

Mention should be made, however, that several methods for the isolation and characterization of ribosomes (Henney and Jungkind, 1969; Hall and Braun, 1977) and polysomes (Brewer, 1972; Schwärzler and Braun, 1977; Adams et al., 1980) from acellular slime molds have been described. Again, undegraded polysomes are hard to obtain from P. polycephalum, particularly from macroplasmodia, and new, simple methods leaving the mRNA intact should be developed.

1. LARGE rRNAS

These have been characterized by several groups using sedimentation velocity and gel electrophoresis. The results have been reviewed in much detail by Melera (1980). Under nondenaturing conditions, the two rRNAs have apparent molecular weights of 1.37×10^6 (26 S) and 0.73×10^6 (19 S). Under denaturing conditions, which give more reliable results, molecular weights are $1.29 \pm 0.03 \times 10^6$ (3.9 kb) and $0.70 \pm 0.02 \times 10^6$ (2.1 kb) (McMaster and Carmichael, 1977). The differences between the molecular weights obtained under the two sets of conditions are rather smaller than in other organisms. Upon denaturation, 26 S rRNA releases 5.8 S rRNA. This holds for P. polycephalum (Hall and Braun, 1977) as for other eukaryotes, in which the molecular weight of 5.8 S rRNA has been determined to be 51,000 (Maden and Robertson, 1974). The size of the large P. polycephalum rRNA lies about halfway between that for bacteria and that for mammals, whereas the small rRNA of P. polycephalum is marginally larger than that of mammals. Apart from its size, not much information is available on *P. polycephalum* rRNA. It has very little secondary structure, as determined by electron microscopy (Gubler *et al.*, 1979). This is in contrast to the well-documented secondary structure of vertebrate rRNA (Wellauer and Dawid, 1973). The base composition of both large rRNA molecules is very similar, with 24% cytosine, 23% adenine, 30% guanine, and 23% uracil (Zellweger and Braun, 1971). No sequence data have so far been published. In *P. rigidum* the two large rRNA molecules were assigned apparent sedimentation values of 26 S and 18 S (Henney and Jungkind, 1969).

In all cells studied so far, the two large rRNA molecules are cleaved from a single precursor which, at least in vertebrates, is more than double the size of the combined 18 and 28 S rRNAs. In *P. polycephalum* several likely precursors have been partly characterized, the largest of which is probably the initial and intact transcript. In earlier pulse-labeling experiments, both Melera and Rusch (1973a) and Jacobson and Holt (1973) demonstrated the formation of very large RNA by *P. polycephalum,* some of which appeared to give mature rRNA after an adequate chase period. A quantitative conversion, however, could not be shown. Melera and Rusch (1973a) observed a sharp peak at about 4.1×10^6 daltons, Jacobson and Holt (1973) a more heterodisperse profile, of which the largest fraction had nearly the same molecular size. Melera (1980) gave a recalculated molecular weight of 4.0×10^6 for the larger precursor. The likely immediate precursor of the mature 26 S rRNA has a molecular weight of about 1.9×10^6 (Melera and Rusch, 1973a) and a base composition indistinguishable from that of mature 26 S rRNA (Zellweger and Braun, 1971).

Experiments by Gubler *et al.* (1980), using a different experimental approach, have led to a more clear-cut picture. RNA was extracted with a method intended to give minimal degradation, then fractionated by gel electrophoresis and characterized by hybridization with nick-translated plasmid DNA following transfer to DBM paper. The largest of the relatively stable rRNA precursors was seen to be 11.8 kb or 3.9×10^6 daltons. Three less stable, still larger precursors contained either IVS 1 (0.5 kb), IVS 2 (1.0 kb), or both of these intervening sequences. The 11.8-kb precursor is probably cleaved in the external transcribed spacer to give a 8.8-kb (2.9×10^6-dalton) precursor still containing 19 S, 5.8 S, and 26 S sequences. In the next step, the internal transcribed spacer is most likely cut to give two precursors, one for the 19 S and one for the 5.8/26 S RNA. Both molecular species need to be further processed. A likely intermediate leading to the large rRNA is 5.3 kb (1.65×10^6 daltons).

The largest precursors in *Physarum* extend about 4 kb beyond the 5' end of the mature 19 S gene. A putative promotor site therefore has to be located about 17.7 kb from either end of the linear palindromic DNA molecule. This location is in very good agreement with studies on the initiation of *in vitro* transcription of *Physarum* rDNA by Sun *et al.* (1979). For details, see Section III,D,1.

Grainger and Ogle (1978) have made interesting observations on the putative first rRNA transcript by electron microscopy. They spread chromatin, in which they could see occasional transcription units of the size expected for ribosomal cistrons. The transcribed lengths of DNA, similar in appearance to those shown by Miller and Beatty (1969), were about 4.2 μm long, only slightly larger than expected from the data reviewed above. In conclusion, several sets of data strongly suggest that a large precursor of rRNA has a molecular weight of about 12 kb or 4×10^6. It must still be shown whether this is really the initial transcript carrying the original 5' triphosphate.

2. tRNA AND OTHER SMALL RNAS

Of the total RNA, tRNA makes up about 13% (Braun *et al.*, 1966). By chromatography and aminoacylation, 44 different tRNA species have so far been identified in *P. polycephalum* (Melera and Rusch, 1973b; Melera *et al.*, 1974). As in other organisms, ribosomes of *P. polycephalum* have also been shown to contain a 5 S RNA. This accounts for about 1% of all RNA (Melera and Rusch, 1973b). Both tRNA and 5 S rRNA are coded for by chromosomal DNA rather than the extrachromosomal rDNA. There are approximately 1000 gene copies per haploid genome for each of the two RNA types (Hall and Braun, 1977). In 1977, Hellung-Larsen and Frederiksen showed the presence of an additional small and presumably stable RNA, with a sedimentation value of 7 to 8 S. This amounts to about 0.1% of the total RNA, and its function is unknown. It appears that *P. polycephalum*, like HeLa cells, may contain some circular RNA made up of a few hundred nucleotides. Its structure and function do not yet seem to have been studied in any detail (Hsu and Coca-Prados, 1979).

3. hnRNA AND mRNA

These two groups of molecules are notoriously difficult to characterize in any organism, primarily because they are intrinsically heterogeneous in size, composition, and function. No single mRNA with a specific function, or its precursor, has so far been studied in detail in an acellular slime mold. Studies on the *in vitro* translation of mRNA for histones are reported to be in progress (Sauer, 1978).

The cytoplasm appears to contain both poly(A)$^+$ RNA and poly(A)$^-$ RNA, though no data have so far been published on the latter (Sauer, 1978). Polysomal poly(A)-containing RNA may represent about 1% of cellular RNA (Sauer, 1978). It is heterogeneous in size, sedimenting at about 8–15 S (Fouquet *et al.*, 1974b; Fouquet and Sauer, 1975). Electrophoresis under denaturing conditions gave an average molecular weight of 0.45×10^6, a value well within the size range given above (Melera, 1980). Most of this RNA hybridizes to unique-sequence DNA (Fouquet and Sauer, 1975). Adams and Jeffery (1978) devoted a detailed study to the poly(A) sequences of the cytoplasmic RNA, most of which is likely to be polysomal. Two types of sequences were discovered, differing in several properties. Short oligo(A) sequences (averaging 26 nucleotides) are metabolically stable, whereas longer poly(A) sequences are turned over and may also be shortened. The oligo(A) sequences do not arise by degradation of poly(A) sequences; in fact, the two may not even be part of the same RNA molecules. The oligo(A) sequences have about the same stability as rRNA; virtually no degradation is observed during an 18-hour period. By contrast, half of the poly(A) sequences disappear within about 4 hours. Superimposed on this degradation, there may be a progressive shortening of the poly(A) sequences. Newly made poly(A) sequences are about 65 nucleotides long, whereas they seem to average only about 50 nucleotides after 6 to 8 hours. The authors suggest that newly made poly(A) sequences progressively lose up to 15 nucleotides. It cannot be ruled out, however, that newly made poly(A) sequences are heterogeneous in length and that the longer sequences are more rapidly degraded than the shorter ones. Inhibitor studies suggest that translation is required for poly(A) sequence destruction.

Not much information is available on hnRNA. Nuclear poly(A) RNA appears to vary in size from 4 to 30 S (Fouquet and Sauer, 1975), with an average molecular weight, as determined under denaturing conditions, of about 0.51×10^6 (Melera *et al.*, 1978). Both results may require revision since no one has yet isolated nuclei to a high degree of purity without degrading RNA. About one-third of the nuclear poly(A) RNA hybridizes to reiterated DNA sequences (Fouquet and Sauer, 1975). Melera *et al.* (1978) have done a preliminary study on polyadenylated RNA using electrophoresis under denaturing conditions, in which the RNA came from fractions enriched either for nuclei or for cytoplasm. After a 45-minute labeling period, the nuclear fraction contained a higher proportion of large RNA molecules (MW $> 1.37 \times 10^6$) than did the cytoplasmic fraction. In a further study, Melera *et al.* (1979) showed that poly(A) RNA from a nuclear-enriched fraction was considerably larger than the poly(A) RNA from a cytoplasm-enriched fraction. With both fractions, an *in vitro* translation system using a rabbit reticulocyte lysate produced a similar spectrum of polypeptides, as analyzed by SDS–polyacrylamide gel electrophoresis. These experiments suggest that part of the polysomal mRNA may derive from a distinctly larger nuclear precursor, as is well established for mammalian cells. In other lower eukaryotes, such as *D. discoideum* and *Achlya bisexualis,* the nuclear precursor of cytoplasmic mRNA appears to be nearly the same size as the mature product (Firtel and Lodish, 1973; Timberlake *et al.*, 1977).

It would be desirable in the future to study specific nucleotide sequences in addition to a mixed population of molecules. The coding sequences for actin might be attractive candidates for such a study since actin is one of the most abundant proteins of *P. polycephalum*. The sequence of *P. polycephalum* actin is similar to that of vertebrate cytoplasmic actin (Vandekerckhove and Weber, 1978). Coding sequences for actin of *D. discoideum* have been cloned in plasmids (Kindle and Firtel, 1978).

III. RNA SYNTHESIS

A. *In Vivo* Transcription

Most of the experiments on *in vivo* RNA synthesis in *Physarum* were done with the aim of gaining a deeper insight into the control mechanisms of the mitotic cycle or of differentiation. These experiments will therefore be critically reviewed in Sections IV and V. At this point, the use of inhibitors to block RNA synthesis in *Physarum* will be reviewed briefly.

Over the past 15 years, several attempts have been made to find a specific inhibitor of *in vivo* RNA synthesis in *P. polycephalum*. The search, however, has not met with striking success. Both actinomycin C and actinomycin D are required in very high concentrations (100 to 250 μg/ml) to inhibit substantially RNA synthesis (Mittermayer *et al.*, 1966; Sachsenmaier *et al.*, 1967). Rather similar results were obtained more recently for daunomycin and toyocamycin (Fouquet *et al.*, 1975b). The same authors found, however, that 50 to 200 μg/ml cordycepin (3'-deoxyadenosine) very substantially inhibited RNA synthesis, particularly if plasmodia were pretreated with the drug before labeling. So cordycepin appears to be the best inhibitor available, but it is far from ideal. Perhaps its effect could be enhanced by using more suitable media. In addition, it might be fruitful to search for derivatives of such antibiotics as actinomycin, since *in vitro* with isolated nuclei, actinomycin is active at very low concentrations, indicating that the antibiotics used so far are unable to penetrate the cell (Mittermayer *et al.*, 1966). It should be possible to find derivatives with better permeability properties. The fungal toxin α-amanitin has no inhibitory effect *in vivo*.

Heat treatment of plasmodia for 10 to 30 minutes at 41°C increases uridine incorporation into RNA, whereas it decreases amino acid incorporation into protein (Bernstam, 1974). The mechanism of this differential effect is not understood. Inorganic cadmium in the concentration range around $10^{-3}M$ considerably reduces RNA synthesis of *P. polycephalum* plasmodia within a few hours (Sina and Chin, 1978). It is not clear from the data presented how specific this effect is, although it is obvious that cadmium strongly distorts the structure of nucleoli, the organelles of rRNA synthesis.

B. RNA Polymerases

1. PURIFICATION AND CHARACTERIZATION

Physarum polycephalum, like all other eukaryotes investigated so far, contains multiple RNA polymerases. Several laboratories have attempted to analyze these enzymes (for reviews, see Sauer, 1978; Melera, 1980).

The presence of two different RNA polymerase species in extracts from iso-

12. RNA Metabolism

lated nuclei was first demonstrated in 1973 (Hildebrandt and Sauer, 1973), and several laboratories have since confirmed and expanded those early findings (Burgess and Burgess, 1974; Gornicki et al., 1974; Weaver, 1976; Smith and Braun, 1978). When a nuclear extract is chromatographed on DEAE–Sephadex, the RNA polymerase activity which elutes first is entirely sensitive to α-amanitin. On the other hand, the RNA polymerase activity eluting in a later peak is resistant to this toxin. This represents an inversion of the situation observed when RNA polymerases of various other eukaryotes are chromatographed on this ion exchanger. Here the α-amanitin-resistant polymerase activity elutes first (and hence is designated "polymerase I"), whereas the α-amanitin-sensitive enzyme follows in a later peak and is designated "polymerase II." For the sake of clarity, the designation of the RNA polymerases from *P. polycephalum* follows that employed for other eukaryotes, so that the α-amanitin-resistant polymerase is designated "polymerase I" and the α-amanitin-sensitive enzyme is "polymerase II," irrespective of their order of elution from the DEAE–ion exchanger. This convention is justified since the α-amanitin-resistant polymerase from *P. polycephalum* is localized in the nucleolar chromatin and is involved in the transcription of rRNA (Davies and Walker, 1977; Seebeck et al., 1979), as are its counterparts in other eukaryotes. Furthermore, a comparison of the salt optima of the RNA polymerases from *P. polycephalum* demonstrates that, again in accordance with the situation in other eukaryotes, the α-amanitin-resistant polymerase has a very much lower salt optimum than the α-amanitin-sensitive enzyme (Burgess and Burgess, 1974; Gornicki et al., 1974; Smith and Braun, 1978).

In addition to RNA polymerases I and II, a third RNA polymerase activity, possibly representing RNA polymerase III, has been discovered after DEAE–Sephadex fractionation of high-salt homogenates from total plasmodia (Hildebrandt and Sauer, 1976a; Ernst and Sauer, 1977). This activity elutes at still higher salt concentrations than do polymerases I and II, and it is resistant to α-amanitin concentrations of up to 100 μg/ml but is sensitive to higher toxin concentrations. This behavior would suggest that the third activity is in fact RNA polymerase III, but due to its instability on purification, no additional information on this enzyme is presently available. A formal demonstration that it is a nuclear polymerase is also lacking at present.

Early attempts to purify RNA polymerases from *P. polycephalum* were based on salt extractions of isolated nuclei (Burgess and Burgess, 1974; Gornicki et al., 1974; Weaver, 1976). Though polymerases from such extracts could be purified to specific activities comparable to those obtained with *E. coli* RNA polymerase, the overall yield was consistently low, a typical figure being 5 μg of purified RNA polymerase from 100 ml of packed plasmodia (Smith and Braun, 1978). An approximately tenfold increase in yield was achieved when total plasmodia, rather than purified nuclei, were used for the initial salt extraction

(Hildebrandt and Sauer, 1973; Smith and Braun, 1978). Since then, Hildebrandt and Sauer (1976b) have very elegantly demonstrated that RNA polymerases I and II reside exclusively in the nucleus in intact plasmodia. The increase in enzyme yield observed when total plasmodia, rather than isolated nuclei, are extracted suggests that considerable leakage of RNA polymerase from the nucleus into the cytoplasm takes place during nuclear isolation.

Fractionation of the two RNA polymerase activities from crude extracts can be achieved by ion-exchange chromatography on DEAE-Sephadex or DEAE-Sepharose. In addition to the ionized groups of the resin, the supposedly inert matrix also appears to be involved in this fractionation step. This may be caused by hydrophobic interactions, since DEAE-cellulose is ineffective in separating RNA polymerases I and II from *P. polycephalum* (Burgess and Burgess, 1974; Hildebrandt and Sauer, 1976a; S. S. Smith, personal communication). This is quite a contrast to other eukaryotic RNA polymerases, in which DEAE-cellulose usually is the material of choice for this purification step. However, a similar observation has been made during the purification of RNA polymerase I from sea urchins (D. Stafford, personal communication).

A further purification step usually applied to DEAE-Sepharose fractionated polymerases I and II is chromatography on phosphocellulose (Hildebrandt and Sauer, 1973; Burgess and Burgess, 1974; Gornicki *et al.*, 1974; Smith and Braun, 1978). Here, polymerase II again elutes before polymerase I. This step usually results in rather poor yields of protein. However, a strong increase in specific activity of the polymerases, notably polymerase I, is observed, leading even to recoveries of polymerase I activities which exceed the input (Burgess and Burgess, 1974). This observation cannot be adequately explained solely by removal of contaminating protein. It suggests that inhibitors were removed which had previously contaminated the enzyme preparation. Hildebrandt *et al.* (1979) have described a nucleolar inhibitor for RNA polymerases, the removal of which during chromatography of the enzyme on a cation exchanger could account for the observed strong stimulation of polymerase I after chromatography on phosphocellulose (see Section III,C,1).

An alternative method for the purification of RNA polymerase II has been established which results in both better yields and greater purity of the enzyme (Smith and Braun, 1978). The crude salt extract from total plasmodia is treated with polyethyleneimine, which leads to the precipitation of, among other proteins, RNA polymerase (Jendrisak and Burgess, 1975). The precipitated polymerase is eluted from the pellet, and the eluate is then treated with ATP and magnesium to precipitate the vast amounts of actin which otherwise seriously contaminate the enzyme preparations (Smith *et al.*, 1979). The actin-depleted supernatant is then fractionated by chromatography on DEAE-Sepharose, followed by heparin-Sepharose. The latter step achieves a dramatic increase in activity which is, in other purification schemes, observed after phosphocellulose chromatography (see above). RNA polymerase prepared in this manner is essentially pure except for a trace of actin and is free of nucleases. The remaining traces of actin can, if necessary, be removed by chromatography on phosphocellulose.

2. SUBUNIT STRUCTURE

The subunit structure of polymerases I and II from *P. polycephalum* is still the subject of investigation. Earlier workers reported an extremely simple subunit structure for polymerase I as well as II. Each enzyme was thought to consist of two polypeptides of molecular weights between 100,000 and 200,000. This view now appears to have been simplistic, probably because insufficient amounts of protein had been loaded onto the gels. More recent investigation, though confirming the presence of the large-molecular-weight subunits identified earlier, has revealed the presence of several additional presumptive subunits. The most recently proposed subunit structure for RNA polymerase I suggests that it consists of five polypeptides of the following molecular weights: 200,000, 135,000, 45,000, 24,000, and 17,000 in the ratio 1:1:1:2:1, respectively (Gornicki *et al.*, 1974). These molecular weight determinations agree well with those reported by other workers for the two large subunits (Hildebrandt and Sauer, 1973; Burgess and Burgess, 1974). A comparison of Gornicki's subunit structure for polymerase I with that proposed by the same laboratory for polymerase II, namely, four polypeptides of molecular weights 175,000, 140,000, 24,000, and 17,000 present in the ratio 1:1:2:1, respectively (Weaver, 1976), suggests that the two polymerases might share some of the low-molecular-weight subunits (24,000 and 17,000). Smith and Braun (1978) have extended the number of putative subunits. Polypeptides of the following molecular weights were found: 215,000, 170,000, 135,000, 38,000, 25,000, 17,000, and 14,000. Such an overall subunit structure is rather similar to that proposed for RNA polymerase II from a variety of plants or from calf thymus (Hodo and Blatti, 1977; Jendrisak and Guilfoyle, 1978). Though the reported accounts of subunits from polymerase II from *P. polycephalum* vary in the number of subunits found, there is general agreement on the molecular weights of those subunits identified so far in the different laboratories. One of the large subunits which is consistently found in polymerase II by all workers (molecular weight about 140,000) might be analogous to a subunit of similar size which is found in RNA polymerase II from all other eukaryotic sources and which is the binding site for α-amanitin (Brodner and Wieland, 1976).

When intact polymerase II is analyzed on nondenaturing gels, two major and various minor subfractions of the enzyme can be detected, though the enzyme preparation is homogeneous when analyzed by ion-exchange chromatography (Smith and Braun, 1978; Smith and Braun, 1981). Two-dimensional gel electrophoresis reveals that each of the two largest subunits found in SDS gels (molecular weights 215,000 and 170,000) probably corresponds to one or the other of the two major subforms of the intact enzymes detected on nondenaturing gels. Subforms of RNA polymerase II have been observed in a variety of organisms (Dezélée et al., 1976; Hodo and Blatti, 1977; Jendrisak and Guilfoyle, 1978), and they supposedly reflect the transcriptional state of the tissue in question (Jendrisak and Guilfoyle, 1978). Heterogeneity of polymerase II from *P. polycephalum*, even though purified from plasmodia supposedly in balanced growth, is also suggested by the broad salt optimum consistently observed (Gornicki et al., 1974; Smith and Braun, 1978). The various subforms of RNA polymerase II could reflect different stages of the cell cycle, or else various subforms might be required for transcription of different gene categories.

3. SUMMARY

The RNA polymerases of *P. polycephalum* closely resemble those found in other eukaryotes. Although little is known about RNA polymerase III, the subunit structure of polymerases I and II have been investigated in some detail. Both enzymes consist of several units. Polymerase I, the enzyme involved in the transcription of rRNA, is completely resistant to α-amanitin. Polymerase II, the enzyme which transcribes RNA destined to become mRNA, is very sensitive to α-amanitin, which inactivates the enzyme presumably by binding to one specific subunit. An analysis of the subunit structure of RNA polymerase II shows heterogeneity in some of the large subunits. These variations are thought to reflect differences in transcriptional activity or specificity.

C. Regulatory Factors

1. INITIATION INHIBITOR

Early work on transcription in *P. polycephalum* has led to the postulation of a nucleolar inhibitor of RNA polymerases. Such a factor was tentatively characterized as an organic polyphosphate molecule (Goodman et al., 1969; Sauer et al., 1969a,b; Goodman, 1972). These results have since been reexamined and considerably expanded (Hildebrandt et al., 1979). Nucleoli are enriched in a polymerase inhibitor. The inhibitor copurifies with DNA during phenol extraction but can be separated from DNA in CsCl equilibrium density gradients due to its higher buoyant density. The inhibitor can be further purified by ion-exchange chromatography on DEAE–Sephadex and by density gradient centrifugation, in which the material sediments homogeneously with a sedimentation coefficient of 4 S. It is composed of acid-labile and acid-resistant phosphate in a ratio of 3:1 and a single carbohydrate component, tentatively identified as glycerol. It does not contain significant amounts of nitrogen. The inhibitor binds tightly to free RNA polymerase, but not to template-bound enzyme or to free DNA, and therefore seems to compete with DNA for the template-binding site of the polymerase. It inhibits homologous RNA polymerase I much more strongly than homologous polymerase II or heterologous RNA polymerase I, and it does not have any effect on homologous DNA polymerase. This at least partial specificity of the inhibitor for polymerase I, together with its nucleolar localization, suggests but by no means proves that this substance might be a negative control element in the regulatory system of rRNA transcription. A possible physiological role for this substance is also suggested by the observed fluctuations of inhibitor concentration *in vivo* during the cell cycle and development (see Sections V and VI), and more information on the structure and function of this interesting substance is eagerly awaited.

2. STIMULATORY PHOSPHOPROTEIN

A large non-histone phosphoprotein (molecular weight 70,000) which acts as a strong positive control element for the transcription of rRNA *in vitro* has been characterized (Kuehn *et al.*, 1979). This protein is highly phosphorylated *in vivo* by a nuclear protein kinase which is dependent on the presence of polyamines and strongly inhibited by cyclic AMP (Atmar *et al.*, 1978). It is localized in the minichromosomes containing the ribosomal genes (Seebeck *et al.*, 1979). In its phosphorylated form, this protein binds to one specific region of rDNA with a binding constant of approximately 10^{-10}. Preliminary mapping of the binding site has shown that it is located upstream from the sequences coding for rRNA and is at least 6000 bp away from the start of those sequences. On dephosphorylation, this protein loses its ability to bind DNA. The effect of the protein on *in vitro* transcription is equally dependent on phosphorylation. The addition of purified phosphoprotein to isolated minichromosomes containing the ribosomal genes results in a fivefold stimulation of transcription, whereas the dephosphorylated protein has no such effect. Analysis of the RNA transcribed *in vitro* in the presence or absence of the 70,000-dalton protein demonstrates that in both cases the same sequences are transcribed, though in the presence of the protein, transcription is five times more efficient.

It is at present unclear whether this stimulation of transcription is brought about by a higher frequency of initiation or by some other effect (e.g. on elongation). The observation that this protein is a strong stimulator of transcription, though it binds at a specific site far removed from the coding sequences, poses an intriguing problem: if this protein-binding site is anywhere near the initiation site of transcription, a very long precursor RNA should be predicted, which, however, has not yet been detected *in vivo* (see Section II,B,1). Alternatively, this protein could facilitate the binding of RNA polymerase molecules to the DNA at a site far away from the onset of transcription. The lined-up polymerases could then slide along the DNA to the coding sequences without transcribing en route and initiate only farther downstream. Unfortunately, too little information is available at present to give substance to any of these possibilities, and certainly many other speculations on the mechanism of such a positive control element for a eukaryotic gene are conceivable. Regulatory proteins, such as the one described above, might in evolution be a highly conserved class of molecules, and preliminary comparison of the chemical composition of this 70,000-dalton protein from *P. polycephalum* has shown striking similarities with a possibly analogous protein isolated from rat nucleoli. These proteins from rat and *P. polycephalum* have the same molecular weight, are both acidic, and have a very similar amino acid composition (James *et al.*, 1977; Kuehn *et al.*, 1979).

3. ELONGATION FACTOR

Although both of the control factors dealt with so far may act at the level of initiation of transcription, a possible elongation factor has been described (Ernst and Sauer, 1977). This elongation factor is found in the material not retained by DEAE–Sephadex during RNA polymerase purification and appears to be a protein with an isoelectric point of about 8. Since it has not yet been purified, its molecular structure has not been determined. The factor stimulates RNA polymerases I and II equally well if the template is double-stranded DNA; no stimulation is observed when either single-stranded DNA is used as template or with *E. coli* RNA polymerase. The extent of stimulation when double-stranded DNA is used as template is strongly dependent on the size of the template, decreasing rapidly with decreasing molecular weight of the DNA. The factor does not have endonuclease activity, nor does it act as an RNase inhibitor under the conditions tested, and from the low salt optimum reported for the stimulatory effect, it appears unlikely to be an unwinding enzyme. Analysis of the transcription products demonstrated an increase in the length of the chains synthesized *in vitro* in the presence of the elongation factor, thus suggesting that this factor might stabilize the transcription complex *in vitro*. A possible *in vivo* role of this factor for transcription is suggested by the fluctuations of factor concentrations with changing transcriptional activities of the plasmodia during the cell cycle and development (see Sections V and VI).

4. SUMMARY

Three factors which regulate transcription have so far been identified. They are an organic polyphosphate which acts as an inhibitor of RNA polymerase I, a presumably proteinaceous elongation factor, and a highly phosphorylated nonhistone protein which specifically stimulates the transcription of rRNA. Whereas the inhibitor and the elongation factors interact directly with RNA polymerase, the stimulatory phosphoprotein binds to specific sites of the rDNA. The mechanisms of action of all three factors are currently being investigated in several laboratories.

D. *In Vitro* Transcription

1. TRANSCRIPTION IN ISOLATED NUCLEI

Physarum polycephalum has long been recognized to afford a particularly suitable system for investigating transcription *in vitro* using either isolated nuclei or nucleoli. The advantages of this organism for this type of study are the high natural synchrony of the mitotic cycle, the possibility of studying transcription during different phases of development, and the ease with which nuclei can be prepared. Early studies demonstrated the transcriptional activity of isolated nuclei (Mittermayer *et al.*, 1966; Cummins and Rusch, 1967) and established that the transcriptional activity of nuclei isolated at different times during the cell cycle approximately follows the fluctuations of the transcriptional activity *in vivo* (Braun *et al.*, 1966; Cummins *et al.*, 1966; Cummins and Rusch, 1967; Grant, 1972; Zellweger and Braun, 1971; see Section V,B,1).

A more detailed investigation of the basic requirements and the action of several transcription inhibitors suggested the existence of at least two different RNA polymerases in nuclei before these enzymes were actually isolated and characterized biochemically (Grant, 1972). Isolated nucleoli were shown to contain an RNA polymerase activity which was largely resistant to the fungal toxin α-amanitin. This enzyme was minimally active at 10 mM $MgCl_2$ and under low salt conditions, and no further stimulation could be achieved by the addition of Mn^{2+} or by raising the ionic strength of the incubation medium. If, however, nuclei instead of nucleoli were analyzed, a second enzyme activity became apparent. This latter activity was sensitive to α-amanitin, and optimal conditions of activity required provision of Mn^{2+} ions and moderately high ionic strength (100 mM KCl). The differential effect of Mn^{2+} ions on the activity of the two polymerases in nuclei, where polymerase II but not I was stimulated, was in apparent contrast to the ionic requirements of the two enzymes purified subsequently (Gornicki *et al.*, 1974), which are both stimulated to a similar extent by Mn^{2+}. However, in the experiments of Grant (1972) with nuclei or nucleoli, addition of Mn^{2+} was always accompanied by the addition of salt to the incubation mixture. Since RNA polymerase I demonstrates optimal activity at a very low salt concentration, the stimulation of activity brought about by the Mn^{2+} ion was probably nullified by the concomitant elevation of the salt concentration to supra-optimal levels.

Davies and Walker (1977) have taken up and considerably expanded these observations. The method of preparing nuclei was shown to be of crucial importance for the overall transcriptional activity of the purified nuclei. Exposure to a high sucrose concentration (1 M) and to 10 mM $MgCl_2$ led to a considerable loss of activity, an observation most readily explained by the leakage of RNA polymerases out of the nuclei during isolation (Hildebrandt and Sauer, 1976b;

see Section III,B,1). Structural integrity of the chromatin was demonstrated to be another important factor, since sonication of isolated nuclei completely abolished transcriptional activity. Overall transcriptional activity in this system has a salt optimum of 100 mM KCl, suggesting an overabundance of RNA polymerase II. This is confirmed by the observation that about 60% of [^{3}H]UTP incorporation can be inhibited by 16.5 μg/ml of α-amanitin. No further inhibition is observed with inhibitor concentrations up to 200 μg/ml, indicating the absence of RNA polymerase III activity (Davies and Walker, 1978). This latter observation, as well as the difficulty of detecting RNA polymerase III during enzyme purification (see Section III,B,1), again suggests that this enzyme is either very easily lost from nuclei during purification or that, in *P. polycephalum*, it is an extremely unstable enzyme which is quantitatively inactivated during the nuclear isolation procedure. Preliminary experiments investigating the possibility of initiating transcription in isolated nucleoli demonstrated a slight inhibition of [^{3}H]UTP incorporation in the presence of the initiation inhibitor rifampicin AF/013 (Davies and Walker, 1978). This observation suggests that reinitiation of transcription may in fact occur in isolated nucleoli.

A potentially very interesting observation by Davies and Walker (1977) is that nuclei, as well as nucleoli, synthesize full-length precursor rRNA molecules and, in addition, are capable of processing these large transcripts (44 S) via the correct intermediate (34 S) to the mature 26 and 19 S rRNA. However, several features of these experiments require further clarification, and a more detailed investigation would certainly be of great interest.

Although the paper states explicitly that only about 40% of the transcription is resistant to α-amanitin, i.e., is ribosomal transcription, the gels presented indicate a virtually complete absence of non-ribosomal transcripts, if one is to follow the argument employed by the authors. Furthermore, gels displaying RNA synthesized either *in vivo* or *in vitro* for various lengths of time indicate a virtually complete absence of transcripts smaller than 19 S, an observation so unusual that it would certainly warrant a closer investigation. The question of RNA processing is difficult to assess from the data presented, since the resolution in the gel system used is rather poor, as the authors themselves acknowledge, and since the very likely possibility of aggregation artifacts has not been taken into account. If, however, the claims of the authors should be substantiated by more rigorous experimentation, the nucleolar transcription system from *P. polycephalum* would certainly be an even more attractive system for investigating the transcription of ribosomal genes in a eukaryotic organism.

Another system for *in vitro* transcription of rRNA has been established by the purification of a minichromosome containing the ribosomal genes and, among other chromosomal proteins, RNA polymerase I but not II or III (Kuehn *et al.*, 1979; Seebeck *et al.*, 1979). In this system the endogenous polymerase transcribes the correct strands of the coding sequences, but in contrast to the observations discussed previously (Davies and Walker, 1977), the size of the transcripts is small. Since RNase does not appear to be a major cause of this small size of the transcripts (G. Bindler and T. Seebeck, unpublished), the discrepancy between the sizes of the transcripts obtained from the two systems remains unresolved. A speculative explanation for this discrepancy would be the loss, during isolation of the minichromosome, of factors which stabilize the transcriptional complex (e.g., Ernst and Sauer, 1977; see Section III,C,3). The transcriptional activity of the minichromosome is positively controlled by a non-histone phosphoprotein (see Section III,C,2). The mechanism of action of this control element still remains to be resolved.

In an attempt to map the initiation site for transcription of the precursor rRNA, Sun et al. (1979) have worked with intact nuclei as their *in vitro* system. Since all experiments were conducted in the presence of α-amanitin, which inhibits RNA polymerases II and III, transcription by polymerase I could be selectively observed. In order to be able to discriminate RNA synthesized as an elongation of preexisting chains from RNA which had been initiated *in vitro*, their reaction conditions contained, among other components, $5'$-(γ-S)-adenosine triphosphate. In the latter compound, an oxygen atom of the γ-phosphate group has been replaced by a sulfur atom. Whenever this analogue is incorporated, during an initiation event, as the first nucleotide of a nascent RNA chain, the modified γ-phosphate group protects this $5'$ terminal triphosphate from degradation by phosphatase and allows the selective recovery of the *in vitro* initiated RNA chains through affinity chromatography on organomercurial–agarose columns. Such isolated, short RNA chains are then hybridized to restriction fragments of rDNA to localize their origin within this DNA molecule, i.e., the point of initiation of transcription. Though the hybridization data are not totally unambiguous, the most likely point of initiation appeared to be about 3.5 kbp upstream from the start of the 19 S gene.

This location is compatible with an rRNA precursor molecule of the size of those actually observed *in vivo* (Jacobson and Holt, 1973; Melera and Rusch, 1973a). These findings have been further corroborated by length determination and R loop mapping of the presumptive primary transcript of the ribosomal genes (Gubler *et al.*, 1980; see Section II,B,1).

In vitro transcription by RNA polymerase II in isolated nuclei or chromatin is at present rather difficult to investigate in detail since no well-defined hybridization probes for individual genes, which could be used for the analysis of *in vitro* transcripts, are yet available. However, specific genes have been identified from a library of cloned DNA from *D. discoideum* (Kindle and Firtel, 1978), and at least some of these genes may be expected to give enough cross-hybridization with the corresponding *P. polycephalum* genes so that they could possibly be used as DNA probes until homologous clones are available. Once such cloned DNA fragments carrying specific genes are available as hybridization probes for identifying *in vitro* transcripts, the analysis of *in vitro* transcription in *P. polycephalum* will enter a truly exciting stage.

Besides the lack of hybridization probes, transcription studies in *P. polycephalum* have been hampered by the impossibility of preparing chromatin without extensively damaging its structure. This problem is due to the extraordinary resistance of isolated nuclei to shear or lysis by EDTA, so that the preparation of chromatin always had to include one step of extensive sonication or vigorous vortexing in the presence of EDTA. This problem has now apparently been solved by a significantly improved technique for the preparation of chromatin which has been reported by Schicker *et al.* (1979). The important step in this procedure is the lysis of the nuclear membrane with lysolecithin in the presence of 5 mM $MgCl_2$. Thus the previously used sonication or vortexing can be avoided and the nuclei do not have to be exposed to EDTA, which by itself disrupts the structure of *P. polycephalum* chromatin so that it can no longer be properly digested by micrococcal nuclease (J. Stalder, personal communication). Chromatin as prepared by Schicker *et al.* (1979) contains endogenous RNA polymerase activities of both type I and II. Salt optima for the two polymerases in chromatin correspond well to those found for purified enzymes. Analysis of the *in vitro* transcripts in denaturing formamide gels demonstrates a broad peak of RNA migrating with 20–30 S, a homogeneous peak at 4–5 S, and a third pronounced peak of material migrating faster than 4 S. Such a profile is in sharp contrast to those presented by Davies and Walker (1977), who could find no evidence for the *in vitro* synthesis of small-molecular-weight RNA molecules. Since these latter authors used nondenaturing gels of rather poor resolving power, the data presented by Schicker *et al.* (1979) may be more reliable. The addition of exogenous, purified RNA polymerase II to isolated chromatin resulted in a doubling of incorporation of [^{3}H]UTP and in a higher proportion of transcripts migrating in the 20–30 S region during electrophoresis.

Such a stimulation of transcription, however, is observed only when the exogenous RNA polymerase is added under low salt conditions, and it is completely abolished when high salt (100 mM KCl) or the initiation inhibitor aurin tricarboxylic acid are present. Once initiation by the exogenous polymerase has taken place under low salt conditions, a subsequent increase of the salt concentration no longer prevents transcription. Experiments using γ-^{32}P-labeled UTP as a precursor have given suggestive evidence of the occurrence of initiation of transcription by the endogenous polymerases, thus supporting the conclusions reached by previous workers (Davies and Walker, 1977), who investigated the effects of the initiation inhibitor rifampicin AF/013. However, both lines of evidence are rather indirect, and none of the numerous pitfalls inherent in either of the two methods have been rigorously excluded. Hence the question of significant, site-specific initiation of transcription of mRNA in *Physarum* is still wide open, as it is in other eukaryotic systems (Grummt, 1978).

2. TRANSCRIPTION IN ISOLATED MITOCHONDRIA

A single paper has appeared to date dealing with *in vitro* transcription in isolated mitochondria from *P. polycephalum* (Grant and Poulter, 1973). Here RNA synthesis is inhibited both by actinomycin D and by rifampicin (50% inhibition by both drugs is exhibited at about 5 μg/ml). The half-life of RNA synthesized *in vitro* is extremely short, about 2–3 minutes. This suggests, unless the rapid turnover is an experimental artifact, that the molecules made *in vitro* are not rRNA, since rRNA would be expected to be much more stable.

Electrophoretic analysis indeed confirms that mitochondrial RNA synthesized *in vitro* migrates rather heterogeneously but exhibits a clear peak migrating more rapidly than the mitochondrial rRNA. No significant amount of RNA synthesized *in vitro* migrates in the positions of ribosomal or of 4 and 5 S RNA. The nature of the mitochondrial RNA synthesized *in vitro* is unclear at present and certainly warrants closer investigation. The lack of interest in the transcriptional capabilities of the mitochondria from *P. polycephalum* is rather surprising since this organism offers the possibility of isolating rather large amounts of mitochondria from a homogeneous cytoplasm. Due to the natural synchrony of the cell cycle and the various life stages exhibited by this organism, the analysis of possible nuclear controls over mitochondrial transcription during the cell cycle or differentiation appears to be a promising approach, and further information on this subject is eagerly awaited.

3. SUMMARY

Transcription has been investigated *in vitro* with both isolated nuclei and nucleoli. During *in vitro* transcription with nucleoli, full-size precursor rRNA is apparently synthesized and processed into mature rRNA. Isolated chromatin from whole nuclei still contains endogenous activity of both RNA polymerases I and II. Transcription of such chromatin can be strongly stimulated by the addition of purified homologous RNA polymerase. For lack of well-defined hybridization probes, detailed investigation of *in vitro* transcription by bulk chromatin has not yet been possible.

E. Chromatin Structure and Transcription

Besides understanding the mechanisms of the transcriptional process on the level of interaction of RNA polymerase with DNA, a strong effort has been made to elucidate the structure of active genes within the chromatin. Studies along these lines are expected to yield information on the role of chromosomal proteins and of higher-order structures of chromatin in the regulation of transcription.

In *Physarum*, the ribosomal genes have become the main object for such studies, due both to their extrachromosomal existence as independent entities and to the fact that rRNA is synthesized at very different rates during various phases of the cell cycle. An overall view of a ribosomal gene from *Physarum* in the process of transcription was presented in an electron micrograph by Grainger and Ogle (1978). The two sets of ribosomal genes of the rDNA molecule are transcribed simultaneously in the form of classic "Miller-type" Christmas trees. Transcription appears to start at a specific site between the symmetry center of the molecule and the coding regions and moves from there toward the respective ends of the molecule, transcribing the ribosomal genes in the order, by inference from the physical map of the rDNA, 19-5.8-26 S. So far, this single picture remains the sole piece of electron microscopic data on these ribosomal genes, and this line of investigation has apparently been completely abandoned for the time being.

Using a biochemical approach, Stalder *et al.* (1979) demonstrated that the genes coding for rRNA are degraded by staphylococcal nuclease with the same kinetics as is the bulk of the nuclear chromatin. On the other hand, size analysis of the cleavage products indicated that ribosomal genes are cleaved to fragments of different size than those obtained from bulk chromatin.

It is unclear from these experiments whether ribosomal genes are cleaved to fragments of random size or to a fragment pattern analogous to the one formed from bulk chromatin, but based on a monomer of different size. In addition, experiments using different techniques seem to indicate that the ribosomal genes are degraded by staphylococcal nuclease to fragments similar in size to those obtained from bulk chromatin (U. Pauli, personal communication). Johnson *et al.* (1978) investigated the distribution of coding and noncoding sequences of rDNA between various nucleosome subpopulations obtained from nuclei digested with staphylococcal nuclease. Ribosomal coding sequences were mainly found in a nucleoprotein particle termed "peak A," sedimenting at 5 S and in mononucleosomes. In contrast, noncoding regions were predominantly found in the oligonucleosome fraction. These experiments indicate that cleavage by staphylococcal nuclease may occur more frequently within the coding region of rDNA, which is assumed to be actively transcribed, than it does in the noncoding, nontranscribed segments of the rDNA molecule. An earlier observation that coding and noncoding segments of rDNA have similar base compositions (Steer *et al.*, 1978; Matthews *et al.*, 1978) supports the concept that the different cleavage patterns are due to a different mode of interaction of chromosomal proteins with the different regions of rDNA.

In higher eukaryotes, DNase I discriminates between genes which are actively transcribed, no matter to what extent, and genes which are inactive (Weintraub and Groudine, 1976). When the ribosomal genes of *P. polycephalum* were similarly probed, Stalder *et al.* (1978) found that they were selectively degraded by DNase I, in good agreement with what had been found with active genes in higher eukaryotes. The susceptibility of ribosomal genes to DNase I remained equally high throughout the whole cell cycle, though rRNA synthesis is virtually absent during mitosis, starts in the early S phase, and continues at an ever-increasing rate into the late G_2 phase, where the rate of synthesis is about fivefold higher than during the early S phase (Hall and Turnock, 1976). These results suggested that DNase I recognizes active genes independently of their momentary frequency of transcription, a conclusion confirmed and extended in many higher eukaryotes (e.g., Stalder *et al.*, 1980).

However, Swofford and Vogt (1980) have repeated these experiments by a different procedure and have obtained contradictory results. They found that ribosomal genes are not selectively degraded by DNase I, nor are the genes coding for cytoplasmic poly(A)-containing RNA. Using the same methodology, the authors were able to detect selective degradation of active genes in mammalian nuclei. Their results lead to the conclusion that the chromatin of lower eukaryotes is intrinsically sensitive to DNase I, regardless of the transcriptional status of individual genes, a tenet held also by Jerzmanowski *et al.* (1979) for *Physarum* and proposed for the situation in yeast chromatin (Lohr and Hereford, 1979). The discrepancy between the data of Stalder *et al.* (1978) and those of Swofford and Vogt (1980) is at present unresolved, but the results can presumably be accommodated on the basis of the different methods used. So far, an attempt to correlate chromatin structure with transcriptional activity in *Physarum* has been made only for ribosomal genes. With the increased availability of cloned probes for many genes of *Physarum*, this aspect of the regulation of transcription will soon be amenable to more thorough analysis than it has been so far.

IV. RNA DEGRADATION

A. *In Vivo* Degradation

Regarding its stability, the RNA of all organisms falls into two distinct categories; one comprises stable RNA, the other labile RNA. Mature rRNA and tRNA are stable, as are several of the small nuclear RNA species, at least during growth. On the other hand, mRNA, hnRNA, and pre-rRNA are quite rapidly cleaved, sometimes even before their synthesis has been completed. In general, the proportion of stable to unstable RNA synthesized at any given moment is about 1:1 in growing cells, whereas the steady-state level of labile RNA amounts to only about 1% of the level of stable RNA. The breakdown or cleavage of unstable RNA is presumably accomplished by several different enzymes, e.g., RNases, nonspecific nucleases, phosphodiesterases, and transphosphorylases.

In *Physarum*, the stability of rRNA and tRNA has been primarily established by the elegant isotope dilution experiments of Turnock and associates (Hall and

Turnock, 1976; Fink and Turnock, 1977). On the other hand, many pulse-labeling experiments have shown the lability of mRNA and hnRNA. They were discussed in Section II,B,3. Since RNA is quite stable in aqueous solution at pH 7, enzymes must be responsible for the cleavage of unstable RNA. None of the enzymes causing this kind of processing, cleavage, or breakdown have so far been identified in *Physarum,* however.

Under certain conditions, not only the labile RNA but also some of the otherwise stable RNA can be degraded. During differentiation of *Physarum* a fraction of the normally stable RNA is degraded, as measured by the decrease in total RNA per culture during spherulation (Chet, 1973). Here again, it is not known which enzymes cause the breakdown or how they come to attack an otherwise stable RNA. It would be interesting to study which, if any, of the enzymes discussed in the following section actually contribute to the *in vivo* cleavage or degradation of RNA.

B. RNA-Degrading Enzymes

Over the past 12 years, several RNA-degrading enzymes of *Physarum* have been isolated or partially purified. Those for which a considerable amount of data are available are listed in Table I; others are briefly mentioned at the end of this section. It seems likely that each of the enzyme activities of Table I depends on a separate protein molecule. This would mean that *Physarum* can elaborate at least nine enzymes capable of cleaving or breaking down RNA. The intracellular enzyme RNase Pp2 and the extracellular enzyme RNase Phy1 might be an exception to this conclusion. They could be identical: the molecular weights given are relatively close, and the heat inactivation data are rather similar. It should be noted that heat inactivations depend a great deal on the purity and concentration of enzymes (Pilly *et al.,* 1978), so that differences in this column between Phy1 and Pp2 do not eliminate the possibility that the two activities are due to the same enzyme. It is striking that eight out of nine enzymes have their optimum in the acid pH range. For the extracellular enzymes, this most likely reflects the fact that the organism grows best around pH 4–5 and that these enzymes may act *in vivo* to solubilize RNA found in the environment. For the intracellular enzymes the low pH optimum may reflect the hydrogen ion concentration in the lysosomes, the organelles in which ingested RNA is thought to be broken down. Most of the intracellular acid nucleases were discovered by Egami's group over 10 years ago (Hiramaru *et al.,* 1969a, 1969b), and it is unfortunate that those studies have not been pursued since.

Of all the nucleases of *Physarum,* RNase Phy1 has been most extensively studied. It is useful in RNA sequencing and is commercially available. The interest in Phy1 stems from its relative aversion to cut next to a cytosine residue, a property observed when it was first described by Braun and Behrens (1969).

TABLE I

Characterization of RNA-Degrading Enzymes from *Physarum*[a]

Enzyme	Source	Substrate	Cofactors	IEP	Opt. pH	MW	Inactivation	Inhib.	Specificity	References
RNase Pp1	Mic.	RNA	None	—	6.7	40	20% 3' 85°C	—	Cuts at G	Hiramura et al. (1969a)
RNase Pp2	Mic.	RNA	None	—	4.5	40	20% 3' 85°C	—	C slower than A, U, G	
RNase Pp3	Mic.	RNA	None	—	5.5	10	30% 3' 85°C	—	C, U slower than A, G	
RNase Pp4	Mic.	RNA	Zn^{2+}	—	4.0	—	40% 1h 37°C	Hg^{2+}	C, U slower than A, G	
Nucl. Pp1	Mic.	RNA/dDNA	None	—	4.5	—	—	—	RNA → 3' NMP	Hiramura et al. (1969b)
Nucl. Pp2	Mic.	RNA/dDNA	Zn^{2+}	—	4.5	—	50% 1h 37°C	Hg^{2+}	RNA → 5' NMP	
Alk. nucl.	Mic.	RNA/dDNA	Zn^{2+}	4.6	8.0	32	50% 20' 80°C	PO_4^{3-}	—	Waterborg and Kuyper (1979)
RNase Phy1	Medium	RNA	None	4.3	4.5	29	50% 80' 60°C	—	C slower than A, U, G	Pilly et al. (1978)
RNase Phy2	Medium	RNA	None	3.8	3.1	32	—	—	Cuts at G	

[a] Nucl., nuclease; Alk. nucl., alkaline nuclease; Mic., microplasmodia; IEP, isoelectric point; dDNA, denatured DNA; Opt. pH, optimal pH for enzyme activity; MW, molecular weight in kilodaltons; Inactivation, percent of inactivation observed following minutes (') or hours (h) of treatment at the indicated temperatures; Inhib., inhibitors; Specificity; C, cytosine; G, guanine; A, adenine; U, uracil; NMP, nucleoside monophosphate.

Subsequently Farr et al. (1972) developed a method for its further purification. In the following years, Phy1 as well as Phy2 has been purified to homogeneity and characterized by Pilly et al. (1978). In a series of thorough experiments on specificity, using primarily dinucleotides, these authors showed that Phy1 cuts CpN much less frequently than UpN, ApN, or GpN. Phy2, on the other hand, is quite similar to RNase T1 of *Aspergillus oryzae* in its specificity. More recent studies on Phy1 have focused on finding variants with different IEPs, determining kinetic parameters K_M, V_{max}, and K_i, and making a preliminary investigation of the active center (A. Bertoli-Proudfoot and J. P. Bargetzi, personal communications). The amino acid composition obtained by these authors is in good agreement with the molecular weight given in Table I.

As mentioned above, several attempts have been made to correlate changes in physiology, in particular differentiation, with changes in the activity patterns of RNA-degrading enzymes. As an analytical tool, isoelectric focusing has been particularly popular, primarily because several different enzyme activities with different isoelectric points can be observed in a single preparation step, starting with a crude extract. Unfortunately, none of these studies has pursued the molecular nature of the enzyme activities, and therefore no firm conclusions about the molecular events involved can be drawn. Brand et al. (1974) have observed three RNase activities in plasmodia of *Physarum*, with a new activity appearing during spherule formation. Yet another activity seems to be associated with amoebae (Brand et al., 1975). Chet et al. (1973) briefly reported on the possible presence of 12 different RNA-degrading activities in crude extracts of *Physarum*. Some of these activities may change during spherule formation. For lack of adequate data, none of the activities mentioned in this section can be correlated with the enzymes purified by Hiramaru et al. (1969a,b). The same conclusion holds for phosphodiesterases, which may also cleave RNA. Several such activities have been demonstrated during both growth and spherulation, but none of the enzymes were further characterized (Hüttermann et al., 1970a,b; Hüttermann and Chet, 1971; Hüttermann, 1972). Somewhat more is known about DNA-degrading enzymes acting either on single-stranded DNA (Waterborg and Kuyper, 1979) or on double-stranded DNA (Polman et al., 1974; Waterborg, 1980). A detailed review of this topic would exceed the scope of an article dealing with RNA metabolism.

V. TRANSCRIPTION IN THE MITOTIC CYCLE

A. *In Vivo* Transcription

The question underlying much of the research done in this area is whether the well-known structural changes in the mitotic cycle are controlled at the level of transcription, or, to formulate a more descriptive question, whether there are quantitative and/or qualitative differences in the types of RNA molecules synthesized at different phases of the mitotic cycle. In *P. polycephalum* some of the biochemical events of the mitotic cycle have been studied in considerable detail. In particular, it is well established that under standard growth conditions chromosomal DNA is replicated only in the first third of interphase (Nygaard *et al.*, 1960) and that each fraction of the genome replicates at a specific time during the S phase (Braun *et al.*, 1965; Braun and Wili, 1969). Also, early-replicating DNA contains fewer repeated sequences than does late-replicating DNA (Fouquet *et al.*, 1974a; Fouquet and Sauer, 1975). In contrast to the main chromosomal DNA, however, rDNA replicates throughout interphase, with the exception of the first hour of the S phase (Zellweger *et al.*, 1972). The second minor DNA fraction, mitochondrial DNA, replicates throughout the intermitotic period, as was found in several other eukaryotic cell types (Sonenshein and Holt, 1968; Braun and Evans, 1969).

1. DIFFERENTIAL TRANSCRIPTION IN THE MITOTIC CYCLE

How does transcription relate to the mitotic cycle? A decrease of RNA synthesis during mitosis was indicated early by Nygaard *et al.* (1960), and later autoradiographic observations of Kessler (1967) showed that very little, if any, RNA is made during metaphase. The arrest of transcription lasts only about 5 minutes under standard growth conditions. These results, obtained by using pulse-labeling techniques, have been confirmed for the two stable RNA types, rRNA and tRNA, by using an elegant isotope dilution procedure (Hall and Turnock, 1976; Fink and Turnock, 1977). In addition, these authors have shown that the rate of stable RNA synthesis clearly increases from early to late interphase. For rRNA it was possible to compute the rate increase more accurately. It appears to rise continuously five- to sixfold from early to late interphase (Hall and Turnock, 1976). The similarity in behavior of both these stable RNA species is particularly interesting, since they are synthesized in a totally different manner. The genes for tRNA are chromosomal, whereas those for rRNA are extrachromosomal (Hall and Braun, 1977). In addition, it is most likely that two different polymerases are required for their synthesis. It is not clear whether the mitotic arrest of synthesis and the gradual rate increase during the intermitotic period are regulated by the same mechanism. This does not seem very likely. The mitotic arrest may be caused by chromosome condensation and inability of the RNA polymerases to move further on their appropriate templates. The interphase rate, on the other hand, is probably controlled both by gene dosage and by some superimposed unknown mechanism. For rRNA the latter is postulated, since the rate of rRNA synthesis rises five- to sixfold over the mitotic cycle, whereas the template only doubles in amount. For the metaphase arrest of transcription, mechanisms other than the one mentioned above can be postulated. One such possibility, which might be tested experimentally, is that RNA polymerase leaves the template during chromosome condensation.

Pulse-labeling experiments with total RNA have revealed a biphasic pattern of uridine incorporation over the mitotic cycle with minima in mitosis and also, unexpectedly, in mid-interphase (Mittermayer *et al.*, 1964; Braun *et al.*, 1966; Sauer *et al.*, 1969b). Many attempts have been made to determine whether the two peaks of apparent transcriptional activity reflect the synthesis of different classes of RNA. There is a great deal of evidence to suggest that differential transcription does occur, as shown by sensitivity toward actinomycin D (Mittermayer *et al.*, 1964), base composition (Cummins *et al.*, 1966), DNA–RNA hybridization under conditions in which multiple copies reanneal (Fouquet and Braun, 1974), fractionation of poly(A)RNA (Fouquet *et al.*, 1974b), and RNA redundancy (Fouquet and Sauer, 1975). In general, the G_2 phase reflects predominantly rRNA synthesis, whereas in the S phase peak other RNA species (presumably hnRNA) appear to be made in the largest amounts. Furthermore, studies of the transcription of unique-sequence DNA strongly suggest that there are qualitative as well as quantitative differences between the RNA populations synthesized during the S and G_2 phases (Fouquet and Braun, 1974). There may be about three times more different sequences transcribed in the S phase than in the G_2 phase. Taking all these results together, it is clear that there is a control over both the type and the quantity of RNA transcribed during the mitotic cycle. This implies the existence of a transcriptional program which is repeated with every cell cycle. There is no compelling evidence to show at what stage of the mitotic cycle the transcription program is reset, but metaphase itself would be a likely candidate for such an event. To investigate this problem further, it would be most helpful to have probes for individual mRNA species. Regarding tRNA, no differential transcription of 20 tRNA families could be detected over the mitotic cycle (Melera and Rusch, 1973b; Melera *et al.*, 1974). In contrast to most nuclear RNA synthesis, mitochondrial RNA seems to be made at a constant rate over the mitotic cycle (Grant and Poulter, 1973).

The intracellular pools of free ribonucleoside triphosphates, the immediate precursors of macromolecular RNA, have been the subject of several studies and appear to fluctuate considerably during the mitotic cycle (Chin and Bernstein, 1968; Sachsenmaier *et al.*, 1969; Bersier and Braun, 1974; Fink, 1975). This topic will not be dealt with further here, since it is outside the scope of this chapter (for a review, see Turnock, 1979).

2. ROLE OF RNA PRIMERS IN DNA REPLICATION

The event by which the S phase of the cell cycle is defined is the replication of chromosomal DNA. There appear to be two close connections between replication and certain functions of RNA. First, Okazaki fragments contain RNA; second, there may be some connection between replication of a particular DNA molecule and its transcription.

The first polydeoxyribonucleotides that are made in cells engaged in DNA synthesis are single-stranded and short. These so-called Okazaki fragments are about 200 nucleotides long in *P. polycephalum* and are subsequently joined to form much larger molecules (Funderud and Haugli, 1977). Interestingly, the transient Okazaki fragments carry a short polyribonucleotide sequence at the 5' end of the nascent polydeoxyribonucleotide (Waqar and Huberman, 1975), as had previously been shown for the corresponding Okazaki fragments in *E. coli* (Sugino *et al.*, 1972; for a review, see Sheinin *et al.*, 1978). These authors injected γ-^{32}P-label deoxyribonucleoside triphosphates into surface plasmodia (Waqar and Huberman, 1973), isolated Okazaki fragments containing a short RNA fragment, and showed that the RNA was joined to the 5' end of the DNA. Any of the four ribonucleotides was found joined to any of the four deoxyribonucleotides, thereby allowing all 16 possible RNA–DNA junctions, although the frequency of different junctions varied up to twentyfold. As a starter, dGTP was used about 10 times more frequently than the other three deoxyribonucleoside triphosphates. From these observations, and similar ones made in other organisms, it may be postulated that the first step in DNA replication is the synthesis of a short polyribonucleotide primer by an RNA polymerase and the subsequent joining to this of deoxyribonucleotides to yield the nascent single-stranded DNA chain. The RNA will obviously have to be removed at a later stage and replaced by DNA. It will be interesting to find out how transcription and subsequent removal of the RNA primers are regulated. Though the existence of an RNA primer for DNA replication has been unambiguously demonstrated, it is unclear at present whether any of the RNA polymerases discussed in Section III,B are involved in the synthesis of the primer or whether this type of transcription is performed by an altogether different enzyme or enzyme system.

3. COUPLING OF TRANSCRIPTION WITH DNA REPLICATION

In *P. polycephalum* there may be a rather close connection between the replication of certain DNA segments and their transcription. This led Sauer (1978) to propose an interesting model of transcriptional control in the cell cycle. As will be shown, this model is still very hypothetical. Rao and Gontcharoff (1969) showed that during the S phase, inhibition of DNA replication by fluorodeoxyuridine leads to a considerable inhibition of uridine incorporation into RNA. Since fluorodeoxyuridine has this effect only during the S phase and not during the G_2 phase, it seems likely that previous replication is required for subsequent transcription of a specific DNA segment. It cannot be totally discounted, however, that the observed effect of fluorodeoxyuridine may be indirectly caused, e.g., by interference with the uptake or metabolism of the labeled uridine used as a precursor for RNA. More recently another inhibitor of DNA synthesis, hydroxyurea, has been used for similar experiments (Fouquet *et al.*, 1975a). In 1-hour pulses given at different phases of the mitotic cycle, hydroxyurea inhibited uridine incorporation into RNA by up to 70%, but only in the early S phase.

Again, it cannot be completely excluded that this striking effect may arise indirectly, as suggested above for fluorodeoxyuridine. However, this objection is largely eliminated by the finding that isolated nuclei of cultures treated *in vivo* with hydroxyurea contain less RNA polymerase activity than nuclei from untreated cultures (A. Hildebrandt and H. W. Sauer, personal communications). RNA from both the S and G_2 phases, labeled in the presence and absence of hydroxyurea, was extracted and partly characterized by RNA–DNA hybridization. The results tend to suggest that the RNA made in the presence of the inhibitor may in part be qualitatively different from that made in its absence, but to establish this firmly, several different types of experiments would be required. It is perhaps surprising that different RNA precursors gave results that were somewhat at variance.

The replication–transcription coupling model makes the attractive suggestion that transcription may be realigned once per mitotic cycle, possibly during mitosis itself (Sauer, 1978). This would allow a simple control of events which occur periodically once per mitotic cycle or per cell cycle. Several observations with extracted RNA polymerase are in agreement with such a model, but they do not prove it (see Section V,B). In this context, studies on specific mRNA species, coded for by nonreiterated genes, would be welcome.

B. *In Vitro* Transcription

1. ISOLATED NUCLEI

Experiments on *in vitro* transcription in *P. polycephalum* led to the observation that the overall transcriptional activity of nuclei isolated during the cell cycle reflected the transcriptional activity of the intact plasmodia *in vivo* (Mittermayer *et al.*, 1966; Cummins and Rusch, 1967; Grant, 1972). Basically, two peaks of transcriptional activity can be discerned when nuclei are isolated during the mitotic cycle. The first peak occurs during the S phase, whereas a second peak becomes apparent during the G_2 phase (Mittermayer *et al.*, 1966; Grant, 1972; Davies and Walker, 1978). When the RNA polymerases involved in these two phases of transcription were investigated, differential activities of polymerases I and II during the cell cycle could be recognized *in vitro*. In the first peak of activity, during the S phase, maximal transcription occurs in the presence of Mg^{2+} ions and with low salt concentrations; adding Mn^{2+} and increasing the salt concentration to 100 mM KCl yields no further stimulation. However, during the second peak, when the overall transcriptional activity is lower than during the first peak, if assayed under low salt conditions, a considerable stimulation of transcription can be achieved by the addition of Mn^{2+} and by raising the salt concentration (Grant, 1972). The stimulation of transcription by Mn^{2+} and salt is completely inhibited by α-amanitin, a potent inhibitor of RNA polymerase II only. The basal activity, which is measured with Mg^{2+} and low salt, is resistant to the toxin. In contrast, the transcriptional activity during the first peak is sensitive to α-amanitin (Davies and Walker, 1977, 1978). These observations correlate well with the properties of purified RNA polymerases (see Section III,B,1), since polymerase I has a low salt optimum and is resistant to α-amanitin, whereas polymerase II has a higher salt optimum and is very sensitive to this inhibitor. Taken together, these observations suggest that during the first peak of activity (S phase), it is chiefly RNA polymerase II which is active, and hence, it is mRNA which is predominantly synthesized. During the second peak (G_2 phase), the predominant activity is that of RNA polymerase I synthesizing rRNA.

This contention was soon supported by the results of analysis of the base composition of the RNA synthesized both *in vitro* (Cummins and Rusch, 1967) and *in vivo* (see Section V,A,1). This demonstrated in both cases that RNA synthesized during the S phase peak of activity has a rather DNA-like base composition (as would be expected for mRNA), whereas RNA synthesized during the G_2 peak has a higher GC content and hence bears a closer resemblance to rRNA. Although the validity of these data has been seriously questioned on technical grounds (Grant, 1972), the final conclusions of this work have gained support from the observation that, *in vivo*, poly(A)-containing mRNA was mainly synthesized during the S phase (Fouquet and Braun, 1974; Fouquet *et al.*, 1975b), whereas the synthesis of rRNA was low during the S phase but predominated during the G_2 phase (Hall and Turnock, 1976).

Since RNA polymerase II can be quantitatively inhibited by low concentrations of α-amanitin, the specific transcription of rRNA, rather than that of the total RNA, can be conveniently studied *in vitro* (Davies and Walker, 1977, 1978). Experiments using isolated nuclei or nucleoli both showed that, again in accordance with the situation *in vivo,* no detectable transcription of rRNA takes place during metaphase. However, *in vitro* transcription reaches its maximal interphase level within 10 minutes after mitosis and, in contrast to what has been observed *in vivo* (Hall and Turnock, 1976), no discernible differences in the rate of transcription of rRNA were observed *in vitro* during the cell cycle. Supportive evidence for this observation is provided by a preliminary experiment with the initiation inhibitor rifampicin AF/013 which showed that, *in vitro,* the rate of initiation of rRNA transcription apparently did not change during the cell cycle. The two sets of observations, *in vivo* and *in vitro,* need not necessarily be conflicting since all kinds of postrna postal processing schemes can be envisaged to account for the apparent contradiction that the rate of initiation seems to be constant during the cell cycle *in vitro,* whereas the rate of transcription of rRNA increases by a factor of five during the cell cycle *in vivo.* But before these findings are evaluated in any depth, it would be desirable to obtain more reliable results *in vitro* in the initiation experiment.

The ability of chromatin or nuclei from different stages of the cell cycle to be transcribed by exogenous RNA polymerase II has been investigated (Schicker *et al.*, 1979). *In vitro* transcription by chromatin or nuclei isolated during the S phase, in which transcription by polymerase II predominates, can be strongly stimulated by the addition of exogenous, homologous polymerase II. In contrast, the stimulation by exogenous enzyme is far less with chromatin or nuclei from the G_2 phase. This observation tentatively suggests that genes which are to be transcribed by polymerase II are inaccessible to the enzyme during G_2, or else that inhibitors are produced during the cell cycle which selectively inactivate either of the two polymerases and which would also inactivate, in G_2 phase chromatin, the exogenous polymerase. Several observations have provided hints that both types of transcriptional control might be operative. The presence of fluctuating concentrations of inhibitors is suggested by the observation that *P. polycephalum* contains approximately equal amounts (i.e., of units of enzyme activity) of polymerases I and II and that these amounts do not change appreciably during the cell cycle when semipurified enzyme preparations are analyzed (Hildebrandt and Sauer, 1976c). Therefore, the differential transcriptional activities observed both *in vivo* and *in vitro* cannot be caused by differential synthesis or rapid turnover of the polymerases, but rather through a modulation of their activity by controlling factors. A possible loss of such presumptive factors during the isolation of nucleoli could also account for the observation by Davies and Walker (1978) that no difference in transcription of rRNA during the cell cycle can be detected with isolated nucleoli (see Section III,D,1).

Other observations point to the possibility that the transcription of mRNA might be coupled with ongoing DNA replication (Fouquet *et al.*, 1975b; for a review, see Sauer, 1978; Section V,A,3), and thus that a particular conformation of a gene, or its surrounding chromatin, might be required to allow transcription by polymerase II. This suggestion is also supported by Schicker *et al.* (1979), who prepared chromatin from plasmodia in the S and G_2 phases, which had been pretreated with 50 m*M* hydroxyurea to block DNA synthesis prior to the preparation of the chromatin. Nuclei were isolated from cultures which had been treated with hydroxyurea at various stages of the cell cycle, and their endogenous activities of RNA polymerase I and II were determined. RNA polymerase I activity was somewhat inhibited by pretreatment with the drug, but the degree of inhibition (about 30%) remained unchanged throughout the cell cycle. In contrast, RNA polymerase II activity was completely abolished in nuclei from cultures which had been treated with hydroxyurea during the S phase, but it was unchanged when hydroxyurea was applied during G_2 (Pierron, 1979). These observations overcome some of the problems associated with previous experiments conducted *in vivo*, in which the influence of the respective inhibitors of DNA synthesis on ribonucleotide pools was difficult to control and therefore led to incorporation data of uncertain meaning. When stimulation of transcription of such chromatin by exogenous polymerase II was tested, none could be achieved, either with S phase or with G_2 phase chromatin, although S chromatin normally can be strongly stimulated (see above).

The evidence outlined so far indicates general agreement that transcription observed *in vitro* with isolated nuclei or nucleoli reflects, at least to a large extent, the *in vivo* state of transcription. *In vitro* transcription systems, therefore, appear to be attractive instruments for analyzing the various factors involved in the regulation of transcription during the cell cycle. A variety of concepts can be envisaged to explain how transcriptional regulation could be brought about, and most probably many different regulatory mechanisms are working in conjunction. An obvious mechanism, namely, that the fluctuation of the numbers of RNA polymerase molecules parallels the observed transcriptional activity, has been made unlikely by the experiments which have determined the amounts of active RNA polymerase during the cell cycle (Hildebrandt and Sauer, 1976c; see above and Section IV,C). The importance of a particular gene or chromatin structure for transcription, e.g., the need for concomitant DNA replication, has been suggested by different lines of experimentation (Fouquet *et al.*, 1975b; Sauer, 1978; Schicker *et al.*, 1979), but direct evidence has not yet been reported and might, in fact, be rather difficult to obtain at present.

2. REGULATORY FACTORS

A third alternative, the presence of regulatory factors of various kinds, is amenable to analysis, and evidence for several such factors has been reported. An organic polyphosphate of nucleolar origin, which is an inhibitor of RNA polymerase I, and to a much lesser extent of polymerase II, has been characterized, and its function in the control of transcription of rRNA during differentiation suggested (Hildebrandt and Sauer, 1977a; Hildebrandt *et al.*, 1979; see also Section III,C,1). Preliminary experiments have indicated that this factor might be involved in the regulation of transcription during the cell cycle (Sauer *et al.*, 1969b), although this possibility still remains to be explored. A similar situation exists with a non-histone phosphoprotein which has been reported to act as a positive control element for the transcription of rRNA (Kuehn *et al.*, 1979). Here again, no information is yet available on its actual role in the regulation of transcription during the cell cycle.

In contrast, an isolated nuclear elongation factor for transcription (Ernst and Sauer, 1977; see Section III,C) has been correlated with transcriptional activity during the cell cycle *in vivo*. The concentration of this factor (measured *in vitro* as a stimulatory effect on RNA polymerase activity) fluctuates during the cell cycle. It reaches a sharp maximum during the peak of transcriptional activity of polymerase II during the S phase, drops to background level (about 50% of peak activity) at the end of S, and remains constant throughout the rest of the mitotic cycle, including mitosis. It is interesting to note that the concentration of this elongation factor is covariant with the activity profile of polymerase II only, although *in vivo,* the factor is capable of stimulating polymerase I as well as polymerase II. In contrast, it has no effect on *in vitro* transcription accomplished with *E. coli* RNA polymerase. A second point of interest is the observation that the activity of this factor, aside from the peak attained during the S phase, remains constant throughout the cell cycle and does not change during mitosis, whereas transcription by both polymerase I and II drops to very low levels during mitosis (see Section V,A).

3. RNA POLYMERASE

Very little is known so far about the importance of subunit modification for the modulation of transcriptional activity in *Physarum*. An analysis of RNA polymerase II from asynchronously growing cultures has shown that this enzyme, although homogeneous upon purification, exhibits a marked heterogeneity in its largest peptide subunit, the size of which can vary from 170,000 to 215,000 daltons (Smith and Braun, 1978; see Section III,B,2). This heterogeneity of the large subunit is reminiscent of the situation observed with RNA polymerases from other eukaryotic sources, in which polymerase IIA (the largest subunit has a molecular weight of over 200,000) appears to be associated with resting, non-transcribing tissue, whereas polymerase IIB (the largest subunit has a molecular weight of less than 200,000) is predominantly found in actively growing tissue (Jendrisak and Guilfoyle, 1978). Therefore, the variation in subunit structure of polymerase II from *P. polycephalum* could well reflect a mixture of enzyme populations from various stages of the cell cycle, and hence of different transcriptional activities.

A further possible regulatory mechanism has been suggested by the observation that in isolated nuclei various classes of polymerase II can be distinguished by the salt concentrations needed to extract them from the nuclei (Hildebrandt and Sauer, 1977b). The bulk of polymerase II is eluted during the preparation of nuclei in hypertonic sucrose solutions (Hildebrandt and Sauer, 1976b). Another 10% of enzyme activity can be solubilized from nuclei with 0.5 M NaCl. A small amount of additional activity can be eluted by 1.5 M NaCl and is considered by the authors to represent initiated enzyme. The relative distribution of total polymerase II activity between these solubility classes appears to fluctuate considerably during differentiation (see Section VI), and therefore it seems appropriate to speculate that differential binding of polymerase to template might constitute an important means of regulating transcriptional activity during the cell cycle.

Further evidence that binding of RNA polymerase II to chromatin is closely correlated with transcriptional activity during the cell cycle has been provided by using [^{3}H]label α-amanitin to titrate RNA polymerase II molecules in various subnuclear fractions (Sauer, 1979). Judging by this technique, RNA polymerase

II is almost quantitatively found in chromatin, and none is in the nucleoplasmic fraction. In contrast, during mitosis, when the rate of transcription is almost zero, RNA polymerase II is excluded from the condensed chromosomes and is found in the nucleoplasm. These observations again suggest that the partition of polymerases between various compartments of the nucleus might be a factor of importance for the overall regulation of transcription.

VI. TRANSCRIPTION IN DIFFERENTIATION

A. *In Vivo* Transcription

Vegetative slime mold plasmodia can differentiate into resistant forms, usually under adverse environmental conditions. Although differentiation has been studied in much less detail with the acellular slime molds than with the cellular slime molds, several investigators have studied the conditions under which sporulation or spherulation occurs and some of the biochemical changes that accompany these processes. The only acellular slime mold that has received any attention regarding transcription is again *P. polycephalum*. Since investigations at the molecular level have advanced little in the last few years, and since the earlier observations have been succinctly reviewed by Chet (1973) for spherulation and by Sauer (1973) for sporulation, we will cover this subject only cursorily.

During the structural reorganization which occurs in differentiation, the RNA and protein contents of cultures decrease, presumably through a breakdown of ribosomes. At the same time, one could expect that new species of RNA have to be made, particularly mRNA for new structural proteins and new enzymes. Experiments with actinomycin D indicated that this might be the case for sporulation, though not for spherulation. The difference may, however, be due entirely to differential penetration of the inhibitor. More direct investigations were undertaken to visualize differences between newly made RNA and old RNA by sizing in either sucrose gradients or polyacrylamide gels. The results were negative, as no clear differences were seen in RNA populations made at various stages of differentiation. Competition hybridization experiments, on the other hand, seemed to indicate that different types of RNA molecules were made at different times. Unfortunately, these experiments gave no quantitative estimates of the number of molecular species concerned. Several reasons for this may be suggested. Hybridization conditions took account only of RNA species made on reiterated DNA sequences. In addition, RNA extraction was not shown to be quantitative and to yield intact molecules. Particularly during differentiation, when RNA-degrading enzymes are known to be activated, such controls are essential to obtain biologically meaningful results. Undoubtedly *P. polycephalum* is a suitable and interesting organism in which to study the role of transcription and its regulation during various stages of differentiation. It will take considerable effort to get some tangible information on the fate of the mixed population of mRNA molecules as well as on individual species coding for proteins with specific functions in differentiation.

B. *In Vitro* Transcription

Differentiation in *P. polycephalum* can be induced by starvation and can proceed via two different pathways. Both types of differentiation seem to be accompanied by changes in the program of RNA transcription. Again, as previously seen in cell cycle studies (see Section V,B), these changes in transcriptional activity are also reflected *in vitro* by isolated nuclei (Hildebrandt and Sauer, 1977b). When nuclei are isolated from microplasmodia after different periods of growth and starvation, the endogenous RNA polymerase II activity is about fivefold lower in plasmodia after 6 days growth, i.e., which have initiated spherulation, than it is in growing plasmodia. A subsequent experiment indicated that this drop in activity does not reflect the absence of polymerase II in nuclei from stationary cultures, but that the enzyme becomes reversibly inhibited upon starvation. When nuclei from both actively growing and stationary cultures were treated with 2% Triton X-100 before RNA polymerase activity was measured, a reverse of the previous observation was obtained. Stimulation of transcription by Triton X-100 was very low in nuclei isolated from actively growing cultures (about 20% stimulation). In contrast, nuclei from old, stationary cultures were stimulated by Triton X-100 by about 400%. Transcriptional activity in nuclei isolated from cultures of different ages therefore reflects rather accurately the decline in transcriptional activity observed *in vivo*. Triton X-100 treatment of such nuclei restores polymerase II activity to the level found in nuclei from actively growing cultures. These findings can be interpreted to show that treatment of nuclei with the detergent reverses the inhibition blocking polymerase II to an increasing degree during the stationary phase and the onset of spherulation. The authors of this study indicate that at least one of the regulatory factors involved in the inhibition had been identified tentatively. Here again, as previously discussed for the events of the cell cycle, regulation of transcription does not appear to operate on the level of turnover of RNA polymerase molecules, but rather via modification of the activity of a constant pool of enzyme protein.

A different picture emerges when the activity of RNA polymerase I is followed through the stationary phase and into spherulation (Hildebrandt and Sauer, 1976b). In these experiments, RNA polymerases were purified from whole plasmodia at different stages of starvation, and relative amounts of enzyme activity were measured after fractionation of the enzymes on DEAE–Sephadex columns. In agreement with the results obtained for RNA polymerase II activity described above, constant amounts of activity of this enzyme were detected in enzyme preparations from actively growing cultures, from starving cultures, and from mature spherules. However, when the activity of RNA polymerase I was measured, a rather different picture was obtained. Whereas during active growth approximately equal amounts of polymerase I and II (in terms of activity) were found, polymerase I dropped to near-background levels during starvation. In mature spherules, however, the level of polymerase I again equalled that of polymerase II and was quantitatively similar to the activity found during active growth. The drop in polymerase I activity during starvation again reflected the drop in transcription of rRNA which is observed *in vivo*. In contrast to the situation observed for polymerase II, the drop in enzyme activity seen *in vivo* was paralleled by the disappearance of soluble enzyme activity found *in vitro*. This drop in polymerase I activity could be achieved by a gradual degradation of the polymerase I population during starvation, followed by resynthesis of the enzyme during maturation of the spherule. Such a course of events would imply a completely different scheme for the regulation of the two RNA polymerases during differentiation. On the other hand, an apparent drop in enzyme activity could also be brought about by an inhibitor which is formed during starvation and which binds reversibly to polymerase I but not to polymerase II. This binding has to be tight enough so that it survives the purification procedure for polymerases. In such a situation, very little enzyme activity could be recovered upon purification, although the overall amount of polymerase I protein in the cell had not changed. During spherule maturation, according to this hypothesis, the inhibitor is released from the enzyme and/or degraded so that full activity of the polymerase is restored.

Despite this point, the role of the polyphosphate inhibitor in the control of rRNA transcription during differentiation is of great interest (Goodman *et al.*, 1969; Sauer *et al.*, 1969b; Hildebrandt *et al.*, 1979; see also Section III,C,1). It is found in the nucleolus, and its concentration increases continuously during starvation. Whereas actively growing cells contain about 0.08 μg of inhibitor per 10^6 nuclei, this amount increases to a maximum of about 1 μg per 10^6 nuclei during prolonged starvation. This is sufficient to inhibit completely all RNA polymerase I found in this number of nuclei when assayed *in vitro*. During maturation of the spherules, the inhibitor concentration decreases, and in mature spherules its concentration has dropped to undetectable levels (Hildebrandt and Sauer, 1977a).

A further modulating element for transcription, which might be implied in the gradual shutoff of RNA synthesis during starvation, is a nuclear elongation factor (Ernst and Sauer, 1977) already discussed in Sections III,C,3 and V,B,2. The activity of this factor (measured as stimulation of RNA polymerase II activity *in vitro*) gradually decreases during starvation to undetectable levels. From the evidence presently available, it is difficult to judge whether its activity, and hence its disappearance, is of consequence for either or both of the two RNA polymerases.

Taking into account all the observations outlined above, a picture of the regulation of transcription during differentiation emerges wherein the overall concentration of potentially active RNA polymerases I and II remains constant throughout starvation. Their respective activities are modulated according to the needs of the hour by reversible interaction of a variety of regulatory factors with one or the other of the enzymes. In the mature spherule, full transcriptional capacity of both polymerases is restored. A very interesting further possibility of modulating the activity of polymerases during differentiation, namely, through chemical modification of subunit peptides or through changes in the subunit composition of these enzymes, has not been investigated to date.

VII. CONCLUDING REMARKS

The information compiled in this chapter illustrates why *Physarum* underwent a remarkable evolution from a modest forest dweller to a pet organism of many laboratories interested in transcription. The observation that in this organism the rDNA is probably the sole genetic content of the nucleolus made *Physarum* a convenient experimental system for functional and structural analysis of this particular set of genes. Analysis of the pattern of RNA transcription, on the other hand, suggested that *Physarum* behaves in this respect much more like higher eukaryotes than do other lower eukaryotes, such as *D. discoideum*. *Physarum* makes mRNA in the form of relatively large precursors, nuclear heterogeneous RNA, and its ribosomal genes are transcribed into a very large precursor, much of which is discarded during processing into mature rRNA. Rather surprisingly, the value of *Physarum* as a differentiating organism has been little exploited, at least in the context of transcriptional regulation. Few recent data are available, and progress generally seems to have been slow in this area. However, this field should gain new attention, with the burgeoning genetics providing new mutants useful in the study of both differentiation and transcription. Further impetus will be gained from the fact that cloned genes are rapidly becoming available as probes for transcription studies.

At the present time, it seems that the investigations on transcription in *Physarum* are ready to move from the groundwork that has been laid in the past few years to the exciting field (or perhaps jungle) of a more refined analysis and deeper understanding of how nature managed to make precise copies long before the advent of xerography. The most interesting stories on this subject are yet to be told.

ACKNOWLEDGMENTS

Research from the authors' laboratory was supported by Grant 3.312.78 of the Swiss National Science Foundation. We are grateful to D. Braun for much editorial help.

REFERENCES

Adams, D. S., and Jeffery, W. R. (1978). Poly (adenylic acid) degradation by two distinct processes in the cytoplasmic RNA of *Physarum polycephalum. Biochemistry* **17,** 4519–4524.

Adams, D. S., Noonan, D., and Jeffery, W. R. (1980). An improved method for the isolation of polysomes from synchronous macroplasmodia of *Physarum polycephalum. Anal. Biochem.* **103,** 408–412.

Atmar, V. J., Daniels, G. R., and Kuehn, G. D. (1978). Polyamine stimulation of phosphorylation of nonhistone acidic protein in nuclei and nucleoli from *Physarum polycephalum. Eur. J. Biochem.* **90,** 29–37.

Bernstam, V. A. (1974). Effects of supraoptimal temperatures on the myxomycete *Physarum polycephalum*. II. Effects on the rate of protein and ribonucleic acid synthesis. *Arch. Mikrobiol.* **95,** 347–356.

Bersier, D., and Braun, R. (1974). Pools of deoxyribonucleoside triphosphates in the mitotic cycle of *Physarum. Biochim. Biophys. Acta* **340,** 463–471.

Brand, G., Hoffmann, W., and Hüttermann, A. (1974). Changes in the specificity of RNA- and protein-degrading enzymes–a possible point of control during differentiation. *Hoppe-Seyler's Z. Physiol. Chem.* **355,** 1181.

Brand, G., Hüttermann, A., and Haugli, F. B. (1975). Differential expression of RNase activities in the life cycle of *Physarum polycephalum. Die Naturwiss.* **62,** 535–536.

Braun, R., and Behrens, K. (1969). A ribonuclease from *Physarum*. Biochemical properties in the mitotic cycle. *Biochim. Biophys. Acta* **195,** 87–98.

Braun, R., and Evans, T. (1969). Replication of nuclear satellite and mitochondrial DNA in the mitotic cycle of *Physarum. Biochim. Biophys. Acta* **182,** 511–522.

Braun, R., and Seebeck, T. (1979). Ribosomal DNA: Extrachromosomal genes of *Physarum. 13th Alfred Benzon Symp. (Copenhagen),* pp. 306–317.

Braun, R., and Wili, H. (1969). Time sequence of DNA replication in *Physarum. Biochim. Biophys. Acta* **174,** 246–252.

Braun, R., Mittermayer, C., and Rusch, H. P. (1965). Sequential temporal replication of DNA in *Physarum polycephalum. Proc. Nat. Acad. Sci. U.S.A.* **53,** 924–931.

Braun, R., Mittermayer, C., and Rusch, H. P. (1966). Sedimentation patterns of pulse-labeled RNA in the mitotic cycle of *Physarum polycephalum. Biochim. Biophys. Acta* **114,** 27–35.

Brewer, E. N. (1972). Polysome profiles, amino acid incorporation *in vitro,* and polysome reaggregation following disaggregation by heat shock through the mitotic cycle in *Physarum polycephalum. Biochim. Biophys. Acta* **277,** 639–645.

Britten, R. J., and Kohne, D. E. (1968). Repeated sequences in DNA. *Science* **161,** 529–540.

Britten, R. J., and Smith, J. F. (1970). The nuclear satellite in slime mold. *Year Book-Carnegie Inst. Washington* **69,** 518–521.

Brodner, O. G., and Wieland, T. (1976). Identification of the amatoxin-binding subunit of RNA polymerase B by affinity labeling experiments: Subunit B 3—the true amatoxin receptor protein of multiple RNA polymerase B. *Biochemistry* **15,** 3480–3483.

Burgess, A. B., and Burgess, R. R. (1974). Purification and properties of two RNA polymerases from *Physarum polycephalum. Proc. Nat. Acad. Sci. U.S.A.* **71**, 1174–1177.
Campbell, G. R., Littau, V. C., Melera, P. W., Allfrey, V. G., and Johnson, E. M. (1979). Unique sequence arrangement of ribosomal genes in the palindromic rDNA molecule of *Physarum polycephalum. Nucleic Acids Res.* **6**, 1433–1447.
Carlile, M. J., and Dee, J. (1967). Plasmodial fusion and lethal interaction between strains in a myxomycete. *Nature (London)* **215**, 832–834.
Chet, I. (1973). Changes in ribonucleic acid during differentiation of *Physarum polycephalum*. In "*Physarum polycephalum:* Object of Research in Cell Biology" (A Hüttermann, ed.), pp. 77–92. Gustav Fischer, Verlag, Stuttgart.
Chet, I., Retig, N., and Henis, Y. (1973). Changes in ribonucleases during differentiation (spherulation) of *Physarum polycephalum. Biochim. Biophys. Acta* **294**, 343–347.
Chin, B., and Bernstein, I. A. (1968). Adenosine triphosphate and synchronous mitosis in *Physarum polycephalum. J. Bacteriol.* **96**, 330–337.
Cummins, J. E., and Rusch, H. P. (1967). Transcription of nuclear DNA in nuclei isolated from plasmodia at different stages of the cell cycle of *Physarum polycephalum. Biochim. Biophys. Acta* **138**, 124–132.
Cummins, J. E., Weisfeld, G. E., and Rusch, H. P. (1966). Fluctuation of ^{32}P distribution in rapidly labeled RNA during the cell cycle of *Physarum polycephalum. Biochim. Biophys. Acta* **129**, 240–248.
Davidson, E. H., Hough, B. R., Amenson, C. S., and Britten, R. J. (1973). General interspersion of repetitive with non-repetitive sequence elements in the DNA of *Xenopus. J. Mol. Biol.* **77**, 1–23.
Davies, K. E., and Walker, I. O. (1977). *In vitro* transcription of RNA in nuclei, nucleoli, and chromatin from *Physarum polycephalum. J. Cell Sci.* **26**, 267–279.
Davies, K. E., and Walker, I. O. (1978). Control of RNA transcription in nuclei and nucleoli of *Physarum polycephalum. FEBS Lett.* **86**, 303–306.
Dezélée, S., Wyers, F., Sentenac, A., and Fromageot, P. (1976). Two forms of RNA polymerase B in yeast: Proteolytic conversion *in vitro* of enzyme B_I into B_{II}. *Eur. J. Biochem.* **65**, 543–552.
Dove, W. F., and Rusch, H. P. (1980). "Growth and Differentiation in *Physarum polycephalum*." Princeton Univ. Press, Princeton, New Jersey.
Ernst, G. H., and Sauer, H. W. (1977). A nuclear elongation factor of transcription from *Physarum polycephalum in vitro. Eur. J. Biochem.* **74**, 253–261.
Farr, D. R., Amster, H., and Horisberger, M. (1972). Isolation and purification of the extracellular ribonuclease from *Physarum polycephalum. Arch. Mikrobiol.* **85**, 249–252.
Fink, K. (1975). Fluctuation in deoxyribo- and ribonucleoside triphosphate pools during the mitotic cycle of *Physarum polycephalum. Biochim. Biophys. Acta* **414**, 85–89.
Fink, K., and Turnock, G. (1977). Synthesis of transfer RNA during the synchronous nuclear division cycle in *Physarum polycephalum. Eur. J. Biochem.* **80**, 93–96.
Firtel, R. A., and Bonner, J. (1972). Characterization of the genome of the cellular slime mold *Dictyostelium discoideum. J. Mol. Biol.* **66**, 339–361.
Firtel, R. A., and Lodish, H. F. (1973). A small nuclear precursor of messenger RNA in the cellular slime mold *Dictyostelium discoideum. J. Mol. Biol.* **79**, 295–314.
Fouquet, H., and Braun, R. (1974). Differential RNA synthesis in the mitotic cycle of *Physarum polycephalum. FEBS Lett.* **38**, 184–186.
Fouquet, H., and Sauer, H. W. (1975). Variable redundancy in RNA transcripts isolated in S and G2 phase of the cell cycle of *Physarum. Nature (London)* **255**, 253–254.
Fouquet, H., Bierweiler, B., and Sauer, H. W. (1974a). Reassociation kinetics of nuclear DNA from *Physarum polycephalum. Eur. J. Biochem.* **44**, 407–410.

Fouquet, H., Böhme, R., Wick, R., Sauer, H. W., and Braun, R. (1974b). Isolation of adenylate-rich RNA from *Physarum polycephalum*. *Biochim. Biophys. Acta* **353**, 313-322.
Fouquet, H., Böhme, R., Wick, R., Sauer, H., and Scheller, K. (1975a). Some evidence for replication-transcription coupling in *Physarum polycephalum*. *J. Cell Sci.* **18**, 27-39.
Fouquet, H., Wick, R., Böhme, R., Sauer, H., and Scheller, K. (1975b). Effects of cordycepin on RNA synthesis in *Physarum polycephalum*. *Arch. Biochem. Biophys.* **168**, 273-280.
Funderud, S., and Haugli, F. (1977). DNA replication in *Physarum polycephalum:* Characterization of replication products made in isolated nuclei. *Biochem. Biophys. Res. Commun.* **74**, 941-949.
Goodman, E. M. (1972). Axenic culture of myxamoebae of the myxomycete *Physarum polycephalum*. *J. Bacteriol.* **111**, 242-247.
Goodman, E. M., Sauer, H. W., Sauer, L., and Rusch, H. P. (1969). Polyphosphate and other phosphorus compounds during growth and differentiation of *Physarum polycephalum*. *Can. J. Microbiol.* **15**, 1325-1331.
Gornicki, S. Z., Vuturo, S. B., West, T. V., and Weaver, R. F. (1974). Purification and properties of deoxyribonucleic acid-dependent ribonucleic acid polymerases from the slime mold *Physarum polycephalum*. *J. Biol. Chem.* **249**, 1792-1798.
Grainger, R. M., and Ogle, R. C. (1978). Chromatin structure of the ribosomal RNA genes in *Physarum polycephalum*. *Chromosoma* **65**, 115-126.
Grant, W. D. (1972). The effect of α-amanitin and $(NH_4)_2SO_4$ on RNA synthesis in nuclei and nucleoli isolated from *Physarum polycephalum* at different times during the cell cycle. *Eur. J. Biochem.* **29**, 94-98.
Grant, W. D., and Poulter, R. T. M. (1973). Rifampicin-sensitive RNA and protein synthesis by isolated mitochondria of *Physarum polycephalum*. *J. Mol. Biol.* **73**, 439-454.
Grummt, I. (1978). *In vitro* synthesis of pre-rRNA in isolated nucleoli. *In* "The Cell Nucleus" (H. Busch, ed.), Vol. V, pp. 373-414. Academic Press, New York.
Gubler, U., Wyler, T., and Braun, R. (1979). The gene for the 26 S rRNA in *Physarum* contains two insertions. *FEBS Lett.* **100**, 347-350.
Gubler, U., Wyler, T., Seebeck, T., and Braun, R. (1980). Processing of ribosomal precursor RNAs in *Physarum polycephalum*. *Nucleic Acids Res.* **8**, 2647-2664.
Hall, L., and Braun, R. (1977). The organisation of genes for transfer RNA and ribosomal RNA in amoebae and plasmodia of *Physarum polycephalum*. *Eur. J. Biochem.* **76**, 165-174.
Hall, L., and Turnock, G. (1976). Synthesis of ribosomal RNA during the mitotic cycle in the slime mould *Physarum polycephalum*. *Eur. J. Biochem.* **62**, 471-477.
Hardman, N., and Jack, P. L. (1977). Characterization of foldback sequences in *Physarum polycephalum* nuclear DNA using the electron microscope. *Eur. J. Biochem.* **74**, 275-283.
Hardman, N., and Jack, P. L. (1978). Periodic organisation of foldback sequences in *Physarum polycephalum* nuclear DNA. *Nucleic Acids Res.* **5**, 2405-2423.
Hardman, N., Jack, P. L., Brown, A. J. P., and McLachlan, A. (1979). Distribution of inverted repeat sequences in nuclear DNA from *Physarum polycephalum*. *Eur. J. Biochem.* **94**, 179-187.
Hardman, N., Jack, P. L., Fergie, R. C., and Gerrie, L. M. (1980). Sequence organization in nuclear DNA from *Physarum polycephalum*. *Eur. J. Biochem.* **103**, 247-257.
Hellung-Larsen, P., and Frederiksen, S. (1977). Occurrence and properties of low molecular weight RNA components from cells at different taxonomic levels. *Comp. Biochem. Physiol.* **58B**, 273-281.
Henney, H. R., Jr., and Jungkind, D. (1969). Characterization of ribosomes from the myxomycete *Physarum rigidum* grown in pure culture. *J. Bacteriol.* **98**, 249-255.

Hildebrandt, A., and Sauer, H. W. (1973). DNA dependent-RNA polymerases from *Physarum polycephalum*. *FEBS Lett.* **35**, 41-44.
Hildebrandt, A., and Sauer, H. W. (1976a). Differential template specificities of nuclear RNA polymerases isolated from *Physarum polycephalum*. *Arch. Biochem. Biophys.* **176**, 214-217.
Hildebrandt, A., and Sauer, H. W. (1976b). Levels of RNA polymerase activity during growth, encystment and germination of *Physarum polycephalum*. *Wilhelm Roux's Arch. Dev. Biol.* **180**, 149-156.
Hildebrandt, A., and Sauer, H. W. (1976c). Levels of RNA polymerases during the mitotic cycle of *Physarum polycephalum*. *Biochim. Biophys. Acta* **425**, 316-321.
Hildebrandt, A., and Sauer, H. W. (1977a). Transcription of ribosomal RNA in the life cycle of *Physarum* may be regulated by a specific nucleolar initiation inhibitor. *Biochem. Biophys. Res. Commun.* **74**, 466-472.
Hildebrandt, A., and Sauer, H. W. (1977b). Discrimination of potential and actual RNA polymerase B activity in isolated nuclei during differentiation of *Physarum polycephalum*. *Wilhelm Roux's Arch. Dev. Biol.* **183**, 107-117.
Hildebrandt, A., Mengel, R., and Sauer, H. W. (1979). Characterization of an endogenous transcription inhibitor from *Physarum polycephalum*. *Z. Naturwiss.* **34C**, 76-86.
Hinnen, A., Hicks, J. B., and Fink, G. R. (1978). Transformation of yeast. *Proc. Nat. Acad. Sci. U.S.A.* **75**, 1929-1933.
Hiramaru, M., Uchida, T., and Egami, F. (1969a). Studies on ribonucleases from *Physarum polycephalum*. Purification and characterization of substrate specificity. *J. Biochem. (Tokyo)* **65**, 693-700.
Hiramaru, M., Uchida, T., and Egami, F. (1969b). Studies on two nucleases and a ribonuclease from *Physarum polycephalum*: Purification and mode of action. *J. Biochem. (Tokyo)* **65**, 701-708.
Hodo, H. G., III, and Blatti, S. P. (1977). Purification using polyethylenimine precipitation and low molecular weight subunit analyses of calf thymus and wheat germ DNA-dependent RNA polymerase II. *Biochemistry* **16**, 2334-2343.
Holt, C. E. (1980). The nuclear replication cycle in *Physarum polycephalum*. *In* "Growth and Differentiation in *Physarum polycephalum* " (W. F. Dove and H. P. Rusch, eds.), pp. 9-63. Princeton Univ. Press, Princeton, New Jersey.
Hsu, M.-T., and Coca-Prados, M. (1979). Electron microscopic evidence for the circular form of RNA in the cytoplasm of eukaryotic cells. *Nature (London)* **280**, 339-340.
Hüttermann, A. (1972). Isoenzyme pattern and *de novo* synthesis of phosphodiesterase during differentiation (spherulation) in *Physarum polycephalum*. *Arch. Microbiol.* **83**, 155-164.
Hüttermann, A., and Chet, I. (1971). Activity of some enzymes in *Physarum polycephalum*. III. During spherulation (differentiation) induced by mannitol. *Arch. Mikrobiol.* **78**, 189-192.
Hüttermann, A., Porter, M. T., and Rusch, H. P. (1970a). Activity of some enzymes in *Physarum polycephalum*. I. In the growing plasmodium. *Arch. Mikrobiol.* **74**, 90-100.
Hüttermann, A., Porter, M. T., and Rusch, H. P. (1970b). Activity of some enzymes in *Physarum polycephalum*. II. During spherulation (differentiation). *Arch. Mikrobiol.* **74**, 283-291.
Jacobson, D. N., and Holt, C. E. (1973). Isolation of ribosomal RNA precursors from *Physarum polycephalum*. *Arch. Biochem. Biophys.* **159**, 342-352.
James, G. T., Yeoman, L. C., Matsui, S. I., Goldberg, A. H., and Busch, H. (1977). Isolation and characterisation of nonhistone chromosomal protein C-14 which stimulates RNA synthesis. *Biochemistry* **16**, 2384-2389.
Jendrisak, J. J., and Burgess, R. R. (1975). A new method for the large-scale purification of wheat germ DNA-dependent RNA polymerase II. *Biochemistry* **14**, 4639-4645.
Jendrisak, J. J., and Guilfoyle, R. J. (1978). Eukaryotic RNA polymerases: Comparative subunit structures, immunological properties, and α-amanitin sensitivities of the class II enzymes from higher plants. *Biochemistry* **17**, 1322-1327.

Jerzmanowski, A., Staron, K., Fronk, J., and Roczko, K. (1979). DNA is less tightly bound to nucleosomal histones in *Physarum polycephalum* as compared to higher eukaryotes. *In* "Current Research on *Physarum*" (W. Sachsenmaier, ed.), pp. 45-50. Univ. of Innsbruck Press, Innsbruck, Austria.

Johnson, E. M., Matthews, H. R., Littau, V. C., Lothstein, L., Bradbury, E. M., and Allfrey, V. G. (1978). The structure of chromatin containing DNA complementary to 19S and 26S ribosomal RNA in active and inactive stages of *Physarum polycephalum*. *Arch. Biochem. Biophys.* **191**, 537-550.

Kessler, D. (1967). Nucleic acid synthesis during and after mitosis in the slime mold *Physarum polycephalum*. *Exp. Cell Res.* **45**, 676-680.

Kindle, K. L., and Firtel, R. A. (1978). Identification and analysis of *Dictyostelium* actin genes, a family of moderately repeated genes. *Cell* **15**, 763-778.

Kuehn, G. D., Affolter, H.-U., Atmar, V. J., Seebeck, T., Gubler, U., and Braun, R. (1979). Polyamine-mediated phosphorylation of a nucleolar protein from *Physarum polycephalum* that stimulates rRNA synthesis. *Proc. Nat. Acad. Sci. U.S.A.* **76**, 2541-2545.

Lohr, D., and Hereford, L. (1979). Yeast chromatin is uniformly digested by DNase I. *Proc. Nat. Acad. Sci. U.S.A.* **76**, 4285-4288.

McMaster, G. K., and Carmichael, G. G. (1977). Analysis of single- and double-stranded nucleic acids on polyacrylamide and agarose gels by using glyoxal and acridine orange. *Proc. Nat. Acad. Sci. U.S.A.* **74**, 4835-4838.

Maden, B. E. H., and Robertson, J. S. (1974). Demonstration of the "5.8S" ribosomal sequence in HeLa cell ribosomal precursor RNA. *J. Mol. Biol.* **87**, 227-235.

Matthews, H. R., Johnson, E. M., Steer, W. M., Bradbury, E. M., and Allfrey, V. G. (1978). The use of netropsin with CsCl gradients for the analysis of DNA and its application to restriction nuclease fragments of ribosomal DNA from *Physarum polycephalum*. *Eur. J. Biochem.* **82**, 569-576.

Melera, P. W. (1980). Transcription in the myxomycete *Physarum polycephalum*. *In* "Growth and Differentiation in *Physarum polycephalum* " (W. F. Dove and H. P. Rusch, eds.), pp. 64-97. Princeton Univ. Press, Princeton, New Jersey.

Melera, P. W., and Rusch, H. P. (1973a). A characterization of ribonucleic acid in the myxomycete *Physarum polycephalum*. *Exp. Cell Res.* **82**, 197-209.

Melera, P. W., and Rusch H. P. (1973b). Aminoacylation of transfer ribonucleic acid *in vitro* during the mitotic cycle of *Physarum polycephalum*. *Biochemistry* **12**, 1307-1311.

Melera, P. W., Momeni, C., and Rusch, H. P. (1974). Analysis of isoaccepting tRNAs during the growth phase mitotic cycle of *Physarum polycephalum*. *Biochemistry* **13**, 4139-4142.

Melera, P. W., Peltz, R., Davide, J. P., and O'Connell, M. (1978). Polyadenylated RNA in the lower eukaryote *Physarum polycephalum*. *Mol. Biol. Rep.* **4**, 229-232.

Melera, P. W., Davide, J. P., and Hession, C. (1979). Identification of mRNA in the slime mold *Physarum* polycephalum. *Eur. J. Biochem.* **96**, 373-378.

Miller, O. L., and Beatty, B. (1969). Visualization of nucleolar genes. *Science* **164**, 955-957.

Mittermayer, C., Braun, R., and Rusch, H. P. (1964). RNA synthesis in the mitotic cycle of *Physarum polycephalum*. *Biochim. Biophys. Acta* **91**, 399-405.

Mittermayer, C., Braun, R., and Rusch, H. P. (1966). Ribonucleic acid synthesis *in vitro* in nuclei isolated from the synchronously dividing *Physarum polycephalum*. *Biochim. Biophys. Acta* **114**, 536-546.

Mohberg, J. (1977). Nuclear DNA content and chromosome numbers throughout the life cycle of the colonia strain of the myxomycete *Physarum polycephalum*. *J. Cell Sci.* **24**, 95-108.

Molgaard, H. V. (1978). rDNA organisation in *Physarum polycephalum*. *In* "The Cell Nucleus" (H. Busch, ed.), Vol. V, pp. 335-371. Academic Press, New York.

Molgaard, H. V., Matthews, H. R., and Bradbury, E. M. (1976). Organisation of genes for ribosomal RNA in *Physarum polycephalum. Eur. J. Biochem.* **68,** 541-549.
Nygaard, O. F., Guttes, S., and Rusch, H. P. (1960). Nucleic acid metabolism in a slime mould with synchronous mitosis. *Biochim. Biophys. Acta* **38,** 298-306.
Pierron, G. (1979). Replication-transcription coupling: Assays of RNA-polymerase in isolated nuclei of *Physarum polycephalum. In* "Current Research on *Physarum*" (W. Sachsenmaier, ed.), pp. 82-86. Univ. of Innsbruck Press, Innsbruck, Austria.
Pilly, D., Niemeyer, A., Schmidt, M., and Bargetzi, J. P. (1978). Enzymes for RNA sequence analysis. *J. Biol. Chem.* **253,** 437-445.
Plaut, B. S., and Turnock, G. (1975). Coordination of macromolecular synthesis in the slime mould *Physarum polycephalum. Mol. Gen. Genet.* **137,** 211-225.
Polman, B. J. J., Janssen, H. M. J., and Kuyper, C. M. A. (1974). Degrading enzymes in the slime mold *Physarum polycephalum:* Evidence for five different enzymes. *Arch. Mikrobiol.* **96,** 119-124.
Rao, B., and Gontcharoff, M. (1969). Functionality of newly synthesized DNA as related to RNA synthesis during the mitotic cycle in *Physarum polycephalum. Exp. Cell Res.* **56,** 269-274.
Sachsenmaier, W., Fournier, D. V., and Gürtler, K. F. (1967). Periodic thymidine kinase production in synchronous plasmodia of *Physarum polycephalum:* Inhibition by actinomycin and actidion. *Biochem. Biophys. Res. Commun.* **27,** 655-660.
Sachsenmaier, W., Immich, H., Grunst, J., Scholz, R., and Bücher, T. (1969). Free ribonucleotides of *Physarum polycephalum. Eur. J. Biochem.* **8,** 557-561.
Sauer, H. W. (1973). Differentiation in *Physarum polycephalum. In* "Microbial Differentiation" (J. M. Ashworth and J. E. Smith, eds.), pp. 375-405. Cambridge Univ. Press, London and New York.
Sauer, H. W. (1978). Regulation of gene expression in the cell cycle of Physarum. *In* "Cell Cycle Regulation" (J. R. Jeter, I. L. Cameron, G. M. Padilla, and A. M. Zimmerman, eds.) pp. 149-166. Academic Press, New York.
Sauer, H. W. (1979). Estimate of RNA pol B in the mitotic cycle of *Physarum polycephalum* obtained from the binding of labelled α-amanitin. *In* "Current Research on *Physarum*" (W. Sachsenmaier, ed.), pp. 87-91. Univ. of Innsbruck Press, Innsbruck, Austria.
Sauer, H. W., Babcock, K. L., and Rusch, H. P. (1969a). High molecular weight phosphorus compound in nucleic acid extracts of the slime mold *Physarum polycephalum. J. Bacteriol.* **99,** 650-654.
Sauer, H. W., Babcock, K. L., and Rusch, H. P. (1969b). Changes in RNA synthesis associated with differentiation (spherulation) in *Physarum polycephalum. Biochim. Biophys. Acta* **195,** 410-421.
Schicker, C., Hildebrandt, A., and Sauer, H. W. (1979). RNA transcription of isolated nuclei and chromatin with exogenous RNA polymerases during mitotic cycle and encystment of *Physarum polycephalum. Wilhelm Roux's Arch. Dev. Biol.* **187,** 195-209.
Schwärzler, M., and Braun, R. (1977). Preparation of polysomes from synchronous macroplasmodia of *Physarum polycephalum. Biochim. Biophys. Acta* **479,** 501-505.
Seebeck, T., and Braun, R. (1980). Transcription in acellular slime moulds. *Adv. Microb. Physiol.* **21,** 1-45.
Seebeck, T., Stalder, J., and Braun, R. (1979). Isolation of a minichromosome containing the ribosomal genes from *Physarum polycephalum. Biochemistry* **18,** 484-490.
Sheinin, R., Humbert, J., and Pearlman, R. E. (1978). Some aspects of eukaryotic DNA replication. *In Annu. Rev. Biochem.* **47,** 277-316.
Sina, J. F., and Chin, B. (1978). Cadmium modification of nucleolar ultrastructure and RNA synthesis in *Physarum polycephalum. Toxicol. Appl. Pharmacol.* **43,** 449-459.
Smith, S. S., and Braun, R. (1978). A new method for the purification of RNA polymerases II (or B)

from the lower eukaryote *Physarum polycephalum*. The presence of subforms. *Eur. J. Biochem.* **82,** 309-320.

Smith, S. S., and Braun, R. (1981). New polypeptide chains associated with highly purified RNA polymerase II or B from *Physarum polycephalum*. *FEBS Lett.* **125,** 107-110.

Smith, S. S., Kelly, K. H., and Jockusch, B. M. (1979). Actin co-purifies with RNA polymerase II. *Biochem. Biophys. Res. Commun.* **86,** 161-166.

Sonenshein, G. E., and Holt, C. E. (1968). Molecular weight of mitochondrial DNA in *Physarum polycephalum*. *Biochem. Biophys. Res. Commun.* **33,** 361-367.

Stalder, J., Seebeck, T., and Braun, R. (1978). Degradation of the ribosomal genes by DNase I in *Physarum polycephalum*. *Eur. J. Biochem.* **90,** 391-395.

Stalder, J., Seebeck, T., and Braun, R. (1979). Accessibility of the ribosomal genes to micrococcal nuclease in *Physarum polycephalum*. *Biochim. Biophys. Acta* **561,** 452-463.

Stalder, J., Groudine, M., Dodgson, J. B., Engel, J. D., and Weintraub, H. (1980). Hb switching in chickens. *Cell* **19,** 973-980.

Steer, W. M., Molgaard, H. V., Bradbury, E. M., and Matthews, H. R. (1978). Ribosomal genes in *Physarum polycephalum:* Transcribed and non-transcribed sequences have similar base compositions. *Eur. J. Biochem.* **88,** 599-605.

Sugino, A., Hirose, S., and Okazaki, R. (1972). RNA-linked nascent DNA fragments in *Escherichia coli*. *Proc. Nat. Acad. Sci. U.S.A.* **69,** 1863-1867.

Sun, I. Y.-C., Johnson, E. M., and Allfrey, V. G. (1979). Initiation of transcription of ribosomal DNA sequences in isolated nuclei of *Physarum polycephalum:* Studies using nucleoside $5'$-[γ-S] triphosphates and labeled precursors. *Biochemistry* **18,** 4572-4580.

Swofford, L. K., and Vogt, V. M. (1981). *Nucleic Acids Res.* (in press).

Timberlake, W. E., Shumard, D. S., and Goldberg, R. B. (1977). Relationship between nuclear and polysomal RNA populations of *Achlya:* A simple eucaryotic system. *Cell* **10,** 623-632.

Turnock, G. (1979). Patterns of nucleic acid synthesis in *Physarum polycephalum*. *Prog. Nucleic Acid Res. Mol. Biol.* **23,** 53-104.

Vandekerckhove, J., and Weber, K. (1978). The amino acid sequence of *Physarum* actin. *Nature (London)* **276,** 720-721.

Vogt, V., and Braun, R. (1976). Structure of ribosomal DNA in *Physarum polycephalum*. *J. Mol. Biol.* **106,** 567-587.

Waqar, M. A., and Huberman, J. A. (1973). Evidence for the attachment of RNA to pulse-labeled DNA in the slime mold, *Physarum polycephalum*. *Biochem. Biophys. Res. Commun.* **51,** 174-180.

Waqar, M. A., and Huberman, J. A. (1975). Covalent linkage between RNA and nascent DNA in the slime mold, *Physarum polycephalum*. *Biochim. Biophys. Acta* **383,** 410-420.

Waterborg, J. H. (1980). Alkaline nucleases in *Physarum polycephalum*. Ph. D. Thesis. Univ. Nijmegen, Nijmegen, Netherlands.

Waterborg, J. H., and Kuyper, C. M. A. (1979). Purification of an alkaline nuclease from *Physarum polycephalum*. *Biochim. Biophys. Acta* **571,** 359-367.

Weaver, R. F. (1976). Structural comparison of deoxyribonucleic acid-dependent ribonucleic acid polymerases I and II from the slime mold *Physarum polycephalum*. *Arch. Biochem. Biophys.* **172,** 470-475.

Weintraub, H., and Groudine, M. (1976). Chromosomal subunits in active genes have an altered conformation. *Science* **93,** 848-856.

Wellauer, P. K., and Dawid, I. B. (1973). Secondary structure maps of RNA: Processing of HeLa ribosomal RNA. *Proc. Nat. Acad. Sci. U.S.A.* **70,** 2827-2831.

Zellweger, A., and Braun, R. (1971). RNA of *Physarum*. I. Preparation and properties. *Exp. Cell Res.* **65,** 413-423.

Zellweger, A., Ryser, U., and Braun, R. (1972). Ribosomal genes of *Physarum:* Their isolation and replication in the mitotic cycle. *J. Mol. Biol.* **64,** 681-691.

Index

A

ALC mutants, 266
acetyltransferase, 350
Achlya, 399
acid phosphatase, 161
act mutants, 238
actidione, see cycloheximide
actin, 165–179, 360, 399
  association with plasmalemma, 171, 195
  polymerization of, 176–177, 195
actinin, *Physarum*, 177
actinomycin C, 400
actinomycin D, 279, 400
activation, enzyme, 91
activator, mitotic, 69, 70–71
actomyosin, 95, 97, 99, 101, 102, 150, 166, 170, 174, 182–183, 192, 360
adenosine diphosphate-ribosyl transferase, 89
adenosine monophosphate, cyclic, 90, 94, 133–134, 177, 190–192, 354–356, 405
adenosine triphosphatase, 101, 166, 181–182, 190
adenosine triphosphate, 129, 137, 166, 174, 177, 180, 192–193, 384
  pyrophosphohydrolase, 182
adenylate cyclase, 89
aequorin, 98, 179, 187
agar, 112–113, 114
aging, 49
alc mutants, 226
alcohols, in chemotaxis, 119–120
alleles,
  mating type, 213, 219
  multiple, 42–43, 213, 219, 231
Allium cepa, 148

aluminum, 127
amanitin, alpha, 89, 400, 401, 407, 408, 421, 424
amino acids, 119
  pools of, 94
aminopeptidase, 238
bis(4-aminophenyl)-1,3,4-oxadiazol, 254
ammonium oxalate, 98
Amoeba proteus, 97, 147, 196
amoeba, 170
amoebo-flagellate stage, 51–52
anaerobiosis, 98
analysis, genetic, 242
anaphase, mitotic stage, 298, 305, 311
aneuploidy, 215, 267
anilinonaphthalene sulfonate, 124–126
anisomycin, 239
apogamy, 34, 214, 217, 220, 224, 230, 236, 244–245, 265–266
apomixis, 31–34
apparatus, mitotic, 309–312
apt mutants, 221–222, 232–234, 247
arrowhead, actin pattern, 172
autoradiography, 302, 305, 308, 356, 373
axe mutants, 237

B

BAO, see bis(4-aminophenyl)-1,3,4-oxadiazol
bacteria, 76–77
Badhamia, 10
benzimidazole carbamate, 268
birefringence, plasmodial fibril, 154, 170, 175
bisbenzimid, 327
bismuth, as cytochemical stain, 290, 299, 302, 305

5-bromo-2'-deoxyuridine, 239, 240, 278, 372, 381
BUdR, see 5-bromo-2'-deoxyuridine
bur mutant, 239

C

C5.1, myxamoebal strain, 215–216
C50, myxamoebal strain, 215–216, 229, 266
CL, culture strain, 216, 217, 223, 227–228, 245, 266, 267, 279
CLd, myxamoebal strain, 217–218, 220, 223, 228, 237, 239, 245
CPF myxamoebal strains, 266
cadmium, 185, 400
caffeine, 75, 101, 180, 195, 242
calcium, 94, 97–101, 127, 133–134, 137, 163–164, 168, 170
   regulation of streaming by, 100–101, 146, 150, 179–196
calmodulin, 181, 182–184, 192
capillitium, 4, 9, 14–16, 50
cat mutants, 226
cell cycle, 61–109, 211, 342–343, 347, 353, 355, 357–358, 380, 416–424
centrioles, 52
Ceratiomyxa fruticulosa, 163
cesium, 123
Chaos carolinensis, 148
chemoreception, 112
   fluorescence techniques in, 124–126
chemotaxis, 98, 111–143, 196
   to carbohydrates, 117
   tests for, 112–115
chlorine, 94
chlorotetracycline, 196
chromatin, 88, 291–292, 351–352, 376–379, 394–396, 422
   isolation of, 410
   nucleolar, 302–304, 325–330, 395–396, 409–410
   replication, 321
   staining for, 288
   subunit structure, see nucleosome
chromatography, ion exchange, 88, 402
chromocenters, 292
chromomycin A_3, 255
chromosomes, 89, 296–300
   counting technique, 257–263
   molecular aspects, 317–369
   numbers of, 35, 253, 259, 262
   replication of, 351
chymotrypsin, 336–338, 356
clear zones, plasmodial, 47
clone, myxamoebal, 214, 216, 224
coated vesicles, 162
Colcemid, 329, 343
colloids, 122
Colonia, culture strain, 214–219, 227–229, 236, 265, 277
columella, 5, 9, 14, 15, 27, 50–51
complementation, genetic, 221, 225, 233–234, 242, 246
contractility, 132, 135–136, 169–172, 174, 186–187, 360
   synchrony of, 187
cordycepin, 400
Craterium, 10
crista, mitochondrial, 158
culture, axenic, 237–238
   characteristics of, 280
   preservation of, 246, 275, 278
   sources of, 279–280
cyanide, 98
cycloheximide, 68, 76, 79, 84, 85, 93, 161, 238, 358, 372–373, 382–383
cytidine, methylation, 327
cytochalasin, 175–176
cytochrome oxidase, 95
cytoplasm, inclusions, structure of, 157–165

D

DAB, see 3,3'-diaminobenzidine
DJ, culture strain, 214, 276
DNA, see deoxyribonucleic acid
daunomycin, 400
density, charge, 116
deoxyadenosine kinase, 90, 360
deoxyguanosine triphosphate, 382
deoxyribonuclease, 291, 304, 395, 412
deoxyribonucleic acid, 61
   arrangement in chromatin, 318–324
   denaturation-renaturation experiments, 375–376
   electron microscopy, 395–396
   foldback sequences, 376, 395
   hybridization with RNA, 379
   in vitro synthesis, 374, 383–384
   in vivo replication, 380

Index

inhibition of synthesis, 70, 78
mitochondrial, 93, 158-160, 373, 376
nucleolar, 325-330, 373, 376, 395-396
nuclear content, 35-38, 213, 217, 231, 254-257
properties of, 394-396
replication of, 351, 371-391
ribosomal, 62, 322-330, 378, 395, 412
satellite, 373, 376
sequencing, ribosomal, 325-326
structure of, 374
synthesis of, 62-63, 67, 71, 76-79, 160, 241
deoxyribonucleic acid polymerase, 88
deuterium, 101
Diachea, 18
3,3'-diaminobenzidine, 291
Dictyostelium, 133, 237, 399
Didymiaceae, Family, 12-17
Didymium, see also specific topics
apogamy in, 216
characteristics of genus, 13-17, 25-27
culture of, 28
isolates studied, 28, 29
life cycle, 26, 27, 265
reproductive cycles, 28-35
dif mutants, 221-222, 227, 232, 247
differentiation, 212, 220
transcription in, 425-426
drugs, 238-239, 279

E

EDTA, see ethylene diamine tetra-acetic acid
EGTA, see ethylene glycol bis(beta-aminoethylether)-N,N'-tetraacetic acid
EMS, see ethylmethane sulfonate
Echinostelium, 19, 20
ectoplasm, 96, 101, 146-148, 157, 163, 169, 174, 192, 195-196
electrophoresis, gel, 81, 330-343, 346-347, 356-357, 360, 377, 396
two-dimensional, 332
Elodea, 148
elongation factor, 406, 423, 427
eme mutant, 239
emetine resistance, 239
encystment, 241
endonucleases, 414-415

endoplasm, 96, 101, 128, 146-148, 157, 163, 169, 175-177, 186, 189, 195
drops of, 171, 176, 180
endoplasmic reticulum, 160, 190
enrichment, method for mutant selection, 240-241
entrainment, stretch, 188-189, 195
enzyme, 62, 70
periodic synthesis in cell cycle, 82-94
RNA-degrading, 414
Escherichia coli, 83
ethidium bromide, 377, 378
ethylene diamine tetra-acetic acid, 193, 196, 288, 305, 329, 410
ethylene glycol bis(beta-aminoethylether)-N,N'-tetraacetic acid, 179, 180, 185, 193, 384
ethylmethane sulfonate (EMS), 226, 242
evolution, 38-41

F

FUdR, see 5-fluorodeoxyuridine
factors, transcription regulatory, 423-424
Feulgen straining, 254
fibrils, plasmodial, 154, 170-171, 176
filtration, gel, 336
fixation, electron microscope, 156, 288
flagellum, 243
fluorescence, nucleosome labeling, 321
5-fluorodeoxyuridine, 70, 279, 372, 419
formamide, 410
fragmin, 177-179
freezing, 246
frequency, nearest neighbor, 374
fructose, 118
Fuligo, 10, 27
fus locus, 218, 228, 235-236, 246
fusion
mammalian cell, 78
myxamoebal, 28-35, 215, 228-231
nuclear, 216, 230
plasmodial, 45-48, 63-64, 72, 75, 215, 228, 235-237, 245-246, 279, 353

G

G_2 phase, 62, 70-71, 373, 374, 378, 422
gad mutants, 224, 226, 228, 232
galactose, 117, 118, 153

galactosidase, beta, 329
genetics, 41-48, 211-247
glucose, 117, 118
glutamate dehydrogenase, 82, 88, 91
glycocalyx, 151, 153, 156, 196
glycogen, 157-158, 311
Golgi apparatus, 156, 161
granules, pigment, 152, 162
griseofulvin, 268, 309
growth, 62
guanidinium chloride, 330
guanosine monophosphate, cyclic, 90, 94
guanylate cyclase, 90

H

HMG proteins, see proteins, high mobility group
heat, effect on plasmodium, 73-74, 267, 360, 400
hemacytometer, 255
het mutants, 226, 227
heteroallelism, 219, 221, 224, 235
heterokaryons, 269
heterothallism, 29-31, 212-214, 216, 224, 232, 234, 243, 263-265
heterozygotes, 34
histones, 88
   acetylation of, 330, 346-351
   arrangement in chromatin, 318-322
   H1, 333-339, 341-346, 357
   H2A, 339-340
   H2B, 340
   H3, 340, 378
   H4, 340-341, 346-351
   interaction with DNA, 345-346
   isolation and analysis, 330-341
   mammalian cells, 343-345
   mitochondrial, 158
   nucleolar, 329
   phosphorylation of, 341-346, 355-357
   postsynthetic modification, 341-354
histone deacetylase, 349
histone kinase, 74-75, 87-88, 322, 341, 345, 353-358
   activation of, 354
   in cell cycle, 357-358
homothallism, 32-34, 214, 216-217
hts mutants, 242-243
hyaloplasm, 148
hydroxyurea, 419-420, 422

I

imz, gene locus, 219, 231, 239
incompatibility, 45-48
Indiana, culture strain, 213, 218, 236, 277
inducer, plasmodium, 217
inhibitor, mitotic, 66
inoculum, suppliers of, 279-280
interphase, mitotic stage, 292, 301-304, 372
ionophore, calcium, 133, 180
isopropyl *N*-(3-chlorophenyl) carbamate, 268
isozymes, 38-41, 84

K

kil locus, 236

L

LU648, myxamoebal strain, 217, 219
LU688, myxamoebal strain, 218, 219
lanthanum, 99, 127, 185
Leicester, culture strain, 214, 273, 278
let locus, 236
lethal interactions, 236
leu mutant, 237
leucine, 237
Licea, 6
light, effect on fruiting, 44-45
   ultraviolet, 242
lipid, in plasmodium, 161
lithium, 123
Lycogala, 6
lyotropic number, 124
lys mutant, 237
lysine, 237
lysolethicin, use in chromatin isolation, 410
lysosomes, 161-414

M

macronucleus, 76
magnesium, 94, 127, 168, 185, 407, 421
maltose, 118
manganese, 407
mannose, 117-118
mapping, genome, 224
mat locus, 218, 223, 225, 230-231, 245, 247, 268
mating, 41-43, 212
   genetic control of, 229-231

Index

mating types, 212–213
McArdle Laboratory strains, 273–275
medium, culture, 274
meiosis, 50, 52, 213, 229, 254, 263, 265
membrane, 61
   effects of alcohol on, 119–120
   fluidity of, 126
   fluorescent probes of, 126–127
   hydrophobicity of, 120–122
   role in chemotaxis, 119–127
   role in streaming, 96, 101–102, 171–172, 187
metaphase, 146, 296–298, 305, 310, 342
6-methylaminopurine, 380
5-methylcytosine, 374, 379–380
N-methyl-N'-nitro-N-nitrosoguanidine (NMG), 225, 227, 239, 242
methylation, 327, 379
Metrizamide, 330, 354
microcyst, 52
microfilaments, 158, 163
microfluorometry, 254–255
microplasmodium, 101, 241, 322, 386, 394, 426
microspectrofluorometry, laser, 255
microspectrophotometry, 36–38, 254–255
microtubules, 162, 309–312
midbody, 312
migration, plasmodial, 113
minichromosome, 304, 307, 329, 405, 408–409
mitochondrion, 373
   division of, 94, 159–160, 374
   involvement with calcium, 180, 190–193
   enzymes, 95
   structure of, 151, 158–160
   transcription in, 411
mitogen, 64
mitosis, 43, 52, 62–63, 74–75, 78, 84, 87, 259, 262, 274, 288–300, 341, 347, 357–358, 372, 377, 416–420
   activator of, 69, 70–71, 75
   advance of, 75
   cessation of streaming during, 146, 193
   delay of, 67, 68, 73
   inhibitor of, 66–69, 75
   models for control of, 64–76
   nucleolus during, 304–307
   perturbations of cycle, 66–76
   thymidine kinase, 360
   transcription during, 416–420

   trigger of, 75, 358
   ultrastructure, 288–312
mobility, electrophoretic, 116
model,
   inhibition of plasmodium development, 231
   limit cycle, 72
   relaxation oscillator, 72
   streaming oscillator, 95–101
mov mutant, 242
mt gene locus, 213, 217, 219–224, 226–232, 243, 244, 247
muscle, skeletal, similarity to plasmodial fibrils, 166, 170, 174, 181, 188, 192
mutagenesis, 214, 220, 238, 242, 268
mutant, 214–215, 217, 218–220, 235, 237, 279
   growth, 241
   isolation of, 244
   leaky, 223
   screening techniques for, 241–242, 244–245
myosin, 96, 166, 172–175, 181–185, 192, 360
   electron microscopy, 166–169
myxamoeba, 51–52, 158, 213
   chromosomes of, 259–263
   crossing of, 245–246
   morphology of colony, 242
   multinucleate, 241–242
   ploidy of, 259–263, 268

N

NMG, see N-methyl-N'-nitro-N-nitrosoguanidine
netropsin, 240, 327
nicotinamide adenine dinucleotide pyrophosphorylase, 89
Nitella, 148
nocodazole, 268
nomenclature, genetic, 247
npf mutants, 221–223, 226–229, 232–234, 244, 247
nuclear envelope, 299, 311
nuclease
   micrococcal, 319, 320, 321, 322, 323, 377, 379
   restriction, 327, 329, 359, 377, 378, 414
   staphylococcal, 412
nucleoid, mitochondrial, 158
nucleolus, 62, 156, 241, 292, 301–308
   chromatin from, 327–329, 378–379
   formation, 307–308

nucleolus (*continued*)
  isolation of, 327, 378
  ultrastructure, 301–304
nucleosome, 318–324, 376–379, 411–413
  model of, 319
  oligomers, 323–324
nucleus, 62–66, 213, 241, 287–312
  deoxyribonucleic acid content of, 254–257, 394–395
  diameter and ploidy, 255
  homogenization, 384–385
  isolation of, 383–384
  removal of, 67
  spreading, method of, 257, 298
  structure of, 151, 156, 317–369
nutrition, 237

O

Okasaki fragments, 381, 382, 385, 419
ornithine decarboxylase, 85, 91
oscillators,
  relaxation, 72
  streaming, 96, 186, 192–195
osmium tetroxide, 158, 288–289
ouabain, 101

P

Paramecium tetraurelia, 76
peak A, nucleosome, 322, 323, 330, 378, 412
pepsin, 304
Perichaena vermicularis, 155, 163, 253, 265
peridium, 8, 14
  dehiscence, 8
pH, 62, 94, 134
phagocytosis, 152
phalloidin, 176
phaneroplasmodium, 6, 146
phosphatase, alkaline, 354
phosphodiesterase, 89, 416
phosphoprotein, 95, 329, 405
phosphorus, 94
phosphotungstic acid, 289, 302, 305, 312
phototaxis, 113
phylogeny, 18
Physaraceae, Family, 4–12
Physarales, *see also* specific genera
  taxonomy of, 4–17
Physarella, 10

Physarum, see also specific topics
  genus characteristics, 5–12
  pigments, 44
  pinocytosis, 153, 196
  plasmalemma, 95, 100–101, 187
    calcium transport across, 100, 190, 195–196
    invaginations of, 154–155, 163–164, 171
  plasmodium, 6, 13, 62, 145–208
    chemotaxis, 111–143, 196
    color, 44–45, 243–244
    development, 212, 216, 231–235
    fibrils in, 154
    formation of, 41–43, 212, 216, 220–223, 245
    fusion of, 63–64, 72, 245
    incompatibility, 45–48
    killing reactions, 45–48, 235
    light microscopy of, 151
    nutrition, 237
    pigments, 44
    ploidy of, 254–257, 263–266
    senescence, 49
    strands, manipulation of, 128–132, 149
    streaming, 96–101, 115–116, 147–150, 169–196
    structure of, 151–169
    types of, 6, 44
    ultrastructure, 155–169
  ploidy, 35–38, 49, 228–229
    factors affecting, 266–268
  polyadenylate, 399, 413
  polylysine, 177
  polymerase,
    DNA, 88, 385
    RNA, 89, 360, 400–404
      inhibitor of, 402, 404–405
      regulation of transcription by, 424
      subunits of, 403–404
  polyphosphate, 423, 427
  polyploidy, 268–269, 278
    with aging, 49
    with heat, 63, 267
  polysome, isolation of, 396
  pools, metabolic, 94
    triphosphate, 94
  pores, nuclear, 95–96, 157
  potassium, 94, 101, 123, 127, 188
  potential,
    membrane, 113–115, 117, 187
    zeta, 117–118
  primer, RNA, 382, 418–419

Index

prophase, mitotic stage, 296, 305
protein,
  contractile, 360
  division, 67, 68
  high mobility group, 319, 359
  in cell cycle, 81
  in replication, 385-386
  ratio to DNA, 74
  synthesis of, 64, 70-71
protein kinase, 405
proteinase K, 304
Protostelida, 19
pseudopodia, 137-138, 147
pyridoxal 5'-phosphate, 85

Q

quadrant test, 113

R

RNA, see ribonucleic acid
RNP, see ribonucleoprotein
RSD, culture strain, 214, 276
rac locus, 218, 247
radiation, gamma, 75
reception,
  chemo-, 111-143
  olfactory, 121-122
recombination, 215, 236
replication, 380-386
  directionality of, 381
  dependence on protein synthesis, 382-383
replicon, 380
respiration, 95, 98
rhamnose, 153
ribonuclease, 304
ribonucleic acid, 61, 396-399
  degradation of, 413-416
  heterogeneous nuclear, 350, 398, 413
  messenger, 71, 92, 93, 350-351, 398, 413, 425
  metabolism of, 393-436
  mitochondrial, 158
  polyadenylated, 399
  ribosomal, 395-398, 407-410, 413
  5S ribosomal, 398
  stains for, 291
  synthesis of, 71, 93, 241, 400-413
    regulation of, 404-406
  transfer, 398

ribonucleic acid polymerase B, 88
ribonucleic acid polymerase I, 401, 403, 405, 407, 422, 424
  in nucleus, 402
ribonucleic acid polymerase II, 360, 401, 403, 407, 408, 421-422, 424, 426-427
  in nucleus, 402
ribonucleic acid polymerase III, 401
ribonucleoprotein, 288, 298
ribosome,
  formation of, 307
  isolation of, 396
  structure, 157-158
rifampicin, 410, 421
rubidium, 123

S

S phase, 62, 373-372, 374, 423
SDS, see sodium dodecyl sulfate
Saccharomyces cerevisiae, 78, 83
sarcoplasmic reticulum, 181, 191-192
Schizosaccharomyces pombe, 77
Schulze-Hardy rule, 122
sclerotium, 246
selfing, 31-32, 214, 216, 224, 226, 235
  revertants, 226-227
Sendai virus, 78
senescence, 49
size, cell, effect on division, 79
slime, 146, 151, 156, 161, 394
  chemical nature of, 153
  histochemistry, 153
sodium, 94, 101, 123, 127, 137
sodium dodecyl sulfate (SDS), 342-343
spectrin, 172
spermidine phosphate, 158, 385
spermine, 384
spherulation, 212, 275, 278, 425-428
spherule, germination of, 278
spindle, mitotic, 43, 52, 309-312
sporangium, 8
  development, 50-51
  in taxonomy, 51
spores, 10, 16
  formation of, 50, 278
  germination of, 229, 259, 278
sporophore, 8, 13, 50-51
sporulation, 161, 212, 215-216, 228-229, 264, 425

stalk, 9
Stemonitis, 6
Stemonitomycetidae, Subclass, 17
Stentor coeruleus, 77
streaming, plasmodial, 96–101, 115, 116,
    147–150, 169–172, 180, 192–196
  analysis by cinemicrography, 185
  and chemotaxis, 128–130
  periodicity of, 96–101, 134–136, 188–196
  regulation of, 180, 185
  shuttle, 147–150, 170, 185, 192–193
  summary of data, 149–150
sucrose, 117–118
sugars, effect on membrane, 123–124
sulfate, 153, 156
sulfur, 94
swarm cell, 51–52, 158, 243
synchrony
  cell, 80
  of nuclear division, 62, 215
  of DNA synthesis, 62

T

telophase, mitotic stage, 299, 307, 312
temperature,
  effect on streaming, 136–138
  effect on pseudopod formation, 137
  effect on mitosis, 73–75
  mutants sensitive to, 239–241, 245
tension, in plasmodial strand, 116
Tetrahymena, 67, 325
thermotaxis, 113
thorium, 127, 137
thymidine kinase, 84–85, 359–360
toyocamycin, 400
transcription, 325, 329, 350–351, 359, 404–413
  coupling to DNA replication, 419
  differential, 417
  inhibition of, 400
  in isolated nuclei, 407–410

in vitro, 407–411, 420–428
in vivo, 400, 416–420
  stimulation of, 405
tropomyosin, 181
troponin, 181, 182

U

ultraviolet irradiation, 67–68, 76, 381
uranyl acetate, 158

V

vacuole
  autophagic, 161
  calcium-sequestering, 180, 188, 190, 195
  contractile, 152, 162
  endocytotic, 161
  food, 161
  in plasmodium, 152–153
valine, 237
valinomycin, 101
vein, plasmodial, 175–177, 185–187, 189
velocity profile, in streaming, 169
vesicles, plasmodial, 154, 190
viscosity, 116

W

water, 123
whi mutant, 243–244
Wisconsin 1, culture strain, 213–214, 217, 218,
    228, 235, 238–239, 243, 274

Y

yeast, cell division in, 78

Z

zygote, 216

CELL BIOLOGY: A Series of Monographs

EDITORS

D. E. BUETOW

*Department of Physiology
and Biophysics
University of Illinois
Urbana, Illinois*

I. L. CAMERON

*Department of Anatomy
University of Texas
Health Science Center at San Antonio
San Antonio, Texas*

G. M. PADILLA

*Department of Physiology
Duke University Medical Center
Durham, North Carolina*

A. M. ZIMMERMAN

*Department of Zoology
University of Toronto
Toronto, Ontario, Canada*

G. M. Padilla, G. L. Whitson, and I. L. Cameron (editors). THE CELL CYCLE: Gene-Enzyme Interactions, 1969

A. M. Zimmerman (editor). HIGH PRESSURE EFFECTS ON CELLULAR PROCESSES, 1970

I. L. Cameron and J. D. Thrasher (editors). CELLULAR AND MOLECULAR RENEWAL IN THE MAMMALIAN BODY, 1971

I. L. Cameron, G. M. Padilla, and A. M. Zimmerman (editors). DEVELOPMENTAL ASPECTS OF THE CELL CYCLE, 1971

P. F. Smith. The BIOLOGY OF MYCOPLASMAS, 1971

Gary L. Whitson (editor). CONCEPTS IN RADIATION CELL BIOLOGY, 1972

Donald L. Hill. THE BIOCHEMISTRY AND PHYSIOLOGY OF *TETRAHYMENA*, 1972

Kwang W. Jeon (editor). THE BIOLOGY OF AMOEBA, 1973

Dean F. Martin and George M. Padilla (editors). MARINE PHARMACOGNOSY: Action of Marine Biotoxins at the Cellular Level, 1973

Joseph A. Erwin (editor). LIPIDS AND BIOMEMBRANES OF EUKARYOTIC MICROORGANISMS, 1973

A. M. Zimmerman, G. M. Padilla, and I. L. Cameron (editors). DRUGS AND THE CELL CYCLE, 1973

Stuart Coward (editor). DEVELOPMENTAL REGULATION: Aspects of Cell Differentiation, 1973

I. L. Cameron and J. R. Jeter, Jr. (editors). ACIDIC PROTEINS OF THE NUCLEUS, 1974

Govindjee (editor). BIOENERGETICS OF PHOTOSYNTHESIS, 1975

James R. Jeter, Jr., Ivan L. Cameron, George M. Padilla, and Arthur M. Zimmerman (editors). CELL CYCLE REGULATION, 1978

Gary L. Whitson (editor). NUCLEAR–CYTOPLASMIC INTERACTIONS IN THE CELL CYCLE, 1980

Danton H. O'Day and Paul A. Horgen (editors). SEXUAL INTERACTIONS IN EUKARYOTIC MICROBES, 1981

Ivan L. Cameron and Thomas B. Pool (editors). THE TRANSFORMED CELL, 1981

Arthur M. Zimmerman and Arthur Forer (editors). MITOSIS/CYTOKINESIS, 1981

Ian R. Brown (editor). MOLECULAR APPROACHES TO NEUROBIOLOGY, 1982

Henry C. Aldrich and John W. Daniel (editors). CELL BIOLOGY OF *PHYSARUM* AND *DIDYMIUM*, Volume I: Organisms, Nucleus, and Cell Cycle, 1982

In preparation

John A. Heddle (editor). MUTAGENICITY: New Horizons in Genetic Toxicology, 1982

Potu N. Rao, Robert T. Johnson, and Karl Sperling (editors). PREMATURE CHROMOSOME CONDENSATION: Application in Basic, Clinical, and Mutation Research, 1982

Henry C. Aldrich and John W. Daniel (editors). CELL BIOLOGY OF *PHYSARUM* AND *DIDYMIUM*, Volume II. Differentiation, Metabolism, and Methodology, 1982.

George M. Padilla and Kenneth S. McCarty, Sr. (editors). GENETIC EXPRESSION IN THE CELL CYCLE, 1982

David S. McDevitt (editor). CELL BIOLOGY OF THE EYE, 1982

Govindjee (editor). PHOTOSYNTHESIS, Volume I: Energy Conversion by Plants and Bacteria, 1982; Volume II: Development, Carbon Metabolism, and Plant Productivity, 1982